SONET/SDH

SONET/SDH

A Sourcebook of Synchronous Networking

Edited by

Curtis A. Siller, Jr.
AT&T Bell Laboratories

Mansoor Shafi
Telecom Corporation of New Zealand

A Selected Reprint Volume
IEEE Communications Society, *Sponsor*

IEEE
COMMUNICATIONS
SOCIETY

The Institute of Electrical and Electronics Engineers, Inc., New York

This book may be purchased from the publisher when
ordered in bulk quantities. For more information, contact:

IEEE Press Marketing
Attn: Special Sales
P.O. Box 1331
445 Hoes Lane
Piscataway, NJ 08855-1331
Fax: (908) 981-9334

Printed in the United States of America

10   9   8   7   6   5   4   3   2   1

ISBN 0-7803-1168-X
IEEE Order Number: PC4457

Library of Congress Cataloging-in-Publication Data

SONET / SDH : a sourcebook of synchronous networking / edited by Curtis
  A. Siller, Jr., Mansoor Shafi ; IEEE Communications Society,
  sponsor.
      p.   cm.
Includes bibliographical references and index.
    ISBN 0-7803-1168-X
    1. Synchronous digital hierarchy (Data transmission)   2. Optical
communications.   I. Siller, Curtis A.   II. Shafi, Mansoor (date).
III. IEEE Communications Society.
TK5105.42.S66   1996
621.382'75— dc20   95-35418
                                                                 CIP

Contents

Preface   ix

Tutorial: Synchronous Optical Network/Synchronous Digital Hierarchy:
   An Overview of Synchronous Networking   1

Section 1   SONET/SDH Overview   17

SONET: Now It's the Standard Optical Network   19
   R. Ballart and Y.-C. Ching (modified from original paper published in *IEEE Communications
   Magazine,* March 1989)
Progress in Standardization of SONET   28
   R. J. Boehm (modified from original paper published in *IEEE LCS,* May 1990)
An Overview of Emerging ITU-T Recommendations for the Synchronous Digital Hierarchy:
Rates and Formats, Network Elements, Line Systems, and Network Aspects   39
   J. Eaves et al.
Architectural and Functional Aspects of Radio-Relay Systems for SDH Networks   47
   G. Richman, M. Shafi, and C. J. Carlisle (modified from original papers published
   in *IEEE GLOBECOM '91* and *IEEE ICC '90*)
Aspects of Satellite Systems Integration in Synchronous Digital Hierarchy Transport Networks   53
   W. S. Oei and S. Tamboli
International Gateway for SDH and SONET Interconnection   62
   C. Hwu and S. Chum (*IEEE GLOBECOM '94*)

Section 2   Deployment Plans and Architecture Issues   73

Defining Network Architecture for SDH Based Networks   75
   A. B. D. Reid (modified from original paper published in *British Telecom Engineering,* July 1991)
Planning the Introduction of SDH Systems in the Italian Telecommunications Network   84
   P. Lazzaro and F. Parente (modified from original paper published in *IEEE GLOBECOM '92*)
France Telecom's Deployment of SDH   91
   G. Bars and D. Bourdeau (an update of "SDH Deployment in the France Telecom Network,"
   IEEE Communications Magazine, Aug. 1990)
SDH Network Evolution in Japan   97
   H. Miura, K. Maki, and K. Nishihata
Plans and Considerations for SONET Deployment   103
   N. Sandesara, R. Richie and B. E. Smith (modified from original paper published in
   IEEE Communications Magazine, Aug. 1990)
SONET Implementation   110
   Y.-C. Ching and H. Sabit Say (*IEEE Communications Magazine,* Sept. 1993)
The Role of SDH/SONET Based Networks in British Telecom   117
   S. Whitt et al. (modified from original paper published in *IEEE ICC '90*)
Planning and Deploying a SONET-Based Metro Network   123
   M. To and J. MacEachern (*IEEE LTS,* Nov. 1991)

Section 3   Survivability and Robust Architectures   129

Self Healing Rings in a Synchronous Environment   131
   I. Haque, W. Kremer and K. Raychaudhuri (modified from original paper
   published in *IEEE LTS,* Nov. 1991)

Feasibility Study of a High-Speed SONET Self-Healing Ring Architecture in Future Interoffice Networks   140
     T.-H. Wu and M. E. Burrowes (*IEEE Communications Magazine*, Nov. 1990)
A Class of Self-Healing Ring Architectures for SONET Network Applications   151
     T.-H. Wu and R. C. Lau (*IEEE Transactions on Communications*, Nov. 1992)
Service Applications for SONET DCS Distributed Restoration   162
     J. Sosnosky (*IEEE Journal on Selected Areas in Communications*, Jan. 1994)
Network Design Sensitivity Studies for Use of Digital Cross-Connect Systems in Survivable
Network Architectures   172
     R. Doverspike, J. Morgan, and W. Leland (*IEEE Journal on Selected Areas in Communications*, Jan. 1994)
The Impact of SONET Digital Cross-Connect System Architecture on Distributed Restoration   182
     T.-H. Wu et al. (*IEEE Journal on Selected Areas in Communications*, Jan. 1994)
Control Algorithms of SONET Integrated Self-Healing Networks   191
     S. Hasegawa et al. (*IEEE Journal on Selected Areas in Communications*, Jan. 1994)
Survivable SONET Networks—Design Methodology   200
     O. J. Wasem, T.-H. Wu and R. H. Cardwell (*IEEE Journal on Selected Areas in Communications*, Jan. 1994)

Section 4   Network Performance   **209**

Improvements in Availability and Error Performance of SONET Compared
to Asynchronous Transport Systems   211
     K. Nagaraj et al. (modified from original paper published in *IEEE ICC '90*)
The Impact of G.826   218
     M. Shafi and P. Smith (modified from original paper published in *IEEE Communications Magazine*, Sept. 1993)
Transmission Performance in Evolving SONET/SDH Networks   226
     J. Gruber, J. Leeson, and M. Green (modified from original paper published in *Telesis*, Dec. 1992)
Traffic Management and Control in SONET Networks   235
     S. Kheradpir et al. (an update of "Performance Management in SONET-Based Multiservice Networks," *IEEE GLOBECOM '91*)
Network Synchronization—A Challenge for SDH/SONET?   242
     M. J. Klein and R. Urbansky (*IEEE Communications Magazine*, Sept. 1993)
SONET Requirements for Jitter Interworking with Existing Networks   251
     R. O. Nunn (*IEEE GLOBECOM '93*)

Section 5   Operations, Administration, and Management   **257**

SDH/SONET—A Network Management Viewpoint   259
     R. F. Holter (modified from original paper published in *IEEE Network Magazine*, Nov. 1990)
SDH Management   266
     J. F. Portejoie and J. Y. Tremel
A Layered Approach to SDH Network Management in the Telecommunications Management Network   275
     L. H. Campbell and H. J. Everitt (modified from original paper published in *IEEE NOMS '92*)
Control and Operation of SDH Network Elements   281
     J. Blume et al. (modified from original paper published in *Ericsson Review*, 1992)
A Synchronous Digital Hierarchy Network Management System   294
     T. Kunieda, S. Sugimoto and N. Sasaki (*IEEE Communications Magazine*, Nov. 1993)
SONET Operations in the TINA Context   301
     W. J. Barr, T. Boyd, and Y. Inoue (modified from original paper published in *IEEE Communications Magazine*, March 1993)

SDH Management Network: Architecture, Routing, and Addressing   308
     H. Katz, G. F. Sawyers, and J. L. Ginger (*IEEE GLOBECOM '93*)

Section 6   SONET/SDH Future   **315**

ATM (Asynchronous Transfer Mode): Overview, Synergy with SDH and Deployment Perspective   317
     J. Legras
An SDH Transmission System for the Transport of ATM Cells   324
     A. Brosio and A. Moncalvo (modified from original paper published in *CSELT Technical Reports*,
     Dec. 1992)
Technologies Towards Broadband ISDN   330
     K. Murano et al. (modified from original paper published in *IEEE Communications Magazine*,
     April 1990)
Optical Fiber Access—Perspectives Toward the 21st Century   337
     A. Cook and J. Stern (*IEEE Communications Magazine*, Feb. 1994)
Realizing the Benefits of SDH Technology for the Delivery of Service in the Access Network   346
     M. Compton and S. Martin
Cost-Effective Network Evolution   352
     T. H. Wu (modified from original paper published in *IEEE Communications Magazine*, Sept. 1993)
Emerging Residential Broadband Communications   362
     D. S. Burpee and P. W. Shumate, Jr. (*Proceedings of the IEEE*, April 1994)

Author Index   **373**

Subject Index   **375**

Editors' Biographies   **391**

Preface

ALMOST 30 years have elapsed since the transition from analog to digital transmission in telecommunications networks. During that period, both transport and switching portions of communications networks have seen the replacement of analog technologies with more reliable and efficient digital technology. The transport network is today in the early stages of yet another wide-ranging change, a transition which consists of:

- Replacing copper cables with optical fiber as the preferred transport medium, today in the trunk network and moving ever closer to the subscriber across the access network
- Deploying improved and more widespread network management capabilities
- Substantially increasing the level of network resilience in response to human-induced or natural calamities
- Deploying network elements and system architectures which are capable of providing a variety of narrowband and broadband services in an integrated fashion.

Of course, such a transition cannot occur overnight. It can only happen gradually, constrained by the need to maintain compatibility with existing infrastructures.

Modern transport networks are based on either the Synchronous Digital Hierarchy (SDH) or Synchronous Optical NETwork (SONET). These closely related standards provide the foundation for transforming the transport network today. The SDH and SONET standards govern interface parameters; rates, formats, and multiplexing methods; and operations, administration, maintenance, and provisioning (OAM&P) of carrier networks. They grew out of work, dating back to the mid 1980's, carried out under the aegis of the American National Standards Institute (ANSI) and the Consultative Committee on International Telegraphy and Telephony, the latter now known as the International Telecommunication Union — Telecommunication Standardization Sector (ITU-T).

This book surveys the many aspects of SONET and SDH. The principal topics treated include:

1) Introductory overview;
2) Architecture and current deployment plans;
3) Network survivability (probably SONET/SDH's most visible "service" enhancement);
4) Network performance;
5) OAM&P; and

6) Future networking issues, notably asynchronous transfer mode (ATM) and synchronous networking closer to the subscriber.

The reader will find a tutorial article which provides a glimpse into the early history of SONET and SDH standards evolution; a review of bit rate, frame format, overhead functions, and multiplexing structures; discussion of basic network topologies; an overview of network management; examination of performance capabilities; and a glimpse into future networking trends. This succinct introduction is augmented with a complete list of applicable ANSI standards, Bellcore technical advisories and recommendations, and ITU recommendations, followed by a glossary of commonly used abbreviations.

Following the tutorial, six chapters, each ranging from six to eight papers, fully develop the themes enumerated above. The 42 papers that make up the body of the book are authored by recognized subject-matter experts from throughout the world. Almost a third of the papers were prepared specifically for this book; another third represent updated, previously published manuscripts; and the remainder are drawn directly from the current literature. Each of the six chapters includes a brief preface and a guide to the papers contained therein.

This book can be read at several levels, depending on the reader's interest. The casual reader will gain a basic understanding of SDH/SONET by reading the opening tutorial. The more serious reader will benefit from perusing the introductory article, then turning to individual chapters for additional information, as interest and needs direct. Even the most serious reader will find a wealth of reference material throughout the collected papers. Whether an administrator, network planner, student, systems engineer, or technologist, every reader will find significant value in this work.

The authors are pleased to acknowledge the fine contributions of authors whose work is presented herein, support from AT&T Bell Laboratories and Telecom of New Zealand, exceptional collaboration with Debbie Graffox and her colleagues at IEEE Press, and the patience of our families.

Curtis A. Siller, Jr.
North Andover, MA, U.S.

Mansoor Shafi
Wellington, New Zealand

SONET/SDH

Synchronous Optical Network Synchronous Digital Hierarchy: An Overview of Synchronous Networking

CURTIS A. SILLER, JR., PH.D.

AT&T BELL LABORATORIES
NORTH ANDOVER, MA 01845, U.S.A.

MANSOOR SHAFI, PH.D.

TELECOM CORPORATION OF NEW ZEALAND
WELLINGTON, NEW ZEALAND

1. INTRODUCTION

SYNCHRONOUS Optical NETwork (SONET) and Synchronous Digital Network (SDH) describe two families of closely related standards that govern interface parameters; rates, formats and multiplexing methods; and operations, administration, maintenance, and provisioning (OAM&P) for high-speed transmission. SONET is primarily a set of North American standards with a fundamental transport rate beginning at approximately 52 Mb/s, while SDH, principally used in Europe and Asia, defines a basic rate near 155 Mb/s. From a transmission perspective, together they provide an international basis for supporting both existing (time-division multiplexed) and new (cell-multiplexed) services in the developed and developing countries.

This tutorial gives the reader a succinct overview of SONET and SDH by providing: 1) glimpses into the early history of their standards evolution; 2) a review of bit rate, frame format, overhead functions, and multiplexing structures; 3) discussion of linear add/drop, ring, and mesh topologies; 4) an overview of network management; and 5) examination of performance capabilities. The concluding section summarizes the benefits of SDH and SONET, with additional comments on future deployment and networking trends. The reader should consider this overview as an introduction to the more detailed treatments that are found in the sections that make up this book and the other cited references.

2. STANDARDS EVOLUTION

Standardization of SONET (and, subsequently, SDH) initially grew out of the AT&T post-divestiture era, beginning in 1984. Up to that point, AT&T had formulated and implemented many of the United States telecommunications standards. And because of its participation in the Consultative Committee on International Telegraphy and Telephony (CCITT), it played a significant role in setting international standards, as well. (CCITT is now known as the International Telecommunication Union—Telecommunication Standardization Sector, ITU-T.)

In that same year, the American National Standards Institute (ANSI), recognizing the need for an optical standard for future broadband communications, accredited standards committee T1-Telecommunications to pursue the investigation of Interexchange Carrier Compatibility Forum (ICCF) proposals for standardizing optical parameters, line formats, multiplexing techniques, and maintenance capabilities. In a rapid series of events, 1985 witnessed three pivotal events: 1) seminal work by R. J. Boehm, Y.-C. Ching, C. G. Griffith, and F. A. Saal [1] of Bell Communications Research in specifying an early version of SONET; 2) T1X1 commencement of the standardization process; and 3) in August of that year, project proposal approval by T1X1.

It should be noted that prior to these initial steps, lightwave equipments were implemented with vendor-specific proprietary signals and multiplexing formats, hence precluding a so-called multivendor "midspan meet," i.e., interworking of multivendor equipment across a single fiber span. Consequently, a principal goal of the SONET standards initiative was to permit attaching different vendors' equipment without loss of functionality. In addition to establishing standardized rates and formats, an additional goal was to strive for a standard that obviated a specific network architecture, a goal largely obtained by selecting a base signal rate and synchronously multiplexing that rate to attain higher speed line rates. Beyond these, the goal today is to permit mixed-vendor intra- and interoffice networking, which includes operation systems and subnetwork management using Open Systems Interconnection (OSI), consistent methods for network restoration, and synchronization (clock stability, jitter/wander).

Returning to the chronology introduced above, 1986 proved an auspicious time in that CCITT showed interest in the SONET work, resulting in 1987 and 1988 conferences that yielded coor-

1

dinated specifications for both the American National standard, SONET, and the CCITT international standard, SDH. Attaining this agreement on the SDH within CCITT required much effort and commitment: the customary four-year plenary cycle was proscribed since the work was brought into CCITT at midcycle. Further, when CCITT first showed interest in the T1X1 work in Summer 1986, procedural impediments arose in that contributions are customarily forwarded to CCITT via U.S. Study Groups only after consensus is reached within the ANSI-accredited body.

A potential impasse was resolved in 1987 when British and Japanese delegations started to take part in T1X1 SONET meetings. They not only served as liaisons into T1X1 from the international standards community, but were influential in their own right by bringing valuable perspectives to the discussion. Separately, Bell Communications Research (Bellcore), showing an interest in internationalizing SONET, informally worked through a number of forums to foster support within T1X1 and CCITT.

Early in 1987, the United States proposed SONET to CCITT for use as the Broadband ISDN (B-ISDN) Network-Node Interface (NNI), the latter having previously initiated study of an NNI for an asynchronous signal hierarchy in mid-1986. This United States proposal expectedly focused on a basic rate at approximately 50 Mb/s (to efficiently transport 44.736 Mb/s DS-3's), although there are no European rates (nominally 2, 34, and 140 Mb/s) near that level. (The Japanese hierarchy follows a mixture of European and North American rates; they, too, had a need for a basic rate of about 50 Mb/s.) This became one point separating North American from European interests, another being bit- versus byte-interleaved multiplexing. Another difference surfaced in July 1987 regarding alternative approaches to framing an NNI signal at 150 Mb/s: the United States perspective was based on a frame 13 rows by 180-byte columns, which reflected a SONET Synchronous Transport Signal Level-3 (STS-3) frame structure; Europe advocated a 9-row by 270-byte column STS-3 frame so as to efficiently transport 2.048 Mb/s signals using 4 columns of 9 bytes, based on 32 bytes/125 μs.

This became the dominant issue for discussion over the next year, complicated by orthogonal North American versus international interests, procedural steps leading to standards approval within the respective bodies, weighing the desire for prompt North American approval of SONET versus the appeal of a global standard, and the occasional lack of consensus within the standards-setting organizations. Finally, T1X1 approved a final standard in August 1988, with CCITT following suit, and a global SONET/SDH standard was established. This global standard is based on a 9-row frame, wherein SONET became a subset of SDH.

Optical parameters for interoffice connections and the basic rates and formats for the SONET signal make up the so-called Phase 1 standards completed in 1988. These include a basic 51.84 Mb/s signal, with overhead signals allocated for data communications channels (DCCs) to convey OAM&P messages; for line system operation (automatic protection switching, error monitoring, framing, and order wires); end-to-end functions

(signal source identification, error monitoring, and far-end error counts); and regenerators (order wire, error monitoring, and framing).

A second group of standards, termed Phase 2, were to include physical interfaces for equipment interconnection within a location, and specification of protocols and messages to control and maintain SONET equipment. Much progress has been attained on both of these two goals. Electrical signals with SONET formats near 52 and 155 Mb/s are now defined, standards have been set for network element (NE)-to-NE communications using an OSI protocol stack within the DCC, and mappings for both Asynchronous Transfer Mode (ATM) and Fiber Distributed Data Interface (FDDI) signals into a 155 Mb/s STS-3C are completed. With attainment of a global SONET standard, there remains interest in OAM&P-related standards (e.g., management messages, alarm surveillance, performance monitoring) that are consistent with ITU-T recommendations. Other SONET Phase 2 activities have included clarification of and requirements for timing and synchronization, clarification of automatic protection switching, and specifications governing overhead usage.

A complete list of SONET and SDH standards and draft standards is provided in the Appendix.

3. SDH/SONET RATES, FORMATS, AND MULTIPLEXING METHODS

3.1 SDH/SONET Rates

Both SONET and SDH formats make use of basic building block signals. For SONET, this corresponds to the Synchronous Transport Signal Level-1 at 51.84 Mb/s. Lower rate payloads are mapped into the STS-1, while higher rate signals are obtained by byte interleaving N frame-aligned STS-1s to create an STS-N. This simple multiplexing approach results in no additional overhead; consequently, the transmission rate of an STS-N signal in precisely N × 51.84 Mb/s, with currently defined rates corresponding to the following values of N: 1, 3, 12, 24, and 48.

Similarly, for SDH systems, the 155.52 Mb/s Synchronous Transport Module Level-1 (STM-1) is the fundamental building block. Again, lower rate payloads are mapped into an STM-1, and higher rate signals are generated by synchronously multiplexing N STM-1 signals to form the STM-N signal. Insofar as the transport overhead of an STM-N is N times the transport overhead of an STM-1, the transmission rate is N × 155.52 Mb/s. As currently defined (ITU-T Recommendation G.707), the SDH transmission rates correspond to N = 1, 4, and 16. (Although not currently recommended, other rates for intermediate and higher values of N could be similarly generated. In particular, work is now progressing to standardize OC-192 and STM-64 signals.)

This information is succinctly presented in the table below, where we further note that concatenation of three STS-1 building block signals (i.e., STS-3C) is equivalent to an STM-1. However, the STS-3 SONET signal is three byte-interleaved (multiplexed) STS-1s, and looks quite different from an STM-1 signal.

SONET/SDH TRANSMISSION RATES

SONET Signal	SDH Signal	Transmission Rate
STS-1		51.84 Mb/s
STS-3	STM-1	155.52 Mb/s
STS-12	STM-4	622.08 Mb/s
STS-24		1244.16 Mb/s
STS-48	STM-16	2488.32 Mb/s

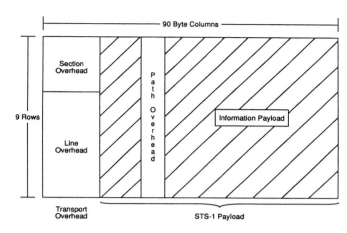

Fig. 2.   SONET STS-1 frame structure.

3.2 SDH/SONET Frame Formats

Both SDH and SONET frames are conveniently depicted as rectangular octet-based units transmitted every 125 μs (8000 frames/s). Because of compatibility between SDH and SONET, the frames are similarly structured, but must differ in dimension to reflect the basic building block transmission rates of 155.52 and 51.84 Mb/s, respectively.

As shown in Fig. 1, the STM-1 frame format is 9 rows of 270 bytes (1 byte = 8 b), or 2430 b/frame, corresponding to an aggregate rate of 155.52 Mb/s. Each octet equates to 64 kb/s, and an octet column of nine rows is 576 kb/s. Transmission occurs from upper left to lower right, as if reading words on a page. The first nine byte-columns of the basic frame are set aside for Section Overhead (SOH); the remaining 261 are for payload mapping. Further, the overhead portion is subdivided into three areas: the first, occupying rows 1–3, is known as Regenerator Section Overhead (RSOH); the second, occupying row 4, contains Administrative Unit (AU) pointers to identify the position of floating payloads; and rows 5–9 are identified as Multiplex Section Overhead (MSOH). In addition to the SOH, a column of Path Overhead (POH) resides within the Information Payload, and can, in fact, spill across two consecutive frames.

Each of these overheads has a functional equivalent in SONET. The specific functions are described generally in Section 5.1, and more specifically with regard to performance in Section

6. Briefly, SDH SOH supports transport capabilities such as framing, error monitoring, and management operations channels (orderwire, monitoring, and control of automatic protection switching, data communication channels, and an unspecified user channel). POH supports performance monitoring, status, signal labeling, a tracing function, and a user channel. This latter overhead is added and dismantled at or near service origination /termination points, and is not processed at intermediary nodes.

As previously mentioned, the SONET frame construct, illustrated in Fig. 2, is similar to that of the SDH frame. Now, the streamline is 9 rows by 90 byte-columns; with a frame repetition again of 8000 frames/s, the transmission rate is 51.84 Mb/s. In the SONET frame, the first three columns comprise Transport Overhead (TOH), the remaining 87 columns carrying payload (including POH, as mentioned in the context of the SDH frame). The TOH is itself divided into two portions: a 3-row Section Overhead (not to be confused with the SDH use of the same term) and a 6-row Line Overhead. The functionality of SONET TOH is akin to the SDH SOH.

The frame structures illustrated in Figs. 1 and 2 are purposefully depicted without offset. Actually, the SONET and SDH payloads can "float" within the respective frames, provided the payload begins on a row that is either the same as or follows the row containing pointers that identify the beginning of the information payload. Further, that the information payload can float reflects phase differences between NEs, i.e., as phase differences among NEs accumulate, the information payload can be moved to the left or right so as to synchronize NEs. Synchronization is described in Section 3.4.

3.3 SDH/SONET Multiplexing Structures

Two different multiplexing/mapping schemes are associated with SDH/SONET. The SONET approach is discussed first because of its comparative simplicity relative to the SDH scheme.

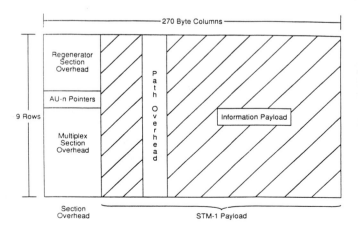

Fig. 1.   SDH STM-1 frame structure.

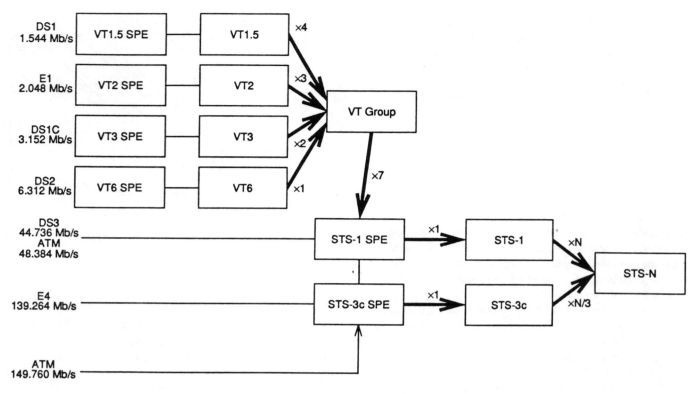

Fig. 3.   SONET multiplexing structure.

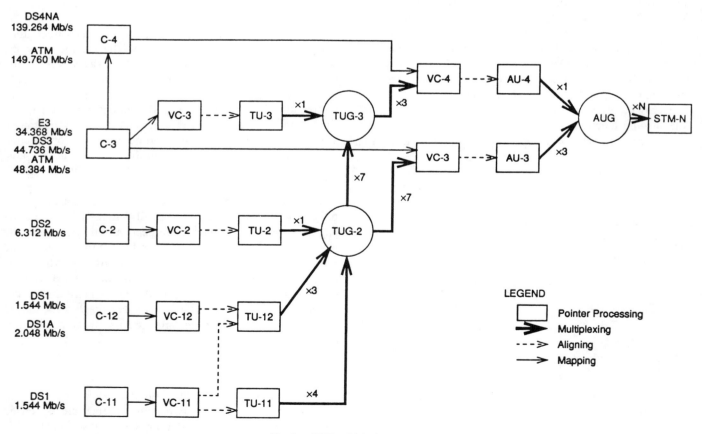

Fig. 4.   SDH multiplexing structure.

Figure 3 depicts the mapping and multiplexing of tributaries (DS1, E1, DS1C, etc.) into an STS-N. As a first step, each low-speed payload is mapped into a Virtual Tributary Synchronous Payload Envelope (VTx-SPE), associated with which are one of four VTs of appropriate size. Each VT has enough bandwidth to carry its respective payload: VT1.5 is carried in three 9-row columns (i.e., 27 bytes), VT2 in four, VT3 in six, and VT6 in 12 columns. As a next step, the VT Group is defined as a 9-row by 12-column payload carrying either four VT1.5s, three VT2s, two VT3s, or a single VT6. Then seven Virtual Tributary Groups (VTGs), making up 84 columns with one POH column and two unused columns, are byte-interleaved to fill out an STS-1 SPE. (VT groups made up of the various VT types mentioned above can be mixed within an STS-1 SPE.) The addition of TOH forms the STS-1 signal, which can be transported optically or electrically. Moreover, multiplexing N STS-1s creates an STS-N. Additionally, by frame aligning STS-N signals with pointer concatenation, an STS-NC signal with a locked ("concatenated") STS-N payload is created. Additional detail on these topics can be found in [2] and [3].

Multiplexing and mapping of SDH frame-formatted signals differs from that of SONET. SDH not only permits different mappings of the same payload, but also requires four hierarchical levels in forming the payload of an STM-1.

Begin by considering Fig. 4, which depicts the SDH multiplexing structure as presented in ITU-T Recommendation G.709. Plesiochronous signals are first mapped into one of five Containers (C) suitable to their individual bandwidth needs. Adding POH then results in a Virtual Container (VC), an SDH signal loosely analogous to an SONET VTG. Here, two types of VCs are defined: higher order VCs (i.e., VC-3 and -4) and lower order VCs (i.e., VC-2, -3, -11, and -12), with VC-3 treated as either higher or lower order. The higher order VCs are mapped to either Administrative Unit Level-3 (AU-3) or AU-4; lower order VCs are mapped to higher order VCs, with Tributary Unit (TU) pointers to locate them. At this point, one AU-4 or three AU-3s are mapped or multiplexed (respectively) to form the Administrative Unit Group (AUG) "SPE of an STM-1." As before, byte-interleaved multiplexing of N STM-1s creates an STM-N. As a final observation, the SDH multiplexing plan carries with it the notion of viewing one or more TUs as a Tributary Unit Group (TUG): no special overhead is associated with a TUG; it exists to facilitate network planning or traffic routing; hence, it becomes the responsibility of SDH network management to properly administer its path.

SDH signals are transported optically, electrically, via terrestrial microwave radio, or satellite.

Comments above make mention of mapping signals of the plesiochronous digital hierarchy (PDH) into SONET Virtual Tributaries or SDH Containers. This is a step of special significance, to the extent that the benefits of synchronous networking accrue only to the traffic types to which it applies. Fortunately, both SONET and SDH stipulate a plethora of suitable signals. For SONET, they are DS1, DS1A, DS1C, DS2, DS3, FDDI, DS4NA, DQDB, and ATM (near 50, 150, 600, and 2400 Mb/s); for SDH, they include DS1 (C-11 or C-12), DS1A, DS2, E3, DS3, DS4NA, and ATM. (ITU-T Recommendation G. 709 specifies the following SDH ATM mappings: 1.600, 2.176, 6.784, N × 6.784, 48.384, 149.76, and N × 149.76 Mb/s.)

Not all signal mappings listed above are depicted in Figs. 3 and 4, and there is much more detail on this subject which is beyond the scope of our survey. Readers should turn to the applicable standards for more information.

3.4 Synchronization

Synchronization is a key aspect of SONET and SDH, and is an underlying basis to the facile hierarchical multiplexing and demultiplexing that the transmission format permits. Synchronization is, first and foremost, predicated on the availability of a strata of clocks which are traceable to one or more extremely precise references, from which local clock is directly available at carrier central offices or loop timed from the signal carried on fiber trunks terminating on NEs.

Even under optimum conditions, however, clock transients, facility interruptions, environmental factors, or NEs themselves can give rise to instantaneous frequency deviations. As a result, the incoming bit rate can depart slightly from the outgoing bit rate, which, over a short interval, can give rise to accumulated phase differences which cause buffer overflows.

SONET and SDH accommodate this by using pointers to offset the phase accumulation due to frequency deviations. The phase difference is accumulated in a buffer until a level is reached indicating that adjustment is necessary. At that point, an information byte is either written to the payload or the buffer is inhibited for one byte interval prior to writing the payload. This is accomplished via pointers in the SDH/SONET overhead which identify the location of the information payload, and incrementing or decrementing that payload. For example, if the incoming signal rate is greater than the outgoing rate, the buffer will overflow, the payload is shifted one byte to the left by virtue of decrementing the pointer value, and the previous payload is reduced by one byte of information.

3.5 SDH and SONET Differences

As noted in the Introduction, SDH and SONET are compatible digital hierarchies. However, differences do exist between them, not only with respect to semantics and the rates of their respective basic building block signals, but also in regard to overhead usage. Those overhead differences have been grouped into two broad categories: format definitions and usage interpretation. Many of the differences are subtle, and a compilation is outside the scope of this survey article. Nevertheless, readers are directed to [3], which presents the relevant information first with respect to Section Line/Path overhead byte definitions for SONET and SDH, and points of difference; and second, using the same format, with regard to performance monitoring information.

4. NETWORK TOPOLOGIES

SONET and SDH, per se, do not afford new network topologies. Rather, SONET/SDH-compliant equipment facilitates some network topologies that would otherwise be economically

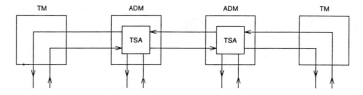

Fig. 5. Linear add/drop configuration.

prohibitive or administratively cumbersome. Here, we review several basic network configurations which are generally dependent upon add/drop multiplexers (ADMs) and digital cross-connect systems (DCSs), with terminal multiplexers (TMs) considered a special case of an ADM. Further, instead of commenting on the evolution of deployment configurations from simple point-to-point systems to hubbing arrangements (as reviewed by Sandesara *et al.* [4]), emphasis is given to ring configurations.

4.1 Linear Add/Drop

The linear add/drop configuration, depicted in Fig. 5, rapidly superseded the initial deployment of simple point-to-point systems, especially as used in the loop versus interoffice portions of

telecommunication networks. Here, a linear chain of connected nodes (outside plant enclosures or central offices) carries traffic among the nodal ADMs, with traffic originating or terminating at any node. As mentioned above, the TMs, which by definition appear at the ends of the chain, are simply special cases of ADMs, the latter permitting adding, dropping, or passing through of traffic in a manner functionally akin to back-to-back multiplexers, although remotely provisionable and more economical. The ADM add/drop/pass-through feature is generally based upon a time-slot assignment (TSA) fabric which permits flexible reassignment of time slots (i.e., time-division multiplexed channels) among the East–West fiber trunks and the local service interfaces.

As customer expectations for improved service reliability increase, the linear add/drop chain gives way to a more survivable network topology. By conceptually folding the linear add/drop chain back on itself (eliminating the TMs [5]), ring architectures are created, as more fully discussed below.

4.2 Ring Architectures

SONET/SDH rings, relying upon ADMs with unidirectional or bidirectional transport, are of two principal types: unidirec-

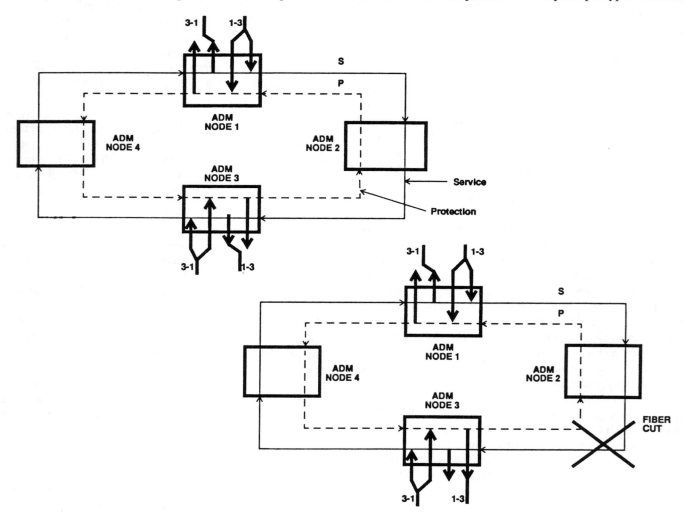

Fig. 6. Two-fiber unidirectional path-switched ring.

tional path-switched rings (UPSRs) and bidirectional line-switched rings (BLSRs). Additionally, these ring architectures are themselves dependent upon basic protection switching, which is briefly described below prior to a more detailed discussion of UPSR and BLSR.

4.2.1 Basic Protection Switching. By their very nature, SONET/SDH fiber optic systems provide high-capacity transport. During network failure, whether electronic or due to a fiber cut, these systems inherently result in substantial service outage, making the need for restoration that much more important. Two basic types of automatic protection switching come into play in these systems: path-protection switching (PPS) and line-protection switching (LPS).

PPS, as the name infers, protects the path level as the low-speed (service) input to ADMs. Referred to as (1+1)-protection, traffic transmitted by the ADM is duplicated and sent out over two independent fibers, the working and protection pair. When a failure occurs, the receiver compares the quality of the dual-fiber input signals and selects the best. Since both copies of the subtrunk channel are available at the receiver, PPS is especially rapid, although the assumption of trunk integrity probably requires consideration of diversely routed fibers.

In LPS, line-layer switching requires that the fiber trunk interconnecting two nodal ADMs be duplicated, with working and protection transmit/receive pairs. In the event of disrupted or degraded signal quality (e.g., loss of signal, LOS, or line-alarm indicating signal, L-AIS), trunk traffic is switched to the protection pair, a protection means designated "1:1." In some instances, one protection fiber pair protects *n* working pairs, "1:*n*," although this approach usually requires some degree of prioritization, with low-priority traffic carried on the protection fibers and shed in the event of a protection switch. Concerns sometimes associated with LPS include service-affecting switching, time-to-recovery (longer than PPS), and expensive fiber and electronics. The availability of the network may be further improved by having the protection fiber pair in separate, diversely routed cable, also known as diverse route protection.

As discussed later in this tutorial, concerns mentioned above with respect to LPS are not particularly significant. For example, although time-to-recovery is longer for LPS than PPS, they both meet the requirement for a 60 ms switch time, measured from onset of failure (e.g., see Bellcore TR-NWT-000253, "Synchronous Optical Network (SONET) Transport Systems: Common Generic Criteria"). Further, we will later observe that PPS used with UPSR is best suited for loop environments, and LPS with BLSR is preferable in interoffice networks.

4.2.2 Unidirectional Path-Switched Rings. Based on the above comments, a UPSR protects traffic on the basis of individual path quality. Further, as UPSR implies, this restoration technique assumes a unidirectional ring that carries service traffic in one direction only, i.e., clockwise or counterclockwise, with the protection fiber carrying traffic in the opposite direction. Consider Fig. 6, where the diagram to the left depicts a normally operating four-node ring, while the illustration on the right shows reconfiguration in response to a fiber cut. Normally, traffic from node 1 to 3 is dual fed at node 1, with traffic carried

clockwise on the service (S) fiber and counterclockwise on the protection (P) fiber. Traffic from node 3 to 1 similarly traverses the ring. With the fiber cut between nodes 2 and 3, traffic is switched internally to node 3, allowing service from node 1 to be carried counterclockwise via node 4 on the P fiber.

4.2.3 Bidirectional Line-Switched Rings. An especially robust form of ring network is attained using LPS in conjunction with bidirectional transmission. Our review initially focuses on four-fiber rings, as shown in Fig. 7; two-fiber BLSRs are subsequently discussed.

Begin by considering interruption of transmission on a transmit and receive service pair between two nodes, as shown on the left of Fig. 7. Such a situation can be due to either an electronics failure (most likely) or a fiber cut. This failure initiates line switching to each of the respective transmit and receive protection pairs. Observe that conventional operation is maintained between those nodes on the ring which do not experience failure; indeed, such "span switching" can occur between other nodal pairs on the ring, and line switching is invoked only to heal the failed service fiber pair.

Most fiber cuts not only take out a service fiber pair, but can sever the protection fiber pair as well. Such an event, depicted to the right in Fig. 7, results in "ring switching," wherein the ADMs on either side of the failed internodal span internally loop traffic back from the failed transmit fiber to the transmit protection fiber, using a longer route around the ring to reach the desired destination ADM. A similar ring switch maintains service integrity in the opposite receive direction.

As mentioned above, BLSRs can also be implemented using only two fibers between adjacent ADMs. These two-fiber BLSRs share service and protection equally between the two fibers, i.e., the capacity of a single fiber (e.g., the clockwise fiber) is given over to half service and half protection. As a result, the aggregate capacity of a four-fiber BLSR is twice that of a two-fiber BLSR. Further, unlike the four-fiber BLSR, span switching is not applicable to two-fiber BLSRs since the latter have only one fiber per direction of transmission. Ring switching in two-fiber BLSRs is, however, fully analogous to the four-fiber ring. If a fiber cut occurs, the ADMs on both ends of the span will initiate a loopback: service traffic previously carried via the two fibers on the interrupted span are routed around the ring in the reverse direction by switching service channels onto the protection channels reserved between the other spans that complete the ring. A second fiber cut is usually service interrupting: the BLSR will partition into two smaller rings with two fiber cuts, while maintaining traffic within the subrings.

4.2.4 UPSR/BLSR: Comparative Benefits. Considerable effort has been expended in understanding the appropriate applicability of UPSRs and BLSRs in telecommunication networks. The following observations seem to find a consensus among network planners:

- Networks in which traffic predominately converges at just one node along a ring, as is often the case in local exchange carrier (LEC) access networks, are best served by UPSRs.
- In contrast, those rings for which traffic demand is more

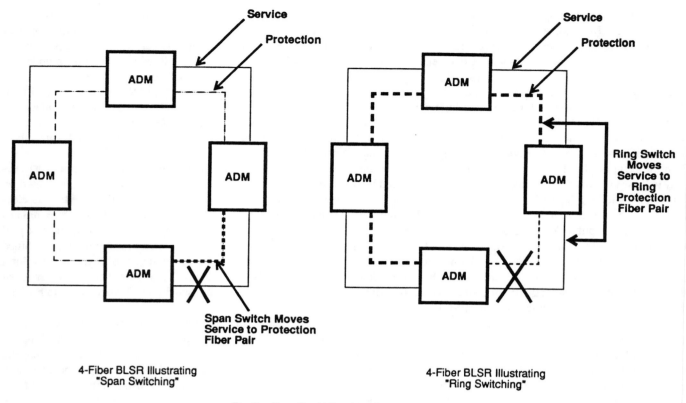

Fig. 7.   Four-fiber bidirectional line-switched ring.

evenly distributed among the nodes, as in LEC interoffice networks and interexchange carrier (IEC) networks, are commonly of the BLSR variety.

- By using both span and ring switching, the four-fiber BLSR affords the more robust (and more expensive) protection means.
- The switch-to-restoration time is less with UPSRs than with BLSRs, with both meeting the 60 ms system switch-time requirement. UPSRs do not require complex rerouting—the traffic is already replicated and available on the protection fiber; switching times are not affected by the ring size or number of ADMs on the ring; and administrative complexity is unaffected by the number or distribution of other fiber failures.
- Regardless of how demand is distributed, UPSR capacity can be no more than the line rate of the interconnecting optical facilities. In a two-fiber BLSR with uniformly distributed traffic, the aggregate capacity is twice the line rate and, therefore, twice that of the corresponding UPSR. As traffic becomes centralized, this capacity must diminish.

The discussion above relates to isolated rings. The subject of multiple rings and the ensuing need to manage interworking among them remains a topic of ongoing investigation, and is partially treated by several papers in this book.

4.3 Mesh Networks

Mesh networks make up a third broad category of network topologies. (Star/hub networks, considered by some as a separate network topology (e.g., see [6]), are viewed here as an example of mesh networks insofar as they, too, rely upon DCSs as the fundamental network element.) These networks are characterized by high levels of connectivity among the individual nodes (for a fully interconnected network with M nodes, there can be as many as $M(M-1)/2$ trunk facilities), and are commonly served by DCSs. In the event of a nodal or facility failure, restoration is achieved by identifying and routing traffic onto a diverse path. Restoration can either be administered centrally or distributed locally to the individual cross-connect network element, and can either depend on prestored contingency events with associated cross-connect tables, or be dynamically controlled by a restoration algorithm [6], [7]. These topics are also areas of active research, some of which are covered elsewhere in this book.

5. NETWORK MANAGEMENT

SDH/SONET is recognized as providing three major benefits to service providers and customers: 1) a standard optical midspan that assures competitive vendor products and is the basis for uniform multivendor equipment assessment; 2) high-speed, flexibly assigned bandwidth to support existing and new applications and services; and 3) comprehensive network management capabilities. This third item, although listed last, has been a consistent vision from SDH/SONET conception through to standards setting and network deployment.

Network management is commonly viewed as embracing five functional areas:

- Configuration management, which deals with administering the physical aspects of a network, i.e., network elements and transmission facilities, with associated bandwidth allocation;
- Fault management, involving alarm monitoring and maintenance procedures, such as protection switching and facility/equipment diagnostics;
- Performance management, which builds upon a variety of quality metrics (e.g., errored seconds, severely errored seconds, and unavailable seconds) that can lead to fault messages;
- Security management, assuring only authorized access to the managed network via passwords and user privileges; and, finally,
- Management of the network management system itself.

SDH/SONET supports each of these function areas via a layered overhead structure, performance monitoring data and maintenance signals, and ample overhead channels. Additionally, SDH/SONET embraces a coherent network management architecture—Telecommunications Management Network, TMN—that capitalizes on the three SDH/SONET attributes mentioned above and individually described below.

5.1 Capabilities Intrinsic to SDH/SONET Signals

As described elsewhere in this overview, the synchronous format and multiplexing structure basic to SONET and SDH provide a layered signal structure and overhead important to network management. Of the four layers—photonic, section, line, and path (SONET terminology)—the latter three have supporting overheads. With regard to the first of these, the section layer provides basic payload performance monitoring, signal framing, a local orderwire channel, and a 192 kb/s DCC. Every network element must process all or part of the section overhead; hence its importance in supporting fundamental network management features.

The line layer supports automatic protection switching and multiplexing functions for the SDH/SONET payload. Protection switching messages are processed by line termination or multiplex equipments; multiplexing management includes payload pointer storage. A 576 kb/s DCC is also provided, as is an express orderwire.

POH provides two overhead functions: a VT/VC overhead supports performance monitoring, status, and signal labeling; a POH spanning information within the synchronized payload provides the same three capabilities, plus a tracing function and user channel.

The comments above make reference to performance monitoring as a standardized aspect of SDH and SONET. Performance monitoring is characterized in terms of monitoring primitives and defects (fundamental measures exercised at each overhead layer) and parameters (as important measures of actual impairment events, parameters are derived from primitives and defects,

and are customarily used to gauge quality or service). These are elaborated in much more detail in ITU-T Recommendation G.784 and ANSI standard T1.231, cited elsewhere in this tutorial and Section 6. Other papers in this book describe these in greater detail: here, we note that the primitives and defects are alarm indication signal, bit-interleaved parity, loss of pointer and loss of signal, remote defect indication, path far-end performance report, and section loss of frame and section severely errored frame. Prominent performance parameters, actually numerical counts of impairments logged over a specific time interval, include errored seconds for section, line, path, and VT/VC path; severely errored seconds associated with the same section, line, and paths; and unavailable seconds for line, far-end line, payload path, far-end payload path, VT/VC path, and far-end VT/VC path.

In addition to performance monitoring, network management depends upon standard maintenance signals which, while reporting on the status of the received payload, require that terminating equipment act on these condition indicators and then report those actions to the management system as status updates. Maintenance signals include alarm indication signal, remote defect indications, remote failure indication, and unequipped indications (status reports for partially equipped network elements).

We gloss over orderwire and user channels, pausing to note that their roles in network management require further study. Instead, focus is directed to DCCs which facilitate passing information among network elements or, with protocol conversion, to operation support systems (OSSs). As noted earlier, the mandatory 192 kb/s section layer DCC must be terminated at every network element; the optional 576 kb/s line layer DCC would only be processed at line terminating elements, and could be used to support digital cross-connect system based network reconfiguration. In both cases, identical OSI protocol stacks have been standardized.

5.2 Telecommunications Management Network

As briefly reviewed above, SDH and SONET provide a variety of channels, signals, and messages amenable to administering the five functional aspects of network management listed earlier in this section. Beyond these, TMN constitutes a structured framework to orchestrate information gleaned from the lower level SDH/SONET signals and messages into a picture of historical data and network status. This coherent view is afforded by OSSs, i.e., computer-based information systems that collect network equipment performance information.

In the past, OSSs (e.g., for alarm surveillance, testing, and provisioning) were connected to NEs via leased lines from each NE to the separate OSSs. The protocols and command language for each OSS–NE interface were unique and generally proprietary. SDH/SONET introduces two modifications. First, a common protocol/language is used, based on the seven-layer OSI stack, defined Common Management Information Service Elements (CMISE), Common Management Information Protocol (CMIP) for message implementation, with data represented using Abstract Syntax Notation 1 (ASN.1). Second, in the SDH/SONET

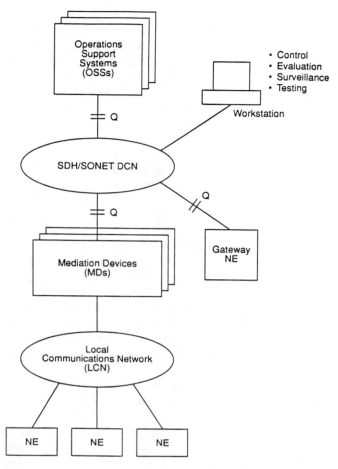

• Control
• Evaluation
• Surveillance
• Testing

Fig. 8.   TMN architecture.

environment, multiple leased lines give way to an SDH/SONET data communications network (SDH/SONET DCN), actually an X.25 packet-switched network (PSN), to interconnect the managed NEs with the multiple OSSs. The usual benefits of a PSN are expected to accrue, namely, sharing a single physical channel among multiple logical channels engenders substantial savings relative to numerous leased lines and their attendant maintenance problems.

Now, consider the high-level TMN architecture illustrated in Fig. 8. The notable entities include multiple OSSs, the SDH/SONET DCN, Mediation Devices (MDs), a Local Communications Network (LCN), and managed NEs. Craft interface the TMN via work stations to the SDH/SONET DCN and/or LCN. Several of these architectural entities are discussed above.

Note the presence of a second, local DCN (i.e., the LCN), intended for use in central office environments, or with specific equipment to support dissemination of collected messages over a small geographical area. Unlike the X.25-based DCN, this LCN is based on Ethernet (IEEE 802.3). Both the DCN and LCN use a seven-layer OSI protocol stack.

MDs are a fundamental aspect of the TMN architecture. Their principal function is to provide protocol conversion between the

SDH/SONET DCN and LCN, thereby facilitating communications among central office equipments and OSSs. An additional prospect is for the MD to handle NE-dependent information processing functions, such as performance monitoring and data storage, and alarm thresholding. This view then gives rise to the concept of subnetworks individually mediated by MDs, so that OSSs begin to view the overall network as managed subnetworks. Indeed, these comments highlight two means for managed NEs to communicate with their OSSs: one has a single NE, referred to as the Gateway Network Element (GNE) with integrated MD functionality, providing the needed protocol conversion and message routing; the second means is by way of the MD, which mediates on behalf of NEs in a subtending network.

Introducing TMN into telephony networks offers many challenges, not the least of which is coadministration of SDH or SONET networks in concert with existing networks. This topic and others associated with TMN and network management are treated in the cited references and several papers in this book.

6. PERFORMANCE ISSUES

SDH/SONET systems are currently being designed and implemented as per the design rules of ITU-T Recommendation G.821. (The discussion here relates principally to SDH. As previously noted, the relationship to SONET is extremely close. Differences between overhead usage in SONET and SDH for performance is highlighted in [3].) This recommendation has been in place for nearly ten years, and has been the basis for the design of PDH digital connections forming part of an ISDN connection. Recommendation G.821 specifies objective values for the error performance parameters for a 27 500 km hypothetical reference path operating at the primary rate of 64 kb/s.

More recently, the ITU-T sector has approved a new Recommendation G.826 which is concerned with the subject of error performance objectives operating at or above the primary rate, e.g., if the end-to-end payload is 2 Mb/s, Recommendation G.826 applies. Consequently, SDH/SONET systems transporting broadband services (video, higher data rates, etc.) will need to be designed according to G.826.

The new performance recommendation is a significant departure from its predecessor, in the sense that objectives are media-independent, block-based, and suitable for performing in-service measurements. Also, the objective values for the new error performance parameters in G.826 translate to much tighter BER requirements than the corresponding values imposed by G.821. The impact of G.826 on the design of transport systems is discussed further in [8].

6.1 Performance Monitoring

The SDH/SONET systems offer a new opportunity to include performance monitoring which was not possible in PDH systems. This is because the overhead bytes in the SDH/SONET frame provide a facility for doing in-service performance measurements. In addition, particular bytes are reserved for media-specific functions. Figure 9 illustrates the Section Overhead (both RSOH and MSOH, SDH nomenclature) byte allocation.

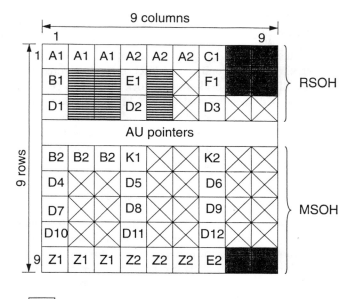

9 columns

RSOH

9 rows

MSOH

AU pointers

  Bytes available for radio specific usage (media specific bytes)

  Bytes reserved for national usage

Bytes reserved for future international standardization

Fig. 9.   Allocation of SOH bytes.

The designation of the SOH bytes is as follows:

- 6 bytes (A1,A2) for frame alignment
- 2 bytes (E1,E2) for order wire channels
- 3 bytes (B2) for multiplex section bit error rate (BER) monitoring
- 1 byte (C1) for STM identification
- 1 byte (B1) for regenerator section BER monitoring
- 1 byte (F1) for user channel
- 2 bytes (K1, K2) for automatic protection switching
- 12 bytes (D1, D2 through to D12) for DCCs
- 6 bytes for national use
- 6 bytes (Z1, Z2) are not yet defined
- 6 bytes for media-specific use
- 26 bytes for future international standardization

In addition, there are Path Overhead (POH, SDH nomenclature) bytes associated with various VCs. They are, for VC-4 and VC-3, a column of 9 bytes, as follows:

- 1 byte (J1) for VC-N path trace
- 1 byte (B3) for bit-interleaved parity (BIP-8)
- 1 byte (C2) for signal label
- 1 byte (G1) for path status
- 1 byte (F2) for VC-4 path user channel
- 1 byte (H4) for multiframe indicator
- 3 bytes (Z3, Z4, Z5) as spare

[1]Remote defect indication (RDI) was previously known as far-end remote fail (FERF).

Likewise, there is a POH for VC-2, VC-12, and VC-11. Details for these can be found in Recommendation G.709.

The overhead bytes are used to determine if compliance with G.826 objectives is attained.

6.2 Performance Measurements

Performance measurements made on STM signals are used to declare the following network events, categorized as either *defects* or *anomalies*.

6.2.1 Defects.   A defect is considered to be a condition under which the network has lost its ability to transport bits. During such a condition, equipment at the receiving end of the path will experience a high BER. Examples of defects are given below:

- Loss of signal (LOS)
- Loss of frame alignment (LOF)
- Multiplex section alarm indication signal (MS-AIS)
- Multiplex section far-end remote defect indication (MS-RDI[1])
- Administrative unit loss of pointer (AU-LOP)
- Administrative unit alarm indication signal (AU-AIS)
- Higher order path far-end remote defect indication (HP-RDI)
- Tributary unit loss of pointer (TU-LOP)
- Tributary unit alarm indication signal (TU-AIS)
- Higher order path trace identifier monitor (HP-TIM)
- Lower order path trace identifier monitor (LP-TIM)

6.2.2 Anomalies.   Anomalies give rise to errored seconds, but do not cause severely errored seconds. Examples of anomalous events are listed below:

- Out of frame alignment (OOF)
- B1 errors
- B2 errors
- B3 errors
- Higher order path far-end block errors (HP-FEBE)
- Lower order path far-end block errors (LP-FEBE)
- BIP-2 errors

Each of the above events is subject to a certain criterion to declare that event. A detailed discussion on the criteria for all the above events is contained in Recommendation O.SDH. A summary of criteria for some key events is contained below.

6.2.3 Criteria for Detecting Defects and Anomalies for Some Events.   The criteria for detecting anomalies and defects are as follows:

- For OOF — Frame alignment is found by searching for the A1 and A2 bytes. The framing pattern searched for corresponds to the overall A1 and A2 bytes. The frame signal is continuously checked with the presumed frame-start position for alignment. An OOF state is declared when at least one bit in A1 or A2 bytes is in error over four consecutive frames.
- For B1 errors — Parity errors contained in byte B1 (BIP-8) of an STM-N signal is monitored. If any of the eight parity

checks fails, the corresponding block is assumed to be in error.

- For B2 errors — The parity contained in byte B2 (BIP-24N) of an STM-N signal is monitored. If any of the N × 24 parity checks fails, the block is declared in error.

- For B3 and BIP-2 errors — The parity errors contained in byte B3 (BIP-8) of a VC-n (n = 3, 4) shall be monitored. If any of the eight parity checks fails, the corresponding block is declared in error. Likewise, for BIP errors, the parity errors in bits 1 and 2 of byte V5 of a VC-m (m = 11, 12, 2), also known as VC-m POH, shall be monitored. If any of the two parity checks fails, the corresponding block is declared in error.

- For LOS — For an electrical interface, an LOS defect is declared when the incoming signal has no transition over a period on N (yet to be decided) time pulses. For an optical interface, an LOS defect is declared when the optical power level of the incoming signal is lower than a specified level.

- For LOF — If an out-of-frame state persists for a specified time (in milliseconds), an LOF defect is declared. The LOF defect is cleared when no OOF is present for a certain period (in milliseconds).

6.3 Compliance with Performance Recommendation

Recommendation G.826 presents a relationship between SDH path performance monitoring and block-based parameters. Similar relationships are also given for PDH and cell-based path performance monitoring. A review of the SDH path performance monitoring is provided below:

The monitoring of the above-mentioned events are used to ensure compliance with performance recommendation (e.g., G.826), but only when the network is in the available state. (An unavailable state period is declared when a period of ten consecutive severely errored seconds is experienced.) The above-mentioned events are also monitored for network management purposes to ensure compliance with G.784.

As previously discussed, Recommendation G.826 uses in-service measurements. These measurements are based on the notion of *errored blocks*, i.e., blocks with one or more bits in error. A block is a set of consecutive bits associated with the path and monitored by means of an error detection code (EDC), e.g., bit-interleaved parity (BIP). The block lengths and associated EDCs for various SDH entities (path, multiplex section, regenerator section) are contained in Recommendation O.SDH. Block sizes for various STM-N signals are also given in G.826.

Based on the block measurements, the following events are identified:

- Errored Block (EB) — A block in which one or more bits is in error

- Errored Second (ES) — A one-second period with one or more errored blocks

- Severely Errored Second (SES) — A one-second period which contains more than 30% errored blocks, or a defect event has occurred

- Background Block Error (BBE) — An errored block not occurring as part of an SES.

Recommendation G.826 gives values of error performance parameters, errored second ratio (ESR), severely errored second ratio (SESR), and background block error ratio (BBER). Compliance with these parameters should be evaluated over a one-month period when the network is available. Various events discussed earlier may give rise to either ESR, SESR, or BBER. A table of these events and the performance parameter effected is contained in Recommendation O.SDH.

7. CONCLUDING COMMENTS

The future of SDH and SONET lies in their ability to demonstrate the benefits of a synchronous multiplexing hierarchy over a plesiochronous multiplexing hierarchy for existing applications, show flexibility in accommodating a new transport modality in ATM, and exhibit resilience by supporting new services and associated network architectures.

7.1 Benefits

The benefits accruing from SDH and SONET were cited earlier as a basis for compatible, less expensive transmission equipment; flexible bandwidth assignment to support existing and new services; and comprehensive network management with the potential for reduced costs. The first of these — often simply alluded to as the so-called midspan meet — assures equipment commonality, a more competitive market, with reduced transmission costs. To this list one might also add increased equipment functionality and a "future proof" network guided by global standards rather than hampered by diverse technology and proprietary implementations.

The second item — flexibly assigned bandwidth — is the necessary foundation for incipient broadband services such as Broadband ISDN. In part, this generic attribute is based on generous overhead channels that permit more rapid network response and operations functions. These channels also allow for remote equipment configuration that can include channel provisioning close to the customer. The final benefit of this triad — improved network management and reduced operations costs — derives from standardized planning and OAM&P that spans multiple equipments, reduced sparing via enhanced equipment functionality (e.g., eliminating back-to-back multiplexing with ADMs), electronic cross-connection of traffic, and the potential to reduce site visits and personnel requirements via intelligent network elements. This last factor is already bringing about fundamental changes in network operation by reducing circuit management on a time scale that is now hours or fractions of an hour (apart from protection switching and restoration, which are yet at least an order of magnitude faster).

Deployment of SONET- and SDH-compliant networks is well underway. Initially, these networks served loop, regional, and metropolitan areas, with transmission equipment serving the distribution portion of loop networks and then connecting main switching exchanges. The current network frontier is to bridge together SDH/SONET islands, and link to compatible long-distance trunk networks. In so doing, we are ever closer to genuine end-to-end connectivity, thereby fulfilling the promise of comprehensive performance management.

7.2 New Networking Applications

Two factors will presently test the promise of SDH and SONET: new network applications and ATM. With regard to the former, we view enterprise networks and broadband services delivery to the residential subscriber as networking trends that afford special promise and challenge to SDH/SONET [9], [10]. Already, large business customers are constructing enterprise networks using synchronous optical transmission across campus environments (i.e., business parks, universities, etc., spanning comparatively small geographic domains, generally serving a single end-user community, and frequently wholly privately owned). These networks are typically data-oriented, and a plethora of "traditional" data networking products with SONET interfaces have come to market. Moreover, these large business customer networks are being linked to local- and interexchange networks, increasingly with the service provider under contract to supply complete OAM&P. Network issues associated with this kind of service capability are partially described in [11].

7.3 ATM

ATM represents a second dimension shaping the future of SDH and SONET. The relationship of ATM to SDH/SONET has been a source of much confusion, with ATM and SDH/SONET sometimes cast in "either/or" adversarial roles. To some extent, this confusion represents a fundamental misunderstanding since, in the OSI model, SDH, and SONET reside in the Physical Level, and ATM is at the Data Link Level. Regardless, the existence of SDH and SONET mappings into Containers and Virtual Tributaries has already been mentioned. Actually, focus should be on the operational complexity of a synchronous network carrying ATM traffic, an issue exemplified in network survivability. This is a topic of active research (e.g., see [12]); however, the issue can be immediately appreciated by recognizing that restoration in the SDH/SONET context (discussed in Section 4) relates to synchronous transfer mode (STM) TDM channels that are coupled to *physical* connections, whereas ATM is associated with *logical* connections.

SDH and SONET are already delivering their promised benefits. Networks using this technology and transmission format will find continued deployment, with islands ultimately bridged to provide a cohesive whole. At the same time, synchronous optical networks will evolve to come even closer to the business customer, deliver broadband services to the residential end-user, and meet the challenges and opportunities of ATM.

ACKNOWLEDGMENT

The authors are pleased to recognize colleagues who have contributed to the preparation of this paper. They include: J. Anderson, S. I. Haque, S. H. Hersey, W. Kremer, M. J. Soulliere, and E. L. Varma of AT&T Bell Laboratories. Appreciation is also extended to Telecom of New Zealand for support of this work. Additionally, we gratefully acknowledge anonymous reviewers whose excellent suggestions improved both this survey and aspects of the book.

We are further grateful for source material used in the preparation of certain sections of this survey, or work of special relevance to the respective topics. They are:

- Section 2: References [2], [13], [14], and [15]
- Section 3: References [2], [3], and [6]
- Section 4: References [4], [5], [6], [7], [16], and [17]
- Section 5: References [18], [19], and [20]
- Section 6: Applicable SDH Recommendations and SONET Standards
- Section 7: References [10] and [12]

In every section, readers' attention is directed to the references cited therein.

APPENDIX

SDH/SONET STANDARDS

The information provided here is believed to be complete. The *status* of these standards is not reported, however, since that information will inevitably change.

SONET Standards

SONET standards have been developed by no less than five ANSI-accredited technical subcommittees, eight working groups, and seven subworking groups. The principal SONET and SONET-affecting standards and draft standards are listed below:

- ANSI T1.101, "Draft Proposed American National Standard for Telecommunications — Synchronization Interface Standard."
- ANSI T1.102, "American National Standard for Telecommunications — Digital Hierarchy — Electrical Interfaces."
- ANSI T1.105, "American National Standard for Telecommunications — Digital Hierarchy — Optical Rates and Formats Specification (SONET)."
- ANSI T1.105.01, "American National Standard for Telecommunications — Synchronous Optical Network (SONET): Automatic Protection Switching."
- ANSI T1.105.02, "Draft Proposed American National Standard for Telecommunications — Digital Hierarchy — Synchronous Optical Network (SONET): Payload Mappings."
- ANSI T1.105.03, "Draft Proposed American National Standard for Telecommunications — Digital Hierarchy — Synchronous Optical Network (SONET): Jitter at Network Interfaces."
- ANSI T1.105.04, "Draft Proposed American National Standard for Telecommunications — Digital Hierarchy — Synchronous Optical Network (SONET): DCC Protocols and Architectures."
- ANSI T1.105.05, "Draft Proposed American National Standard for Telecommunications — Digital Hierarchy — Synchronous Optical Network (SONET): Tandem Connection Maintenance."

- ANSI T1.105.06, "Draft Proposed American National Standard for Telecommunications — Digital Hierarchy — Synchronous Optical Network (SONET): Physical Layer Specifications."
- ANSI T1.106, "American National Standard for Telecommunications — Digital Hierarchy Optical Interface Specifications, Single Mode."
- ANSI T1.107, "American National Standard for Telecommunications — Digital Hierarchy — Formats Specifications."
- ANSI T1.107a, "Supplement to American National Standard for Telecommunications — Digital Hierarchy — Formats Specifications."
- ANSI T1.107b, "Supplement to American National Standard for Telecommunications — Digital Hierarchy — Formats Specifications."
- ANSI T1.117, "American National Standard for Telecommunications — Digital Hierarchy Optical Interface Specifications, Short Reach."
- ANSI T1.119, "American National Standard for Telecommunications — Synchronous Optical Network (SONET): Operations, Administration, Maintenance, and Provisioning (OAM&P) Communications."
- ANSI T1.119.01, "Draft Proposed American National Standard for Telecommunications — Digital Hierarchy — Synchronous Optical Network (SONET): Operations, Administration, Maintenance, and Provisioning (OAM&P) Communications Protection Switching Fragment."
- ANSI T1.204, "Draft Proposed American National Standard for Telecommunications — Operations, Administration, Maintenance and Provisioning (OAM&P) — Lower Layer Protocols for Telecommunications Management Network (TMN) Interfaces."
- ANSI T1.208, "American National Standard for Telecommunications — Operations, Administration, Maintenance and Provisioning (OAM&P) — Upper Layer Protocols for Telecommunications Management Network (TMN) Interfaces Between Operations Systems and Network Elements."
- ANSI T1.214, "Draft Proposed American National Standard for Telecommunications — Operations, Administration, Maintenance and Provisioning (OAM&P) — A Generic Network Model for Interfaces Between Operations Systems and Network Elements."
- ANSI T1.214a, "Supplement to American National Standard for Telecommunications — Operations, Administration, Maintenance and Provisioning (OAM&P) — A Generic Network Model for Interfaces Between Operations Systems and Network Elements."
- ANSI T1.215, "Draft Proposed American National Standard for Telecommunications — Operations, Administration, Maintenance and Provisioning (OAM&P) — Fault Management Messages for Interfaces Between Operations Systems and Network Elements."
- ANSI T1.229, "American National Standard for Telecommunications — Operations, Administration, Maintenance and Provisioning (OAM&P) — Performance Management

Functional Area Services for Interfaces Between Operations Systems and Network Elements."
- ANSI T1.231, "Draft Proposed American National Standard for Telecommunications — Layer 1 In-Service Digital Transmission Performance Monitoring." ANSI T1.5D, "Draft Proposed American National Standard for Telecommunications — Network Performance Parameters for Dedicated Digital Services — Specifications."
- T1M1.2/87-37R2, "Functional Requirements for Fiber Optic Terminating Equipment."

In addition to the standards and draft standards listed above, Bellcore has several technical advisories (TAs) and technical reports (TRs) which warrant vendor consideration for equipment deployment among North American Bell Operating Companies. These include:

- TA-NWT-000233, "Wideband and Broadband Digital Cross-Connect Systems Generic Criteria."
- TA-NWT-000253, "Synchronous Optical Network (SONET) Transport Systems: Common Generic Criteria."
- TA-TSY-000496, "SONET Add-Drop Multiplex Equipment (SONET ADM) Generic Criteria for a Self-Healing Ring Implementation."
- TA-NWT-000782, "SONET Digital Trunk Interface Criteria."
- TR-NWT-000253, "Synchronous Optical Network (SONET) Transport Systems: Common Generic Criteria."
- TR-NWT-000303, "Integrated Digital Loop Carrier System Generic Requirements, Objectives, and Interface" and supplement, "IDLC System Generic Requirements, Objectives, and Interface: Feature Set C — SONET Interface."
- TR-NWT-000496, "SONET Add-Drop Multiplex Equipment (SONET ADM): Generic Criteria and Supplement 1."
- TR-NWT-000917, "SONET Regenerator (SONET RGTR)."

SDH Standards

Standardization of the Synchronous Digital Hierarchy commenced with the European Telecommunications Standards Institute (ETSI), a forum wherein options preferred by Europe were discussed. International standardization progressed within the ITU-T (formerly CCITT) and ITU-R (formerly CCIR) sectors: CCITT study groups 15 and 18, and CCIR study groups 4 and 9 were involved in preparing the relevant SDH standards. A taxonomy of standards and draft recommendations prepared thus far are given below:

- G.707, "Synchronous Digital Hierarchy Bit Rates."
- G.708, "Network Interface for the Synchronous Digital Hierarchy."
- G.709, "Synchronous Multiplexing Structure."
- G.774.04, "SDH Management of Subnetwork Connection Protection for the Network Element View."
- G.780, "Vocabulary of Terms for Synchronous Digital Hierarchy Networks and Equipment."
- G.781, "Structure and Recommendations on Multiplexing Equipment for the Synchronous Digital Hierarchy."

- G.782, "Types and General Characteristics of Synchronous Digital Hierarchy Multiplexing Equipment."
- G.783, "Characteristics of Synchronous Digital Hierarchy (SDH) Multiplexing Equipment Functional Blocks."
- G.784, "Synchronous Digital Hierarchy (SDH) Management."
- G.803, "Architectures of Transport Networks Based on the Synchronous Digital Hierarchy."
- G.825, "The Control of Jitter and Wander Within Digital Networks Which Are Based on SDH."
- G.826, "Error Performance Parameters and Objectives for International Constant Bit Rate Digital Paths At or Above the Primary Rate."
- G.831, "Management Capabilities of Transport Networks Based on the Synchronous Digital Hierarchy."
- G.832, "Transport of SDH Elements on PDH Networks: Frames and Multiplexing Structures."
- G.957, "Optical Interfaces for Equipment and Systems Relating to the Synchronous Digital Hierarchy."
- G.958, "Digital Line Systems Based on the Synchronous Digital Hierarchy for Use on Optical Fiber Cables."
- G-SHR 1, "SDH Protection: Rings and Other Architectures."
- G-SHR 2, "SDH Protection Interworking."
- G.ATME 1, "Types and General Characteristics of ATM Equipment."
- G.ATME 2, "Functional Characteristics of ATM Equipment."
- ITU-R, "Architectures and Functional Aspects of Radio-Relay Systems for SDH-Based Networks."
- ITU-R, "Transmission Characteristics and Performance Requirements of Radio Relay Systems for SDH-Based Network."
- O.SDH, "Equipment to Assess Error Performance on STM-N SDH Interface."

GLOSSARY

ADM — Add/Drop Multiplexer
AIS — Alarm Indicating Signal
ANSI — American National Standards Institute
ASN — Abstract Syntax Notation
ATM — Asynchronous Transfer Mode
AU-n — Administrative Unit Level-n
AUG — Administrative Unit Group
B-ISDN — Broadband Integrated Services Digital Network
BBE — Background Block Error
BBER — Background Block Error Ratio
BER — Bit-Error Rate
BIP — Bit-Interleaved Parity
BLSR — Bidirectional Line-Switched Rings
C — Container
CCITT — Consultative Committee on International Telegraphy and Telephony
CMIP — Common Management Information Protocol
DCC — Data Communications Channel

DCN — Data Communications Network
DCS — Digital Cross-Connect System
EB — Errored Block
EDC — Error Detection Code
ES — Errored Second
ESR — Errored Second Ratio
FDDI — Fiber Distributed Data Interface
FEBE — Far-End Block Errors
FERF — Far-End Remote Fail
GNE — Gateway Network Element
HP — Higher (Order) Path
ICCF — Interexchange Carrier Compatibility Forum
ITU-T — International Telecommunication Union — Telecommunication Standardization Sector
LCN — Local Communications Network
LEC — Local-Exchange Carrier
LOF — Loss of Frame
LOP — Loss of Pointer
LOS — Loss of Signal
LP — Lower (Order) Path
LPS — Line-Protection Switching
MD — Mediation Device
MS — Multiplex Section
MSOH — Multiplex Section Overhead
NE — Network Element
NNI — Network-Node Interface
OAM&P — Operation, Administration, Maintenance, and Provisioning
OC — Optical Carrier
OOF — Out of Frame
OSI — Open Systems Interconnection
OSS — Operations Support System
P — Protection (Fiber)
PDH — Plesiochronous Digital Hierarchy
POH — Path Overhead
PPS — Path-Protection Switching
PSN — Public Switched Network
RDI — Remote Defect Indication
RSOH — Regenerator Section Overhead
S — Service (Fiber)
SDH — Synchronous Digital Hierarchy
SES — Severely Errored Second
SESR — Severely Errored Second Ratio
SOH — Section Overhead (functionality differs between SONET and SDH)
SONET — Synchronous Optical NETwork
SPE — Synchronous Payload Envelope
STM — Synchronous Transfer Mode
STM-n — Synchronous Transport Module Level-n
STS-m — Synchronous Transport Signal Level-m
STS-mC — Synchronous Transport Signal Level-m, Concatenated
TIM — Trace Identifier Monitor
TM — Terminal Multiplexer
TMN — Telecommunications Management Network
TOH — Transport Overhead

TSA — Time-Slot Assignment
TU — Tributary Unit
TUG — Tributary Unit Group
UPSR — Unidirectional Path-Switched Ring
VC — Virtual Container
VT — Virtual Tributary
VTG — Virtual Tributary Group

References

[1] R. J. Boehm et al., "Standardized fiber optic transmission systems — A synchronous optical network view," *IEEE J. Select. Areas Commun.*, vol. SAC-4, pp. 1424–1431, Dec. 1986.

[2] R. Ballart and Y.-C. Ching, "SONET: Now it's the standard optical network," in *SONET/SDH: A Sourcebook of Synchronous Networking*, C. A. Siller, Jr. and M. Shafi, Eds. New York: IEEE Press, 1996.

[3] T1 Technical Report No. 36, "A technical report on a comparison of SONET (Synchronous Optical NETwork) and SDH (Synchronous Digital Hierarchy)," document T1X1.2/93-024R2, prepared by T1X1.2 (Digital Transmission Network Architecture) of ANSI Accredited Standards Committee T1 (Telecommunications).

[4] N. Sandesara, R. Ritchie, and B. E. Smith, "Plans and considerations for SONET deployment," in *SONET/SDH: A Sourcebook of Synchronous Networking*, C. A. Siller, Jr. and M. Shafi, Eds. New York: IEEE Press, 1996.

[5] I. Haque, W. Kremer, and K. Raychaudhuri, "Self-healing rings in a synchronous environment," in *SONET/SDH: A Sourcebook of Synchronous Networking*, C. A. Siller, Jr. and M. Shafi, Eds. New York: IEEE Press, 1996.

[6] S. P. Ferguson, "Implications of SONET and SDH," *Electron. Commun. J.*, vol. 6, pp. 133–142, June 1994.

[7] J. Baudron, A. Khadr, and F. Kocsis, "Availability and survivability of SDH networks," *Elec. Commun.*, issue I/4, pp. 339–348, 1993.

[8] M. Shafi and P. Smith, "The impact of G.826," *IEEE Commun. Mag.*, vol. 31, pp. 56–62, Sept. 1993.

[9] K. Lynch and K. Kreager, "Digital solutions for regional interconnection," *Commun. Technol.*, pp. 32, 34, 36, 38, 40, July 1994.

[10] M. Compton and S. Martin, "Realizing the benefits of SDH technology for the delivery of services in the access network," in *Proc. IEEE Int. Conf. Commun.*, May 1994, pp. 1071–1076.

[11] C. A. Siller, Jr. *et al.*, "Synchronous optical network (SONET) evolution and its implication for data communications," in *Proc. 16th Biennial Symp. Commun.*, Kingston, On., Canada, May 1992, pp. 49–54.

[12] T.-H. Wu, "Emerging technologies for fiber network survivability," *IEEE Commun. Mag.*, vol. 33, pp. 58–59, 62–74, Feb. 1995.

[13] "Up close with SONET: An overview of Sonet standards, uses, and market trends," *TE&M*, pp. 46, 47, 51–53, Aug 15, 1995.

[14] Y.C. Ching and G. W. Cyboron, "Where is SONET?," *IEEE LTS*, vol. 2, pp. 44–51, Nov. 1991.

[15] R. J. Boehm, "Progress in standardization of SONET," in *SONET/SDH: A Sourcebook of Synchronous Networking*, C. A. Siller, Jr., and M. Shafi, Eds. New York: IEEE Press, 1996.

[16] G. W. Ester, "Let survivability ring," *TE&M*, pp. 46, 48, 50, Apr. 15, 1993.

[17] S. Kingsley, "Choices in SONET ring architecture," *Business Commun. Rev.*, pp. 61–65, June 1994.

[18] R. Holter, "SDH/SONET — A network management viewpoint," in *SONET/SDH: A Sourcebook of Synchronous Networking*, C. A. Siller, Jr. and M. Shafi, Eds. New York: IEEE Press, 1996.

[19] J. Blume *et al.*, "Control and operation of SDH network elements," in *SONET/SDH: A Sourcebook of Synchronous Networking*, C. A. Siller, Jr. and M. Shafi, Eds. New York: IEEE Press, 1996.

[20] J. E. Jakubson, "Managing SONET networks," *IEEE LTS*, vol. 2, pp. 5–6, 9–13, Nov. 1991.

Section 1
SONET/SDH Overview

BELL Communications Research (Bellcore) commenced standardization of Synchronous Optical Network (SONET) in the mid 1980s. With the rapid deployment of fiber, it was considered necessary to develop standards for optical interfaces which would permit the operation of transmission equipment sourced from multiple vendors. Later, this standardization process was expanded to also include signals transported under the European and Japanese hierarchies. Consequently, the standardization sector of the International Telecommunications Union (known as CCITT at the time) was charged in the late 1980's with the development of appropriate standards for signals following the Synchronous Digital Hierarchy (SDH). Although it was intended that SDH would integrate the North American, European, and Japanese plesiochronous digital hierarchies (PDH) and allow a method of mapping any one of these PDH signals into an SDH signal, this objective could not be fully achieved. Today, there are two main de facto synchronous signal standards: namely, SONET, covering North American bit rates, and SDH, covering Asian and European bit rates.

Both CCITT and ANSI have adopted many standards covering various aspects of SONET/SDH equipment. An understanding of these standards is crucial to operators, network planners, and designers so that equipment compatible with world standards may be deployed in their respective networks. This section provides the reader with an overview of the CCITT and ANSI standards.

The first two papers, entitled "SONET: Now It's the Standard Optical Network" by R. Ballart and Y.-C. Ching, and "Progress in Standardization of SONET" by R. J. Boehm, describe the ANSI (SONET) standards and discuss basic signal formats, frame structures, multiplexing, etc.

The third paper, entitled "An Overview of Emerging ITU-T Recommendations for the Synchronous Digital Hierarchy: Rates and Formats, Network Elements, Line Systems, and Network Aspects" by J. Eaves et al., provides a succinct overview of all the CCITT SDH-related recommendations which have been approved to date, and also gives a flavor of those currently being worked on. Two additional CCITT recommendations, covering architecture and network management, are respectively covered in Sections 2 and 5 of this book.

Although the initial thrust of SONET/SDH was "fiber oriented," there are many parts of the transmission networks which are based on digital microwave radio and/or satellite systems. The fourth paper, "Architectural and Functional Aspects of Radio-Relay Systems for SDH Networks" by G. Richman, M. Shafi, and C. J. Carlisle, describes the appropriate CCIR[1] recommendations for SDH/SONET-compliant radio relay systems. The fifth paper in this section, "Aspects of Satellite Systems Integration in Synchronous Digital Hierarchy Transport Networks" by W. S. Oei and S. Tamboli, shows how satellite systems can be used to transport an SDH payload and evaluates the impact of the latter on satellite system design.

For SDH/SONET to fully capitalize on its inherent compatibility, it must facilitate international networking among SDH and SONET networks. To that end, C. Hwu and S. Chum, in "International Gateway for SDH and SONET Interconnection," mention some of the differences between SDH and SONET, describe a generic international gateway in terms of traffic types and carrier services, and then comment on approaches to a management network which encompasses both environments.

The above body of papers serves as a useful, general introduction to the "basics" of SDH/SONET, the capabilities of SDH/SONET network elements, appropriate standards with which compliance is required, and issues associated with international networks at the SDH/SONET gateway.

[1] International Radio Consultative Committee (CCIR). This work is now transferred to the International Telecommunication Union — Telecommunications Standardization Sector (ITU-T).

SONET: Now It's the Standard Optical Network

RALPH BALLART AND YAU-CHAU CHING

SONET (Synchronous Optical Network) is the name of a newly adopted standard, originally proposed by Bellcore (Bell Communications Research) for a family of interfaces for use in operating telephone company (OTC) optical networks. With single-mode fiber becoming the medium of choice for high-speed digital transport, the lack of signal standards for optical networks inevitably led to a proliferation of proprietary interfaces. Thus, the fiber optic transmission systems of one manufacturer cannot optically interconnect with those of any other manufacturers, and the ability to mix and match different equipment is restricted. SONET defines standard optical signals, a synchronous frame structure for the multiplexed digital traffic, and operations procedures.

SONET standardization began during 1985 in the T1X1 subcommittee of the ANSI-accredited Committee T1 to standardize carrier-to-carrier (e.g., NYNEX-to-MCI) optical interfaces. Clearly, such a standard would also have an impact on intracarrier networks and, for that reason, has been a subject of great interest for many carriers, manufacturers, and others. Initial T1 standards for SONET rates and formats and optical parameters have now been completed. The history and technical highlights of the SONET standard are the subject of this paper.

Since it began in the postdivestiture environment, SONET standardization can be thought of as a paradigm for the development of new transmission signal standards. Bellcore's original SONET proposal was not fully detailed because all the technical questions were not yet answered. However, some aspects of the proposal have carried through the entire process and are now part of the final standards. These include:

- The need for a family of digital signal interfaces, since the march of technology is going to continually increase optical interface bit rates.
- The use of a base rate SONET signal near 50 Mb/s to accommodate the DS3 electrical signal at 44.736 Mb/s.
- The use of synchronous multiplexing to simplify multiplexing and demultiplexing of SONET component signals, to obtain easy access to SONET payloads, and to exploit the increasing synchronization of the network.
- Support for the transport of broadband (> 50 Mb/s) payloads.
- Specification of enough overhead channels and functions to fully support automation of facility and equipment operations and maintenance.

[1] As defined in CCITT, corresponding signals are plesiochronous if their signal instants at nominally the same rate, any variation in the rate being constrained with specified limits.

As standardization progressed, two key challenges merged, the solution of which gave SONET universal application. The first was to make SONET work in a plesiochronous[1] environment and still retain its synchronous nature; the solution was the development of payload pointers to indicate the phase of SONET payloads with respect to the overall frame structure (see "SONET Signal Standard — Technical Highlights"). The second was to extend SONET to become an international transmission standard, and thereby begin to resolve the incompatibilities between the European signal hierarchy (based on 2.048 Mb/s) and the North American hierarchy (based on 1.544 Mb/s). Toward the latter goal, the International Telegraph and Telephone Consultative Committee (CCITT, which changed its name to ITU-T in 1992) standardization of SONET concepts began in 1986 and a first Recommendation (standard) was completed in June 1988.

This paper will not present a full technical picture of the national and international SONET standards. Instead, we will concentrate on those aspects of the standard and the standardization process that are of particular interest. In the next section, a brief and instructional history of the SONET standard is presented. As philosopher George Santayana said, "Those who cannot remember the past are condemned to repeat it." We will then discuss key technical aspects of the SONET standards. Lastly, additional work to advance the implementation of SONET is given in the final section.

A HISTORY OF SONET IN T1 AND CCITT

The standardization of SONET in T1 started in two different directions and in three areas. First, the Interexchange Compatibility Forum (ICCF), at the urging of MCI, requested T1 to work on standards that would allow the interconnection of multiowner, multimanufacturer fiber optic transmission terminals (also known as the midfiber meet capability). Of several ambitious tasks that ICCF wanted addressed to ensure a full midfiber meet capability, two were submitted to T1. A proposal on optical interface parameters (e.g., wavelength, optical power levels, etc.) was submitted to T1X1 in August 1984 and, after three and a half years of intensive work, resulted in a draft standard on single-mode optical interface specifications [1]. The ICCF proposal on long-term operations was submitted to T1M1, and resulted in a draft standard on fiber optic systems maintenance [2].

In February 1985, Bellcore proposed to T1X1 a network approach to fiber system standardization that would allow not only the interconnection of multiowner, multimanufacturer fiber optic transmission terminals, but also the interconnection of fiber optic network elements of varying functionalities. For example, the standard would allow the direct interconnection between several optical line-terminating multiplexers, manufactured and owned by different entities, and a digital cross-connect system. In addition, the proposal also suggested a hierarchical family of digital signals whose rates are integer multiples of a basic module signal, and suggested a simple synchronous bit interleaving multiplexing technique that would allow economical implementations Thus, the term Synchronous Optical NETwork (SONET) was coined. This proposal eventually led to a standard on optical rates and formats [3]. For the remainder of this paper, the focal point is the history and highlights of the rates and formats document. However, one should always be reminded that the rates and formats standard is only one part of the inseparable triplet, i.e., optical interface specifications, rates and formats specifications, and operations specifications.

As it turns out, the notion of a network approach and simple synchronous multiplexing had been independently investigated by many manufacturers. Some of them were already developing product plans, thus complicating the standards process. With the desire of the network providers (i.e., the OTCs) for expedited standards, SONET quickly gained support and momentum. By August 1985, T1X1 approved a project proposal based on the SONET principle. Because the issues on rates and formats were complex and required diligent but timely technical analyses, a steady stream of contributions poured into T1X1. Several ad hoc groups were formed and interim meetings were called to address them. The contributions came from over 30 entities representing the manufacturers and the network providers alike.

In the early stage, the main topic of contention was the rate of the basic module. From two original proposals of 50.688 Mb/s (from Bellcore) and 146.432 Mb/s (from AT&T), a new rate of 49.920 Mb/s was derived and agreed on. In addition, the notion of a virtual tributary (VT) was introduced and accepted as the cornerstone for transporting DS1 services. By the beginning of 1987, substantial details had been agreed on and a draft document was ready for voting. Then came CCITT.

The SONET standards were first developed in T1X1 to serve the U.S. telecommunications networks. When CCITT first expressed its interest in SONET in the summer of 1986, major procedural difficulties appeared. According to the established protocol, only contributions that have consensus in T1X1 could be forwarded, through U.S. Study Group C, to CCITT. As a result, some aspects of U.S. positions in CCITT appeared to lack flexibility without input from T1X1. Additionally, the views of other administrations in CCITT were not thoroughly understood in T1X1. There were also differences in schedule and perceived urgency. CCITT runs by a four-year plenary period and their meetings are six–nine months apart, while T1 approves standards whenever they are ready and its technical subcommittees meet at least four times a year. While T1X1 saw the SONET standard as a way to stop the proliferation of incompatible fiber optic transmission terminals, no such need was perceived by many other nations whose networks were still fully regulated

and noncompetitive.

The procedural difficulties were partially resolved when representatives from the Japanese and British delegations started to participate in T1X1 meetings in April 1987. These representatives not only gave to T1X1 the perspectives of two important supporters of an international SONET standard, they also served as a conduit between T1X1 and CEPT, the European telecommunications organization.

Separately, interests in an international SONET also gained support in the U.S. Spearheaded by Bellcore, informal discussions in search of an acceptable solution took place in a variety of forums, and contributions in support of this standard were submitted to both T1 and CCITT. Many of these informal discussions had the highest level of corporate support from several U.S. companies, including manufacturers and network providers.

In July 1986, CCITT Study Group XVIII began the study of a new synchronous signal hierarchy and its associated Network-Node-Interface (NNI). The NNI is a nonmedia specific network interface, and is distinct from the User-Network-Interface (UNI) associated with Broadband ISDN. The interaction between T1X1 and CCITT on SONET and the new synchronous hierarchy was fascinating to the participants, and probably will alter the way international standards are made in the future. The U.S. wanted an international standard, but not at the price of scrapping SONET or seriously delaying an American national standard upon which OTC networks were planned. The CCITT was not used to working so quickly on so complicated an issue, but was concerned about being supplanted by T1 in the development of new standards.

The U.S. first formally proposed SONET to CCITT for use at the NNI in the February 1987 Brasilia meeting; this proposal had a base signal level near 50 Mb/s. Table 1 shows that the European signal hierarchy has no level near 50 Mb/s and, therefore, CEPT wanted the new synchronous hierarchy to have a base signal near 150 Mb/s to transport their 139.264 Mb/s signal.

Thus, the informal European response was that the U.S. must change from bit interleave to byte interleave multiplexing to provide a byte organized frame structure at 150 Mb/s. However, there was still no indication from many administrations that an international standard was either desirable or achievable. It took T1X1 three months and three meetings to agree to byte interleaving, and the results were submitted to CCITT as a new T1X1 draft standard document. Thus, T1X1 never gave up the responsibility to develop a SONET standard for the U.S. and, while conceding changes to CCITT wishes, progress was made in other areas of the U.S. standard.

TABLE 1. CCITT RECOMMENDATION G.702, ASYNCHRONOUS DIGITAL HIERARCHY BIT RATES (IN Mb/s)

Level	North America	Europe	Japan
1	1.544 (DS1)	2.048 (CEPT-1)	1.544
2	6.312 (DS2)	8.448	6.312
3	44.736 (DS3)	34.368	32.064
4	139.264 (DS4NA)	139.264	97.728

After CCITT met again in Hamburg in July 1987, a formal request was made to all administrations to consider two alternative proposals for an NNI specification near 150 Mb/s. The U.S. proposal was based on the SONET STS-3 frame structure; the STS-3 frame cold be drawn as a rectangle with 13 rows and 180 columns of bytes. CEPT proposed, instead, a new STS-3 frame with 9 rows and 270 columns. (Commonly referred to as the 9 row/13 row debate, this prompted one amateur poet to chide that neither conforms to the correct SON(N)ET format of 14 lines.) An NNI near 150 Mb/s received unanimous support because it was assumed that future broadband payloads would be about that size. A North American basic module near 50 Mb/s could be easily derived in both proposals, with a frame structure of either 13 rows and 60 columns or 9 rows and 90 columns.

The Europeans wanted a 9 row frame structure to accommodate their 2.048 Mb/s primary rate signal. This signal has 32 bytes per 125 μs, but in the 13 row proposal could only be accommodated in the most straightforward way using three 13 byte columns or 39 bytes. The Europeans decried this waste of bandwidth, and refused to consider any alternative and more efficient mapping of the 2.048 Mb/s signal into the 13 row structure. Their 9 row frame structure could carry the U.S. 1.544 Mb/s primary rate signal (requiring about 24 bytes/125 μs) in 3 columns of 9 bytes and the 2.048 Mb/s signal in 4 columns of 9 bytes.

The CEPT 9 row proposal called for changes in both the rate and format in the U.S., just as T1X1 was about to complete the SONET standard. However, the request also carried an attractive incentive from a CEPT subcommittee, which stated in a letter that these were the only changes necessary for an international agreement. In addition, the text of the CEPT proposal was based largely on the T1X1 draft document so that it was complete. Therefore, after the Hamburg meeting, there was tremendous international pressure on the U.S. to accept the 9 row proposal. After some intense debates in T1X1, the U.S. agreed to change.

Unfortunately, the CEPT proposal did not have unanimous support from all CEPT administrations. While some administrations were anxious to get an international standard, a few became concerned that the 9 row proposal still favored the U.S. DS3 signal over the CEPT 34.368 Mb/s signal. A CEPT contribution to the November 1987 CCITT meeting stated that it was too early to draft Recommendations on a new synchronous hierarchy. Little progress was made at that meeting, and the international SONET standard was in serious jeopardy.

Many T1X1 participants were upset at the apparent change in CEPT's position. Since there were no alternative proposals from CCITT at its November meeting, T1X1 decided to approve the two SONET documents for T1 letter balloting. However, the balloting schedule was deliberately set such that it fell between the CCITT meeting at the beginning of February 1988 and the T1X1 meeting at the end of February 1988. This scheduling allowed a last ditch attempt for an international agreement. In CCITT, a mad rush to rescue the international standard also took place. In addition to a series of informal discussions, a pre-CCITT meeting was held in Tokyo to search for a compromise. Under the skillful helmsmanship of Mr. K. Okimi of Japan, the CCITT meeting in Seoul proposed an additional change to the U.S. draft

standards. The new proposal called for a change in the order that 50 Mb/s tributaries are byte multiplexed to higher SONET signal levels. It also put more emphasis on the NNI as a 150 Mb/s signal by including optional payload structures to better accommodate the European 34.368 Mb/s signal. The U.S. CCITT delegates eventually viewed this proposal as a minor change to the U.S. standards (minor to the extent that equipment under development would probably not require modification) and agreed to accept it. An extensive set of three CCITT Recommendations was drafted and approved by the working party plenary. The U.S. acceptance of these changes was predicated on the understanding that no additional changes of substance would be considered in approving the final version of the Recommendations.

In February 1988, T1X1 accepted the new changes at its meeting in Phoenix. T1 default balloting based on the change was completed in May, and the final passage of the standard occurred on August 8. Editorial corrections to CCITT Recommendations [4]–[6] were completed in June during the Study Group XVIII meeting and approved in October 1988. Under the title of Synchronous Digital Hierarchy (SDH), an international SONET standard was born!

SONET SIGNAL STANDARD — TECHNICAL HIGHLIGHTS

In this section, we describe the technical highlights of the American national standards related to SONET. We use U.S. rather than CCITT terminology, although everything described is consistent with both the American national standards and the CCITT Recommendations. Since the first publication of this article in 1989, much progress has also been made in T1X1, as reflected in the updating of [3] as [7], approval of [8], and other pending standards.

SONET Signal Hierarchy

The basic building block and first level of the SONET signal hierarchy is called the Synchronous Transport Signal — Level 1 (STS-1). The STS-1 has a bit rate of 51.84 Mb/s, and is assumed to be synchronous with an appropriate network synchronization source. The STS-1 frame structure can be drawn as 90 columns and 9 rows of 8-b bytes (Fig. 1). The order of transmission

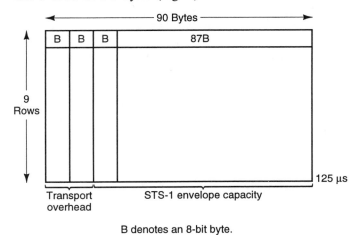

B denotes an 8-bit byte.

Fig. 1.   STS-1 frame.

Transport overhead			Path overhead
Framing A1	Framing A2	(STS-1 ID) C1	Path trace J1
BIP-8 B1	Orderwire E1	User F1	BIP-8 B3
Data com D1	Data com D2	Data com D3	Signal label C2
Pointer H1	Pointer H2	Pointer action H3	Path status G1
BIP-8 B2	APS K1	APS K2	User channel/DQDB F2
Data com D4	Data com D5	Data com D6	Indicator H4
Data com D7	Data com D8	Data com D9	Growth DQDB Z3
Data com D10	Data com D11	Data com D12	Growth Z4
Growth/sync status Z1	Growth/FEBE Z2	Orderwire E2	Tandem connection Z5

Section overhead (rows 1–3), Line overhead (rows 4–9).

Fig. 2. Transport and path overhead byte designations.

of the bytes is row by row, from left to right, with one entire frame being transmitted every 125 μs. (A 125 μs frame period supports digital voice signal transport since these signals are encoded using 1 byte/125 μs = 64 kb/s.) The first three columns of the STS-1 contain section and line overhead bytes (see the following subsection). The remaining 87 columns and 9 rows are used to carry the STS-1 synchronous payload envelope (SPE); the SPE is used to carry SONET payloads including 9 bytes of path overhead (see next section). The STS-1 can carry a clear channel DS3 signal (44.736 Mb/s) or, alternatively, a variety of lower rate signals such as DS1, DS1C, and DS2.

STS-1 is defined as a logical signal internal to SONET equipment; the Optical Carrier—Level 1 (OC-1) is obtained from the STS-1 after scrambling (to avoid long strings of ones and zeros and allow clock recovery at receivers) and electrical to optical conversion. The OC-1 is the lowest level optical signal to be used at SONET equipment and network interfaces. In addition, electrical equivalent signals to STS-1 and STS-3 have also been standardized in T1.102 [8].

SONET Overhead Channels

The SONET overhead is divided into section, line, and path layers; Fig. 2 shows the overhead bytes and their relative positions in the SONET frame structure. This division clearly reflects the segregation of processing functions in network ele-

ments (equipment) and promotes understanding of the overhead functions. The section layer contains those overhead channels that are processed by all SONET equipment, including regenerators. The section overhead channels for an STS-1 include two framing bytes that show the start of each STS-1 frame, an STS-1 identification byte, an 8-b Bit-Interleaved Parity (BIP-8) check for section error monitoring, an orderwire channel for craft (network maintenance personnel) communications, a channel for unspecified network user (operator) applications, and 3 bytes for a section level data communications channel (DCC) to carry maintenance and provisioning information. When a SONET signal is scrambled, the only bytes left unscrambled are the section layer framing bytes and the STS-1 identification bytes.

The line overhead is processed at all SONET equipment except regenerators. It includes the STS-1 pointer bytes, an additional BIP-8 for line error monitoring, a 2 byte Automatic Protection Switching (APS) message channel, a 9 byte line data communications channel, bytes reserved for future growth, and a line orderwire channel. The path overhead bytes are processed at SONET STS-1 payload terminating equipment; that is, the path overhead is part of the SONET STS-1 payload and travels with it. The path overhead includes a path BIP-8 for end-to-end payload error monitoring, a signal label byte to identify the type of payload being carried, a path status byte to carry maintenance signals, a multiframe alignment byte to show DS0 signaling bit phase, and others.

Multiplexing

Higher rate SONET optical transmission signals are obtained by first byte interleaving N frame aligned STS-1s to form an STS-N (Fig. 3). Byte interleaving and frame alignment are used primarily to obtain a byte organized frame format at the 150 Mb/s level that was acceptable to the CCITT; as discussed below, frame alignment and byte interleaving also help an STS-N to carry broadband payloads of about 150 or 600 Mb/s. All the section and line overhead channels in STS-1 #1 of an STS-N are used; however, many of the overhead channels in the remaining STS-1s are unused. (Only the section overhead framing, STS-1 ID, and BIP-8 channels and the line overhead pointer and BIP-8

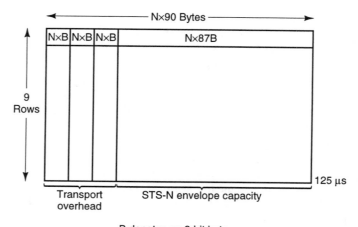

B denotes an 8-bit byte.

Fig. 3. STS-N frame.

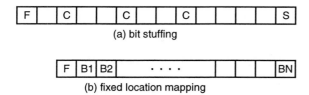

(a) bit stuffing

(b) fixed location mapping

Fig. 4.   Comparison of bit-stuffed and fixed location mappings.

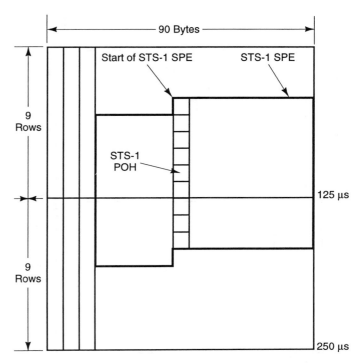

Fig. 5.   STS-1 SPE floating inside STS-1 envelope capacity.

channels are used in all STS-1s in an STS-N.) The STS-N is then scrambled and converted to an Optical Carrier—Level N (OC-N) signal. The OC-N will have a line rate exactly N times that of an OC-1. The American national standard allows only the values N = 1, 3, 12, 24, 48, with OC-192 under consideration in 1993.

SONET STS-1 Payload Pointer

Each SONET STS-1 signal carries a payload pointer in its line overhead. The STS-1 payload pointer is a key innovation of SONET, and it is used for multiplexing synchronization in a plesiochronous environment and also to frame align STS-N signals.

Pointers and Multiplexing Synchronization

There are two conventional ways to multiplex payloads into higher rate signals. The first is to use positive bit stuffing to increase the bit rate of a tributary signal to match the available payload capacity in a higher rate signal. As shown in Fig. 4(a), bit-stuffing indicators (labeled C) are located in a fixed position with respect to signal frame F, and indicate whether the stuffing bit S carries real or dummy data in each higher level signal frame. Examples of bit stuffing are the multiplexing of four DS1 signals into the DS2 signal and the multiplexing of seven DS2 signals into the asynchronous DS3 signal. Bit stuffing can accommodate large (asynchronous) frequency variations of the multiplexed payloads. However, access to those payloads from the high-level multiplexed signal must first be destuffed (real bits separated from the dummy bits), and then the frame pattern of the payload must be identified if complete payload access is required.

The second conventional method is the use of fixed location mapping of tributaries into higher rate signals. As network synchronization increases with the deployment of digital switches, it becomes possible to synchronize transmission signals into the overall network clock. Fixed location mapping is the use of specific bit positions in a higher rate synchronous signal to carry lower rate synchronous signals; for example, in Fig. 4(b), frame position B2 would always carry information from one specific tributary payload. This method allows easy access to the transported tributary payloads since no destuffing is required. The SYNTRAN DS3 signal is an example of a synchronous signal that uses fixed location mapping of its tributary DS1 signals. However, there is no guarantee that the high-speed signal and its tributary will be phase-aligned with each other. Also, small frequency differences between the transport signal and its tributary signal may occur, due to the synchronization network failures or at plesiochronous boundaries. Therefore, multiplexing equip-

ment interfaces require 125 μs buffers to phase-align and slip (repeat or delete a frame of information to correct frequency differences) the tributary signal. These buffers are undesirable because of the signal delay that they impose and the signal impairment that slipping causes.

In SONET, payload pointers represent a novel technique that allows easy access to synchronous payloads while avoiding the need for 125 μs buffers and associated slips at multiplexing equipment interfaces. The payload pointer is a number carried in each STS-1 line overhead (bytes H1, H2 in Fig. 2) that indicates the starting byte location of the STS-1 SPE payload within the STS-1 frame (Fig. 5). Thus, the payload is not locked to the STS-1 frame structures as it would be if fixed location mapping were used, but instead floats with respect to the STS-1 frame. (The STS-1 section and line overhead byte positions determine the STS-1 frame structure; note in Fig. 5 that the 9-row-by-87-column SPE payload maps into an irregular shape across two 125 μs STS-1 frames.)

Any small frequency variations of the STS-1 payload can be accommodated by either increasing or decreasing the pointer value; however, the pointer cannot adjust to asynchronous frequency differences. For example, if the STS-1 payload data rate is high with respect to the STS-frame rate, the payload pointer is decremented by one and the H3 overhead byte is used to carry data for one frame (Fig. 6).

If the payload data rate is low with respect to the STS-1 frame rate, the data byte immediately following the H3 byte is nulled for one frame and the pointer is incremented by one (Fig. 7). Thus, slips and their associated data loss are avoided while the phase of the STS-1 synchronous payload is immediately known by simply reading the pointer value. Thus, SONET pointers com-

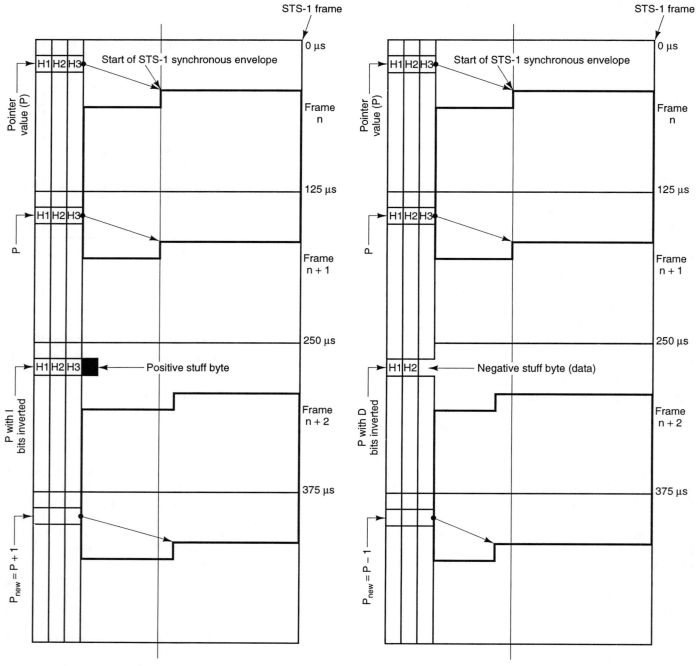

Fig. 6. Positive STS pointer justification operation.

Fig. 7. Negative STS pointer justification operation.

bine the best features of positive bit stuffing and the fixed location mapping methods. Of course, these advantages come at the cost of having to process the pointers; however, pointer processing is readily implementable in today's Very Large Scale Integration (VLSI) technologies.

Broadband Payload Transport with Payload Pointers

As discussed above, STS-1 payload pointers can be used to adjust the frequencies of several STS-1 payloads in multiplexing to the STS-N signal level. As this is done, the various STS-1 section and line overhead bytes are frame-aligned. In Fig. 8, two hypothetical and simplified SONET frames (A and B) are out of phase with respect to the arbitrary, outgoing (multiplexed) SONET signal phase. By recalculating the SONET pointer values and regenerating the SONET section and line overhead bytes, two phase-aligned signals (A and B) are formed. A and B can then be byte-interleaved to form a higher level STS-N signal. As shown, this can be done with minimum payload buffering and signal delay.

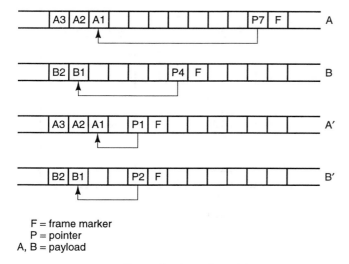

F = frame marker
P = pointer
A, B = payload

Fig. 8. Frame alignment using pointers.

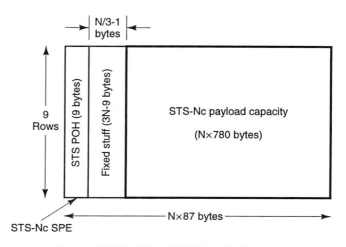

Fig. 9. STS-Nc SPE and STS-Nc payload capacity.

With frame alignment, the STS-1 pointers in an STS-N are grouped together for easy access at an OC-N receiver using a single STS-N framing circuit. If it is desired to carry a broadband payload requiring, for example, three STS-1 payloads, the phase and frequency of the three STS-1 payloads must be locked together as the broadband payload is transported through the network. This is easily done by a "concatenation indication" in the second and third STS-1 pointers. The concatenation indication is a pointer value that indicates to an STS-1 pointer processor that this pointer should have the same value as the previous STS-1 pointer. Thus, by frame aligning STS-N signals and using pointer concatenation, multiple STS-1 payloads can be created. The STS-N signal that is locked together in this way is called an STS-Nc, where "c" stands for concatenated. Figure 9 shows the payload capacity for an STS-Nc SPE. Allowed values of STS-Nc are STS-2c, STS-3c, STS-6c, STS-9c, etc. For broadband User-Network Interfaces (UNI), STS-3c and STS-12c are of particular interest.

As discussed in the section on the history of SONET standards, the Europeans had no interest in using the SONET STS-1 signal. Instead, they were interested in using a base signal of about 150 Mb/s to allow transport of their 139.264 Mb/s electrical signal and for possible Broadband ISDN applications. As the above discussion shows, the technical solution to this problem is the use of the STS-3c signal. In the U.S., we can continue to think of this signal as three concatenated STS-1 signals. In Europe and the CCITT, the STS-3c is considered as the basic building block of the new Synchronous Digital Hierarchy and is referred to as the Synchronous Transport Module — Level 1 (STM-1). In T1.105-1991, broadband signals that can be mapped into STS-3c include DS4NA (139.264 Mb/s), FDDI (125 Mb/s), and ATM and DQDB cells (53 bytes).

SUB-STS-1 PAYLOADS

To transport payloads requiring less than an STS-1 payload capacity, the STS-1 SPE is divided into payload structures called virtual tributaries (VTs). There are four sizes of VTs: VT1.5, VT2, VT3, and VT6, where each VT has enough bandwidth to carry a DS1, CEPT-1 (2.048 Mb/s), DS1C, and DS2 signal, respectively. Each VT occupies several 9-row columns within the SPE. The VT1.5 is carried in three columns (27 bytes), the VT2 in four columns (36 bytes), the VT3 in six columns (54 bytes), and the VT6 in 12 columns (108 bytes).

A VT group is defined to be a 9-row-by-12-column payload structure that can carry four VT1.5s, three VT2s, two VT3s, or one VT6. Seven VT groups (84 columns), one path overhead column, and two unused columns are byte-interleaved to fully occupy the STS-1 SPE. Figure 10 shows the STS-1 SPE configured to carry 28 VT1.5s. VT groups carrying different VT types can be mixed within one STS-1. As discussed in the section on history, the ability of the 9 row format structure to flexibly carry both the 1.544 and 2.048 Mb/s signals was a necessary step in reaching an international agreement on SONET.

Two different modes have been adopted for transporting payloads within a VT. The VT operating in the "floating" mode improves the transport and cross-connection of VT payloads. A floating VT is so called because a VT pointer is used to show the starting byte position of the VT SPE within the VT payload structure. In this sense, the operation of the VT pointer is directly analogous to that of the STS-1 pointer, and has the same advantages of minimizing payload buffers and associated delay when mapping signals into the VT. Fig. 11 shows conceptually how the STS-1 and VT pointers are used to locate a particular VT payload in an STS-1. The other VT mode is the "locked" mode. The locked VT does not use the VT pointer, but instead locks the VT payload structure directly to the STS-1 SPE. (Of course, the STS-1 SPE still floats with respect to the STS-1 frame.) The locked mode improves the transport and cross-connection of DS0 signals by maintaining the relative phase and frequency of DS0 signals carried in multiple locked VTs. When VT-organized, each STS-1 SPE carries either all floating or all locked VTs.

More than one specific payload mapping is possible with each of the VT modes described above. Asynchronous mapping is used for clear channel transport of nominally asynchronous signals using the floating mode of operation; conventional positive

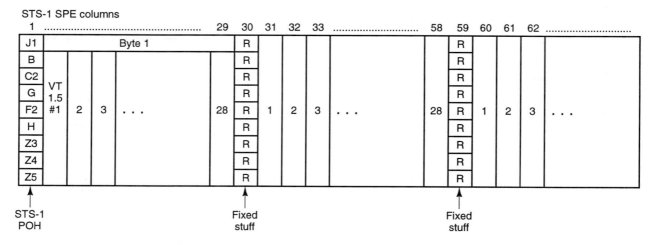

STS-1 SPE columns

J1	Byte 1				R						R				
B					R						R				
C2	VT				R						R				
G	1.5				R						R				
F2	#1	2	3 ...	28	R	1	2	3 ...	28	R	1	2	3 ...		
H					R						R				
Z3					R						R				
Z4					R						R				
Z5					R						R				

STS-1 POH   Fixed stuff   Fixed stuff

Fig. 10. STS-1 SPE configured for 28 VT1.5s.

bit stuffing is used to multiplex these signals into the VT SPE. "Byte synchronous" mappings have been defined in both the locked and floating modes for the efficient, synchronous transport of DS0 signals and their associated signaling; conventional fixed position mappings are used to carry the DS0s in the VT SPE (floating mode) or VT (locked mode). "Bit synchronous" mappings are used in both the locked and floating modes for the clear channel transport of unframed, synchronous signals. The VT mappings that have been defined in the current version of the American National Standard are given in Table 2.

Automatic Protection Switching

With the high capacity of any SONET transport system, automatic protection switching (APS) for a failed facility to improve system availability and reliability becomes a necessity. Two bytes (K1, K2 in Fig. 2) in SONET frame overhead are allocated for this purpose. A simple bit-oriented protocol has been established, which allows 16 different types and priority of APS. In addition, the protocol allows for one-for-N protection where N can be as high as 16. One-plus-one protection is also allowed, in which case the traffic is present on both working and protection fibers; the receiver decides which to choose.

In a recent effort, APS has been extended to cover ring applications [9]. In this case, K1 and K2 bytes indicate the protection status of a ring of SONET nodes. Up to 16 nodes can be accommodated by this protocol. The protocol allows for full traffic recovery in 50 ms for single-fiber failures and partial recovery for multiple failures.

Data Communication Channels

SONET equipment is assumed to have sufficient processing power to provide alarm surveillance, performance monitoring, and memory administration capabilities. These capabilities necessitate the communication between SONET equipment and operations systems and among SONET equipment. The data communications channels (DCCs) are established for this communications purpose. Two DCCs have been defined. Section DCC (D1–D3 in Fig. 2) has a capacity of 192 kb/s, while the line DCC (D4–D12) has a capacity of 576 kb/s. Reference [7] established an OSI protocol suite as the standard protocol over DCC, with CMISE at the application layer. Current work in T1 [10] has established the information models for termination fragments, automatic protection fragments, and performance monitoring fragments for SONET equipment. It should be noted that a number of operations systems still use Transaction Language

TABLE 2. SUB-STS-1 MAPPINGS

Mappings	VT (Virtual Tributary) Modes	
	Floating	Locked
Asynchronous	DS1, CEPT-1, DS1C, DS2	Not Defined
Byte Synchronous	DS1, CEPT-1	DS1, CEPT-1, SYNTRAN
Bit Synchronous	DS1, CEPT-1	DS1, CEPT-1

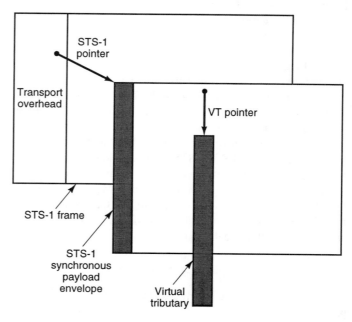

Fig. 11. STS and VT pointers to locate VT bytes.

1 (TL1) as an interim solution to communicate with SONET equipment. However, this subject is beyond the scope of T1X1 standards.

Optical Parameters

The SONET optical interface parameters were developed in parallel with the SONET rates and formats. The optical parameters specified in the American national standard include spectral characteristics, power levels, and pulse shapes. The current optical specifications extend up to OC-48. The intent of the first optical interface standard is to provide specifications for "long reach" fiber transmission systems, i.e., systems using lasers. A second standard was approved in 1990 to address "short reach" specifications for fiber transmission systems based on LEDs and low-power loop lasers [11]. There is a general consensus in T1X1 that these standards should be updated to reconcile the differences with the international standard in CCITT G.958 [12]. However, such a task has not been completed in T1.

CONCLUSION

The Synchronous Optical Network concept was developed to promulgate standard optical transmission signal interfaces to allow midsection meets of fiber systems, easy access to tributary signals, and direct optical interfaces on terminals, and provide new network features. The basic SONET format can transport all signals of the North American hierarchy up to DS4NA, and also future broadband signals. SONET is now an American national standard and a CCITT transmission signal hierarchy standard. The second phase of SONET T1 standardization fully specified the data communications channel protocol and information models, and specified short reach SONET optical interfaces for use in intraoffice and loop applications.

SONET represents a successful test case for standards-making in the postdivestiture environment. Of course, the ultimate test for any standard is the development and deployment of products and services that are compliant with the new standard. For specific implementations and their associated operations support features, requirements beyond those contained in the standard are often needed. Bellcore has issued a series of Technical Advisories giving additional requirements for SONET multiplexes, digital cross-connect systems, and other fiber equipment. The first field trial of SONET equipment occurred in 1989, and midfiber meet of equipment from different manufacturers was demonstrated in 1990. SONET deployment is expected to dominate the optical fiber equipment market for OTC network, and is expected to have a great impact for the end-user fiber equipment market as well.

References

[1] ANSI T1.106-1988, "American National Standard for Telecommunications — Digital Hierarchy Optical Interface Specifications, Single Mode."

[2] T1M1.2/87-37R2, "Functional Requirements for Fiber Optic Terminating Equipment."

[3] ANSI T1.105-1988, "American National Standard for Telecommunication—Digital Hierarchy Optical Rates and Formats Specification."

[4] CCITT Recommendation G.707, "Synchronous Digital Hierarchy Bit Rates."

[5] CCITT Recommendation G.708, "Network Node Interface for the Synchronous Digital Hierarchy."

[6] CCITT Recommendation G.709, "Synchronous Multiplexing Structure."

[7] ANSI T1.105-1991, "American National Standard for Telecommunication—Digital Hierarchy Optical Rates and Formats Specification (SONET)."

[8] ANSI T.102-1993, "American National Standard for Telecommunications — Digital Hierarchy — Electrical Interfaces."

[9] ANSI T1.105.01-1993, "American National Standard for Telecommunication — Synchronous Optical Network (SONET): Automatic Protection Switching."

[10] ANSI T1.119-1993, "American National Standard for Telecommunication—Synchronous Optical Network (SONET): OAM&P Communications."

[11] ANSI T1.117-1990, "American National Standard for Telecommunications—Digital Hierarchy Optical Interface Specifications, Short Reach."

[12] CCITT Recommendation G.958, "Digital Line Systems Based on the Synchronous Digital Hierarchy for Use on Optical Fiber Cables."

Progress in Standardization of SONET

RODNEY J. BOEHM

Abstract—Before divestiture, most of the telecommunication standards for the United States were developed and implemented by AT&T. In addition to this, most of the participation in international standards bodies, i.e., CCITT, was handled by AT&T. This was perfectly acceptable to the industry because AT&T controlled so much of the North American telecommunications market. When other manufacturers wanted to supply equipment to different Bell Operating Companies, they would request specifications from AT&T and implement equipment to that standard. Divestiture brought about a very different standardization process. The Exchange Carrier Standards Association (ECSA) Committee T1 was formed to fill most of the needs for telecommunication standards.

One of the projects T1 has undertaken is to standardize an optical interface commonly referred to as SONET (Synchronous Optical Network). This interface has received not only national but international attention, and is in various stages of being completed. Initial documents were released in 1988 which detailed a great deal of the necessary interface specification. In addition, CCITT recommendations released at that time contain a similar level of detail concerning this interface. In an effort to release the SONET specifications, some of the necessary specifications were left incomplete, and work since that time concentrated on finishing the standard. Additionally, there sometimes exists confusion in the terms used in CCITT and ECSA Committee T1 which refer to the same specification. In this paper, we will explain the difference and similarities between the North American SONET signal and the CCITT Synchronous Digital Hierarchy (SDH) signal. Also, we will describe the continuing work and how this standard and subsequent revisions/additions to this standard will affect equipment deployment.

1. INTRODUCTION

BEFORE the North American telecommunications industry went through divestiture, most standards for telecommunication were developed by AT&T and distributed to other vendors interested in building equipment to those specifications. When the Bell System was broken into the corporations we know today, this standardization process needed to change. To fill the place of a single organization setting standards, a committee in the Exchange Carrier's Standards Association was established. This is known today as Committee T1 (T1 referring to the fact that this was the first standardizations committee for telecommunications established in ECSA).

At the same time, a new company was being formed to provide the Regional Bell Operating Companies (RBOCs) the same support Bell Laboratories had supplied them before divestiture. This company is known as Bellcore. One of its major jobs, but by no means the only one, was to establish Technical Requirements for equipment needed by the RBOCs and aid in the standardization process. Since many of the existing specifications were generated by AT&T and had not formally been standardized, Bellcore had a great deal of work to accomplish in a very short time.

Even though existing interfaces needed to be standardized, some people were attempting to establish specifications for in-terfaces not yet dreamed of. One of the interexchange companies approached some of the RBOCs and asked to establish an interface optically. Since, at that time, all optical systems had proprietary frame formats and optical parameters, there was not a standard manner in which to accomplish this. Two methods were finally agreed to — one short term and the other longer term. The short term method, and definitely less preferable to the companies involved, was to agree to purchase the same system on both ends of the fiber. The other was to start a standards project in the newly formed T1 committee to specify a standardized optical interface.

The RBOCs sought help from Bellcore (referred to as the Central Services Organization at the time while Judge Green decided if this new company could use the Bell name) to initiate the standardization effort. Mr. Yau Ching and I were assigned the task of getting the optical interface project started. We knew some form of a proposal needed to be developed, therefore, we established and documented the basic concepts behind SONET, which stands for the Synchronous Optical Network.

This was the skeleton of a proposal, and did not contain all of the details necessary for a supplier to build a system with a standardized interface. We knew that establishment of an optical interface in the telecommunications network would have far ranging effects and would be with us for many years. We also knew that many people were working on optical systems and could provide valuable input to this project. Therefore, we decided to bring the framework of a proposal to the standards body and work out the details of the interface in open forums to enlist the support and intelligence contained in our industry.

In 1988, after three years of very hard work, over 400 contributions by numerous corporations, many compromises, gallons of coffee, hundreds of thousands of airline miles, and personal time spent away from families, T1X1 developed the initial SONET standard that not only accomplished the initial goals, but also went beyond this and became an interface specification that will allow us to build today's networks and transition easily to tomorrow's networks. As will be shown later, this interface also was adopted as an international telecommunication standard that can be used in all countries. This project illustrates the standardization process at its best because dedicated people helped to take an idea and develop it into a specification that goes beyond its initial application into something that will affect the telecommunications network for years to come.

While the initial effort was monumental, it is a credit to Committee T1, and specifically those involved in T1X1, that they continued to work to bring the SONET standards to the stage they are today. Much more is known about this interface and how it will impact the network. This has led to many new ap-

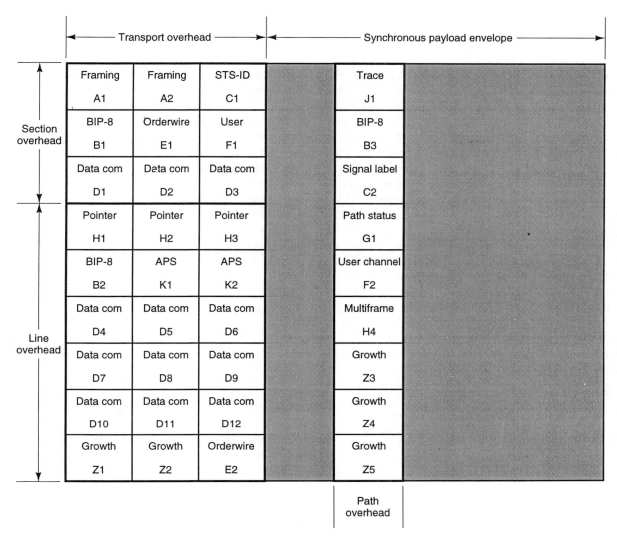

Fig. 1. The SONET frame.

plications and complications that have caused us to extend the standard even further than was even dreamed of initially.

1.1 Project Goals

Initially, we established very simple goals for this project — that is, to be able to specify an optical interface which would allow optical midspan meets between different suppliers' equipment. This would be analogous to the DS3 or DS1 interfaces so prevalent in the network today. We quickly realized that selecting a single interface rate would be impossible because so many different applications existed. In addition, we recognized it would be a difficult and time-consuming process to establish even a single interface. Therefore, a method was needed to allow establishment of a family of interfaces at the same time. In this manner, a number of interfaces would be established simultaneously, thus avoiding the time-consuming process of defining them one at a time.

We chose to specify a base signal, complete with a format, and a multiplexing method that would allow us to create a number of interfaces from the base signal. After a compromise was reached with CCITT, a base signal with a rate of 51.84 Mb/s and frame format shown in Fig. 1 was chosen. The multiplexing method chosen was to synchronously byte interleave the basic signals together to form higher rate interfaces. Using this technique, a simple one-step multiplex and demultiplex process can be designed to provide direct access to the desired basic signal in a higher rate interface. An analogy of stacking blocks on top of each other to make a taller tower is a very appropriate one. The term synchronous in SONET came from this multiplexing technique. Now that we had a base signal and a multiplexing technique, we could create a family of interfaces which could satisfy all information transport needs. Algorithms were established to allow multiplexing up to 256 base signals together, which would create interfaces from 51.84 Mb/s to 13.27 Gb/s.

1.2 Accomplishments in the Initial Document Release

Most of the initial goals were accomplished with the release of the initial SONET specifications, ANSI T1.105-1988 and ANSI T1.106-1988 [2]. While the numbers are somewhat confusing,

29

the specifications are actually simple. ANSI T1.105-1988 contains specifications for the rate and format of the SONET interface. This is analogous to other format specifications for other interfaces — such as the DS1 or DS3 format. A SONET interface, however, is different in some very important ways which will be covered in more detail later. Also contained in this document are instructions for mapping different payloads into the interface, multiplexing techniques, and overhead specifications. The T1.105 document was reissued in 1991 as ANSI T1.105-1991 [1]

ANSI T1.106-1988 contains the optical parameter specification needed to complete the interface specification. This is analogous to the electrical parameter specification for a DS1 or DS3 signal. These specifications were released in this manner to allow different optical parameters to be specified for different rate signals in different applications. ANSI T1.106-1988 specifies parameters for longer distance spans in single-mode fiber. This was given the highest priority by the committee. Another specification, described later in more detail, will specify a much shorter distance interface which can use significantly less expensive optical components.

As mentioned before, most of the goals were met by these two documents. One of the major items not contained in these initial standards is the specification of the embedded operations channels called Data Communication Channels (DCC). Two channels are allocated bandwidth in the overhead. One is the Section DCC and the other is the Line DCC. These will be covered in further detail in another section.

Because these channels will be used to communicate operations support information from one network element to another, some have maintained that the basic goal of defining a midspan meet interface has not been achieved. In fact, all of the necessary specifications for reliable information transfer have been completed.

It is the protocol stack and messages necessary to trade operations information that were not complete at that time. This situation is analogous to connecting one vendor's M13 to another. Transport of DS1s can take place reliably; however, commands to loop back a DS1 or notifying the other multiplexer that the far end has an alarm cannot be accomplished because no standard method has been established to perform this function.

Therefore, while a midspan meet can be accomplished, many of the features in transmission equipment used by network providers are not present. This situation was to be corrected by the SONET Phase II document scheduled for release in December 1989. After a great deal of work, only the protocol specification for the DCCs was released. This will be covered in more detail in a subsequent section.

1.3 Decision to Phase the Standard

One could legitimately question the wisdom of releasing the SONET specification in phases. The reason this approach was taken can be seen by looking at the normal development cycle needed to design telecommunication equipment. Since signals must be processed at very high speeds and cost is a primary consideration, most multiplexing and demultiplexing functions must be accomplished in Application Specific Integrated Circuits (ASICs). Design time for these can be quite lengthy.

In order to allow suppliers to start developing the needed ASICs, T1X1 released the SONET specification which detailed the multiplexing and demultiplexing functions first. For the same reason, the optical parameter specification was released in the same time frame. The specification of the DCCs was left until later phases because most of the changes to equipment would be in software which could be updated much more easily than hardware changes. In addition, we determined that specification of these channels would take a couple of years beyond the time when the initial rate, format, and optical specifications were available.

By phasing the standard, manufacturers could start designing equipment and network providers could start deploying equipment knowing that as the standard was completed for the DCCs, only software, or at most a processor card, would need to be updated.

1.4 Differences in the SONET Standard and Other Interface Specifications

Most transmission signals have been designed to carry only a limited set of payloads. For instance, the DS1 was designed to carry 24 voice signals at 64 kb/s. Other types of payloads are now being mapped into the basic DS1 format, but they must maintain some of the signal structure inherent in the DS1 format. We knew that the SONET interface must be able to carry all of the North American and CEPT signals; however, we also knew that the signal structure must allow transport of signals and services not defined with characteristics not known.

The concept of layering the basic signal was introduced in the standards body to allow the SONET interface to be structured properly. Figure 1 illustrates this by showing the major layers. A SONET interface is basically broken into two parts, with divisions within each of these parts. The two major parts are the synchronous payload envelope and transport overhead. A synchronous payload envelope (SPE) is the portion of the signal that can be structured to efficiently carry signals with different characteristics and formats. For instance, the SPE is structured one way to directly carry DS1s and another way to carry DS3s. Included within the SPE is Path overhead that is used to carry information about the signal (such as performance checks, a signal structure indication, etc.) from end to end through the entire transmission system. This allows complete end-to-end checks.

Transport overhead is the other major division within the basic signal. This is overhead necessary for transport through the network. It is broken into two parts — Section overhead and Line overhead. Section overhead is that overhead necessary for reliable communication between network elements such as terminals and regenerators. A minimum amount of overhead is placed here to allow regenerators to be constructed as cost effectively as possible. Some of the functions contained here are framing, high-level performance monitoring, a Data Communications Channel (DCC), and an orderwire channel. More will be presented about the DCC in later sections.

Line overhead was established to allow reliable communica-

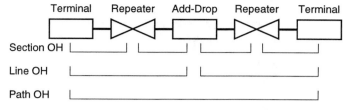

Fig. 2. Section, line, and path overhead usage.

tion of necessary information between more complicated network elements such as terminals, digital cross-connects, multiplexers, and switches. Some of the functions contained in this layer include pointers for frequency justification, automatic protection switching channels, a much larger DCC, a lower level performance monitoring check, and an express orderwire.

By structuring the overhead in this manner, network elements need only access the information necessary, and thus cost can be avoided in some network elements because not all of the overhead must be processed. Figure 2 illustrates the layers that different network elements need to process. For a complete discussion of the overhead and functions of each byte, see [1].

In addition to layering the interface to allow signals to be transported with different characteristics, we established a method of linking several basic signals together to form a transport channel with greater than 51 Mb/s carrying capacity. This capability became very important in the establishment of an international optical specification which is described in more detail in a later section. Establishing this procedure will allow the SONET interface to transport BISDN, FDDI, IEEE 802.6, and other signals that have not been defined which will require a larger bandwidth channel than 51 Mb/s.

1.5 Subsequent Standards Work

The rest of this document will concentrate on other related standards work involving the SONET interface. First, we will cover the international standardization effort that took place in CCITT and some of the different terms used in this standard which have counterparts in the ANSI document. Next, we will concentrate on the SONET Phase II effort completed in 1991, the short reach optical parameter specification, other related standards efforts, what is going on currently in SONET standardization, and some of the network impact of a standard that will be released in phases.

2. CCITT STANDARDIZATION EFFORT

In the summer of 1986, CCITT expressed an interest in defining an optical interface similar to the work going on in the United States. We proposed the same signal structure developed in the U.S. Because international compatibility was not one of the initial project goals in the U.S., the signal structure was not particularly suited to transport of a CEPT hierarchy. We had based our interface on signals with granularity of 51 Mb/s, which the rest of CCITT did not favor. Instead, granularity of 155 Mb/s was preferred. In addition to this, the original format was not well suited to transport 2.048 Mb/s signals.

Everyone very much wanted an international standard which would finally provide a common hierarchy for the three hierarchies — North American, Japanese, and CEPT. After numerous compromises on both sides, a Synchronous Digital Hierarchy (SDH) was defined which would satisfy all of the needs of countries with different hierarchies. One of the compromises involved changing a number of terms from the North American SONET specification to ones recognized by CCITT SDH recommendations. Both specify the same signal; however, it is sometimes difficult to tell because of terminology. Here, we will explain some of the different terms and how they are related to each other. For a complete definition of all the terms and how they are used, please see [3], [4], and [5].

2.1 Administrative Units

We made very little progress in international standards until the concept of Administrative Units (AUs) was introduced. Basically, an AU is a bundle of bandwidth used to manage a telecommunications network. In the U.S., most networks are managed in DS1s and DS3s. In other words, most transmission networks are managed at the DS1 and DS3 level. This is the reason digital cross-connect systems are being introduced to switch signals at these rates.

In order for a network provider to manage a network at an administrative unit, the transmission signal must be capable of being broken into these size bandwidth bundles very economically. This is the reason the original SONET signal was optimized at approximately 51 Mb/s (DS3 size bundles) and why such careful work was put into the concept of virtual tributaries (DS1 size bundles).

CEPT countries do not manage their networks at DS1s and DS3s. Instead, their networks were managed at 2.048 and 34 or 139 Mb/s. Clearly, a 51 Mb/s signal would not accommodate management at these levels.

Fortunately, we had developed an algorithm which allowed us to link a number of the basic signals together to form a concatenated signal. The basic signal in North America is called an STS-1 (Synchronous Transport Signal — level 1). When three are multiplexed together, it is called an STS-3, and when three are concatenated, it is called an STS-3c. Concatenation specifies that these signals are to be considered as a single signal, and transported that way through the network. The algorithm established actually allows any number of STS-1s to be concatenated. When these signals are linked together, it forms a transport channel in multiples of 51.84 Mb/s. Using the STS-3c signal as a base allowed CCITT and the U.S. to reach a compromise which satisfied everyone.

Figure 3 illustrates a 51 Mb/s administrative unit and how an STS-3 is formed from this. Notice pointers are used to indicate the beginning of the SPE. A good pointer description can be found in [6]. Figure 4 illustrates a 155 Mb/s administrative unit. Note that, because of the concatenation algorithm, establishing a 51 Mb/s AU with pointers also establishes a 155 Mb/s AU. Figure 5 illustrates 34 Mb/s AU containers. Note that CCITT has established only a 155 Mb/s base interface; therefore, the 34 Mb/s AU is administered by decoding four pointers located

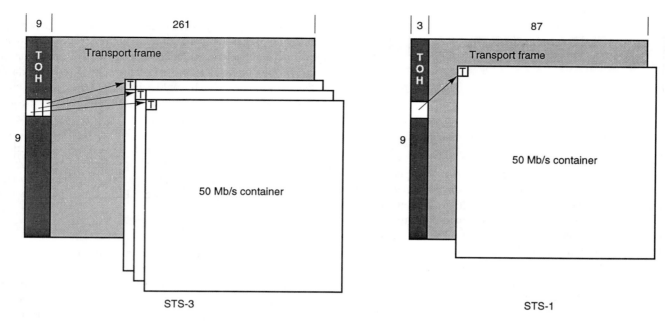

Fig. 3.   51 Mb/s AUs.

in a special location in a 155 Mb/s frame. Because of this, choosing 34 Mb/s AUs also establishes a 155 Mb/s AU. CCITT has established the name of the basic signal at 155 Mb/s as a Synchronous Transport Module — level 1 (STM-1), which is completely equivalent to the STS-3c established in the U.S.

By observing the needs of different countries and different hierarchies, we were able to establish base signals that would allow network providers the ability to administer the SONET network in the same manner as their existing network, saving untold dollars invested in network management systems. In addition, we were able to establish a common meeting point at which both hierarchies could join and above which there would be common interfaces.

The final concept incorporated into administrative units is the ability to carry one type of AU in a country optimized for the other type. Figure 6 illustrates nested signals. A nested signal is a set of AUs contained within a 155 Mb/s AU which is transported through a network. Using nested signals, a CEPT Hierarchy country can carry three AU-3s through their network using a 155 Mb/s AU without constructing a special network to manage it. A North American Hierarchy based country can transport four 34 Mb/s signals in the same manner. Nested signals are to be used to transport bulk traffic (e.g., DS3s or 34 Mb/s) through the network because lower level traffic (DS1s, 2.048 Mb/s, etc.) would be transported in virtual tributaries.

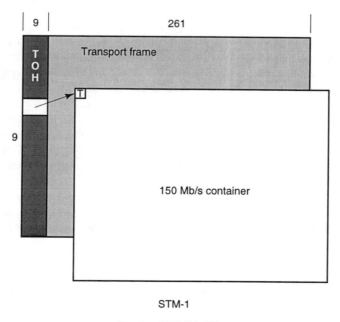

Fig. 4.   155 Mb/s AU.

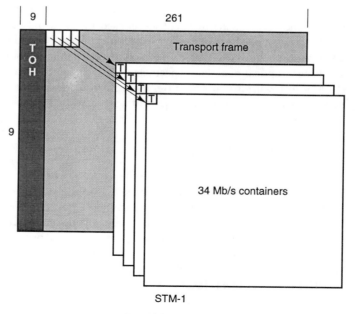

Fig. 5.   34 Mb/s AUs.

32

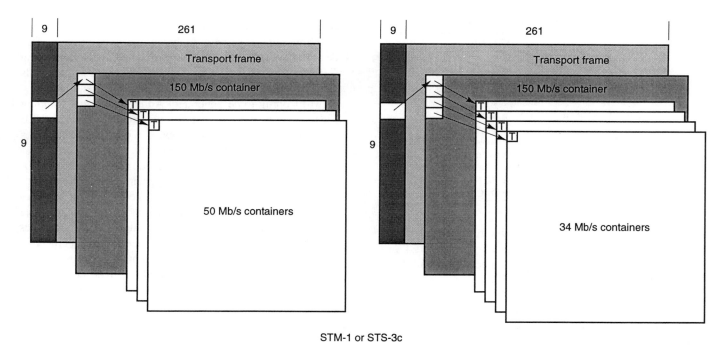

9	261

Transport frame

150 Mb/s container

50 Mb/s containers

9

9	261

Transport frame

150 Mb/s container

34 Mb/s containers

9

STM-1 or STS-3c

Fig. 6.   Nested signals.

2.2 VTs and TUs

Mentioned before, North American Hierarchy networks are managed at DS1 and DS3 rates. We established AUs to manage a DS3, but we needed to define another easily manageable container for DS1s. In SONET, these are called Virtual Tributaries (VTs). Basically, a VT is a container into which one can place a sub-DS3 signal for efficient management. There are several different types for different size signals: VT1.5, VT2, VT3, VT6 for DS1, 2.048 Mb/s, DS1C, and DS2 signals. In addition to this, we have established an algorithm which will allow VTs to be concatenated together to create a variable size transport pipe.

A VT Group is a set of VTs that has been grouped together to carry like VTs. VT groups were established because it was recognized that 4 DS1s or 3 2.048 Mb/s signals could be contained in a VT6. By establishing VT groups, a single VT mapped STS-1 signal can transport a mixture of DS1 and 2.048 Mb/s signals, something no other hierarchy can accomplish easily.

CCITT established a signal identical to a VT called a Tributary Unit (TU). In addition, a TUG (or Tributary Unit Group) was established to correspond to a VT group. Even though their terms are different, they refer to the same signal format. For further information on CCITT definitions, see [4], and for SONET, see [1].

2.3 Results

The most significant outcome of this work is that now one transmission interface hierarchy can be established throughout the world. Starting at 155 Mb/s, the hierarchy of the world telecommunication networks becomes common. Because they still need to transport the existing asynchronous hierarchies, payload structures will be different, but at least the transmission and cross-connecting equipment can be common.

3. SONET Phase II

We achieved many of our initial goals for SONET standardization when the first specifications were published in 1988. However, we still did not have a standard for communicating operations support information across the data communications channels (DCCs). Operations support information includes commands (such as DS1 loop backs), alarm information, provisioning information (such as what options to set in a particular interface card), surveillance information (such as performance monitoring data), and other information needed to maintain a network element. We had established the bandwidth allocated for the DCCs (192 kb/s for the Section DCC and 576 kb/s in the line DCC), but there was insufficient time to reach consensus on the protocol and messages to be used over these channels. To many, one of the significant advantages of a SONET interface was the availability of the DCC channels which would be used to construct an operation support network at the same time the transport network is constructed. Therefore, standardization of the DCC became the primary goal of the SONET Phase II effort.

T1M1 has primary responsibility to develop protocols and messages for use in operation support networks in telecommunications interfaces. We requested that they develop a standard for SONET, and a joint ad hoc group was formed to aid in standardization. T1M1 at the time was completely involved in the development of generic interfaces for all OS to NE and NE to NE interfaces, and was organized to accomplish this. When our request arrived to develop a specification for a single interface, it caused a massive redirection of resources and involved many of the acting working groups in T1M1. The effort and time dedicated to our request were very significant, and a great deal of credit goes to T1M1 for responding to our request with the spec-

TABLE 1. SONET PHASE II ADDITIONS

Item	Description
Text changes	Editorial in nature
Timing & synchronization	Clarifications and requirement relaxation
APS	Clarification
Mapping additions	Added an ATM and DS4NA mapping
Overhead usage	Specification when to use certain overhead
DCC protocol	Specifies the protocol used on the Line and Section DCC

ifications developed.

Because the work was fragmented across several working groups and many people interested in only the SONET interface had to attend numerous meetings where only a portion of the time could be dedicated to SONET, a decision was made to complete the specification of the protocol in T1M1 and move the final message specification back to T1X1.5 where all of the time could be dedicated to this task. Obviously, a great deal of coordination is necessary between these groups to ensure that standards for the SONET specific messages and generic standards do not diverge. This recent change is significant because it will allow us to dedicate the time and resources necessary to complete the specification in a timely manner.

Even though specification of the DCCs was the primary goal of the SONET Phase II effort, there were a number of other items which needed to be cleaned up. The standard ANSI T1.105-1991 [1] contains all of the changes for SONET Phase II (see Table 1).

3.1 Results

Changes which had the least effect on equipment designs are the minor editorial and text changes. One significant addition was inclusion of the term SONET in the standard. Until this point in time, the term SONET had been avoided for a variety of reasons. When most of these reasons were found to be unjustifiable, we included the term to allow someone to identify the standard easier in a search. Another change included relaxing the synchronization requirements for network elements. After careful examination of pointer justification movements, stuffing techniques, and desynchronizer phase lock loop characteristics, it was found that the specification could be relaxed somewhat while maintaining required jitter performance. It should be noted that subsequent work in this area has resulted in a new standard being developed to cover synchronization. It completed the voting stage in 1994 and is now a standard.

Changes were made to the automatic protection section which clarified many points and provided a method of communicating switch status (such as lock out) via a standard means. Part of this included requiring the section DCC to be present in the line side interface or providing an alternate method of communicating information through the OS network. As in the synchronization area, new work on APS caused an entirely new document to be developed specifically to cover this important topic. The new structure of the T1.105 series of documents can be found at the end of this paper.

Two new mappings were added and clarifications to the byte synchronous mappings were included. The new mappings include a DS4NA (139.264 Mb/s) mapping and an ATM mapping for BISDN. An ATM mapping is significant because the details concerning cell size in CCITT have settled sufficiently to allow an ATM mapping to be generated. Additionally, this mapping will also be used with very slight modifications for the IEEE 802.6 mapping. This can be accomplished because the cell sizes for BISDN and IEEE 802.6 are the same.

Another clarification added to the document is a table describing the usage of various overhead bytes in different situations. Table 2 lists line side and drop side interfaces, and specifies which overhead bytes are to be present at the interface. Notice that there are both electrical and optional drop side interfaces which correspond to the electrical interface parameters recently agreed to by T1X1.4 and the short reach optical parameter. One of the main drivers behind this table was to restrict the use of DCCs on the drop side interface. Recent contributions to T1M1 have indicated that communication of operation support information within an office or location can be much more cost effectively accomplished with a Local Communication Network or LCN. This will be described in Section 6 on future standards work.

3.2 DCC Specification

In previous sections, we discussed the goals of completely specifying the DCCs and the method used to accomplish this. Because of underestimation of the effort required and the organization of the work, we were only able to specify the protocol stack used on the DCCs. This means that we can now transfer data reliably over these channels, but we do not yet have a common language which will allow us to communicate. The protocol stack is contained in Fig. 7, and more information about choices made within each layer is contained in [1].

Is it useful to have the protocol specified without message? The answer is yes because now hardware can be built which will implement the protocol. When the messages are specified, then software can be updated to include the messages. The only risk is that sufficient memory or processing power be included when the hardware is developed.

Layer	Name	Protocol
7	Application	ISO 9595-2, 9596-2 (CMISE) X.217, X.227 (ACSE) X.219, X.229 (ROSE)
6	Presentation	X.216, X.226 X.209, (ASN.1 basic encoding rules)
5	Session	X.215, X.225
4	Transport	ISO 8073, 8073-PDAD2 (TP4)
3	Network	ISO 8473 (ISO IP)
2	Data link	LAPD
1	Physical	DCC

Fig. 7.   DCC protocol stack.

TABLE 2. OVERHEAD SPECIFICATION

Signal	Drop Side		Line Side (Optical)		
	Electrical	Optical	Protection	Line 1	Line 2-N
A_1 - A_2	R	R	R	R	R
B_1	NA	NA	R	R	R
D_1 - D_3	NA	NA	R	R	NA
C_1	R	R	R	R	R
E_1	NA	NA	OPT	OPT	NA
F_1	NA	NA	OPT	OPT	NA
B_2	R	R	R	R	R
D_4 - D_{12}	NA	NA	OPT	OPT	NA
E_2	NA	NA	OPT	OPT	NA
H_1 - H_3	R	R	R	R	R
K_1, K_2 (APS)	NA	Note 1	R	NA	NA
K_2(6-8) Line AIS	NA	NA	R	R	R
K_2(6-8) Line FERF	R	R	R	R	R
H_1, H_2 (Path AIS)	R	R	R	R	R

Note 1: For further study.

4. SHORT REACH OPTICAL INTERFACE SPECIFICATION

So much emphasis has been placed on the standardization of the DCCs that it is easy to overlook other very important work. One of these is the issuance of the Short Reach Optical Parameter Standard ANSI T1.117-1990. Work on this specification started at the same time as the DCC specification. The goal for this document was to assign optical parameters to an interface that is optimized for shorter distances (somewhere around 2 km). By restricting the distance, lower cost components can be used which will allow a lower cost interface to be established.

Main uses for this specification include interfaces between different network elements inside a central office and short interfaces to customers. If these interfaces are used inside an office (drop side), then certain overhead need not be present (e.g., Section and line DCC, orderwire, etc.). The reason is to minimize the cost of providing an optical interface between equipment inside an office by using only those overhead bytes necessary. See Table 2 for the compete specification.

If, on the other hand, the short reach parameters are applied to an interface to a customer, then it is a line side interface and most of the overhead must be present to ensure reliable data transfer. These two cases illustrate the need to keep the optical parameter and format specifications separate.

5. OTHER RELATED STANDARDS ACTIVITIES

Because we have specified SONET to be a transport signal in the public network that is capable of transporting many different services of varying capacity and structure, a great deal of interest has been generated in developing mappings for new services. Local area networks can be built or a data network can be constructed inside a single location or campus to satisfy a particular need. When these networks must be connected through the public network, many times some type of conversion from the data format or rate must be performed to meet the existing public network standards. When SONET equipment is deployed, it can allow us to create bandwidth channels through the public network more easily and structure the SPE to more efficiently transport the service.

Broadband ISDN is one service which will require a large bandwidth channel through the public network. A great deal of effort is being placed on the final definition of the BISDN service. Several issues are already resolved. Two of the most important are the Asynchronous Transfer Mode (ATM) cell size, 5 byte header and 48 byte information field, and how it will be transported in the public network. An STS-3c was chosen to carry the BISDN signal because it offers a 155 Mb/s channel and is identical to the international interface at STM-1. Mentioned in a previous section, the BISDN ATM mapping is included in the SONET Phase II supplement. By carrying the BISDN signal as a SONET payload, we can allow ATM to perform the functions it does best, supplying variable bandwidth to a customer, and the synchronous frame of the SONET signal to perform the functions it does best, providing a large bandwidth channel that is maintainable in the network.

A related activity is the IEEE 802.6 standardization effort. A great deal of work by numerous people has ensured that the important characteristics of the IEEE 802.6 standard are identical to the BISDN standard. Because of this, the IEEE 802.6 signal can be carried in a SONET STS-3c payload in much the same manner as the BISDN signal. In fact, only a very minor modification to the mapping, involving allocation of additional bandwidth for a control channel needed in the IEEE 802.6 signal, is necessary. Therefore, a method of transport for this signal in the public network will exist when SONET equipment is deployed.

The X3T9 Committee has undertaken a project to map the FDDI interface into a SONET payload. Again, the reason for this is to provide a method of easily transporting this signal through the public network. The mapping is under consideration currently, but will most likely map the 125 Mb/s FDDI signal into the 155 Mb/s payload of an STS-3c. More analysis ensured that this method is the proper approach.

These are some examples of what is currently going on in related standards bodies to utilize some of the capabilities available in a SONET signal. Others will surely follow as more peo-

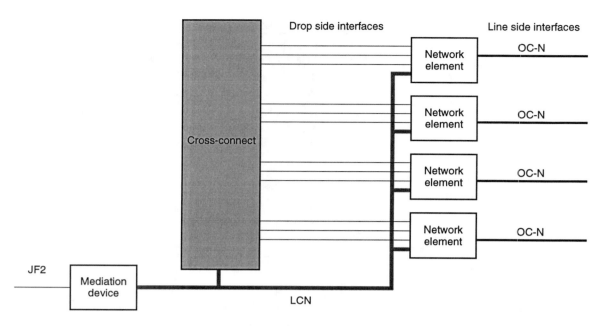

Fig. 8. LCN deployment.

ple look at methods of obtaining bandwidth channels through the network and more SONET equipment is deployed. Several capabilities included in the SONET format to facilitate this are the ability to link VTs together to create bandwidth channels in multiples of 1.5 Mb/s and the ability to link STSs together to form channels in multiples of 51 Mb/s. This, coupled with payload transparency through SONET equipment and extensive performance monitoring, will ensure that additional applications of the interface will be found.

6. SONET STANDARDS IN THE FUTURE

As we have shown, not all of the goals of Phase II were met, and other work must be accomplished. The standard, in fact, will have numerous other changes in the future. These will involve mostly software and hardware changes which will need to be implemented to allow equipment to be backward compatible.

6.1 North American Activities

The main activities in North America will involve standardization of the DCCs and related network element to operation support (NE to OS) interfaces.

As we saw in previous sections, we were able to standardize the protocol stack for the DCC, but not the messages. Actually, two steps are involved in standardizing the messages. The first is to specify an information model. In its simplest terms, an information model is an agreement about the characteristics of a network element that one wants to communicate to another party. This is how many parameters will be communicated, how one can describe performance monitoring, and so on. After this agreement is reached, then the actual message can be constructed which will transfer the information. This work is being developed in the T1X1.5 Working Group. They have released their first document, ANSI T1.119-199X, which will be published

very soon.

Because of the complexity of the problem, message sets will most likely be released in several phases. The first set will accomplish surveillance and alarm functions. Other sets will include provisioning, etc. This is the reason it is so important to deploy a network element which can have the software updated as the standard evolves, and why we specified the hardware affecting portions of the interface first. Most commercial software packages require updating every couple of years, and the SONET equipment will most likely follow the same course.

A related but separate standards activity has recently been started in T1M1. This is to define a Local Communication Network (LCN) to communicate DCC information inside a central office or on the drop side of equipment. It was recognized that the DCCs provided a great deal of data transport capability, but currently no interfaces existed between NEs and OSs which would allow full benefit to be gained from this capability. Therefore, an LCN protocol has been standardized which uses an IEEE 802.3 Ethernet Connection over twisted pair wire for the lower layers and the same upper layers as the DCCs. Figure 8 illustrates a situation where an LCN can be used for NE to NE and NE to mediation device communication. The standard calls for a 10 Mb/s connection.

The purpose for defining an LCN in this manner is to allow a multidrop data network to be created which could transfer a large amount of operations data over a short time. In addition, because the upper layers are the same as the DCC, the software necessary to implement the DCC and messages standardized for the DCC can be used for the LCN.

In addition to concentrating on OS and DCC issues, future work will include additional optical parameters for interfacing above 600 Mb/s, additional mappings for new services, and add/drop and ring automatic protection systems.

6.2 New SONET Standard Structure

In order to make the SONET standards more recognizable and to allow users of the standard to be sure they have the documents needed, T1X1 has started the process of reissuing the entire SONET standard in a family of documents. Table 3 lists the documents in process.

7. NETWORK IMPACT OF STANDARDS EVOLUTION

We have seen in the previous sections that the SONET standard will evolve over a number of years. A natural concern for organizations deploying equipment is when the standard will be complete enough to start deployment. Actually, we have developed this standard in a method to minimize the impact of evolution, and deployment has started already. In fact, SONET equipment is considered just another generation equipment available for deployment. In this section, we will examine some of the situations that will occur with SONET deployment and how these can be planned for. In addition, we will look at the developing OSS interfaces, and how these will affect deployment of some of the important capabilities embedded in SONET equipment.

7.1 DCC Evolution

Other sections in this paper have pointed out the fact that the SONET Phase I standard specified the interface in sufficient detail to allow reliable information transport from one point to another. This includes all of the North American Hierarchy as well as the CCITT specified hierarchy. What was left for Phase II was completion of specification of the DCC channel. We wanted to completely specify protocol and message sets to be used over these channels. As we have seen, we were only able to complete the protocol specification, and the standardized message sets will need to be added to the equipment at a later date. The reason for releasing the specification in this manner was to allow hardware development to be completed and update the necessary software when the messages become available.

As illustrated in Fig. 9, most network elements are constructed with a microprocessor to process messages on the DCC. Since the protocol is now specified to allow these processors to communicate reliably over the DCCs, when a new message set is defined, the software must be updated to include this message set if this interface requires it.[1] This is analogous to obtaining updates or new releases of software for a word processor or spreadsheet routine. In both cases, care must be taken to ensure backward compatibility. How the update is accomplished largely depends on the hardware architecture chosen by the equipment supplier. One could replace EPROMs or load the new software into memory via a craft terminal. As will be illustrated later, eventually we will download software to network elements from central Operation Support Systems. Until a message set is standardized, vendor specific messages can be used to allow the equipment to function as needed. It is in the best interest of all parties to complete the standardization of messages in order to reduce reliance on vendor-specific messages.

[1] Note that if the equipment is functioning reliably and adequately without the new message set, then it may not be necessary to include the new set unless other circumstances require it.

TABLE 3

Standard	Topic	Standard	Topic
T1.105	Base SONET standard	T1.105.06	Physical layer specifications
T1.105.01	Automatic protection switching APS	T1.119	Base operations, administration, maintenance, and provisioning standard
T1.105.02	Tributary mappings		
T1.105.03	Synchronization and jitter		
T1.105.04	DCC protocols and architectures	T1.119.01	APS information model extensions
T1.105.05	Tandem connection maintenance	T1.119.02	Network level information model extensions

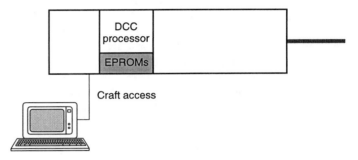

Fig. 9.   Software update procedure.

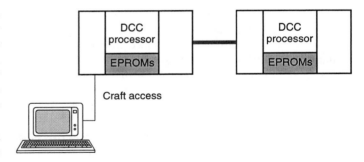

Fig. 10.   Point-to-point systems update.

Figure 10 illustrates the situation when two network elements are connected in a link. Here again, software can be updated by replacing EPROMs or by downloading software into one element and using the DCC to load it into the other end.

Figure 11 illustrates the procedure that we would like to arrive at in the future. In this case, many SONET network elements are connected in a network, and they are not all from a single supplier. In fact, not all of the elements are the same type of equipment. Some of them are terminals, hubs, add/drop elements, and DCS elements. This network has been connected with the LCN, which can be used to facilitate downloading message sets and software updates from a central location. When a vendor supplies an update, it can be loaded into the OS system and addressed to the appropriate network element. This way, a number of elements can be updated, some of them not from the same supplier, when new generics of software are released.

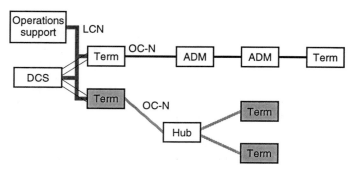

Fig. 11.   SONET network.

As can be seen in these examples, networks can be built with SONET equipment and then updated with messages as they are developed. Of course, the actual method of furnishing and implementing the update will be dependent on the supplier of the equipment, but the standard can be constructed to allow this to happen. It is important to realize that some of this capability, specifically the messages to allow downloading updates to other network elements, must be standardized first, but this should be a task of high priority in the committee because it greatly eases deployment of the equipment into the network.

7.2 Operations Support Interfaces

One of the most common difficulties facing a network element supplier, and network element buyers, is selection of OSS interfaces which need to be supplied. The most common interface in use today is parallel contact closures. These will only supply alarm information from the element. Because micro-processors have become so inexpensive and network elements have become so intelligent, furnishing a network element of today's generation with a parallel alarm interface is like having a stock ticker that tells you only if the Dow Jones Average is up or down. Useful information if you need this to invest, but all of the detail is lost.

Other interfaces are standardized or are in the process of being standardized (e.g., TBOS, TABS, G2, G4, and LCN). Because some of the currently available interfaces cannot supply the detailed information network management systems require, many vendor-specific interfaces have been deployed as well. The question quickly becomes one of a chicken and egg. Since we are building SONET equipment with a very sophisticated DCC interface which will allow us to perform many needed functions and allow us access to a great deal of information about that element, which interface should be designed? The answer is very clear, but not always satisfying: one must build all of them because network management systems are already in place to manage the equipment currently in the network and it is very expensive to change them.

Many other applications will be found for this interface in the future. We have only started to scratch the surface of what is possible.

This point is very important. The specification of DCC messages and protocol is independent of the interface back to the OS. When SONET equipment is deployed that contains a DCC implemented according to the standard, the network element can supply needed information through whatever OSS interface is present. If it is parallel alarms, then one will have the most sophisticated communication path of alarms possible. What is important is that when a more sophisticated interface is deployed to gather more information, the network element is ready, and most likely only a card that implements the OS interface will need to be deployed.

Fortunately, many of the embedded OS systems have migrated to the point where an X.25 interface, using TL1 (Transaction Language 1) messages, is available to communicate with network elements. This interface provides a very rich set of commands and alarm notifications which allows these OS systems to take full advantage of the SONET DCC capabilities. While this is not the final standard solution, it is only necessary for the OS to be programmed to understand any vendor-specific messages.

SONET network elements can and will exist in any of these environments. It is up to the network builder to decide what is the best OSS interface to be deployed. It is up to the suppliers of the equipment to ensure that these interfaces are supplied cost effectively.

8. CONCLUSIONS

As we have seen in this paper, the SONET standard has been divided into stages in order to allow equipment suppliers to deliver and network builders to deploy hardware in the network. Because the standard has been developed in stages, it makes the job a little more difficult However, sufficient detail exists in the standards to allow graceful migration as more of the standard is developed. This is because the later stages of the standard will affect only the messages used to communicate network management information from one network element to another, and these can be supplied by software updates.

Additionally, we have shown how this interface can be used to build networks that can efficiently transport services today and tomorrow without knowing the structure of these future signals.

References

[1] ANSI T1.105-1991, "American National Standard for Telecommunications — Digital Hierarchy Optical Rates and Formats Specification," 1991.
[2] ANSI T1.106-1988, "American National Standard for Telecommunications — Digital Hierarchy Optical Interface Specifications, Single Mode." 1988.
[3] CCITT Recommendations G.707, "Synchronous Digital Hierarchy Bit Rates."
[4] CCITT Recommendation G.708, "Network Node Interface for the Synchronous Digital Hierarchy."
[5] CCITT Recommendation G.709, "Synchronous Multiplexing Structure."
[6] R. Ballart and Y.-C. Ching, "SONET: Now it's the standard optical network." IEEE Commun. Mag., Mar. 1989.

An Overview of Emerging ITU-T Recommendations for the Synchronous Digital Hierarchy: Rates and Formats, Network Elements, Line Systems, and Network Aspects

JOHN EAVES, GILLES JONCOUR, MICHAEL STEINBERGER, AND TIM WRIGHT

INTRODUCTION

In 1988, the International Consultative Committee for Telephone and Telegraph (CCITT) approved Recommendations G.707, G.708, and G.709 [1]–[3], which specify the rates and formats for the Synchronous Digital Hierarchy (SDH). In the 1989–1992 CCITT study period, new Recommendations were developed for synchronous multiplexers and cross-connects, synchronous optical fiber line systems, SDH management, and the network aspects of SDH. Significant progress was made, with several important Recommendations achieving accelerated approval in 1990. The progress at that time was surveyed in a paper by Balcer, Eaves, Legras, McLintock, and Wright [4]. Considerable work remained, however, and revised versions of these Recommendations were approved in 1993. These Revised Recommendations are better integrated with each other, and more complete. In this respect, they represent a great deal of progress in our insight into the interaction between signal formats, network elements, optical line systems, and network architectures in the context of a managed (i.e., manageable) telecommunications network. More work remains, however, and these Recommendations remain under active development and refinement in the 1993–1996 study period of the International Telecommunications Union — Telecommunications Sector (ITU-T), which was formerly the CCITT. This paper reviews the present status of these Recommendations, and identifies some of the areas for future study.

SDH RATES AND FORMATS: G.707, G.708, G.709

Following the work carried out in the U.S. on SONET, discussions, which would later lead to the definition of the Synchronous Digital Hierarchy (SDH), were initiated in 1986 under CCITT SG XVIII. They resulted, in 1988, in the approval of the G.707, G.708, and G.709 Recommendations. These documents contain the specifications of the basic principles of SDH. The hierarchical bitrates, multiplexing principles, and multiplex elements are defined there.

Studies on SDH multiplexing have been pursued since 1988, and the initial content of G.708 and G.709 has evolved. Their second version was approved in 1990, and the latest one is dated March 1993. Since the beginning of the current study period, their maintenance has been under the province of ITU-T SG 15. A new version of G.707, G.708, and G.709 should be submitted to approval of ITU-T SG 15 in 1995.

The G.707, G.708, and G.709 Recommendations define a worldwide Network Node Interface (NNI) having the following advantages.

- It allows simple interworking between hierarchies based on 1.5 and 2 Mb/s.
- The interface is applicable to any kind of equipment that may be found in a transmission network (multiplexer, cross-connect, line system).
- The principles can be applied to any physical medium (optical fiber, radio relay).
- It takes into account the current PDH hierarchy while being fully open to future applications.

Recommendation G.707

G.707 contains the specifications of the bitrates forming the Synchronous Digital Hierarchy. Above the STM-1 level (155.520 Mb/s), which corresponds to the lower level of the hierarchy, two other levels, namely STM-4 and STM-16, are defined. They correspond, respectively, to bitrates of 622.080 and 2488.320 Mb/s. An STM-64 level (9953.280 Mb/s) will be specified in the next version of the Recommendation.

Recommendation G.708 and G.709

These two Recommendations, specifying the Network Node Interface for the SDH and the SDH multiplexing structure, respectively, can be considered together. They mostly define the multiplex elements and the way they are to be arranged together to form STM-N frames.

Containers (C-n). For their transport through an SDH network, the client signals are adapted into containers (C-n). This mapping, as it is called, consists, for instance in the case of plesiochronous 2.048 Mb/s, of a bit justification process in order to adapt them in containers (C-12) having a fixed size. Mappings are currently defined for all the PDH signals (except the 8448 kb/s) and for ATM cells. Concerning those, only a mapping in a C-4 (equivalent to a 140 Mb/s) is currently defined. The next

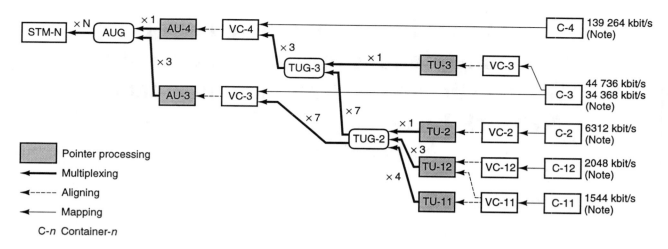

NOTE – G.702 tributaries associated with containers C-x are shown. Other signals, e.g., ATM, can also be accommodated.

Fig. 1. SDH multiplexing structure.

version of G.708 will define ATM mappings in lower order C-n.

Virtual Containers (VC-n). A VC-n consists of the association of a C-n and its corresponding path overhead (POH). The VC-n is the entity which is to be transparently transported (including multiplexing and cross-connection) in an SDH network.

Overhead Bytes. Parts of the frame are allocated to the transport of OA&M information. In accordance with the SDH layered approach, two levels of overhead bytes have been defined: one at the section level (SOH), the other one at the path level (POH). The overhead bytes carry information for equipment and network supervision and management. Among others, they contain parity check information and various embedded data channels transporting alarm and control messages.

The SOH consists of 9×8 bytes, whereas the VC-4/3 POH is constituted of only 9 bytes. The identification of new functionalities has recently led to the extension of the POH of the VC-2/12/11 from 1 to 4 bytes. Some network applications associated with overhead bytes are still under discussion (Tandem connection monitoring, Path trace, . . .).

Pointers. A synchronous multiplexing method requires phase alignment of the signals within a node. A physical phase alignment in a frame buffer was not thought to be an adequate solution for SDH. A pointer mechanism has therefore been invented. The value of a pointer indicates the position of a VC-n within the STM-N payload. A positive-null-negative byte justification mechanism is associated with the pointer in order to handle the frequency differences that may exist between different network node clocks. Thus, a VC-n is allowed to float within the payload. Two levels of pointer have been defined: one at the AU-4/3 level (associated to VC-4/3), and one at the TU-2/12/11 level (associated to VC-2/12/11).

Multiplexing structure. The G.708 multiplexing structure takes into account the various multiplexing routes adopted throughout the world, including Europe, North America, and Japan, and also addresses the necessary interworking aspects. The initial multiplexing structure was revised in 1990 in order

to reflect ETSI decisions, and to allow all types of mixed payloads.

The payload of an STM-1 can either be a single VC-4 or three VC-3s (used mostly in North America and Japan). A VC-4 can either contain a C-4 or a multiplex of lower order VC-n. Similarly, a higher order VC-3 can either contain a C-3 or lower order VC-ns. Within a VC-4/3 or, hence, within the payload of an STM-N frame, many combinations (including mixed ones) of VC-n are allowed by the multiplexing structure. Concatenation of m VC-n is allowed for the transport of signals which do not fit into a single virtual container. The multiplexing structure is shown in Fig. 1.

Frame structure. Independent of the STM-N bitrate, the frame is defined as a 125 μs block. This frame, usually shown as a 9 row per N $\times$ 270 column structure, comprises three different areas: the payload, the Section Overhead (SOH), and the AU-4 pointers. The SOH is itself divided in two parts: the Multiplex Section Overhead (MSOH) and the Regenerator Section Overhead (RSOH). The latter is the only part of the frame in which information can be overwritten in a regenerator. A synchronous byte (or column) interleaving of STM-1 payloads leads to the constitution of an STM-N payload. The frame structure is shown in Fig. 2.

In addition, G.708 contains the description of a frame structure for a 51.840 Mb/s interface for use in radio or satellite digital sections.

SYNCHRONOUS EQUIPMENT: RECOMMENDATIONS G.781, G.782, AND G.783

Currently, the range of multiplex equipment recommended for the Plesiochronous Digital Hierarchy (PDH) is specified in ITU-T Recommendation G.702 [5]. Listed in this Recommendation are all of the types of equipment which can be interconnected at each of the layers of the PDH. For each type of equipment listed, there is another Recommendation which specifies the equipment characteristics.

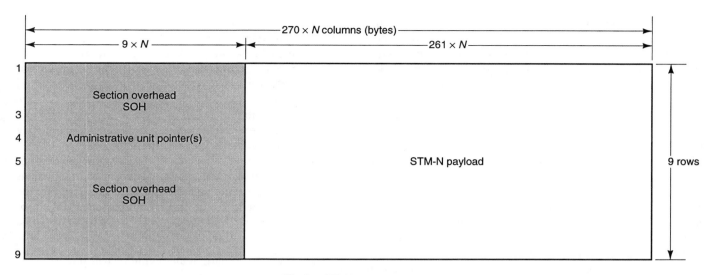

Fig. 2.   STM-N frame structure.

In May 1990, Recommendations G.781, G.782, and G.783 [6]–[8] were approved via accelerated procedure. These Recommendations describe the characteristics of multiplex equipment for SDH. The goal of these documents included:

1. Allow for the much richer variety of multiplex equipment types which are possible in SDH.
2. Define the equipment characteristics with enough detail and precision that functioning, manageable networks could be assembled using equipment from many suppliers.
3. Define the equipment characteristics in a way which does not unduly constrain the choice of implementation.

SDH makes a wider variety of equipment types possible by including many more mappings than in PDH, and by supporting mixed mappings within a single payload. Thus, multiplex equipment may add/drop traffic from mixed payloads using any of several different combinations of mappings, in addition to converting from one mapping to another, and multiplexing from PDH to SDH. For example, it would be possible to multiplex 42 2048 kb/s signals and one 44 736 kb/s signal into a single STM-1 signal, add/drop several of the 2048 kb/s signals at one intermediate node, drop the 44 736 kb/s signal at a second intermediate node, and then completely demultiplex the signal at a terminal node. The equipment at each of the four nodes just described could be considered to be a different equipment type.

To make practical use of the flexibility SDH offers, it must be possible to manage the resulting SDH network. SDH makes this task easier by including much more complete overhead information than was available in PDH. To be useful, however, this overhead must be generated, detected, and reported in an entirely consistent manner by all SDH equipment. This is especially true for network fault and performance information. Thus, much more detail and precision are required in the description of equipment behavior.

To achieve widespread adoption, equipment Recommendations should not place undue constraints on physical implemen-

tation or design innovation. They must encourage a multivendor environment, and not become a barrier to developments which may be of value to the telecommunications industry. Thus, it would be inappropriate for a Recommendation to directly address details of internal implementation in even the most general terms.

The approach adopted in G.781-3 was to define a *functional reference model* which describes the externally observable behavior of the equipment as the behavior exhibited by an interconnection of functional blocks. The functional reference model for a particular equipment therefore consists of a definition of the interconnection of the functional blocks, the definition of the interface between functional blocks, and the definition of the input/output characteristics of each functional block.

It is important to note that any implementation which exhibits precisely the same externally observable behavior as the functional reference model is compliant with the Recommendations. It is definitely *not* necessary for the implementation to have the same internal structure as the functional reference model. In fact, great care was taken to make sure that all of the characteristics defined for the functional blocks would be externally observable, so that the model does not contain any more detail than necessary.

Following the 1990 approval, several of the functional blocks were defined in greater detail, and the flexibility of the functional reference model approach was exploited to define the characteristics of SDH cross-connect equipment as well as multiplex equipment in the same Recommendations. The latter step involved merging G.781-3 with Draft Recommendations G.sdxc 1–3 on SDH cross-connect equipment characteristics into a revised version of G.781-3 [9]–[11] which is reorganized and expanded. These Recommendations now refer to "SDH equipment" rather than "SDH multiplexers" or "SDH cross-connects." This is appropriate since there is very little distinction between the latest generation of intelligent multiplexers and cross-connects, aside from system capacity.

Revised Recommendations G.781, G.782, and G.783 completed the accelerated approval procedure in September 1993. G.781 has changed very little, and remains a guide to the overall structure of the Recommendations, with guidelines on the choice of options in them. G.782 [12] describes SDH equipment at the applications level. It introduces the functional blocks used in the functional reference model, and then shows how these functional blocks can be used to describe equipment which supports various network applications. Examples are given of various possible multiplex and cross-connect equipment types. Although these equipment types are numbered, they are included primarily for illustrative purposes, and are not intended to restrict the range of capabilities which may be included in equipment offerings. G.782 also describes general equipment characteristics, such as transit delay and availability/reliability, which cannot be ascribed to a single functional block.

Recommendation G.783 [13] provides a detailed description of each functional block and its interfaces. Each description includes information on the transmission, management, and synchronization behavior of the functional block, as appropriate. Transmission behavior includes mapping and demapping of payloads, insertion and detection of overhead, and insertion of maintenance signals such as Alarm Indication Signal (AIS) and unequipped signals. Management behavior includes the insertion of management information into overhead, and the detection and reporting of transmission fault and performance information to the management system. Synchronization behavior includes the derivation of clock timing from input signals, the distribution of clock timing to other functional blocks, and the use of clock timing to generate transmission signals. To aid in this multidimensional description, separate reference points are identified for transmission, management, and synchronization interfaces to each functional block.

A key feature of the SDH signal format is the generous provision of bytes for equipment and network management purposes; and the processing of these bytes is a large part of the behavior of a functional block. The result of these overhead processes is a comprehensive array of management information which must be condensed before being made available to the management system. Section 4 of G.783 describes a "Synchronous Element Management Function," where the information is filtered and presented to the management system in the form of managed objects. The treatment of managed objects, and SDH equipment management in general, is the subject of Recommendation G.784 [14].

Future Equipment Recommendations

One of the major growth directions for SDH equipment Recommendations is in the area of network protection applications. In this regard, a new series of Recommendations is being developed to describe the equipment aspects of SDH ring protection applications. These applications will definitely include path-switched rings and line-switched rings. In addition, there is interest in more generalized versions of these applications in the context of subnetwork connection protection and trail protection, as well as detailed improvements to be made to the Multiplex

Section Protection application already defined in G.783.

Another possible direction is the further refinement of the functional modeling approach. As new applications are defined, new functional blocks are needed to describe the equipment behavior properly. Thus, the functional modeling approach utilized presently in G.783 may need to be extended to accommodate these new applications more effectively. For example, new functional modeling methods being developed within ETSI [15] may be used to improve the existing recommendations.

Finally, there are still a number of detailed problems associated with the detection and reporting of transmission anomalies and defects. While some of these problems are posed by the emerging applications, others are problems associated with applications which have been recognized for a long time. These problems include:

- The prevention of spurious alarms during path provisioning.
- Detection of path trace identifier mismatch.
- Detection and modeling of signal label mismatch.

SDH OPTICAL LINE SYSTEMS: RECOMMENDATIONS G.957 AND G.958

Prior to the development of the Synchronous Digital Hierarchy (SDH), the ITU-T (previously CCITT) had developed several specifications for the Plesiochronous Digital Hierarchy (PDH), based on both the 1.5 and 2 Mb/s hierarchies. Among these were Recommendations G.955 and G.956 [16], [17] for digital line systems for use on optical fiber cables. The purpose of this Recommendation was to provide sufficient specifications to achieve "longitudinal compatibility" on elementary cable sections of different optical line systems, i.e., the possibility of parallel installation of systems designed by different manufacturers on the same optical cable. The parameters specified to meet this requirement are the minimum values of the maximum attenuation and dispersion achievable by the system on a regenerator section length for a given level of performance [e.g., a bit error ratio (BER) of 10^{-10}]. Under Rec. G.955, the PDH-based system designer has freedom to specify component characteristics of the system such as line code and bit rate, operating wavelength, transmitter power range, receiver sensitivity and overload, and operation and maintenance features consistent with the system performance design objective. Interconnection at an administrative boundary generally required that the two network operators interface at an electrical hierarchical level or, to interface via an optical signal, choose the same manufacturer of the optical system.

Following approval of the SDH hierarchy, multiplexing structure, and payload mappings in 1988, the ITU-T undertook the development of new Recommendations which could achieve "transverse compatibility" on elementary cable sections, i.e., the possibility of mixing different manufacturers' multiplexer and regenerator equipment within a single optical section. In 1990, the ITU-T approved Recommendations G.957 and G.958 [18], [19] which enable transversely compatible SDH optical systems.

Recommendation G.958

Recommendation G.958 provides information and specifications on the following aspects of digital synchronous optical line systems (SLS):

- Applications — SLS can be used for interoffice and intraoffice applications links. Currently, the Recommendation covers only interoffice applications; the simplifications in several system aspects for intraoffice applications are for further study.
- Transmission medium — General system aspects related to the type of fiber used, wavelength ranges, and type of transmitters are presented.
- System design considerations — The various concepts associated with the design and installation of SLS are discussed, and Rec. G.957 that is the basis for transverse compatibility for the optical interfaces used in SLS is summarized.
- Regenerators — The functions of regenerators of SLS associated with overhead processing, line scrambling, operations and maintenance, and corresponding interfaces are described or specified.
- Jitter performance — To achieve transverse compatibility, jitter characteristics of regenerators of different designs have to be compatible. This means that a regenerator from one manufacturer must not produce more jitter than can be tolerated by the regenerator of another manufacturer. Depending on the technology used for the regenerator timing recovery circuit (e.g., narrow-band phase locked loop or wide-band passive filter), two fundamental types of regenerators can be defined. Rec. G.958 specifies jitter generation, transfer, and tolerance values for two fundamental types of regenerators as well as the conditions for interworking among them that ensure satisfactory performance in mixed installations.
- Error performance — Error performance of SLS is related to the specifications of Rec. G.821[20] which implies an optical section error performance of the BER not exceeding 1×10^{-10}. This value is used in Rec. G.957 to define appropriate optical interface parameters (e.g., receiver sensitivity).
- Operations, Administrations, and Maintenance (OA&M) — The standardization of OA&M features of SLS is an important aspect of transverse compatibility. Both forward compatibility with the general principles of the Telecommunications Management Network (TMN) and the need for line systems to provide standalone management functions for performance monitoring, fault location, and alarm generation for early installations where connection to a TMN is not possible have been considered. These aspects are addressed mainly in Rec. G.784. Rec. G.958 contains the definition of parameters that should be monitored for alarm surveillance or fault localization which are specific to the optical synchronous line systems.
- Consecutive Identical Digit (CID) Immunity — Due to the details of the SDH frame structure, digital sequences in STM-N systems may contain strings of marks or spaces that remain after scrambling. A test method which may be used to assess the ability of the SLS receivers to tolerate these signals is described.
- Safety — In some countries, national regulations make it necessary to provide for an Automatic Laser Shutdown (ALS) facility of the transmitter in case of fiber cable break. This capability is optional; but when provided, it is necessary to implement the mechanism in a way that ensures transverse compatibility. The corresponding requirements are contained in the Recommendation.

Recommendation G.957

The intent of Rec. G.957 is to provide specifications for SDH optical interfaces for several applications so that manufacturers have some freedom in equipment implementation, but network operators have assurance that equipment built to the specifications communicates properly at the optical level in a multi-supplier installation. The scope of this Recommendation covers optical parameters for devices and systems appropriate for typical applications expected in current and near-term future networks. In particular, Rec. G.957 does not include specifications for wavelength division multiplexed (WDM) systems or necessarily for systems employing optical amplifiers. These more advanced technologies are considered in new Recommendations being developed in the ITU 1993–1996 study period.

The enormous range of applications and implementation possibilities provided by fiber optic transmission systems presents a significant challenge in specifying optical parameters for transversely compatible SDH systems. For each of the SDH hierarchical levels, Rec. G.957 describes interface characteristics for three broad application categories covering intraoffice interconnection with typical distances less than about 2 km, short-haul interoffice interconnections with typical distances to about 15 km, and long-haul applications with typical distances of 40–60 km or more. The specifications are further defined by the type of single-mode fiber used in each application category.

Owing to the predominant deployment of single-mode fiber in current telecommunications networks worldwide, optical parameters for SDH interfaces are specified according to the characteristics of the three types of fiber described in Recs. G.652, G.653, and G.654 [21]–[23]. Conventional dispersion-unshifted fiber, specified in Rec. G.652, with nominal 1310 nm sources applies in all three of the application categories. Although G.652 fiber is dispersion-optimized for operation in the 1310 nm region, it can also be used in the 1550 nm window; this possibility is covered by Rec. G.957 for interoffice installations where coarse wavelength division multiplexing may be desired.

For long-haul installations that may be dispersion-limited with G.652 fiber, Rec. G.957 recognizes the use of dispersion-shifted single-mode fiber as specified in Rec. G.653. This fiber is dispersion-optimized for use in the 1550 nm region where attenuation is also low. Finally, for long-haul interoffice applications, Rec. G.957 also recognizes the possible use of attenuation-optimized single-mode fiber as specified in Rec. G.654 with nominal 1550 nm sources.

In developing Rec. G.957, it was agreed to balance the needs of both suppliers and network operators by adopting the following guidelines:

- To minimize the number of distinct implementations, SDH optical parameter specifications should allow as much device commonality as technically and economically possible across the various applications.
- To avoid constraining innovation in technology and implementation choices, particular technologies such as laser or LED and PINFET or APD receivers should not be directly required for the various applications.
- To provide a direct connection between the SDH optical interface specifications and simple design approaches for SDH optical spans, a simple optical span model consisting of a transmitter segment, a receiver segment, and an optical path segment can be used. Optical parameters should be specified for each of the model segments.

The specifications developed in Rec. G.957 for the transmitter segment include spectral characteristics, the allowed range of mean launch power, the minimum value of the extinction ratio, and general transmitter pulse-shape characteristics. For LEDs and multilongitudinal mode (MLM) lasers, spectral width is specified as the maximum rms spectral width accounting for modes 20 dB below the peak mode. For single-longitudinal mode (SLM) lasers, the spectral width is specified as the maximum full width of the central wavelength, measured 20 dB down from the central wavelength. In addition, to control mode partition noise in SLM systems, a minimum value is specified for the SLM laser side mode suppression ratio. Transmitter pulse shapes are characterized somewhat indirectly through specifying the mask of the eye diagram of the optical transmit signal.

The optical path is specified, in part, by the allowed attenuation range in each application considered in the Recommendation. Intraoffice applications are defined by the attenuation range 0–7 dB. Short-haul interoffice applications are characterized by the attenuation range 0–12 dB. Finally, the long-haul interoffice applications are specified by attenuation ranges that vary SDH hierarchical level: 10–28 dB for STM-1, 10–24 dB for STM-4, and 10–20 dB for STM-16 (extension to 24 dB is currently being studied in ITU-T SG 15). To provide network operators some flexibility in designing their networks, an overlap of 2 dB is provided in the attenuation range specifications between the intraoffice and the short-haul interoffice applications and between the short-haul and long-haul interoffice applications.

For systems that are considered dispersion-limited, Rec. G.957 indicates the maximum allowable dispersion (in ps/nm) of the optical path. Where specified, this value is consistent with typically 1 dB (2 dB in selected cases) power penalty estimated from simple models for the dispersion effects of intersymbol interference and, for MLM lasers, mode partition noise. The possible effect of laser chirp in long-haul applications is not quantified, but is implicitly considered to be small. As a result, transverse compatibility of high-speed long-haul systems cannot be guaranteed by the current specifications of Rec. G.957. For applications considered to be particularly sensitive to reflection-induced degradations, Rec. G.957 specifies the maximum value

of the optical return loss of the cable plant itself.

Parameters specified for the receiver segment include the minimum acceptable value for the receiver sensitivity and receiver overload. For those systems considered sensitive to reflections, the maximum value of the receiver reflectance is given. Finally, the receiver is required to tolerate an optical path penalty of 1–2 dB to account for the total degradation due to reflections and the combined effects of dispersion, resulting from intersymbol interference, mode partition noise, and laser chirp.

SDH Network Aspects: Recommendations G.803 and G.831

The 1988 versions of Recommendations G.707, G.708, and G.709 covering the NNI and the early draft Recommendations covering SDH equipment and management were prepared before there was a common understanding of the architectural principles and management capabilities of SDH-based transport networks. This was recognized as an important omission by Study Group XVIII, and in June 1989, work commenced on the preparation of two new Recommendations dealing with the network aspects of SDH:

- G.803 [24]: Architectures of transport networks based on the SDH
- G.831 [25]: Management capabilities of transport networks based on the SDH.

These were finally approved in early 1993. They offer users of SDH guidance and information on the nature of SDH-based transport networks and on the management features which they provide. It was recognized that operators have differing aspirations for SDH with differing deployment strategies. However, there are some common principles which must be observed — principles such as Virtual Container (VC) payload independence, VC path transparency, and migration to direct support of connectivity services in the VCs. A common understanding of these and other issues gives operators sufficient confidence to start early deployment of stand-alone, vendor-specific networks while allowing for evolution to geographically widespread, multioperator, multivendor configurations. Such configurations should permit, for example, VC path connectivity to be established automatically across global, multioperator networks.

Recommendation G.803

Telecommunications networks are complex arrangements which can be described in a number of different ways depending upon the particular purpose of the description. G.803 describes the network in terms of the architecture of its transport functions (i.e., those functions which deal with transferring information between locations) in the context of SDH. The functional architecture is also applicable to networks based on other technologies including PDH and ATM. In order to simplify the description and yet allow for a wide variety of topologies and equipment types, a transport network model is used based on two concepts, each of which allows a high degree of recursiveness:

- Layering in which the transport network is decomposed into a number of independent transport network layers with a client/server association between adjacent layers.
- Partitioning in which each network layer can be divided in a way which reflects the internal structure of that layer.

G.803 goes on to use the function architecture to describe examples of actual topologies, structures, and network elements used in SDH-based networks. It also uses the architecture to describe the various types of performance and event monitoring, as well as the techniques for enhancing the availability of transport networks, such as protection.

In addition to functional architecture, G.803 also describes:

- The architectural aspects of the distribution of timing within SDH-based networks including the evolutionary scenarios from existing synchronization networks, synchronization robustness, and PDH/SDH interworking implications.
- The applications of the various primary rate mappings into the VC-11 and VC-12, and the selection criteria.
- Some of the choices which need to be made when introducing SDH-based networks.

Recommendation G.831

The SDH NNI and the functionality of the various overheads offer SDH-based networks significant potential for remote management. G.831 describes the management goals of SDH-based networks and the facilities that must be provided within such networks to achieve these goals; it is not concerned with the structure or design of the management systems themselves. The primary management functions relevant to SDH-based networks are to:

- Set up VC paths between layer network access points automatically on request and across interoperator boundaries.
- Continuously monitor performance of paths while in-service, and validate compliance with service commitments.
- Maintain paths to the required availability, restoring failed paths automatically if needed.
- Provide for simple remote maintenance of the fabric of the network, including the identification and location of faulty equipment.
- Generate resource utilization information to support routing and internetwork accounting between operators, and to support planning and cost accounting within a domain.
- Support a range of ancillary management functions such as inventory and planning.

Although these functions may be implemented in differing ways by network operators, there are certain principles (which are given in G.831) which will ensure that the management functions will work correctly in a global, multioperator context. Essential to many of these management functions is a unique means of identifying any access point to a layer network. The scheme is based on the format given in Recommendation E.164 [26].

New Recommendations in the 1993–1996 Study Period

During the 1993–1996 ITU study period, specifications for transversely compatible systems are anticipated to be completed in the following areas:

- System aspects and interface specifications for STM-64 (10 Gb/s) SDH line systems.
- System aspects and interface specifications for single-channel SDH line systems using Optical Fiber Amplifiers (OFAs). This will provide specifications to increase the range of interconnect distances currently covered in Rec. G.957, especially for STM-4 and STM-16.
- Functional characteristics and interface specifications for line systems employing WDM techniques.

The enabling technology common to these new areas is the commercial availability of optical fiber amplifiers, in particular, Erbium Doped Fiber Amplifiers (EDFAs). For STM-64, use of EDFAs will likely be required to achieve interconnect distances reaching 40 km or more, thereby maintaining backward compatibility with optical line systems previously designed according to Recs. G.955 and G.957. The use of EDFA technology for systems operating at STM-4 and STM-16 may allow optical spans of 100 km, making regenerators unnecessary in many installations. WDM systems, operating in the flat region of the EDFA passband, will likely provide a cost-effective method for increasing the capacity of a single transport system as well as a high degree of transparency to the constituent signal format.

Among the important technical issues that will need to be addressed in developing these new recommendations will be:

- Definition of new optical interface parameters to enable transverse compatibility for STM-64 systems (e.g., chirp coefficient, specifications to control nonlinear effects including stimulated Brillouin scattering).
- Dispersion accommodation techniques for long-haul, high bitrate systems (e.g., modulation technique and/or use of dispersion compensating fiber).
- Selection of a frequency standard and wavelength allocation scheme for WDM systems.
- Implementation of a supervisory channel function in WDM systems.
- Safety aspects of systems using OFAs.

One of the major growth directions for SDH equipment recommendations is in the area of network protection applications. In this regard, a new series of recommendations is being developed to describe the equipment aspects of SDH ring protection applications. These applications will definitely include path-switched rings and line-switched rings. In addition, there is interest in more generalized versions of these applications in the context of subnetwork connection protection and trail protection, as well as detailed improvements to be made to the multiplex section protection application already defined in G.783.

Another possible direction is the further refinement of the functional modeling approach. As new applications are defined, new functional blocks are needed to describe the equipment behavior properly. Thus, the functional modeling approach utilized presently in G.783 may need to be extended to accommodate these new applications more effectively. For example, new functional modeling methods being developed within ETSI[15] may be used to improve the existing recommendations.

Finally, there are still a number of detailed problems associated with the detection and reporting of transmission anomalies and defects. While some of these problems are posed by the emerging applications, others are problems associated with applications which have been recognized for a long time. These problems include the following:

- The prevention of spurious alarms during path provisioning.
- Detection of path trace identifier mismatch.
- Detection and modeling of signal label mismatch.

References

[1] CCITT Recommendation G.707, "Synchronous Digital Hierarchy Bit Rates," *Blue Book*, 1988.

[2] CCITT Recommendation G.708, "Network Node Interface for the Synchronous Digital Hierarchy," *Blue Book*, 1988.

[3] CCITT Recommendation G.709, "Synchronous Multiplexing Structure," *Blue Book*, 1988.

[4] R. Balcer, J. Eaves, J. Legras, R. McLintock, and T. Wright, "An overview of emerging CCITT Recommendations for the Synchronous Digital Hierarchy: Multiplexers, line systems, management, and network aspects," *IEEE Commun. Mag.*, pp. 21–25, Aug. 1990.

[5] ITU-T Recommendation G.708, "Network Node Interface for the Synchronous Digital Hierarchy," Mar. 1993.

[6] ITU-T Recommendation G.709, "Synchronous Multiplexing Structure," Mar. 1993.

[7] CCITT Recommendation G.702, "Digital Hierarchy Bit Rates," *Blue Book*, 1988.

[8] CCITT Recommendation G.781, "Structure of Recommendations on Multiplexing Equipment for the Synchronous Digital Hierarchy (SDH)," May 1990.

[9] CCITT Recommendation G.782, "Types and General Characteristics of Synchronous Digital Hierarchy (SDH) Multiplexing Equipment," May 1990.

[10] CCITT Recommendation G.783, "Characteristics of Synchronous Digital Hierarchy (SDH) Multiplexing Equipment Functional Blocks," May 1990.

[11] ITU-T Recommendation G.781, "Structure of Recommendations on Equipment for the Synchronous Digital Hierarchy (SDH)," Sept. 1993.

[12] ITU-T Recommendation G.782, "Types and General Characteristics of Synchronous Digital Hierarchy (SDH) Equipment," Sept. 1993.

[13] ITU-T Recommendation G.783, "Characteristics of Synchronous Digital Hierarchy (SDH) Equipment Functional Blocks," Sept. 1993.

[14] ITU-T Recommendation G.784, "Synchronous Digital Hierarchy (SDH) Management," Sept. 1993.

[15] ETSI Recommendation ETS DE/TM-1015, "Transmission and Multiplexing (TM); Generic Functional Requirements for SDH Transmission Equipment, Part 1: Generic Processes and Performance," Version 1.0, Nov. 1993.

[16] CCITT Recommendation G.955, "Digital Line Systems Based on the 1,544 kbit/s Hierarchy on Optical Fiber Cables," *Blue Book*, 1988.

[17] CCITT Recommendation G.956, "Digital Line Systems Based on the 2,048 kbit/s Hierarchy on Optical Fiber Cables," *Blue Book*, 1988.

[18] CCITT Recommendation G.957, "Optical Interfaces for Equipments and Systems Relating to the Synchronous Digital Hierarchy," July 1990.

[19] CCITT Recommendation G.958, "Digital Line Systems Based on the Synchronous Digital Hierarchy for Use on Optical Fiber Cables," July 1990.

[20] CCITT Recommendation G.821, "Error Performance of an International Digital Connection Forming Part of an Integrated Services Digital Network," *Blue Book*, 1988.

[21] CCITT Recommendation G.652, "Characteristics of a Single-Mode Optical Fiber Cable," *Blue Book*, 1988.

[22] CCITT Recommendation G.653, "Characteristics of a Dispersion-Shifted Single-Mode Optical Fiber Cable," *Blue Book*, 1988.

[23] CCITT Recommendation G.654, "Characteristics of a 1,500 nm Wavelength Loss-Minimized Single-Mode Optical Fiber Cable," *Blue Book*, 1988.

[24] ITU-T Recommendation G.803, "Architectures of Transport Networks Based on the Synchronous Digital Hierarchy (SDH)," Mar. 1993.

[25] ITU-T Recommendation G.831, "Management Capabilities of Transport Networks Based on the Synchronous Digital Hierarchy (SDH)," Mar. 1993.

[26] CCITT Recommendation E.164, "Numbering Plan for the ISDN Era," *Blue Book*, 1988.

Architectural and Functional Aspects of Radio-Relay Systems for SDH Networks

GEOFF RICHMAN
BRITISH TELECOM
UNITED KINGDOM

MANSOOR SHAFI AND CHRIS J. CARLISLE
TELECOM NEW ZEALAND LIMITED
WELLINGTON, NEW ZEALAND

Abstract—The synchronous digital hierarchy is a standard that has been defined by the International Telecommunications Union — Telecommunications Standardization Sector (ITU-T), formerly International Telephone and Telegraph Consultative Committee (CCITT), to substantially replace the current plesiochronous digital hierarchy. The ITU-T standards cover fiber optic transmission systems and multiplex equipment. Since there will also be a need for radio-relay systems to carry SDH traffic, the CCIR has established standards for radio-relay systems that can transport SDH signals.

The paper is based on the International Telecommunications Union — Radiocommunications Standardization Sector's (ITU-R), formerly International Radio Consultative Committee (CCIR), Recommendations for SDH, and discusses the functionality of SDH radio and possible architectures for employing SDH radio in SDH networks. The SDH radio equipment is described in terms of functional blocks.

1. INTRODUCTION

The majority of transmission networks in the world at present are plesiochronous. Tributaries at the same level in the multiplex hierarchy have nominally the same bit rate, but are not synchronized by the same clock. Multiplexing to a higher level requires bit stuffing, and demultiplexing requires an inverse process. Therefore, individual tributaries are not easily accessible within a higher level bit stream.

In a synchronous transmission network, all transmission systems are synchronized to a master clock. Asynchronous tributaries are mapped into a synchronous container using bit stuffing. The synchronous digital hierarchy (SDH) allows lower level tributaries to be accessed directly from a higher level bit stream.

The basic transmission unit for the SDH is the synchronous transport module STM-N; currently N can equal 1, 4, or 16, but higher levels can easily be defined as required. An STM-1 frame is shown in Fig. 1 as a block of 9 bytes by 270 bytes, which is transmitted at the rate of 8000 frames per second. STM-1 corresponds to the lowest SDH bit rate of 155.52 Mb/s.

The STM-1 frame contains a payload and a section overhead (SOH). The payload contains tributary data, path overheads, and pointers. The SOH contains a regenerator section overhead (RSOH) and a multiplex section overhead (MSOH). The mapping of asynchronous tributaries into an STM-1 frame is discussed in ITU-T Recommendation G.709 [1].

The main benefits of the SDH compared to the PDH are:

- Reduced costs due to the simplification of multiplex equipment

- Greater network flexibility due to software control of the function of network elements
- Enhanced operation, administration, and maintenance facilities using the overhead provided in SDH frames
- Increased network compatibility, both internationally and between manufacturers
- Simple migration to higher bit rates.

Although the SDH was originally specified only for fiber optic transmission, the widespread use of digital radio-relay systems (DRRS) means that there is a need for STM-1 signals to be transported by digital radio-relay systems. The ITU-R Interim Working Party (IWP) 9/5 (later to be called Task Group 9/1) met in Florence in October 1990, in Kobe in June 1991, and in Geneva in November 1992 to draft two recommendations [2], [3] on the application of DRRS to SDH networks. These recommendations are entitled: "Architectures and Functional Aspects of Radio-Relay Systems for SDH-Based Networks" and "Transmission Characteristics and Performance Requirements of Radio-Relay Systems for SDH-Based Networks." This paper discusses the first recommendation.

DRRS to transport STM-1 (155.52 Mb/s) and sub-STM-1 (51.84 Mb/s) signals will be discussed. It is also possible to design DRRS to transport higher level STM-N or N × STM-1 using dual-polarization co-channel operation, higher level signal constellations, or higher transmission bandwidths (see ITU-R Recommendation 751). The use of DRRS to transport STM-1 signals is discussed in Section 2, and the use of DRRS

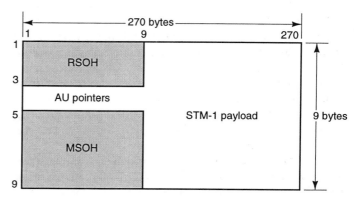

Fig. 1. STM-1 frame structure.

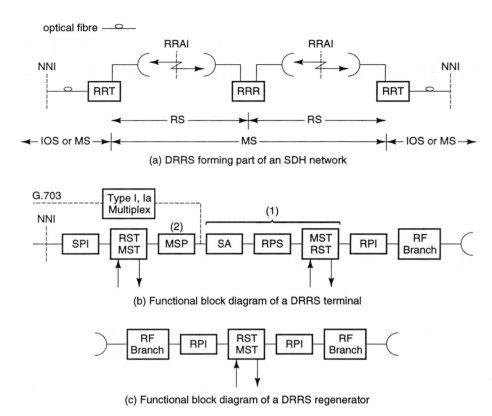

(a) DRRS forming part of an SDH network

(b) Functional block diagram of a DRRS terminal

(c) Functional block diagram of a DRRS regenerator

Fig. 2.   Functional block diagrams of a DRRS for the SDH.

to transport sub-STM-l signals is discussed in Section 3. Operations, Administration, and Maintenance aspects are discussed in Section 4.

2. RADIO-RELAY SYSTEMS TO TRANSPORT STM-1

A DRRS forming part of an SDH network is shown in Fig. 2(a). The logical functions of the radio-relay terminal (RRT) and radio-relay regenerator (RRR) will be described in Section 2.1 in terms of functional blocks. The network node interface (NNI) is the standard interface between network nodes; it can be either electrical or optical. The radio-relay air interface (RRAI) is the air interface between two radio stations. It is defined for architectural purposes only, and is not subject to international standardization.

DRRS for networks based on the SDH will form either a multiplex section (MS) or a regenerator section (RS) as shown in Fig. 2(a). In a multiplex section, both the RSOH and the MSOH within the STM-l signal are accessible, but in a regenerator section, only the RSOH is accessible. An interoffice section is a multiplex section between two transmission stations. An intraoffice section (IOS) is a section within a transmission station, which may not have full RSOH and MSOH termination. The sections before and after a radio multiplex section can be either intraoffice sections or interoffice sections.

Radio-relay systems have been granted the use of media specific bytes in the SOH by ITU-T SGXVIII. If this option is taken, the multiplex sections will be media dependent; otherwise, they will be media independent. Media-independent multiplex sec-

tions will only be possible if all media use the MSOH for the same overhead functions. If the MSOH is media dependent, then the MSOH must be recomputed when multiplex section protection (MSP) switching is activated in a multimedia protection architecture.

2.1 Functional Block Diagrams

The logical functions of an SDH radio can be represented by functional blocks (as used in ITU-T Recommendations G.782 and G.783). The functional blocks do not necessarily correspond to any physical partitioning or implementation, but are used to simplify and generalize the description.

The functional block diagram of a radio-relay terminal (RRT) following an interoffice section and a radio-relay regenerator (RRR) are shown in Figs. 2(b) and (c).

The functions of the blocks in the diagrams are as follows:

• *SDH Physical Interface (SPI)* — Converts between an internal logic level STM-N signal and an STM-N line interface signal.

• *Regenerator Section Termination (RST)* — Generates/terminates the RSOH of the STM-N signal and scrambles/descrambles the SOH (except for row 1).

• *Multiplex Section Termination (MST)* — Generates/terminates the MSOH. MST* means that the MST functionality is optional.

• *Multiplex Section Protection (MSP)* — Provides capability for branching the signal onto another line system for protection purposes.

- *Section Adaptation (SA)* — Provides adaptation of higher order paths into administrative units (AUs), assembly and disassembly of AU groups, byte interleaving, and pointer generation and processing.
- *Radio Protection Switching (RPS)* — Provides switching between working and protection channels. This function is discussed in more detail in Section 2.2.
- *Radio Physical Interface (RPI)* — Includes typical radio-relay functions like scrambling/descrambling, error correction coding, modem, up/down conversions, power amplification (including automatic transmit power control), filtering, etc.
- *RF Branching* — Provides RF branching between working and protection channels.

If the RRT follows an intraoffice section, the (MSP) functionality is removed and the RST/MST block following the SPI block may have reduced functionality. The logical sequence of the three blocks marked (1) may be implementation independent, and the functional blocks marked (2) are optional.

A type I or type Ia multiplexer [4] may be used so that the DRRS can carry plesiochronous G.703 traffic. When an SDH signal is available for the radio to transport, the multiplexer is no longer required.

2.2 Radio Protection Switching

Multiplex section protection (MSP), defined in ITU-T Recommendation G.782, is not suitable to provide the improvement of transmission quality required for radio-relay systems. It takes place when multiplex section parity bytes (B2 bytes) are found to be corrupted; this usually happens when the BER is as high as 10^{-3}. In contrast, radio protection switching (RPS) is usually initiated when the BER is about 10^{-7}, so that RPS will have taken place well before the BER reaches 10^{-3}. On the other hand, MSP is suitable for network restoration where switching may be required between different routes. Therefore, both multiplex section protection and radio protection will be used. However, if RPS is used, the RPS action must not cause a network switch. Preferably, the RPS action time must be nested within the MSP switching time.

One possible RPS arrangement is shown in Fig. 3 for M working channels and N protection channels. The switching operation may cause a loss of synchronization on the protection channels if the STM-1 signals on all channels are not synchronized in frequency and phase. Synchronization of the working and protec-

tion channels is performed using the SA function to synchronize all channels to the frame synchronization clock (FSC). There are other possible RPS arrangements. For instance, the SA function can be performed after the RPS function, so that switching is performed at the VC-4 level. In Fig. 3, switching is done at the STM-1 level.

Radio protection switching is errorless, and is usually done via the radio frame complementary overhead (RFCOH). Although the RPS may switch well before the B2 parity bytes are corrupted, all the parity bytes (B1 and B2 bytes) must be recalculated at the beginning of each switching section. Since the parity bytes are recalculated at the beginning of each radio switching section, the radio switching sections should be SDH multiplex sections.

2.3 Applications

Three possible applications of SDH radio-relay systems in SDH networks are shown in Fig. 4. The diagrams do not nec-

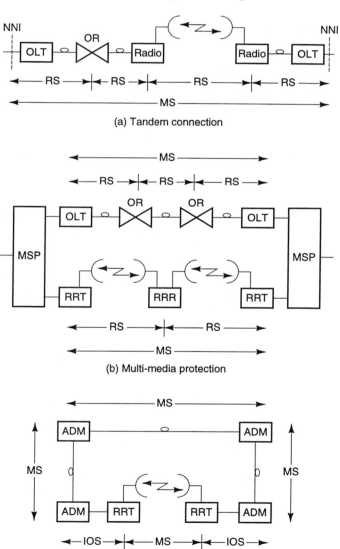

(a) Tandem connection

(b) Multi-media protection

(c) Ring connection

Fig. 4.   Applications of SDH DRRSs in SDH networks.

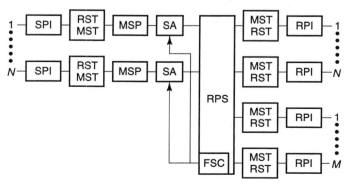

Fig. 3.   Radio protection switching arrangement.

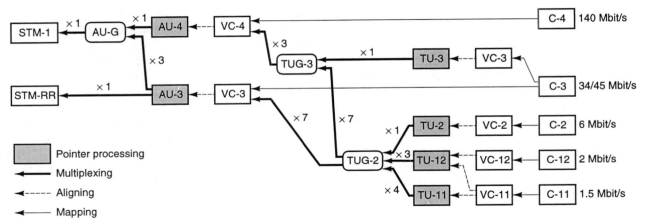

Fig. 5.   Sub-STM-1 multiplexing routes.

essarily represent the physical implementation for the applications; this depends on the way manufacturers package their SDH products.

Additional abbreviations in these diagrams are: OLT (optical line terminal) and OR (optical regenerator).

Fig. 4(a) shows an SDH DRRS forming a tandem connection with a fiber optic transmission system (FOTS). This application could be used for the purpose of "lead in" to a metropolitan area, or for crossing terrain that is not suitable for the laying of fiber cable. In this case, the MS may be media dependent; the choice is left to the network planner.

Fig. 4(b) shows the use of an SDH DRRS and an SDH FOTS in a multimedia protection configuration. Although not necessary, it is desirable for the MSOH to be media independent if multimedia protection is employed. This avoids the need to recompute the MSOH when MSP is activated.

Another use of SDH radio could be to close an optical fiber ring, as shown in Fig. 4(c). Add/drop multiplexers (ADMs) are used to add/drop traffic at the nodes on the ring. The optical line terminal functionality is assumed to be incorporated in the ADMs, as it will be in practice.

3. RADIO-RELAY SYSTEMS TO TRANSPORT SUB-STM-1

When an STM-1 signal is partially filled, there is an opportunity for medium capacity radio-relay systems to transport only part of the STM-1 signal with the necessary overhead. This may result in savings of cost, radio spectrum, and reduced modulation complexity.

To enable medium capacity radio systems to benefit from the advantanges of SDH, a sub-STM-1 interface, corresponding to a bit rate of 51.84 Mb/s, has been standardized to enable the transport of partially filled STM-1 frames by medium capacity radio or satellite. However, the sub-STM-1 rate is not a standard level of the SDH.

The multiplexing routes for sub-STM-1 signals are shown in Fig. 5, together with the multiplexing routes for STM-1 signals.

The radio-relay reference point (RRRP) is a functional reference point within a sub-STM-1 radio system where the synchronous transport module for sub-STM-1 radio-relay (STM-RR) is assembled.

To form an STM-RR from an AU-4 based signal, the STM-1 signal must be decomposed to the TUG-2 level. The seven TUG-2s to be transported by the sub-STM-1 radio can then be formed into an STM-RR signal.

To form an STM-RR signal from an AU-3 based STM-1 signal, the STM-1 signal must be decomposed to the AU-3 level. The AU-3 to be transported by the sub-STM-1 radio can then be formed into an STM-RR signal.

3.1 Functional Block Diagrams

The functional block diagram for a sub-STM-1 DRRS terminal is shown in Fig. 6. An STM-RR signal is formed from an STM signal by the STM-RR MUX block, which will incorporate a number of standard SDH functional blocks. The section adaptation function for sub-STM-1 radio-relay (SA-RR) is analogous to the SA described in Section 2, and provides for the adaptation of the paths to be transported. In particular, the SA-RR function processes the pointer to indicate the phase of the first column of the VC-3 relative to the STM-RR SOH and assembles/disassembles the complete STM-RR frame.

The regenerator section termination for sub-STM-1 radio-relay (RST-RR) and the multiplexer section termination for sub-STM-1 radio-relay (MST-RR) are analogous to the RST and MST of ITU-T Recommendation G.782.

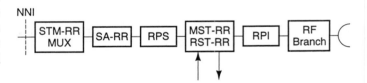

Fig. 6.   Sub-STM-1 radio relay terminal.

4. Operations, Administration, and Maintenance

Operations, administration, and maintenance facilities are transported in the SDH by the SOH bytes [5]. The SOH bytes provide channels for monitoring and control of operations, administration, and maintenance functions. The allocation of SOH bytes complies with the SOH byte allocations contained in ITU-T Recommendation G.708, but also makes use of some media specific bytes. The SOH for sub-STM-l signals consists of one third of the STM-1 SOH. Depending on the implementation, media-specific bytes can be transported in the SOH or in the RFCOH.

The RSOH could be duplicated in the radio frame complementary overhead (RFCOH) to avoid having to regenerate individual bits in the repeaters. This could lower the cost of implementing SDH for DRRS, but is not standardized.

The following functions may make use of media-specific bytes or may be carried in the RFCOH.

• *Early warning RPS activation* — Error detection can be used to indicate the presence of adverse propagation conditions and to activate "early warning switching" in RPS equipment. This is required to obtain error-free switching. A transmission capacity in the range 32–64 kb/s is required to transport early warning RPS activation information between terminals.

• *Automatic transmitter power control (ATPC)* — ATPC can reduce nodal interference between radio-relay systems. It can also be used to improve linearity or increase the dynamic range of multilevel modulation radio equipment. DRRS typically have a 35 dB thermal noise fade margin. However, this margin is not needed for about 90% of the time. During these quiescent periods, it is possible to achieve high-quality transmission with about 15 dB reduction in power, while minimizing the interference to other systems.

• *RPS information and control channel* — A transmission capacity of about 64 kb/s is required for fast hitless RPS control.

• *Propagation monitoring* — One byte has been proposed for propagation monitoring.

• *Auxiliary functions* — SDH radio-relay equipment may make provisional use of all media-specific bytes for future international standardization to perform auxiliary functions, e.g., wayside traffic, propagation data collection, and temporary data/voice channels for special maintenance purposes.

SDH radio-relay systems will be part of an overall managed

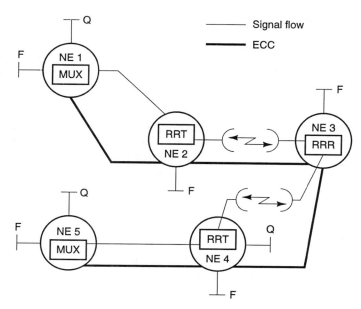

Fig. 8.   SDH management subnetwork.

telecommunications network, based on the telecommunication management network (TMN), as shown in Fig. 7. ITU-T Recommendation G.784 [5] allows the SDH management network (SMN) to consist of various managed SDH subnetworks. SDH radio-relay systems will be managed within an SDH management subnetwork (SMS).

Details of an SMS, showing a radio system interconnected to multiplex and optical line equipment (MUX), are shown in Fig. 8. The RRT and RRR may be regarded as network elements (NE) to be managed. As such, they may have an F or a Q interface, and be linked to other NE and other SMS via an architecture described in ITU-T Recommendation G.784. Usually, communication between NE in an SMS is via the embedded control channel (ECC) or a local communications network. The ECC uses the data communication channel (DCC) as its physical layer. There are two DCC within the SOH. Bytes Dl–D3 form a 192 kb/s channel, which is accessible by all network elements. Bytes D4–D12 form a 576 kb/s channel, which is not accessible at regenerators.

Figure 9 shows a functional block diagram of the manage-

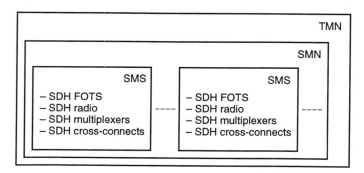

Fig. 7.   SDH DRRSs forming part of a TMN.

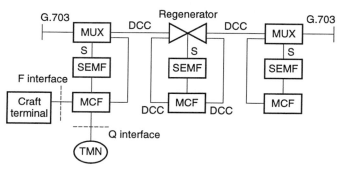

Fig. 9.   Network management interfaces.

ment interfaces for an SDH multiplex section. Three interfaces are provided for the communication of messages. The DCC provides message communication channels between network elements. Communication with the TMN is via the Q interface, and communication with a craft terminal is via the F interface.

The synchronous equipment management function (SEMF) converts performance data and implementation-specific hardware alarms into object-oriented messages for transmission on the DCC and/or a Q interface. It also converts object-oriented messages related to other management functions for passing across the S reference points. The message communication function (MCF) receives and buffers messages from the DCC, Q, and F interfaces, and the SEMF.

The craft interface should have access to the following facilities:

- Control and monitoring of protection switches
- Provisioning of tributaries
- Performance monitoring
- Fault history check
- Alarm surveillance
- Setting of performance thresholds
- Management facilities in remote network elements within the same maintenance entity
- Installation functions.

Radio-specific alarms and a standardized message set have to be defined within the Q protocols defined by the ITU-T. Examples of primitives for radio specific alarms are:

- Transmitter status
- Receiver status
- RPS status
- Modulator/demodulator status.

5. FURTHER WORK

ITU-R Recommendations 750m and 751 have now been approved. However, there are several possible areas where further work can be undertaken. These areas include:

- *Development of a sub-STM-1 interface* — It has been recognized that efficient transmission of low-capacity signals by radio or linking to ADSL systems would be enabled if interfaces in the range from 1.5 to 6.9 Mb/s were standardized. These interfaces would extend the SDH management system into the access network offering a potential reduction in O&M costs. This concept has already been introduced into the ITU-R Recommendations on SDH radio systems, but further work has to be undertaken.

- *Transport of ATM traffic on SDH-compatible radio systems* — Mappings for ATM cell-based signals onto SDH optical line systems is underway, and the ITU-R needs to continue its work by studying similar mappings onto SDH-compatible radio systems.

- *In-service performance monitoring* — ITU-T Recommendation G.826 specifies performance objectives that are more stringent than those derived from Recommendation G.821. Further work should be carried out to monitor the performance of SDH radio systems to verify compliance with Recommendation G.826.

6. CONCLUSIONS

Future digital radio-relay systems will need to carry STM-1 payloads if SDH is to achieve high penetration in transmission networks. The ITU-R has produced recommendations for SDH radio-relay systems to transport STM-1. A sub-STM-1 rate of 51.84 Mb/s has also been defined to allow medium-capacity radio systems to transport synchronous payloads and access the SOH bytes. The sub-STM-1 rate is not a standard SDH interface.

In addition to the SOH bytes defined by the ITU-T, SDH radio may use media-specific bytes for radio specific functions. The SOH bytes will allow SDH radio systems to form part of a total managed telecommunication network.

7. ACKNOWLEDGMENTS

The authors wish to acknowledge the Telecom Corporation of New Zealand and British Telecommunications for providing time and resources to undertake this work.

References

[1] ITU-T, "Recommendation G.709," ITU-T Study Group XVIII, Geneva, 23–25 May 1990.

[2] ITU-R "Recommendation 750: Architectures and Functional Aspects of Radio-Relay Systems for SDH-Based Networks," ITU-R Task Group 9/1, Kobe, Japan, 17–28 June 1991.

[3] ITU-R, "Recommendation 751: Transmission Characteristics and Performance Requirements of Radio-Relay Systems for SDH-Based Networks," ITU-R Task Group 9/1, Kobe, Japan, 17–28 June 1991.

[4] CCITT, "Draft Recommendation G.782: Types and General Characteristics of Synchronous Digital Hierarchy (SDH) Multiplexing Equipment," ITU-T Study Group XV, 16–27 July 1990.

[5] CCITT, "Draft Recommendation G.784: Synchronous Digital Hierarchy (SDH) Management," ITU-T Study Group XV, Geneva, 16–27 July 1990.

Aspects of Satellite Systems Integration in Synchronous Digital Hierarchy Transport Networks

W. S. OEI AND S. TAMBOLI

INTERNATIONAL TELECOMMUNICATIONS SATELLITE ORGANIZATION
3400 INTERNATIONAL DRIVE N.W.,
WASHINGTON, DC 20008, U.S.

Summary—This paper discusses functional and architectural aspects of satellite systems in SDH transport networks. The satellite system design is reevaluated in the light of SDH networking principles, with the objective of ensuring compatibility with SDH while deriving SDH benefits to satellite systems. Adaptations to SDH standards are identified for use internal to the satellite systems acting as SDH subnetworks or synchronous network elements. Three SDH network scenarios accommodating satellite traffic matrices with thin, medium, and thick routes are presented. Integration aspects of satellite systems in SDH networks are highlighted, including system design and OAM requirements relative to international standards.

1. INTRODUCTION

Synchronous digital transport networks based on the Synchronous Digital Hierarchy (SDH) are rapidly being deployed in industrially developed countries, and it can be expected that developing countries will "leap-frog" to this new technology. SDH, which was motivated by the U.S. SONET technology [1], is a set of new digital multiplex standards comprising hierarchical bit rates, multiplexing structures, transport network and its enhanced network management, and Operations, Administration, and Maintenance (OAM) [4]–[9]. The advantages and features of SDH relative to the prevalent Plesiochronous Digital Hierarchy (PDH) network structures are adequately addressed in companion papers, as well as other referenced papers [1], [2].

SDH is primarily geared towards optical fiber transmission systems. In operational networks, however, other transmission media such as radio relay and satellite systems may be required for network interconnections and digital sections within synchronous digital networks. As a consequence of widespread SDH deployment, the need to interconnect various SDH networks, some of which may be geographically dispersed, is imminent. INTELSAT and other satellite systems in the Fixed Satellite Services (hereafter referred to as FSSs) have traditionally served as global and regional network interconnect systems. Given its inherent capability to provide wide geographical coverage, satellite systems can play an important role by serving as an interconnect system for various national SDH transport networks, complementing terrestrial long distance systems. For the interconnection of some SDH networks, satellite systems may be the only viable transmission medium, either as a transitional measure or because terrestrial means may not be economical. It follows that in order to facilitate integration in synchronous network infrastructures, FSS must be capable of transporting SDH signals or their derivatives. In this integrated network role, FSS can benefit from the SDH features such as simplified (de)multiplexing and cross-connection, standard OAM facilities, and commonality of network management.

2. REEVALUATION OF FSS DESIGN AND APPLICATION OF SDH TECHNIQUES

Bandwidth Efficiency

Relative to fiber-optic systems, FSSs are bandwidth limited. System cost considerations lead to the need for bandwidth-efficient transport of SDH signals. This requirement is met by choosing SDH satellite connection bit rates matched to the operational traffic route size. As is the case with international cable systems, the prevalent satellite connections are at multiples of the primary rate (2.048 Mb/s), as this type of connection is increasingly used for carrying compressed DCME international traffic. The lowest level of the standard SDH multiplex hierarchy is STM-1 (155.52 Mb/s). The possibility to transport "low bit rate" SDH signal elements or sub-STM-1 signals (< 155.52 Mb/s) is imperative, and methods to derive these sub-STM-1 signals need to be developed. These methods should be designed such that the transport of sub-STM-1 signals within the satellite network is transparent to the external SDH networks.

Multidestination

Multidestination is commonly used in satellite systems to increase space segment utilization efficiency. More importantly, multidestination operation brings to the system users cost savings in terms of earth station communications equipment, particularly on the transmit side (transmitters, modulators), potentially including even earth station antennas. Multidestination digital operation has so far been somewhat hampered with PDH due to its relatively complex multiplexing techniques involving bit-justification and framing for each multiplexing step above the primary level [15], [16]. The advent of SDH with its simplified (de)multiplexing methods now enables cost-effective multidestination digital operation [3]. Satellite traffic grooming and destination sorting at baseband can be done efficiently with SDH. However, SDH network elements as currently specified are strictly designed for single destination or point-to-point oriented terrestrial cable systems. For multidestination use in FSS-SDH, they need to be modified for asymmetrical connection topologies, i.e., one transmit link may have multiple corresponding receive links. However, the asymmetry should be internal to the satellite network to ensure symmetrical appearance to the external interconnected SDH networks.

Open System Specification

The technical and economic benefits of FSS-SDH to the system users are maximized by developing international standards that serve as open system specification. The set of specifications includes satellite specific requirements and maximum functional and equipment commonality with the SDH international standards. These standards are developed mostly by the CCITT, now called the Telecommunications Standardization Sector (TSS) of the ITU [4]–[13]. Study Group 9 of the CCIR (now the Radio communications Sector (RS)) has developed SDH recommendations for digital microwave radio relay systems [23], and similar recommendations for FSS are being developed by Study Group 4 [24].

Network Management

Network management systems for satellite and terrestrial systems have traditionally been segregated. SDH transport network resource management and its commonality across network operator domains is a prerequisite to enhanced end-to-end network service provisioning. Functional integration of FSS in SDH requires concurrent SDH network management capabilities and elements in the FSS. The FSS should use compatible management protocols and interfaces to maximize commonality with the rest of the SDH Management System (SMS). With these initial steps, the possibility of incorporating FSS, together with the SDH transport networks of which it is part, in the future comprehensive concept of the Telecommunication Management Network (TMN) [22] is ensured.

3. SDH Concepts Applied to FSS

The basic set of SDH standards is given in [4]–[9]. Some adaptations of the standards are necessary for application in the satellite environment.

Network Node Interface (NNI) and Satellite Equipment Interface (SEI)

Recommendation G.708 specifies the Network Node Interface (NNI) for SDH. The basic multiplexing structure at the NNI is the Synchronous Transport Module (STM-N). The first level STM, termed as STM-1, has a bit rate at 155.52 Mb/s. Higher level hierarchical bit rates of N × STM-1, where N is currently 4 and 16, are also defined [4]. As these bit rates are excessive for radio systems, Recommendation G.708 acknowledges a frame structure at a bit rate of 51.84 Mb/s (STM-0) for use in low/medium capacity radio and satellite digital sections. STM-0 does not represent an SDH hierarchical level at the NNI, but it can be used internally within radio systems for transport of SDH payloads. To distinguish the corresponding

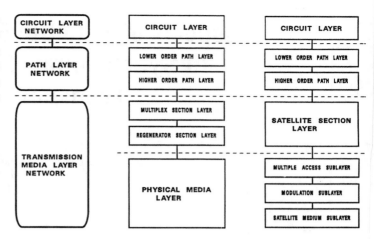

Fig. 1.   Functional layers of SDH transport networks.

section interfaces from the NNI, they are referred to as Equipment Interfaces (EI), or Satellite Equipment Interfaces (SEI) for FSS. For compatible SDH network interconnection, FSS-SDH networks and network elements are still required to have NNIs at STM-N. Compatible interfacing through NNI ensures "normal" SDH network appearance of the FSS-SDH to the interconnected SDH transport networks.

To obtain further system efficiencies and flexibility in matching the section bit rates to satellite route sizes, additional low SEI bit rates — Satellite STM-2n (SSTM-2n) and SSTM-1k — are necessary, which will be discussed in further detail later.

Functional Layering

The concepts of functional layering and network partitioning have been devised to aid the functional description, design, and structured management of SDH transport networks [9]. Figure 1 is an illustration of SDH transport network layers and their adaptation to the FSS environment.

- The Physical Media Layer is subdivided into three FSS sublayers: Satellite Medium sublayer (intermediate and radio frequencies), (De)Modulation sublayer (baseband to/from intermediate frequency), and Multiple Access sublayer. The latter is particularly important from the point of view of implementation of various FSS-SDH network architectures discussed in this paper.
- The SDH Regenerator Section (RS) layer and the Multiplex Section (MS) layer, taken together, is termed as the Satellite Section layer.
- The Path layers of the SDH transport network model are retained in the FSS layered model. This signifies adherence to the principles, definition, and control of connectivities in the path layer networks as required in [12]. Within the FSS, higher order path connections may either be assembled/disassembled or transparently cross-connected and carried through, while lower order path connections are handled transparently.
- As is the case for the interconnected SDH networks [9], the Circuit layer in the FSS is less of a concern to the man-

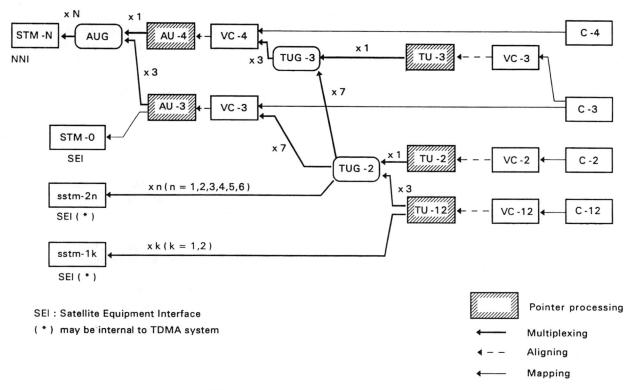

Fig. 2. SDH multiplexing tree adapted to FSS.

agement of FSS-SDH path connections. The Circuit layer will be used to transparently carry a wide variety of possible client network services, e.g., PSTN, ISDN, B-ISDN/ATM, Frame Relay, SMDS, and applicable telecommunication services such as voice, n × 64 kb/s data, video.

Multiplexing Tree

In Fig. 2, the standard SDH multiplexing structure, as specified in Recommendation G.709, is adapted to the FSS environment through the addition of two new multiplexing "paths" shown as n × TUG-2 (n = 1, ···, 6) and k × TU-12 (k = 1, 2). The corresponding multiplex structures, SSTM-2n and SSTM-1k, include appropriate Satellite Section Overhead. These new structures allow a range of additional low section bit rates to be defined for synchronous SDH payload transport over the FSS. This provides internal system flexibility in route dimensioning, and thus efficient satellite bandwidth utilization.

Satellite-Specific OAM Functions

The lower multiplex structures SSTM-2n and SSTM-1k have satellite section overhead functions such as framing, section error monitoring, Data Communications Channel (DCC), and backward alarms (FERF, FEBE), including their multidestination addressing. Although they are internal and specific to FSS-SDH, commonality with SDH SOH functions is strived at to facilitate management commonality. Efficient overhead capacity allocation is necessary in view of the low bit rate of the payloads.

In the STM-1 frame, 6 bytes have been allocated in the RSOH for transmission media specific application [5]. For FSS, these bytes could be used, for example, for system-specific additional OH functions such as section layer backward alarm (FERF, FEBE) addressing in multidestination operation. However, in the STM-0 frame structure, there is no provision for media-specific bytes. This can be solved by means of special allocation of overhead capacity in the underlying Multiple Access sublayer, accompanied by special internal system management messages across the SEI.

Physical Media layer functions specific to satellite transmission are to be placed in a Satellite Complementary Frame Overhead (SCFOH). Example overhead functions are multiple access sublayer frame synchronization, order-wires, scrambler/descrambler synchronization, modem alarm handling for asymmetrical links, and possible future automatic uplink power control and FEC indications for early warning site diversity routing.

Network Management

For maximum commonality with terrestrial SDH management networks, the FSS-SDH network management facilities use the Embedded Control Channel (ECC) protocols, management functions, and communications defined in [8]. The ECC protocol comprises the seven-layer OSI protocol stack of which, in the case of STM-1 and STM-0, the DCC_R (192 kb/s) and DCC_M (576 kb/s) serve as the Physical Layer. For satellite sections operating at these relatively high bit rates, delay-sensitive parameters of the lower layer protocols, such as the Data Link Layer implemented based on LAPD, require proper parameter values to be selected, negotiated, or managed by the ECC management functions to ensure adequate data performance (throughput).

For the new multiplex structures, the expected low DCC bit rates will be the limiting factor for the data throughput.

4. ARCHITECTURAL ISSUES

A number of architectures of FSS-based SDH networks or network elements can be constructed, each characterized by the following attributes:

• *Satellite Route Size* — Satellite resource efficient transmission requires the selection of a section bit rate sufficient to accommodate the size of the traffic route "spanned" by the section. The range of possible bit rates is:

• STM-1: 155.52 Mb/s
• STM-0: 51.84 Mb/s
• SSTM-2n: $n \times$ TUG-2 (6.912 Mb/s) + SSOH (n = 1, $\cdots$, 6)
• SSTM-1k: $k \times$ TU-12 (2.304 Mb/s) + SSOH (k = 1, 2)

• *Type of Multiple Access* — Two common methods of satellite transponder sharing are Frequency Division Multiple Access (FDMA) and Time Division Multiple Access (TDMA). A mixed form of FDMA and TDMA, known as Low-Rate TDMA, is also used in a few satellite systems. The choice of the appropriate transponder multiple access method is classical to satellite communications, and is fundamental in the design of the satellite link and the specification of earth station terminal equipment. Generally, FDMA is well suited for a few moderate to thick sized routes, whereas TDMA is more economical for a network with a large number of thin to medium sized routes, and a multiplicity of correspondents for each user in the community.

• *Number of Section Destinations (Section Topology)* — Single destination per section or efficient traffic distribution by multidestinational digital satellite sections can each be used with either FDMA and TDMA. The ground communications equipment savings due to multidestination operation alone is higher with FDMA, provided that traffic routes can be destination-sorted economically at baseband level. On the other hand, the opportunity of multidestination operation is greater with TDMA based on the premise that TDMA is the better choice for larger numbers of correspondents per user.

• *Integration of SDH and Multiple Access (MA) Functions* — In terms of earth station equipment implementation, two approaches can be followed. The SDH multiplex function and the FSS Multiple Access functions can either be segregated in different equipment units, or integrated into a single piece of equipment. For SDH multiplex structures already defined by an existing standard, i.e., STM-1 and STM-0, it is advantageous to adopt a segregated equipment approach. This allows off-the-shelf SDH multiplex equipment to be used, at least to support single destination (point-to-point) SDH section connections. For derived SDH signals such as SSTM-2n and SSTM-1k with a wide range of bit rates and payload types, an integrated approach is favorable since it would obviate the need to standardize new SEI signal structures and physical interface characteristics.

• *FSS Network Role in the SDH Managed Transport Network* — The role of FSS-SDH in the overall SDH transport network architecture is important from the viewpoint of network management. The most straightforward roles are FSS acting as a point-to-point Regenerator Section (RS) or a point-to-point Multiplex Section (MS) in the SDH. In the first case, FSS can access the RSOH and the MSOH, and the entire payload must be transported transparently across the satellite section. If the FSS serves as an MS of the SDH, access to the RSOH and MSOH is possible. In either case, there are no intermediate points with MSOH termination/generation and SDH payload rearrangements (e.g., VC cross-connections). If SDH payload rearrangements are carried out within the FSS, the FSS takes the role of an SDH network element (e.g., cross-connect, multiplexer) at a network node. In reality, the functions of this network element are spread across a wide area (the FSS coverage), involving multiple earth stations, with the terrestrial SDH networks connected to the baseband multiplex equipment via standard NNIs. Within the wide area network element, the FSS can have its own media-dependent multiplex sections for internal FSS synchronous payload transport.

5. NETWORK SCENARIOS

Three example FSS-SDH network scenarios particularly suited for an international interconnect network such as the INTELSAT system are described in the following. Other scenarios, either mixed forms or to suit different networking needs, are possible but not presented.

FSS-SDH Network Element Scenario for Thin/Medium Routes

This scenario is suited for an FSS network with relatively thin to medium sized routes with wide geographical traffic distribution. Such a traffic matrix is typical in the current INTELSAT system, in which a growing percentage of traffic is bundled in primary groups of 2.048 Mb/s (DCME traffic) distributed across a moderately large number of nodes (earthstations) in the FSS network. The system characteristics of this FSS-SDH network element, shown in Fig. 3, are:

• Internal FSS Section Bit Rate:
  $k \times$ TU-12 (k = 1,2) + SSOH,
  $n \times$ TUG-2 (n = 1, $\cdots$ 6) + SSOH,
• Multiple Access: TDMA
• Section Topology: Single-destination and multidestination
• Implementation: Integrated SDH functions and TDMA functions
• Network Role: Wide-area SDH cross-connect network element

The choice of TDMA with integrated implementation is appropriate for the range of relatively low bit rate sections with different combinations and numbers of cross-connected TU-12s and TUG-2s, dictated by traffic requirements. Frequent changes in bandwidth and connectivity requirements can be accommodated by implementing burst time plan changes through the network management system.

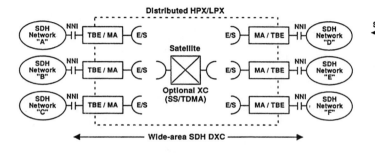

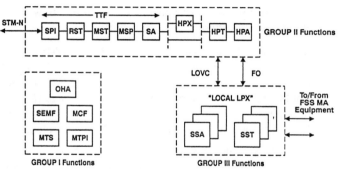

Fig. 5. Generic functional blocks in TBE.

TBE: Terrestrial Baseband Equipment
MA: Multiple Access (TDMA)
DXC: Digital Cross-Connect
XC: Cross-Connect On-Board

Fig. 3. Thin route scenario.

As shown in Fig. 3, an important element of the wide-area cross-connect is the FSS Terrestrial Baseband Equipment (TBE). Additional cross-connect functions at the Satellite Medium sublayer may be performed on-board the satellite to restore inter-beam connectivities of a multitransponder satellite system, e.g., the INTELSAT SS-TDMA system. The TBEs are connected to the various national SDH transport networks via the standard NNI. The integration of the SDH and TDMA functions into a single piece of equipment is illustrated in Fig. 4.

The functions of the TBE can be described in terms of "generic functional blocks" of SDH multiplex equipment defined in [7]. The TBE generic functions are subdivided in three functional groups, shown in Fig. 5. Group I comprises functions common to SDH multiplex equipment, i.e., Overhead Access (OHA), Multiplexer Timing Physical Interface (MTPI), Multiplexer Timing Source (MTS), and the two management related functional blocks Synchronous Equipment Management Function (SEMF) and Message Communications Function (MCF).

Group II functions at the terrestrial interface side to ensure full compliance across the NNI. They comprise standard SDH equipment functional blocks necessary to reach the VCs at the appropriate level (i.e., LOVC). The functions are SDH Phys-

ical Interface (SPI), Regenerator Section Termination (RST), Multiplex Section Termination (MST), Multiplex Section Protection (MSP), and Section Adaptation (SA), collectively called Transport Terminal Function (TTF). Intermediate HOVC pointer processing and cross-connection, if necessary, are accomplished by the Higher Order Path Cross-Connect (HPX) function, followed by Higher Order Path Termination (HPT) and Higher Order Path Adaptation (HPA). The result of Group II functions is the generation of Lower Order VCs (LOVCs) and their Frame Off-sets (FO).

Group III comprises local cross-connect and FSS transmission functions. The wide-area cross-connect function is basically the combination of the FSS network internal transmission, including its multiple access system, and the Group III local cross-connect function (LPX). For transmission over the FSS, the reduced set of TTF in Group III comprises Satellite Section Adaptation (SSA) and Satellite Section Termination (SST) functions. The latter contains functions similar to standard SDH SOH functions, albeit at reduced capacity, e.g., satellite section error performance monitoring and FEBE backward error reporting for the individual satellite sections. Each satellite link correspondence is separately given in these facilities. Group III functions would be implemented asymmetrically in multidestination operation.

For system efficiency, other satellite SOH functions such as an FERF for backward failure event reporting, the Data Communication Channel, and Order-Wire Channels are provided in the multiple access common terminal shared by all satellite digital sections (route associations) originating/terminating at the TDMA terminal (Fig. 4). The centralized TDMA system management, including Burst Time Plan (BTP) management, can be linked to the SDH transport network management system for enhanced flexibility and responsiveness to service/resource provisioning.

FSS-SDH Network Element Scenario for Medium Routes

In the INTELSAT system, the demand for PDH digital connections in the range of 34–45 Mb/s for aggregated thin route traffic and digital video is likely to increase in the future. Moreover, at these bit rates, there are operational requirements for temporary restoration of submarine cables. Thus, in view of the planned conversion to SDH, it is desirable to include standard SDH features on "medium rate" digital connections. This re-

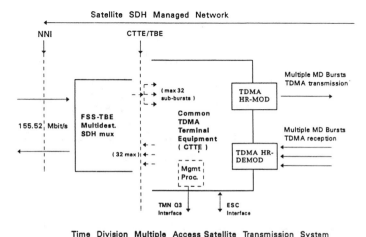

Fig. 4. Integrated TDMA-SDH terrestrial baseband equipment.

sults in the medium route scenario, depicted in Fig. 6, with the following characteristics:

- Internal FSS Section Bit Rate: STM-0 at 51.84 Mb/s
- Multiple Access: FDMA
- Section Topology: Single-destination and multidestination
- Implementation: Separated SDH functions and FDMA functions
- Network Role: Wide-area distributed SDH multiplexers

In this scenario, the FSS network element operates with one satellite section internal bit rate of 51.84 Mb/s, for which Frequency Division Multiple Access (FDMA) is the suitable method for (homogeneous) transponder sharing. Similar to the previous case, standard SDH functional blocks within the Terrestrial Baseband Equipment (TBE) are essential for SDH compatibility across the NNI and also SEI. For the STM-0 frame structure, the availability of standardized SOH functions and byte allocations [5], [23] justifies the segregation of SDH baseband multiplex equipment from FDMA functions, as shown in Fig. 7.

SDH multiplexing and cross-connect functions to be performed within the TBE depend on the following operational factors:

- LOVC destination sorting for multidestination distribution
- Multipoint HOVC (digital video) distribution
- STM-N/sub-STM-1 interworking requirement

The typical size of operational FSS traffic routes would be too small to warrant a separate 51.84 Mb/s point-to-point digital section for each pair of correspondents. However, in this scenario it is assumed that the total traffic level of the routes on the transmit side of all local multiplex equipment is sufficient to fill the STM-0 payload capacity, but destined to a number of different correspondents. This operational scenario justifies multidestination operation, extending the concept and asymmetry into the SDH Terrestrial Baseband Equipment. An asymmetrical configuration (point-to-multipoint in the transmit direction and multiple single points in the receive direction) at the satellite section layer and at the VC-3 path sublayer is required for multidestination distribution of the individual point-to-point path connections at either TUG-2 or TU-2s/VC-2s (containing 6.312 Mb/s PDH signals), and also at TU-12/VC-12 (containing 2.048 Mb/s PDH signals). In addition, 34/45 Mb/s digital video distribution and its management would also be facilitated by multipoint networking at the VC-3 path level. The SDH multiplex equipment needs internal "destination" filtering of various received signal tributaries and associated POHs to reconcile the multipoint asymmetry. The FSS internal section layer multipoint topology across the 51.84 Mb/s SEIs also requires destination addressing of section layer OAM related information and alarms in the backward direction (FERF and FEBE).

Similar to the previous case, the TBE functional elements can be subdivided in three general functional groups. The Group I common functions are essentially the same as in the previous case. Aside from TTF, Group II includes HPC, HPT functions, and possibly additional HPA functions. The Group III functions include the multipoint/multidestination receive side filtering at

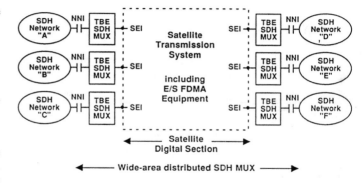

SEI: Satellite Equipment Interface (@51.84 Mbit/s)

Fig. 6. Medium route scenario.

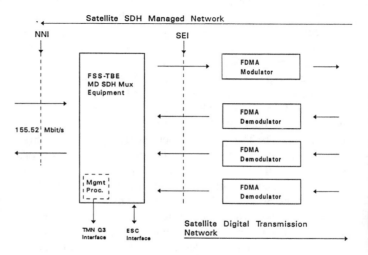

Fig. 7. Segregated FDMA-SDH terrestrial baseband equipment.

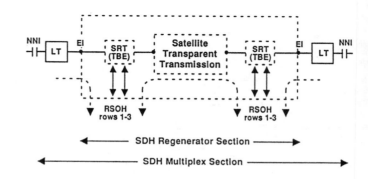

EI: Equipment Interface
SRT: Satellite Regenerator Terminal (in TBE)
LT: Line Terminal

Fig. 8. Thick route scenario.

the VC-3 level, and for each destination, HPT, HPA, and standard TTF towards the 51.84 Mb/s FSS interface. Due to multidestination operation, Group III functions are implemented asymmetrically.

Point-to-Point Regenerator Section for Thick Routes

In the INTELSAT system, there are very few instances of point-to-point satellite transmission links at 140 Mb/s. Currently, the main application at this bit rate is undersea cable restoration. However, in the future, it is anticipated that new applications such as High Definition TV as well as the need for SDH compatible cable restoration will require satellite system transport capability at STM-1 bit rate. Figure 8 depicts a satellite link constituting a single point-to-point SDH Regenerator Section of the SDH transport network. In general, such a scenario could apply to both 155.52 Mb/s (STM-1) and 51.84 Mb/s (STM-0) bit rates. The characteristics of this FSS-SDH scenario are:

- Internal FSS Section Bit Rate: STM-1 at 155.52 Mb/s
                                      STM-0 at 51.84 Mb/s
- Multiple Access: Single carrier per transponder or FDMA
- Section Topology: Single-destination point-to-point
- Implementation: Separated SDH functions and MA functions
- Network Role: Regenerator section in the SDH network

By acting as an SDH Regenerator Section, the FSS can access the standard RSOH functions, i.e., the BIP-8 error monitoring, DCC, and voice order-wire channels. The FSS TBE, representing a Satellite Regenerator Terminal (SRT), can use elements of CCITT-defined SDH regenerators including the optical terrestrial interfaces [13].

6. SYSTEM DESIGN ISSUES

Transmission Performance and Link Engineering

Transmission error performance objectives for international digital "paths" at the primary rates and above, and the allocation of the end-to-end objectives over portions of the connections, are specified in TS Recommendation G.826 [14]. Since this recommendation covers both PDH and SDH transmission networks, the design of the satellite portions of FSS-SDH network scenarios presented in this paper would have to meet the satellite link objectives derived from G.826. It is anticipated that Forward Error Correcting (FEC) will be needed in all cases to attain G.826 performance objectives, requiring additional power and/or bandwidth to be allocated to the satellite carriers.

Performance Monitoring

Error performance evaluation tools are important operational elements for integrated network management and OAM. The SDH multiplex structures provides a range of performance monitoring facilities based on the Bit Interleaved Parity - N (BIP-N) technique [5], which are summarized in Table 1.

Generally, VC path error monitoring is used on an end-to-end basis, whereas section monitoring is used in the SDH transport

TABLE 1. BIP ERROR MONITORING CAPABILITIES

	STM-0	STM-1
Regenerator Section Layer	BIP-8	BIP-8
Multiplex Section Layer	BIP-8	BIP-24
VC-3/4 Path	BIP-8	BIP-8
VC-1/2 Path	BIP-2	BIP-2

network between network elements at nodes and between regenerators. The comprehensive use of the SDH error monitors will facilitate unambiguous characterization of the performance of the network and its constituent parts. The FSS-SDH functional network elements discussed in this paper have access to one or more of these standardized SDH error performance monitoring facilities accessible at the SDH terrestrial baseband equipment. When used in both directions (over the satellite sections and as terrestrial section terminations), they contribute to the operational integration of FSS in SDH managed transmission networks.

Protection Switching

An important feature of SDH networks is the improved network availability and service restoration in case of equipment and transmission failures. An Automatic Protection Switching (APS) capability and associated K1/K2 bytes in the Multiplex Section Overhead (MSOH), as described in [7] Annex A, permit fast restoration at the Multiplex Section layer. In a 1-for-N protection arrangement, the APS protocols enables a switchover to a Protection line to be executed within 50 ms upon detection of a failure on a Working line. FSS systems acting as SDH regenerator sections at STM-1 ensure transparency to all Multiplex Section Overhead (MSOH) functions including the K1 and K2 bytes. This permits, in principle, multimedia (cable restoration by satellite) automatic network level protection, with the Multiplex Section Protection (MSP) functions located at the extreme (terrestrial) ends of the interoffice multiplex sections. Service restoration by satellites, in the event there is a failure of the protected fiber optic cable system, could be provided with much shorter service interruptions. However, spare transponder capacity to support STM-1 bit rate would have to be maintained as idle reserve or hot-standby, which may be prohibitively expensive. Economic and operational tradeoffs of maintaining a significant amount of satellite capacity for protection switching need to be examined in further detail. The possibility of carrying preemptible traffic in the spare transponder could be considered, although this will prolong service restoration time.

Considerations of satellite resource efficiency tend to also discourage the use of APS-like facilities and protocols at the Satellite Section Layer within the FSS network. In practice, improvement of FSS transmission quality and availability is accomplished by equipment level protection in the Multiple Access and Modulation sublayers. In addition, link margins possibly combined with automatic transmit power control techniques are provided at the Satellite Medium sublayer as protection against link quality degradation due to radio-path fading. Unlike radio relay systems with multipath dispersive fading, satellite systems experience performance degradation on a frequency nonselective basis across a given operational band, rendering 1-for-N radio channel link protection schemes not useful.

FSS can be part of a more complex SDH network protection arrangement with restoration at higher- and/or lower-order VC path level. In such arrangements, the protection switches are placed at the path termination points in line terminal equipment located outside the FSS subnetwork. The required transparency to pertinent VC paths is provided in all FSS scenarios described earlier. To the FSS, path level protection requires partial system capacity reservation, while other paths within the multiplex structure may be used for normal traffic.

Timing and Synchronization

One of the early applications of SDH transport networks is to carry PDH payload signals and to deliver them to PDH networks at the SDH-PDH network boundaries. Network limits for PDH signal phase variations (jitter/wander) at the hierarchical interfaces are specified in [19], [20]. SDH transport networks should not generate or transfer more phase variations than those allowed for existing PDH multiplexer/demultiplexer equipment [15], [16] and line systems [21].

The SDH payload timing quality is a function of the payload mapping process, pointer processing, and SDH network synchronization quality [9]. SDH network timing variation is caused by impairments in the synchronization reference clock (frequency offset, clock noise) and regenerator jitter, and the NNI upper limits of jitter and wander are given in [10]. Payload phase variation (jitter) at the desynchronizer output are mainly attributed to pointer adjustment jitter and payload mapping jitter. Pointer adjustments can be minimized by proper design of the SDH network synchronization. Clocks slaved to a G.811 Primary Reference Clock (PRC) [17] must meet relevant performance objectives, i.e., limits set in [13] for regenerator slave clocks and in [11] for cross-connect/ADM slave clock, respectively. In addition, the synchronization network topology must provide for adequate back-ups in case of clock failure.

Preliminary studies have been conducted on the combined effects of SDH network synchronization and the pointer processing in G.783 type multiplex equipment having a clock stability specified for a worst case network synchronization scenario. The results show that the pointer phase variation statistics are upper bounded with increasing number of cascaded pointer processing nodes, and thus there appears to be no practical constraint on the number of AU and TU pointer processing nodes that can be cascaded. Furthermore, with the pointer processor buffer spacings specified in [7], it is shown that pointer adjustments at the TU-1 level are extremely rare events [9].

Digital FSSs use Doppler buffers to fully compensate for the timing effect of satellite diurnal motion. These buffers are part of the digital baseband equipment, and are usually combined with (optional) plesiochronous buffers properly sized such that, together with the system high accuracy (1×10^{-11}) PRC, the controlled slip rate for PDH plesiochronous networking is held within limits set in [18]. When SDH networks and network elements are integrated in existing synchronization hierarchies,

occasional pointer adjustments will occur at the boundaries of the independently synchronized digital networks. This SDH synchronization mode, referred to as "pseudo-synchronous" mode (i.e., not all clocks in the network are slaved to the same G.811 PRC) is the normal mode of operation for the international and interoperator SDH networks [9].

Network Management

The management objectives of SDH transport networks [12] are:

- to set up VC paths automatically on request and across multiple operator boundaries,
- to maintain the paths to a high degree of availability; if necessary, with proper automatic restoration arrangements,
- to perform continuous in-service monitoring of allocated paths, and validation of compliance,
- to remotely maintain/reconfigure the network and localize faulty equipment,
- to record resource utilization for routing and billing.

Common use of standard SDH management concepts, principles, communication channels, and protocols by all systems, including FSS, is necessary for meeting these objectives and also future TMN objectives of integrated network management systems, while economically benefitting from cost savings through the use of common equipment. The paths to be managed may be given Access Point Identifiers based on the format of Recommendation E.164 (ISDN Numbering Plan), and country code allocation as per Recommendation E.163 (PSTN International Numbering Plan) [12]. The SDH Path numbering principles may need to consider international FSS subnetworks with path terminations points as separate "countries."

7. CONCLUSIONS

The integration of satellite systems in SDH transport networks generates architectural and system design requirements that are unique to satellite systems. Foremost among these are the need for STM-0 and other sub-STM-1 SDH connections, and multidestination operation for system efficiency and cost savings in satellite earth stations. The engineering solutions developed to meet these requirements will result in FSS-SDH networks and network elements which lend themselves to common SDH network management, and in meeting international functional and performance standards. The early identification of these unique requirements and international standardization of engineering solutions will make the benefits of "economies of scale," as well as the operational benefits of proper and orderly integration of satellite systems in SDH transport networks, available to the system users.

8. ACKNOWLEDGMENT

This paper is based on studies being conducted within the INTELSAT management. Views expressed are those of the authors. The authors wish to thank Messrs. D. K. Sachdev, J. F. Phiel, and G. P. Forcina from INTELSAT for their encouragement and valuable comments during the preparation of this paper.

9. APPENDIX: ABBREVIATIONS

APS	Automatic Protection Switching
AU	Administrative Unit
BIP-N	Bit Interleaved Parity-N
DCC	Data Communications Channel
ECC	Embedded Control Channel
FEBE	Far End Block Error
FERF	Far End Receive Failure
FO	Frame Offset
HOVC	Higher Order Virtual Container
HPA	Higher Order Path Adaptation
HPC	Higher Order Path Connection
HPT	Higher Order Path Termination
HPX	Higher Order Path Cross-Connect
LOVC	Lower Order Virtual Container
LPA	Lower Order Path Adaptation
LPX	Lower Order Path Cross-Connect
MCF	Message Communications Function
MSOH	Multiplex Section Overhead
MSP	Multiplex Section Protection
MST	Multiplex Section Termination
MTPI	Multiplexer Timing Physical Interface
MTS	Multiplexer Timing Source
NNI	Network Node Interface
OHA	Overhead Access
POH	Path Overhead
PDH	Plesiochronous Digital Hierarchy
RSOH	Regenerator Section Overhead
RST	Regenerator Section Termination
SA	Section Adaptation
SDH	Synchronous Digital Hierarchy
SEI	Satellite Equipment Interface
SEMF	Synchronous Equipment Management Function
SPI	SDH Physical Interface
SRT	Satellite Regenerator Terminal
SSA	Satellite Section Adaptation
SST	Satellite Section Termination
STM	Synchronous Transport Module
SSTM	Satellite Synchronous Transport Module
STS-1	(SONET) Synchronous Transport Signal — L1
TTF	Transport Terminal Function
TU	Tributary Unit
TUG	Tributary Unit Group
VC	Virtual Container

References

[1] R. Ballart and Y. C. Ching, "SONET: Now it's the standard optical network," *IEEE Commun. Mag.*, pp. 8–15, Mar. 1989.

[2] Various Authors, Special Issue on SDH, *IEEE Commun. Mag.*, Aug. 1990.

[3] P. T. Thompson, R. Silk, R. Samuel, and D. Aldous, "Exploitation of SDH features to provide multidestination satellite communications," *IEE Conf. Publ. 381*, Nov. 1993.

[4] *CCITT Rec. G.707*, "Synchronous Digital Hierarchy Bit Rates," Geneva, 1991.

[5] *CCITT Rec. G.708*, "Network Node Interface for SDH," Geneva, 1991.

[6] *CCITT Rec. G.709*, "Synchronous Multiplexing Structure," Geneva, 1991.

[7] *CCITT Rec. G.783*, "Characteristics of SDH Equipment Functional Blocks," Nov. 1992 version.

[8] *CCITT Rec. G.784*, "SDH Management," Geneva, 1991.

[9] *CCITT Rec. G.803*, "Architectures of Transport Networks based on SDH," Geneva, 1992.

[10] *CCITT Rec. G.825*, "The Control of Jitter and Wander within Digital Networks which are Based on the Synchronous Digital Hierarchy (SDH)," Helsinki, 1993.

[11] ITU-T (draft) Rec. G.81s, "Timing Characteristics of Slave Clocks Suitable for Operation of SDH Equipment," Jan. 1993 version.

[12] *CCITT Rec. G.831*, "Performance and Management Capabilities of Transport Networks Based on the SDH," Geneva, 1992.

[13] *CCITT Rec. G.958*, "Digital Line Systems Based on the Synchronous Digital Hierarchy for Use on Optical Fibre Cables," Geneva, 1990.

[14] *ITU-T Rec. G.826*, "Error Performance Parameters and Objectives for International Constant Bit Rate Digital Paths at or above the Primary Rate," July 1993 version.

[15] *CCITT Rec. G.743*, "Second Order Multiplex Equipment Operating at 6312 kbit/s and Using Positive Justification," *CCITT Blue Book, Vol. III-Fascicle III.4*, Nov. 1988.

[16] *CCITT Rec. G.751*, "Digital Multiplex Equipment Operating at the Third Order Bit Rate of 34368 kbit/s and the Fourth Order Bit Rate of 139264 kbit/s and Using Positive Justification," *Blue Book, Vol. III-Fascicle III.4*, Nov. 1988.

[17] *CCITT Rec. G.811*, "Timing Requirements at the Output of Primary Reference Clocks Suitable for Plesiochronous Operation of International Digital Links," *CCITT Blue Book, Vol. III-Fascicle III.5*, Nov. 1988.

[18] *CCITT Rec. G.822*, "Controlled Slip Rate Objectives on an International Digital Connection," *CCITT Blue Book, Vol. III-Fascicle III.5*, Nov. 1988.

[19] *CCITT Rec. G.823*, "The Control of Jitter and Wander within Digital Networks which are Based on the 2048 kbit/s Hierarchy," Helsinki, 1993.

[20] *CCITT Rec. G.824*, "The Control of Jitter and Wander within Digital Networks which are based on the 1544 kbit/s Hierarchy," Helsinki, 1993.

[21] *CCITT Rec. G.921*, "Digital Sections Based on the 2048 Mbit/s Hierarchy," *CCITT Blue Book, Vol. III-Fascicle III.5*, Nov. 1988.

[22] *CCITT Rec. M.30*, "Principles for a Telecommunications Management Network," *CCITT Blue Book, Vol. IV-Fascicle IV.l*, Nov. 1988.

[23] *CCIR Rec. 750*, "Architectures and Functional Aspects of Radio-Relay Systems for SDH-Based Networks," Geneva, 1991.

[24] *CCIR Doc. 4B/Temp/20 (Ref. 2)*, "Draft New Report on Work Toward New Recommendations on FSS-SDH Transport Subnetworks," amended June 1993.

International Gateway
For SDH and SONET Interconnection

Cannon Hwu and Stanley Chum

Sprint Communications, 1 Adrian Court, Burlingame, California 94010, USA

Tel : 1-415-375-4383 Fax : 1-415-375-4079

Abstract

The ITU-T SDH and the ANSI SONET are compatible but not identical standards for synchronous digital networks. There is significant complexity when carriers have to handle traffic originated from and terminated in different countries of different implementation preferences of SDH or SONET, or transit this traffic across one's network.

This paper analyzes the diversity of the traffic and the management of international networks and services in considering the appropriate architectural designs of an international network and the SDH/SONET gateway. As an example, this paper presents a field trial plan to be conducted in 1994-95 by Sprint Communications. Relevant information and partial results of the trial will be presented at the GlobeCom '94 Conference.

1.   Introduction

The concept of ANSI SONET (Synchronous Optical Network) was introduced by Bell Communications Research (Bellcore) in mid-1980s and quickly emerged as a promising higher level digital hierarchy beyond the then existing DS-3 (44.736 Mbits/s) in the US.

Meanwhile in ITU-T (formerly CCITT), development of the standards for Broadband ISDN in Study Group XVIII suggested that a next generation of digital hierarchy at levels higher than E3 (34.368 Mbits/s) would be necessary. Such a hierarchy would be synchronous in nature and should provide for a common hierarchical transport platform unifying the world telecommunications community. Much effort and compromises had been made by all parties involved since then, and ITU-T Recommendations for SDH (Synchronous Digital Hierarchy) were published during the 1989 - 1992 Study Period.

SONET and SDH are compatible, but not identical, digital hierarchies. They have identical transmission bit rates and framing structures, but they differs in the multiplexing and mapping of tributary payloads into higher digital levels. Both define similar sets of overheads and functions, but the usage and definitions of some overheads vary. Today two major camps emerge from ITU-T SDH standards in the world telecommunications community, namely, the ETSI (European Telecommunications Standards Institute) and the US SONET preferences. The gaps between these two preferences are narrowing recently, but not completely eliminated. Therefore, in practices, there is significant complexity when carriers have to handle international traffic originated from and terminated in different countries of different SDH implementation preferences, or transit this traffic across one's network.

Today, SONET is being deployed in the US, while SDH is implemented in Europe and Asia at a rapid pace. For international traffic, SDH-based submarine fiber optic cable systems are scheduled for services by mid-1995. International carriers such as Sprint Communications now focus their attention in ensuring a smooth interworking between the two preferences where differences still exist. We believe that it will be very unlikely in practices for SDH networks to be modified or adapted to pass SONET traffic, due to the fact that most parts of the world will be using SDH standards except noticeably USA and Canada. Therefore, our discussion in this paper will focus on how to pass SDH traffic via a SONET network.

In section 2, the differences relevant to international interworking between the two preferences are reviewed. Section 3 describes a generic SDH/SONET international gateway for which a diversity of traffic types and patterns are determining factors. Alternative approaches in the design of a SDH management network which oversees the integrity of international digital services spanning across SDH and SONET environments are also discussed. As an example, a field trial plan of SDH/SONET international gateway is presented in Section 4. With SDH, coupled with an appropriate SDH/SONET gateway, the wish of having a common

Reprinted from *IEEE GLOBECOM '94 Conf. Rec.*, pp. 725–734, Dec. 1994.

hierarchical transport platform unifying the world telecommunications community is becoming a virtual reality.

2. Differences Between SDH and SONET

Differences exist between the SDH and the SONET standards. An ANSI T1 Committee document, T1X1.2/93-24R2, provides a comprehensive discussion on these differences. Table 1 shows the transmission rates supported by SDH and SONET respectively. Table 2 shows the payloads and their containers, together with their respective payload mappings under SDH and SONET standards. Figure 1 shows the SDH and SONET multiplexing structure.

Table 1 SDH / SONET Transmission Rates		
SDH Signal	SONET Signal	Transmission Rates
STM-1 STM-4 STM-16	STS-1 STS-3 STS-12 STS-24 STS-48	51.840 Mbit/s 155.520 Mbit/s 622.080 Mbit/s 1244.160 Mbit/s 2488.320 Mbit/s

Figure 1 and Table 2 show that SDH and SONET are compatible in mapping payloads of E4 and above, where SONET's STS-3c is equivalent to SDH's STM-1. However, for mapping lower capacity payloads (i.e., DS1, E1, ..., E3, DS3), SONET provides one unique mapping path for each payload (note : SONET does not support E3), while SDH allows at least two alternative paths for each payload (Note 1 : four possible paths for mapping DS1. Note 2 : SDH does not support DS1C). Mapping paths of SDH indicated by bold-typed brackets in Table 2 are compatible to those of SONET.

In the SDH world, ETSI has decided to use AU-4 mapping for STM-1 while Japan will use AU-3 mapping for STM-1. Fortunately, Japan also decides to use only AU-4 mapping at international boundaries, and thus from a SONET operator's perspective, there will be no difference between the traffic from Europe or Japan. Furthermore, interworking between SDH and SONET would present the least complication if they meet at the tributary interface level but not on the "line" side (i.e., no mid-span meet). The SDH's STM-1 and the SONET's STS-3c appear to be the most suitable interface for international SDH/SONET interworking.

An essential feature of SDH (and SONET) is its embedded capabilities of handling OAM&P (Operations, Administrations, Maintenance and Provisioning) information. These capabilities are made possible by the overhead bytes built in the SDH (and SONET) frames, for which the overhead bytes (for STM-1 and STS-3c) are shown in Table 3.

The terminology and definitions of these overhead bytes are somewhat different between SDH and SONET. The methods of performance monitoring using these overhead bytes also vary. Many of these differences will have no or little effect on the design of an international gateway for SDH/SONET interworking because most of the overhead bytes (Regenerator Section Overheads and Multiplex Section Overheads) are terminated at the international boundaries. The following highlights those overhead bytes which are still of concerns.

The original definitions of the C1 byte for STM-1 or STS-1 identifications were different between SDH and SONET. The C1 byte has now been redefined by ITU-T to be used for Section Trace. ANSI is likely to adopt this change as well. If this is the case, there will be no incompatibility problem for interworking. Furthermore, equipment designed for the original C1 definitions which ignores the C1 byte at the receivers will present no backward compatibility issue.

The "ss" bits in the Pointer byte H1 are different between SDH and SONET. SDH defines the "ss" bits as "01" and can set them to other values to indicate the payload mappings of AU-4, AU-3 and TU-3. SONET always sets the "ss" bits as "00" which remains undefined, and thus multiple values of "ss" bits (which are generated by SDH equipment) cannot be carried in an STS-3c format. These bits shall be ignored on international boundaries.

Due to the fundamental differences in speech encoding (SDH uses A-Law while SONET uses u-Law), the E1 and E2 bytes for orderwires have to be decoded and accessed via 4-wire analog circuits at the international boundaries.

The Z1 byte (called S1 in SDH) allocation for Synchronization Messages is compatible between SDH and SONET. The actual messages are different (due to the differences in stratum clock levels in different regions) and have not been firmly defined. These differences shall not affect the interconnection of the SDH/SONET networks at the "drop" side interface.

Table 2 SDH / SONET Payloads and Mappings

Payload	SDH					SONET			
	Container	Actual Payload Capacity	Payload and POH	Mapping AU-3/AU-4 Based		Container SPE	Actual Payload Capacity	Payload and POH	Mapping
DS1 (1.544)	VC-11	1.648	1.728	**(AU-3)**,AU-4		VT1.5	1.648	1.664	STS-1
	VC-12	2.224	2.304	AU-3,AU-4					
E1 (2.048)	VC-12	2.224	2.304	**(AU-3)**,AU-4		VT2	2.224	2.240	STS-1
DS1C (3.152)						VT3	3.376	3.392	STS-1
DS2 (6.312)	VC-2	6.832	6.912	**(AU-3)**, AU-4		VT6	6.832	6.848	STS-1
E3 (34.368)	VC-3	48.384	48.960	AU-3,AU-4					
DS3 (44.736)	VC-3	48.384	48.960	**(AU-3)**,AU-4		STS-1	49.536	50.112	STS-1
E4 (139.264)	VC-4	149.760	150.336	**(AU-4)**		STS-3c	149.760	150.336	STS-3c
ATM (149.760)	VC-4	149.760	150.336	**(AU-4)**		STS-3c	149.760	150.336	STS-3c
ATM (599.040)	VC-4-4c	599.040	601.344	**(AU-4)**		STS-12c	599.040	601.344	STS-12c
FDDI (125.000)	VC-4	149.760	150.336	**(AU-4)**		STS-3c	149.760	150.336	STS-3c
DQDB (149.760)	VC-4	149.760	150.336	**(AU-4)**		STS-3c	149.760	150.336	STS-3c

Note 1 : **(AU-n)** indicates compatible mapping to SONET   Note 2 : Numbers are in Mbit/s unit

Table 3   Overheads in a SDH's STM-1 (and SONET's STS-3c) Frame

Framing A1	Framing A1	Framing A1	Framing A2	Framing A2	Framing A2	STS-1 ID C1	STS-1 ID C1	STS-1 ID C1		Trace J1
BIP-8 B1			Orderwire E1			User F1			for STS-3c only, not included in STM-1	BIP-8 B3
Data Comm D1			Data Comm D2			Data Comm D3				Sig Label C2
Pointer H1	1001 ss11	1001 ss11	Pointer H2	1111 1111	1111 1111	PtrAction H3	Ptr Action H3	Ptr Action H3		Path Stat G1
BIP-8 B2	BIP-8 B2	BIP-8 B2	APS K1			APS K2				User F2
Data Comm D4			Data Comm D5			Data Comm D6				Multi-Frame H4
Data Comm D7			Data Comm D8			Data Comm D9				Growth Z3
Data Comm D10			Data Comm D11			Data Comm D12				Growth Z4
Growth Z1	Growth Z1	Growth Z1	Growth Z2	Growth Z2	Growth Z2	Orderwire E2				Growth Z5

◄————————————— Transport Overhead —————————————►   ◄—————— Payload Capacity ——————►

Path Overhead

By definition, path overheads (POHs) are delivered from end to end, and are transparent to the transport facility. Thus the differences of POHs between SDH and SONET should not impact on the international interworking. However, in order for the POHs to be properly interpreted at the receiving end, the path originates from a SDH equipment must be terminated in a SDH equipment; the same holds true in the case of SONET.

3. SDH / SONET International Gateway

3.1 Types of International Traffic :

Today, carriers in US and Canada are deploying SONET standards in their networks. A basic and representative model of SONET deployment is shown in Figure 2, which has a multi-rings architecture (SONET Rings) with self-healing protection, add-drop multiplexer (ADM) and SONET access digital cross-connect systems (SA-DCS). Similar models are also being employed in other parts of the world where SDH standards are used. Incompatibility of the SDH and SONET standards becomes a critical concern to international carriers spanning across the two different camps.

There are five key types of international traffic:

(1) International public switched voice circuits : this type of traffic is normally routed to an international switching center (ISC) of a specific international carrier where a gateway digital switch will process this traffic and will pass it onto a domestic interexchange carrier (IEC) and/or a local exchange carrier (LEC).

(2) International dedicated transit (IDT) : this type of traffic will be digital in nature; it originates from a number of SDH-based countries, passes through the SONET-based US/Canada continent, and continues onward to other SDH-based countries.

(3) Digital private lines traffic : this type of traffic originates from a number of SDH-based countries and terminates at various locations within the continent US/Canada; and vice-versa.

(4) Network management information traffic : the automation of the OAM&P functions of one's telecommunications networks is a major feature and advantage of the SDH/SONET technology; management information related to OAM&P will be communicated within a carrier's SDH (or SONET) network and/or other adjacent facilities of the carrier via the overhead bytes of the synchronous frame structures. For international carriers, management information may need to be communicated across different networks of different designs, i.e., network management information flows between a SDH network and a SONET network.

(5) Digital Switched Circuits : This includes, for examples, ATM (Asynchronous Transfer Mode), frame relay, etc.

3.2 SDH Cablehead :

Today, consortia are formed for the initial stages of SDH submarine cable systems deployment due to the huge amount of capital investment involved. A single international carrier will not have large enough trans-oceanic traffic volume to justify a dedicated cable system, nor even a dedicated channel such as a STM-1 channel for each country it serves. Thus within a single channel (e.g., STM-1 or STM-4) of a SDH submarine cable system, there may be traffic mixed from different countries, for different international carriers, and/or with different tributary payloads. Generally, the cablehead at the landing location of a submarine cable system will separate the traffic among different carriers and deliver them to the respective carriers. The delivery from the cablehead to a specific carrier will likely take place in a form of SDH channels (e.g., STM-1 or STM-4), and different types of traffic mentioned in section 3.1 above (which belong to that particular carrier) will still be mixed inside these channels. The SDH cablehead operators will doubtfully perform any further grooming on behalf of the carriers due to cost and strategic concerns of different parties.

From the point of view of an international carrier, the interface facility between the carrier and the cablehead must be able to accept the signals from the cablehead and transform them into some suitable forms that can be carried and processed appropriately along its SONET's network. We will call such an interface facility a "SDH/SONET International Gateway". Our discussion in this paper will focus around the scenario where a relatively small traffic volume is incurred for an international carrier of concern. Specifically, for digital private lines services, a traffic volume of a few STM-1's for a carrier is considered as "small" in this paper. This is probably true for most international carriers in the early stages of SDH deployment. A more detailed discussion for each type of traffic is given below.

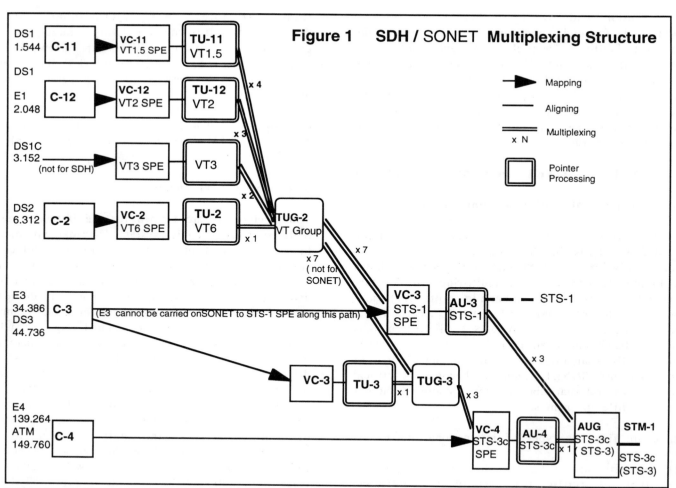

Figure 1   SDH / SONET Multiplexing Structure

DS1 1.544 — **C-11** → **VC-11** VT1.5 SPE — **TU-11** VT1.5

DS1 / E1 2.048 — **C-12** → **VC-12** VT2 SPE — **TU-12** VT2 — x 4

DS1C 3.152 (not for SDH) → VT3 SPE — VT3 — x 3

DS2 6.312 — **C-2** → **VC-2** VT6 SPE — **TU-2** VT6 — x 2 — x 1 — **TUG-2** VT Group

x 7 (not for SONET)

E3 34.386 / DS3 44.736 — **C-3** — (E3 cannot be carried onSONET to STS-1 SPE along this path) → **VC-3** STS-1 SPE — **AU-3** STS-1 ---- STS-1

→ **VC-3** — **TU-3** — x 1 — **TUG-3** — x 3 — x 7 — x 3

E4 139.264 / ATM 149.760 — **C-4** → **VC-4** STS-3c SPE — **AU-4** STS-3c — x 1 — **AUG** STS-3c (STS-3) — **STM-1** STS-3c (STS-3)

Mapping ▶

Aligning ——

Multiplexing ══ x N

Pointer Processing ☐

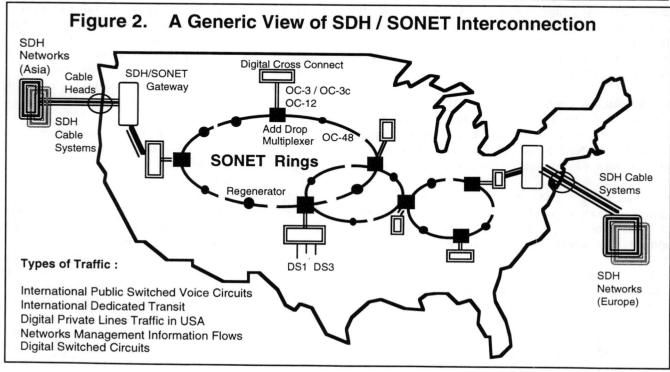

Figure 2.   A Generic View of SDH / SONET Interconnection

SDH Networks (Asia)

Cable Heads

SDH/SONET Gateway

SDH Cable Systems

Digital Cross Connect

OC-3 / OC-3c OC-12

Add Drop Multiplexer   OC-48

SONET Rings

Regenerator

DS1 DS3

SDH Cable Systems

SDH Networks (Europe)

Types of Traffic :

International Public Switched Voice Circuits
International Dedicated Transit
Digital Private Lines Traffic in USA
Networks Management Information Flows
Digital Switched Circuits

3.3 International Public Switched Voice

For public switched voice traffic from Europe, voice circuits (64 Kbits/s each) are bundled in an E1 (2.048 Mbits/s) tributary payload entity at the origins. These traffic originated from different SDH-based countries could be carried in various STM-1 channels of a submarine SDH cable. When they arrive at the US, they go to one destination commonly called ISC (International Switching Center) of an individual international carrier (sometimes two' or more redundant ISCs for diversity protection). This type of traffic traditionally is carried via PDH (Plesiochronous Digital Hierarchy) cables but may migrate to SDH cables in the future. The gateway must be able to separate this type of traffic from the other types which are all embedded in the incoming STM-1s (or STM-4s), and re-pack them into a suitable form for transport to and interface with the ISC.

handled. They will be first separated from other types of traffic at the gateway location, and subsequently grouped into a number of SONET channels to be transported across the SONET network to another SDH/SONET gateway, where inverse functions are to be performed before this IDT traffic is put back onto other SDH systems to other countries.

3.5 Digital Private Lines Traffic

Due to the potential diversity of the tributary payload types and volume from Europe and Asia, and due to the fact that they will be terminated (added-dropped) at various locations throughout the US/Canada continent, routing and capacity engineering to handle this type of traffic in a SONET network would be a substantial task. The design of an appropriate gateway must couple closely with one's approach in the above

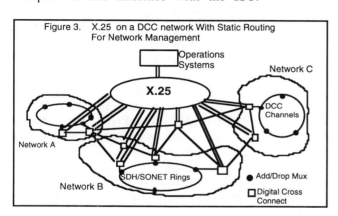

Figure 3. X.25 on a DCC network With Static Routing For Network Management

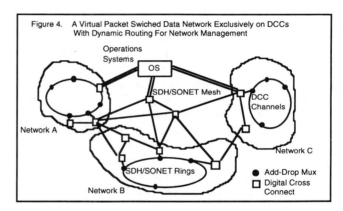

Figure 4. A Virtual Packet Swiched Data Network Exclusively on DCCs With Dynamic Routing For Network Management

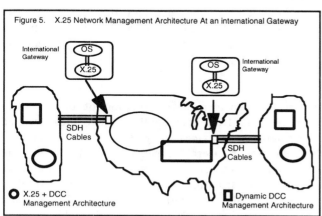

Figure 5. X.25 Network Management Architecture At an international Gateway

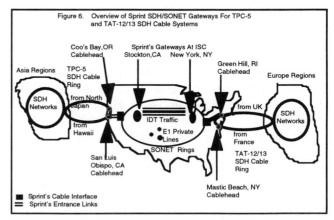

Figure 6. Overview of Sprint SDH/SONET Gateways For TPC-5 and TAT-12/13 SDH Cable Systems

3.4 International Dedicated Transit (IDT)

Payloads on IDT could be any tributaries supported by SDH standards, e.g., E1, E3, ATM, etc. The types of payloads and their respective traffic mix and volume will have little effect on how IDT shall be

engineering. The following two extreme scenarios can be used to illustrate this point. For the simplicity of this illustration, let us assume in this section that E1's are the only tributary payloads.

(1) If the E1 traffic volume is large and the number of add-drop destinations is small (says, 10 cities

and about 10 to 15 E1's per city), a simple grooming function at the gateway will allow these E1's to be bundled in a number of dedicated STS-1 SPE envelopes (in the SONET world). Each envelope may then be transported straight through to its intended destination via the SONET network.

(2) If there are a large number of destinations involved and if only several E1's are added-dropped at each location, a grooming function is certainly still necessary at the gateway location. However, the methods of packaging in the grooming process could vary significantly depending on the architecture of the SONET network itself. The digital cross connect capability which forms a part of the overall architecture of one's SONET network will determine the efficiency of the packaging approach chosen. The capability of a SONET digital cross connect architecture can be characterized by its levels of cross connects, i.e., at VT1.5, VT2, VT6 and/or STS-1 SPE envelopes levels. It is obvious that an E1 payload cannot be packaged into a VT1.5 envelope. The cross connect at the STS-1 SPE level provides a rather poor overall efficiency of only 52%, assuming an equal probability of distribution of E1's ranging from 1 to 21 in a STS-1 SPE entity to each destination. The VT6-based cross connect provides a good overall efficiency of 87%. The VT2-based cross connect has an 100% efficiency but it will incur deployment of VT2-based switching fabric and software at most junctions of the SONET network. The economic of this approach is questionable since it bears no benefit to the US domestic SONET-based traffic. VT6 offers an attractive choice and compares favorably to VT2 because relatively less switching fabric and software are involved. Cross connect at STS-1 SPE level is at parity with that of VT6 only if the number of E1's to each destination is more than 14, for which 86% efficiency can be obtained.

For those carriers whose SONET networks have already been built or specific types of equipment/architecture have already been committed to, and if as such does not include a VT6 cross connect capability or upgrade-ability, alternative approaches must be considered to address the efficiency issue. For digital private lines services at bit rates higher than E1, the issue of transport efficiency is relatively less stringent.

3.6. Network Management Information Traffic

The overhead bytes and especially the Data Communication Channels (DCC), embedded in a SDH/SONET network constitute a set of logical packet switched data networks for the communications of management information. Overhead bytes other than DCCs will be used for the flows for those management information standardized in SDH/SONET. Management information specific to a particular carrier could be communicated on the logical packet switched data network made up by the DCCs. Initially, one alternative is to use a X.25 packet switched network architecture on the DCCs physical facility of the SDH/SONET network (and possibly complemented by other existing asynchronous transmission facilities) with a static routing scheme. Another alternative in the future could be using the DCCs facility exclusively as a logical network with dynamic routing scheme, for which a new standard protocol stack will be required. Figures 3 and 4 illustrate the above two alternatives.

For international carriers whose services and traffic are spanning across a multiple of networks (either SDH, SONET and/or even PDH), the flows of management information across networks depend on the alternative designs chosen for individual networks involved. The regenerator section overheads and the multiplex section overheads are terminated at the international boundaries. Path overheads (POH) are delivered from customer to customer. The transport facility should be transparent to POH, and it is made possible for as long as the POHs are embedded in the specific payload envelopes which are cross connected as individual entities.

Most of the management information of one network do not need to flow into another. Only those related to international services may be required and are subjected to bilateral arrangement between the international carriers and the involved networks owners. This situation will significantly simplify the design of the international SDH/SONET gateway from a network management prospective. Due to the uncertainty of the network management architecture of each involved countries, a X.25 architecture at the gateway would be a feasible choice, as illustrated in Figure 5. However, since all traffic (both payloads and network management information) coming from the submarine SDH cables into the gateway location of a particular carrier are embedded in a number of STM-1 channels, the gateway device must segregate and handle them accordingly.

3.7 Digital Switched Circuits

Examples of digital switched circuits included those offered by frame relay and/or ATM technologies. Generally, these types of services would require a network of frame relay switches and/or ATM switches, linked together by a synchronous transmission network of the carrier offering these services. For international services of these types, a transmission link between two switches may involve both SDH and SONET facility. Therefore, the SDH/SONET gateway must be able to pass and maintain the integrity of the ATM/Frame Relay tributary payloads. The interface of STM-1 and STS-3c between SDH and SONET shall be appropriate for this interconnection, similar to that for the case of IDT (International Dedicated Transit).

4. A Field Trial Plan

Anticipating the growth potential of the trans-oceanic telecommunications traffic well into the 21st century, new submarine fiber optic cables are being deployed in both the Pacific and the Atlantic Oceans. Specifically, SDH-based submarine cable systems, namely, the TPC-5 and the TAT-12/13, are scheduled for services by mid-1995. Both TPC-5 and TAT-12/13 will employ a synchronous 4-fibers ring architecture with 1-for-1 protection along the bi-directional ring. The transmission bit rate on each fiber will be 5 Gbits/s, i.e., equivalent to 2 STM-16's. (Note : the protocols are vendor proprietary, i.e., not STM-32 as such.) DCS equipment at the VC-4 cross connect level will be installed at the cableheads, which will provide a STM-1 (optical or electrical) interface to participating carriers.

Sprint Communications Corporation is a major participant of the consortia. Initial Sprint's shares of these SDH cable capacities are of the order of several STM-1's. The nature of a small traffic volume for digital private line services presents to Sprint constraints in the design of the required SDH/SONET gateway and the related network routing architecture, as discussed in section 3.5. Initially, only E1 digital private lines are expected. Meanwhile Sprint is accelerating her pace in the deployment of SONET in US, and a significant portion of the targeted SONET infrastructure would

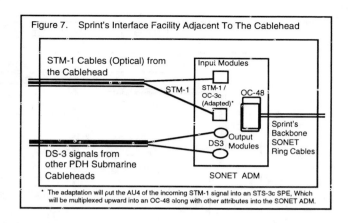

Figure 7. Sprint's Interface Facility Adjacent To The Cablehead

STM-1 Cables (Optical) from the Cablehead

Input Modules

STM-1

STM-1 / OC-3c (Adapted)*

OC-48

Sprint's Backbone SONET Ring Cables

Output Modules

DS3

DS-3 signals from other PDH Submarine Cableheads

SONET ADM

* The adaptation will put the AU4 of the incoming STM-1 signal into an STS-3c SPE, Which will be multiplexed upward into an OC-48 along with other attributes into the SONET ADM.

be completed in the similar time frame, i.e., by 1995-96. Growth in Sprint international network and services is expected, and considerations of such need to be included in today's design of the gateway and its associated architecture.

Sprint's SONET is essentially of a ring architecture as depicted in Figure 2. Accesses to Sprint SONET rings will be via SONET access digital cross connect facility (SA-DCS). Interface modules to SA-DCS include DS1, DS3, E1, STS-1, STS-3 (STS-3c), OC-1, and OC-3 (OC-3c). Due to economic consideration for the bulk of the domestic USA traffic, the digital cross connect at the VT6 level is currently not supported by the network DCS and the access DCS.

With the above services and constraints in mind, various gateway and network architecture are being considered. The following presents a field trial plan being scheduled by Sprint in 1994-95.

Sprint's facility interfacing the SDH cable systems will be installed in a site adjacent to the cableheads in San Luis Obispo, CA and Green Hill, RI respectively, as shown in Figure 6. The entrance links between the cablehead and the Sprint interface will be via redundant STM-1 optical fibers. Specific equipment types inside the interface are shown in Figure 7. For emergency backup and future network migration purposes, some DS3 channels from PDH submarine cables will be re-routed through this facility into Sprint SONET network. At the international boundary, the STM-1's from the cablehead will be connected to the OC-3c input ports of the SONET ADM equipment inside the interface facility, as suggested in Section 2 of this paper. All traffic as is will be hauled via an OC-48 channel from this SONET ADM to the nearest Sprint's ISC where the SDH/SONET gateway equipment will be housed, i.e., in Stockton, CA and

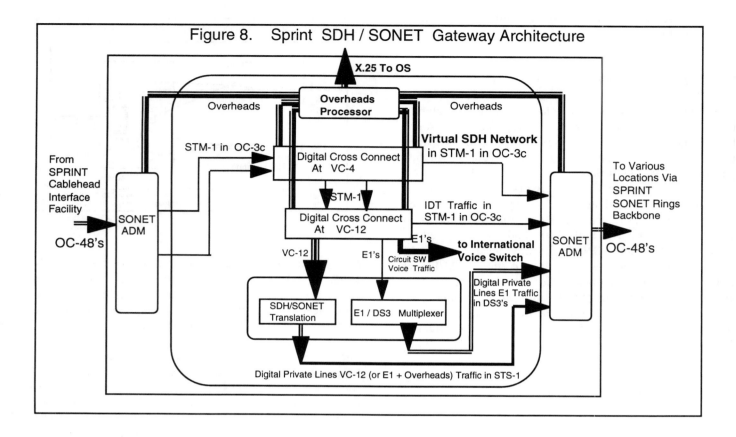

Figure 8. Sprint SDH / SONET Gateway Architecture

New York, NY respectively. Note that the transport between the interface facility and the gateway is also protected due to the SONET ring nature of the Sprint network.

Figure 8 shows a schematic block diagram of the SDH/SONET gateway being pursued by Sprint for the field trial. Two digital cross connect equipment (i.e., at VC-4 and VC-12) will be used to segregate the incoming traffic (the reverse direction is understood and is not shown here for simplicity) of different types, and to regroup them into appropriate bundles. The IDT traffic are packed into independent OC-3c payloads. ATM traffic (if any) could be handled in a similar fashion as IDT. Circuit switched voice traffic are separated from the other types of traffic by the SDH-based DCS's and are presented as E1's to the international voice switch at the ISC.

Digital private lines (currently E1 only) can be either packed into DS3's or STS-1. They will be inserted into the SONET network at the SONET ADM as shown in Figure 8. The former (i.e., DS3) cannot include any end-to-end path overheads for the customer, while the latter (i.e., STS-1) can maintain the integrity of POH end-to-end, but will require a

SDH/SONET translation device in the multiplexing and mapping processes from VC-12 to STS-1. The digital private lines (E1's) now embedded in various DS3 (or STS-1) will flow along the SONET rings and will be routed by the SONET DCS's at appropriate junctions of the SONET network. Since initially Sprint's SONET DCS can only cross connect at the STS-1 (and above) levels [note: the VT1.5 level is irrelevant here for the E1 traffic], the following alternate designs will be trialed to address the issue of transport efficiency.

(1) **Virtual SDH Network** : A definite number of STS-3c channels along appropriate routes of the SONET network can be allocated to form a virtual SDH ring structure. The number of channels will be sufficiently large to accommodate the accumulated total volume of the add-drop E1 traffic, (and with some room for growth). With this architecture, a "small scale" designed SDH-baesed DCS and/or ADM equipment (at the E1 or VC12 level or other levels when needed) can be installed in or close to the customer premise. "Small scale" will provide the benefits of versatility and economic of capital investment. A "virtual" SDH ring structure provides an opportunity of a clean and independent OAM&P environment for this type of international traffic; this is particularly critical since in the

initial stage of the SDH/SONET interconnection, the Sprint SONET network itself is still in its infant stage and its OAM&P functions and practices may not be maturely in place. This architecture also simplifies the processing requirements en-route of the path overheads that are to be terminated at the customer premises. Furthermore, a virtual SDH architecture would make feasible the provisioning of future higher levels digital private lines services with end-to-end POH.

(2) **Partitioning** : A partition in the form of geographical cells of suitable sizes for the required add-drop sites can be made such that the relevant digital private lines are grouped accordingly at the gateway locations. This grouping will increase the fill (of E1's for example) in each DS-3 or STS-1 payloads which will be routed by the network DCS's (cross connect at the STS-1 level) and delivered to the corresponding cells. Individual E1's will be demultiplexed from the STS-1 at each cell, and will be send to the end users via suitable local transport facilities. An advantage of this approach is that initial capital equipment investment is smaller comparing to the former approach. However, it may be more complicated especially in the re-partitioning processes when capacity expansion is needed. In this architecture, network management functions for this type of traffic are mostly decentralized. This may present more complexities in the operations practices since international digital private line services may become more dynamic in terms of both capacities and destinations (i.e., the trend of bandwidth on demand).

A X.25 packet switched architecture will be used at the SDH/SONET gateways for the communications of network management information to the related operations systems (OS's). In Figure 8, the overheads processor (OP) together with the overheads paths between the DCS's and the OP is essentially a server on a local area network bus for the OS. One function of the DCS's at the gateway is to be able to access to the overhead bytes of the incoming STM-1 frames, extract and then load these bytes on the bus.

Equipment for the above trial are being acquired/developed in the first half of 1994, a feasibility test at the Sprint Advanced Technology Laboratories, California, will take place before the field trial begins by the end of 1994.

5 . Conclusion

This paper has considered the differences between SDH and SONET, the international digital services requirements, and the possible constraints dictated by the already existing or committed network architecture of individual carriers. Various alternatives are discussed from the prospective of the traffic types, scales and scope. This paper also specifically addresses the concern of the transport efficiency under the scenario of a small volume of add-drop digital private lines.

A field trial plan of Sprint has been presented in this paper and due attention has been paid to the targeted service schedules of the TPC-5 and the TAT-12/13 submarine SDH-based cable systems. The trial will provide insights of the feasibility of interworking between STM-1 and STS-3c interfaces and issues related to network architecture and management. At the time this paper is sent to GlobeCom'94 for press, specific test data are not at hand. However, verbal reports of the then available results will be made at the conferences.

References :

ITU-T Rec. G.803, Architecture of Transport Networks Based On SDH, 1991.
ITU-T Rec. G.708, Network Node Interface for SDH, 1991.
ITU-T Rec. G.709, Synchronous Multiplexing Structure, 1991.
ITU-T Recommendation G.784, SDH Management, 1990.
ANSI T1X1.2/93-24R2, "A comparison of SONET and SDH".
H Katz, G. Sawyers, and J. Ginger : "SDH Management Network : Architecture, Routing and Addressing", Globecom 1993, pp. 223.

Section 2
Deployment Plans and Architecture Issues

EXISTING telecommunication network architectures are optimized for a public switched telephone network (PSTN) service. However, new services such as broadband video services of various types, LAN interconnection, high-speed data services, etc., are beginning to appear, and consideration must be given to suitable network architectures. The development of SDH/SONET, and its ability to provide flexible networking and new network topologies, offer a unique opportunity to build a network architecture capable of efficiently transporting and managing existing and new services. Consequently, this section is devoted to architecture issues, and contains papers which demonstrate experience to date in the use of SONET and SDH technology and the manner in which it is deployed.

While most of the standardization work relates to generation of standards for network node interfaces, the ITU-T[1] has done considerable work in defining some key architectural principles which need to be considered in the deployment of an SDH network. The explanation of these principles is the subject of the first paper, "Defining Network Architecture for SDH Based Networks" by A. Reid.

This introductory paper is followed by several others on the experience of various telephone companies in the deployment of SDH. Specifically, the second paper, "Planning the Introduction of SDH Systems in the Italian Transmission Network" by P. Lazzaro and F. Parente, looks at the evolution of SDH in Italy, and presents the SDH deployment strategy and target architectures at the trunk and local levels of the Italian network. The subsequent paper, "France Telecom's Deployment of SDH" by G. Bars and D. Bourdeau, presents the existing France Tele-

com network, a target SDH network structure, and a deployment strategy which will lead to a realization of the target network architecture and also take full advantage of the SDH network management abilities to realize a Telecommunications Management Network (TMN).

The fourth paper, "SDH Network Evolution in Japan" by H. Miura et al., describes the evolution of Nippon Telegraph and Telephone's network from plesiochronous digital hierarchy to an SDH network. Network architectures which lend themselves to enhanced network management are then discussed.

"Plans and Considerations for SONET Deployment" by N. Sandesara et al. describes plans for SONET deployment in the United States by the Regional Bell Operating Companies (RBOCs), and also shows how a SONET network will act as a platform for providing broadband services. A related paper, "SONET Implementation" by Y.-C. Ching and H. S. Say, then examines the status of SONET deployment, and asks if it meets the original expectations of systems developers.

The next paper, entitled "The Role of SDH/SONET-Based Networks in British Telecom" by S. Whitt et al., describes the advantages and possible disadvantages of SONET/SDH-based networks and presents a network structure, control structure, and management philosophy that could be utilized within British Telecom's networks.

The final paper of this section, "Planning and Deploying a SONET-Based Metro Network" by M. To and J. MacEachern, shows, with the aid of examples of generic network architectures which can be realized with SONET network elements, how SONET can be deployed in a metropolitan network.

The above body of experience will act as an invaluable guide to network planners who are anticipating the introduction of SDH/SONET capabilities into their networks.

---

[1] International Telecommunication Union — Telecommunications Standardization Sector (formerly CCITT).

Defining Network Architecture for SDH Based Networks[1]

ANDY REID[2]

Abstract—The international standardization of the Synchronous Digital Hierarchy has caused a significant rethink in the way the architecture of telecommunications networks is defined. A set of concepts has been developed in the ITU-TS[3] which allow all the features of telecommunications networks including those based on SDH to be described, modeled, designed, dimensioned, and managed. These fundamental concepts, while originally developed for SDH, are applicable to any telecommunications network, in particular those designed to provide broadband services.

1. INTRODUCTION

The international work on the definition of the Synchronous Digital Hierarchy (SDH) in ANSI[4], ETSI[5], and ITU-TS is based on several key principles which were developed in the standards bodies prior to the definition of the standards. Some of these principles are technological, for example, the use of optical fiber and the latest integrated circuit technologies. However, most of the principles are architectural. These were based on network operators' experience in operating the existing Plesiochronous Digital Hierarchy (PDH) and included the goal of a single world standard [1], the provision of generous management capacity for performance monitoring at every bit rate, and a strict adherence to the layering principles described below. While the original priority of the standards bodies was the generation of standards for the SDH network node interfaces and equipment specifications, ITU-TS subsequently returned to some of the key architectural principles and formalized them. This work has resulted in a number of new and very important concepts for all telecommunications and are recorded in ITU-TS Recommendations G.803 and G.831. While all the principles in these Recommendations are reasonably straightforward, G.803 especially is, of necessity, a set of formal definitions, and there is not a great deal of explanation in the text. Moreover, it can be quite difficult to picture what is really happening in a network. This means that using the architecture does require a certain amount of learning; this is essential for any network operator or manufacturer who is to successfully deploy SDH.

SDH is often contrasted to ATM. A helpful perspective is that if ATM is a revolution in switching technology, SDH is a revolution in network architecture.

The first and most essential point is that *the architecture describes network functions and **not** boxes*. A manufacturer has considerable freedom in the way the functions are packaged into boxes, and the operator has considerable freedom in where the boxes are sited. Neither impacts the functionality of the network.

2. NETWORK ARCHITECTURE

Architecture is a widely used term. Most notably, architecture is applied to buildings, although telecommunications operators have used the term for telecommunications networks. It is possible to draw out the common understanding of architecture which can be applied to both building and networks, and that is that *architecture is concerned with a set of descriptive tools and their use in the design process.*

There are two important aspects to architecture, whether it is the architecture of buildings or networks. Firstly, the descriptive tools may be used to evaluate and test different designs without constructing them. Secondly, the descriptive tools may be used to describe a design so that those who construct it and maintain the construction afterwards correctly understand the architect's intent in the design. Both of these aspects of architecture are essential to the design of any telecommunications network so that it is fully optimized for the right balance of cost and functionality.

3. NETWORK ARCHITECTURE ISSUES RAISED BY SDH

In the past, network architecture has applied predominately to the PSTN. When most operators undertook their original digitalization programs, a network architecture was developed based on digital exchanges, PDH higher order multiplexing and line systems based on a variety of technologies including optical fiber, microwave radio, satellite, pair cable, and coaxial cable. The sort of network architecture used then is shown in Fig. 1. BT, along with all network operators, has used these types of pictures to describe their network. The diagram illustrated in Fig. 1 of the companion paper, "The Role of SDH/SONET-Based Networks in British Telecom, " shows the picture BT has used in the past.

SDH presupposed the use of technology which allows the economic manufacture of digital cross-connects (DXC) which can cross-connect bit rates from 2 Mb/s up to 2.5 Gb/s. The DXC replaces multiplexers and Digital Distribution Frames (DDFs) and is very much smaller in physical size. This means the DXC, when compared to the multiplexers and DDFs, will be more reliable, require less maintenance, use less power and accommodation, and will cost significantly less. More importantly, the

---

[1]This is an updated version of a paper originally published in the *BT Engineering Journal*, July 1991.

[2]Network Strategy Department, BT, London.

[3]International Telecommunications Union — Telecommunications Standardization Section (formally CCITT).

[4]American National Standards Institute.

[5]European Telecommunications Standards Institute.

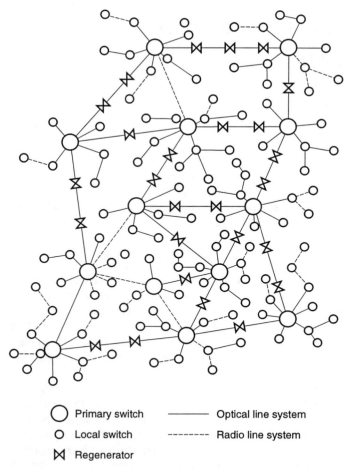

⬡	Primary switch	———	Optical line system
◯	Local switch	- - - - -	Radio line system
⋈	Regenerator		

Fig. 1. Example of traditional PSTN network architecture.

from Internet to possible future networks based on optical wavelength division multiplexing (WDM) and wavelength routing. It takes into account existing architectural frameworks including that shown in Fig. 1 and the ISO seven-layer stack. ITU-TS has also agreed that this transport network architecture is the basic framework for the development of broadband networks including B-ISDN.

4. GENERIC TRANSPORT NETWORK ARCHITECTURE

It is often the case when describing any new set of concepts that a certain amount of redefinition of terms is necessary. Terms have been used loosely in the past, and a more precise definition is now necessary. This has been the case with transport network architecture, and much of the time in ITU-TS has been taken up with the definition of these terms; Recommendation G.803 contains 39 definitions. Considerable effort has been made to avoid confusion with existing usage of these terms, and while the definitions themselves are now considered to be stable, it is possible that some of the terms may overlap with existing use of these terms. The English language has a very rich vocabulary, but even this was insufficient to avoid the need for some redefinitions! The terms used in this paper are those defined by ITU-TS.

This paper gives a brief presentation of transport network architecture and introduces the important concepts and terms. Some of the definitions may appear arbitrary, but this is due to constraints on the length of this paper rather than any lack of rigor in the transport network architecture. Further reading material can be found in the Bibliography.

The Layer Network, Access Points, and Trails

In the past, the term network has been used loosely to cover several concepts including all that shown in Fig. 1. The term network is now used in a very general way to include all the things that a network operator must put in place in order to provide telecommunications services including equipment, software systems, and human processes. The *transport network* is defined to be the collection of logical functions in the network which convey user information between distant locations. Transport network architecture, therefore, is concerned with the architectural aspects of the transport network.

A key feature of the transport network is its ability to set up communication between any *access points* at distant locations when so requested. Not all access points, however, will be of the same type, and successful communications will only take place between access points of the same type. A *layer network* is the logical entity which can connect together access points of the same type.

The type of access points, and hence the layer network, is defined by its *characteristic information*. Characteristic information has a defined format which is formed by the access point in one direction of transmission and is interpreted by the access point in the other direction of transmission. It will contain both user information and a certain amount of standard overhead which can normally be used for management functions like

DXC can be remotely managed, which means that it can facilitate rapid circuit provision without any of the existing circuit provision effort, again significantly improving quality and reducing costs. The remote management also means that the DXC can be used as part of a network restoration scheme to further improve network availability.

Most network operators are beginning to find that the PSTN is not the dominant network service it was in the past. New services based on information technology, for example, LAN interconnect, high-speed interprocessor links, video services of various types, remote CAD/CAM, etc., now account for a significant proportion of line system capacity, and this is certain to increase. SDH is designed to provide a general transport network infrastructure which can support new networks specifically for these new services like Broadband ISDN (B-ISDN), Switched Multi-Megabit Data Service (SMDS), and Frame Relay (FR), as well as supporting some of the services directly.

None of these features of SDH can be adequately described by the network architecture indicated in Fig. 1, and so a new network architectural framework was required. ITU-TS has called this *transport network architecture*. While transport network architecture was originally developed for SDH, it is generic in nature, and can be applied to any telecommunication network

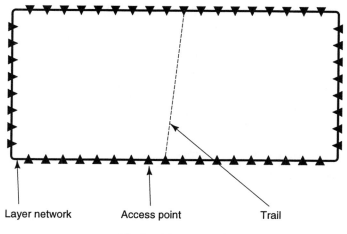

Fig. 2   A layer network.

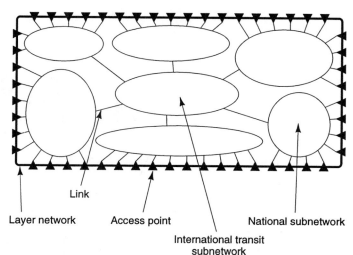

Fig. 3   Partitioning of a layer network.

performance monitoring. Examples of this overhead are frame alignment words in PDH and path overhead in SDH.

The layer network can set up communication between access points, and this specific communication is called a *trail*. These concepts are illustrated in Fig. 2.

Partitioning

In order to set up trails in an economic way, a layer network will normally consist of a number of nodes (exchanges or DXCs) which are linked together in a defined topology. Normally, the layer network will be global in extent, as all possible access points worldwide are included, which will make the overall topology of the layer network virtually impossible to define as a single structure. It is necessary to *partition* the layer network into more controllable parts.

The importance of the partitioning concept is that it allows a part of a layer network to be considered as a single entity by the rest of the layer network and the internal structure of this *subnetwork* is hidden. This is significant both for reducing the complexity of the management and control of the layer network, and because it allows individual network operators the freedom to change and optimize their subnetwork without affecting the rest of the layer network.

A first level of partitioning might be to divide the layer network into an international transit subnetwork and national subnetworks. The result is a number of subnetworks which are linked together in a defined topology as shown in Fig. 3.

The international transit subnetwork could then be further partitioned into the subnetworks provided by the international transit operators. Each national subnetwork could be further partitioned into a trunk transit subnetwork and regional subnetwork. Partitioning can continue, and the end result of partitioning will be to end up with the original nodes and links with the nodes equating to subnetworks.

The partitioning process will produce a tree structure of transit subnetworks, and so it is possible to define levels in the tree structure as shown in Fig. 4. These levels are significant for several applications of the partitioning concept; for example, capacity consolidation, trail setup, and restoration all make use

of these partitioning levels. It can also be seen that each step in the levels of partitioning results in a simple equation:

$$subnetwork = subnetworks + links + topology.$$

While the partitioning concept has not been formalized in the past, extensive use has been made of it in switched networks, especially for numbering and routing. The fields of the PSTN numbering scheme define the levels of the partitioning tree and the step-by-step routing makes full use of the hiding feature of partitioning.

So far, this description allows for the partitioning of the topology of the layer network. It is equally possible to partition the individual trails formed across the layer network. A trail is made from a *trail termination* function at each end, which generates and terminates the characteristic information, together with a *network connection* as illustrated in Fig. 5. The network connection can be partitioned into *link connections* and *subnetwork connections*. The link connection is formed across a link, while the subnetwork connection is formed across a subnetwork. The subnetwork connection can be further partitioned into subnetwork connections and link connections in the same way as the subnetwork, giving the following equation:

$$subnetwork\ connection = subnetwork\ connections$$
$$+ link\ connections.$$

Layering

The other key concept of transport network architecture is *layering*. A layer network is made up from its subnetworks and links. The links are fixed and generally cannot be partitioned. Since they are not partitioned, the most economic way of providing the links is by bundling a number of link connections together and transporting them on a trail of a higher capacity. For example, 30 64-kb/s connections can be bundled together and carried on a 2 Mb/s trail. The bundling function is called *adaptation* and normally will involve multiplexing. This is illustrated in Fig. 6.

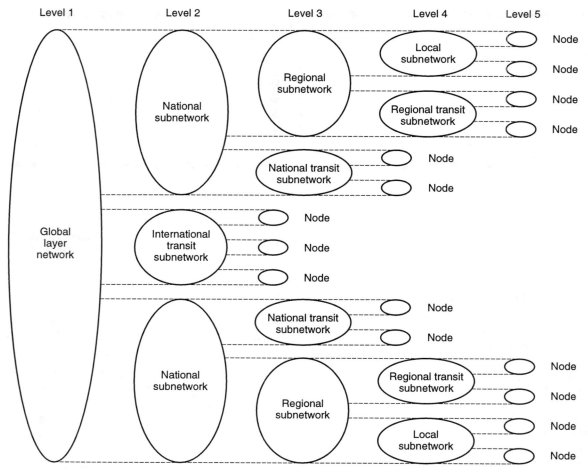

Fig. 4   Partitioning of a layer network.

The concept of layering allows a view of a one-layer network serving another, or of a one-layer network being a client of another — a *client/server association* is formed between the two one-layer networks. This association provides a formal description of the way layer networks interact and allows layer networks to be designed for each other, but independent of each other, as illustrated in Fig. 7. Fig. 1 might suggest that the 140 Mb/s layer network has only a point-to-point topology; however, a correct use of the client/server association allows an independent topology to be developed.

While the topology of any layer network can be independent, the architectural components are the same. This means that the same management and control features can be used for every layer network, giving significant simplifications.

Note that it is possible to equate the layer network to the OSI seven-layer stack. *The OSI use of the term layer, however, is significantly different from that used in layer network.* The transport of client layer links is the application of the layer network — OSI layer 7. The trail is the OSI transport layer — OSI layer 4. The network connection is the OSI network layer — OSI layer 3. The link connection is the OSI link layer — OSI layer 2, and is an application of another layer network.

5. APPLICATIONS OF TRANSPORT NETWORK ARCHITECTURE

Designing and Sizing Network Topology

It may be noted that the existing notion of network architecture has centered on the PSTN, and hence the 64 kb/s layer network. Traditionally, it has been the only layer network which has had a well-developed topology, and the transport layers have been viewed primarily as point-to-point links between exchanges.

This has meant that the existing PDH transport network has been built up on a route-by-route basis. When digital capacity was required between two centers, especially for exchange modernization, line systems of the appropriate size have been planned and installed. Sometimes, there has been capacity available on some part of the route from existing line systems, and so line systems have been required only for parts of the route with a through connection at 8 or 34 Mb/s (taking the example of the ETSI PDH hierarchy). This has meant that the layer networks at 2, 8, 34, and 140 Mb/s each lack a good structure as they have not been designed as individual networks. Applying the transport network architecture, it is possible to interpret each bit rate of the PDH as its own layer network, with its own topology which

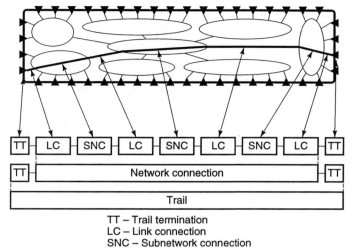

TT – Trail termination
LC – Link connection
SNC – Subnetwork connection

Fig. 5   Partitioning of a trail.

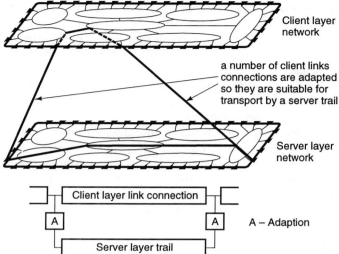

Client layer network

a number of client links connections are adapted so they are suitable for transport by a server trail

Server layer network

A – Adaption

Fig. 6   Layering and the client/server association.

can be sized and optimized separately. This equally applies to 1.5, 6, and 45 Mb/s of the North American PDH hierarchy and to 1.5, 6, 32, and 97 Mb/s of the Japanese PDH hierarchy.

With SDH, this optimization of each layer network is very important as each layer network is used by a number of client layer networks. Each client layer network will have differing requirements of the server layer network, and so the server layer network should be optimized to suit all requirements. For example, the VC-12 [3] layer network is used by both the PSTN (64 kb/s) layer network and a 2 Mb/s Leased Line layer network. The differing requirements from each can be added together to form a general VC-12 transport requirement which can then be used as the basis for developing an optimized topology for the VC-12 layer network. The VC-4 layer network will have VC-12, VC-2, VC-3, broadcast TV, and broadband ISDN as client layer networks. Again, the requirements from all of these layer networks can be added together and an optimized topology developed.

One essential element allowed by the application of layering to SDH is that each layer network can have an independent topology. The topology of the VC-12 network can be independent of both the PSTN and the VC-4 network as illustrated in Fig. 7.

Developing and Defining Routing Algorithms

As layer networks are normally global, and the topology can be complex, the development of algorithms which can work out the routing for a trail is an important aspect of the overall design of the layer network. Several factors need to be taken into account in the development of routing algorithms:

- efficient utilization of the network topology;
- network operators will not necessarily allow another network operator to control routing through its network;
- network failure and network congestion may make certain preferred routings impossible;
- different network operators may wish to use different routing algorithms.

The partitioning concept allows the routing algorithm to be

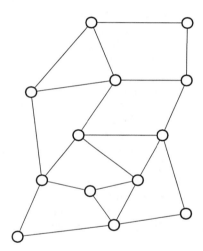

Possible VC-4 layer network topology

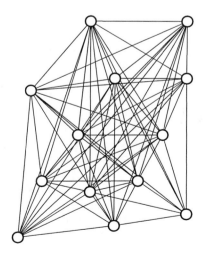

Possible VC-12 layer network topology

Fig. 7   Independence of topology in different layers.

79

broken down into a number of stages. Routing across a subnetwork can be the sole responsibility of that subnetwork. As one network operator's part of the layer network can be regarded as a subnetwork, the network operator can choose a routing algorithm that suits its network topology. However, routing between different network operators' subnetworks must be agreed on a multilateral basis. This applies particularly to the international transit subnetwork illustrated in Fig. 3.

The ability to define this international transit subnetwork as a collection of subnetworks, each belonging to an international transit network operator linked together in a defined topology, allows suitable algorithms to be defined. Currently, international signaling systems, which are based on the partitioning concept, define and use such algorithms for the PSTN and ISDN. European network operators have agreed that international routing algorithms for SDH payloads should be developed for Europe, and these may well be extended to all countries.

Deriving Network Performance Parameters

Some network performance parameters depend on the topology of the network. These include availability and blocking probabilities. A significant factor in the definition of these parameters is the way the topological aspects of the parameter are taken into account and how the topology is divided across an interoperator boundary.

This is illustrated in Fig. 5 where a trail is formed from connections across several network operators' domains. SDH control will make possible the restoration of the trail as a whole rather than by the restoration of individual connections, and so the availability of the trail cannot be simply related to the availability of each connection. An application of the partitioning concept allows a network operator's domain to be adequately defined, thus allowing a definition of accessibility and availability to be related to the complete boundaries between network operators' domains including all the points on the boundary that might be used in a restoration.

Simulating Network Availability

Several factors have made network simulation for the purposes of assessing availability performance a difficult process. Firstly, it has been very difficult to relate the effects of restoration at higher bit rates with restoration at lower bit rates and how they may interact. For example, BT currently uses a 140 Mb/s restoration scheme called ASDSPN[4] which will restore capacity following line system failures. The PSTN exchanges also take action when a failure is detected on the 2 Mb/s switch port. Ultimately, the PSTN customer can also take action to restore the call by clearing down and dialing again. The interaction between these two restoration mechanisms has been difficult to model.

The application of both the layering and partitioning concepts allows a complete model of the network to be constructed in which the interactions are properly formalized and are variables of the model. Simulations of the network can then be analyzed and a restoration strategy developed which gives the best availability at the lowest network cost. This is illustrated in Fig. 8.

This work is especially relevant to SDH as most of the layer networks will be capable of offering restoration: $1 + 1$ or 1 for N protection at the multiplex section layer, network restoration at the VC-4, VC-3, VC-2, VC-12, and VC-11 layers, as well as the example given above where the PSTN user can clear down and redial.

Deriving the Equipment Functional Reference Blocks

One of the first applications of the transport network architecture was in the development of the ITU-TS SDH equipment Recommendations, G.781, G.782, G.783, and G.784 [5].

Using some of the techniques already identified above which use the transport network architecture, a network operator can develop a transport network functional design. This process will automatically identify all the functional components required,

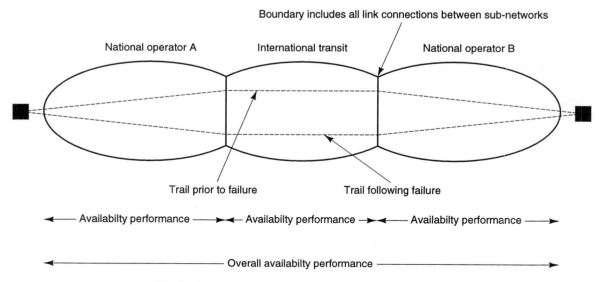

Fig. 8   The use of partitioning to define availability performance.

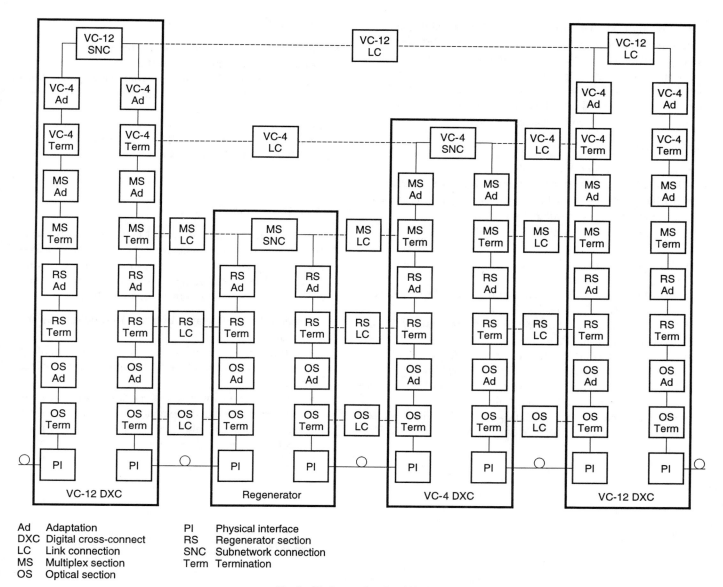

Ad	Adaptation	PI	Physical interface
DXC	Digital cross-connect	RS	Regenerator section
LC	Link connection	SNC	Subnetwork connection
MS	Multiplex section	Term	Termination
OS	Optical section		

Fig. 9   Equipment functional blocks.

and therefore the functionality required in the equipments. The emphasis on the functionality is important with SDH equipments as the technology makes possible high levels of functional integration in one equipment. ITU-TS has recognized this and developed the equipment recommendations such that the functionality is specified as well as the external interfaces. This allows manufacturers to develop equipment with high levels of functional integration without losing a precise specification as shown in Fig. 9. Network operators can therefore build a transport network with many features which is tailored precisely to the network operator's requirements, but without large numbers of equipments. The small number of equipments means that the network will be inherently more reliable with good availability and low maintenance cost.

Deriving the Management Information Model

Remote management of equipment has long been a goal of network operators, although this has necessitated a management

interface on equipments. In order to reduce the cost of this interface, both network operators and manufacturers have sought to agree on an international standard for this interface. One of the most difficult aspects of developing this interface has been the answer to the general question, "What is the generic functionality that requires management?" The transport network architecture provides the ideal answer as it describes not just the functional components requiring management, but it also gives all the relationships between the functional components; an essential part of management. Within ITU-TS, the transport network architecture has been developed in close cooperation with the management interface.

6. A VISION OF THE TRANSPORT NETWORK IN 2000

The development of layer networks will play an essential role in the evolution of the transport network over the next ten years.

With growing customer demands for higher and higher bandwidths and increased competition at all levels of telecommunications, the ability to provide the right amount of capacity for a customer at the lowest possible network cost is vital to any network operator. It seems likely that this will necessitate the development of a number of layer networks, each of which is optimized for the transport of particular clients:

Circuit Layers —

- 64 kb/s layer network for ISDN services and some 64 kb/s layer network management and control links (signaling links);
- B-ISDN/Virtual Circuit layer network for the variable bit rate and integrated broadband services.

Path Layers —

- VC-11 layer network for transport of 64 kb/s links and existing 1.5 Mb/s leased lines;
- VC-12 layer network for transport of 64 kb/s links and existing 2 Mb/s leased lines;
- VC-2 layer network for the transport of broadcast TV links, and B-ISDN/Virtual Path links;
- VC-3 layer network for the transport of LAN-LAN links and existing North American 45 Mb/s leased lines;
- B-ISDN/Virtual Path layer network for B-ISDN/Virtual Channel links and management and control links;
- VC-4 layer network for the transport of VC-12 links, VC-2 links, VC-3 links, B-ISDN/Virtual Path links, and HDTV links.

Transmission Media Layers —

- Multiplex Section layer network for the transport of VC-4 links and management and control links;
- Regenerator Section layer network for the transport of Multiplex Section links and management and control links (particularly relating to regenerators);
- Possible Optical Section layer network based on wavelength division multiplexing (WDM) and wavelength routing to carry Regenerator Section links.

Management and Control Layers —

- A single management and control network conformant to international signaling standards (ITU-TS Signaling System No. 7), the telecommunications management network (TMN), and ISO management standards for all layer networks with three main groups of client applications:

  - basic subnetwork connection control which sets up and restores trails;
  - customer and service management, for example, billing, virtual private networks (VPNs), mobility, guaranteed diversity, etc.;
  - planning and maintenance support, for example, event and performance management with repair scheduling, inventory control, planning tools to predict requirements for new equipment in the network.

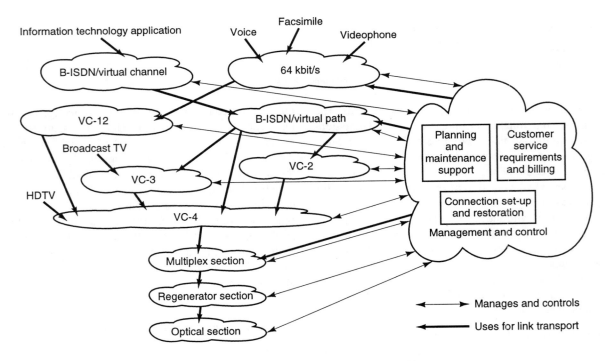

Fig. 10   A view of layer networks in the year 2000.

These interrelated layer networks are illustrated in Fig. 10. It is seen that the management and control network is an integral part of the transport network architecture and controls all layer networks. This is possible because all layer networks are based on the same architectural principles.

This architecture gives all the essential properties that BT, along with other network operators, are looking for as the world of telecommunications moves from its historic focus on telephony and telephony-derived services. The key properties BT has identified are flexibility, quality of service, monitoring, reduced costs, future proofing, and support of broadband and are described in the companion paper, "The Role of SDH/SONET-Based Networks in British Telecom." All of these depend principally on the architecture and not the technology, and can only be fully realized by correct adherence to the architecture. The vision painted above meets all these requirements, and whichever services take off in the future, such a set of networks is able to quickly adapt to suit rapidly changing needs.

7. CONCLUDING REMARKS

The transport network architecture described in this article is generic to telecommunications, and was first derived to ade-quately describe some of the features of SDH. It will, however, form a complete descriptive tool for broadband networks, and will be essential to network operators worldwide if the ever-increasing complexity of network architecture is to be managed both by software systems and a finite human resource.

References

[1] K. R. Harrison, "The new CCITT synchronous digital hierarchy: Introduction and overview," *British Telecommun. Eng. J.*, vol. 10, part 2, July 1991.
[2] ITU-TS Recommendation G.803, 1993.
[3] T. C. Wright, "SDH multiplexing concepts and methods," *British Telecommun. Eng. J.*, vol. 10, part 2, July 1991.
[4] M. J. Schickner, "Service protection in the trunk network, Part 3 — Automatically-switched digital service protection network," *British Telecommun. Eng. J.*, vol. 7, part 2, July 1988.
[5] W. R. Balcer, "Equipments for SDH networks," *British Telecommun. Eng. J.*, vol. 10, part 2, July 1991.

Bibliography

Architectures of Transport Networks Based on the Synchronous Digital Hierarchy, ITU-TS Recommendation G.803, ITU Geneva, 1993.

Performance and Management Capabilities of Transport Networks Based on the SDH, ITU-TS Recommendation G.831, ITU Geneva, 1993.

M. Sexton and A. Reid, *Transmission Networking: SONET and the Synchronous Digital Hierarchy*. Boston and London: Artech House, ISBN 0-89006-551-9.

Planning the Introduction of SDH Systems in the Italian Transmission Network

P. LAZZARO AND F. PARENTE

TELECOM ITALIA — SIP — DIREZIONE GENERALE, ROMA, ITALY

1. INTRODUCTION

The technology progress and the deployment effort produced around optical transmission put recently into evidence the intrinsic limitations of the existing transmission network, based on the Plesiochronous Digital Hierarchy of multiplexing (PDH). That raised the interest of many Telecom Operators towards an accelerated introduction of new transmission and multiplexing systems, based on the Synchronous Digital Hierarchy (SDH).

The SDH standard, in fact, offers several important advantages, and among them:

- easy access to tributary signals and simplified multiplex–demultiplex operations;
- facilitated drop-insert and cross-connect functions;
- large amount of overhead bytes easily accessible for powerful operation and management of equipment and network;
- interworking between the American and the European hierarchies and extended compatibility of equipment of different manufacturers.

Different approaches to the above change have been proposed, with the common target of an advanced transmission network architecture, with integrated transport capabilities for all TLC services, and with more effective management features through centralized support systems.

This paper, starting from the evolution of the Italian digital network, presents concepts and guidelines on which are based the current deployment strategy and the target structures of SDH network, both at trunk and local loop level; moreover, the relevant equipment functionalities and the management issues are addressed. Finally, the network dimensioning and planning methodology is outlined and an application example is shown.

2. EVOLUTION OF THE DIGITAL NETWORK

To meet the growing and diversified needs of the market with the highest standard of service quality and the most advanced network facilities, the Italian PNO is at present engaged in a complex reorganization and modernization process at both the technological and operational levels. The main strategic guidelines are:

- further development of the basic telephone service, in terms of penetration and traffic growth;
- improvement of the quality of offered services;
- deployment of new technologies in order to broaden the range of telecom services and to improve the network management efficiency.

To achieve these objectives, a key role is played by the network digitalization. The evolution of the Italian network in the past decade has been actually characterized by the massive introduction of digital technologies in both transmission and switching facilities, so that a platform is being made up allowing in the early 1990s the provision of a vast set of services, including Intelligent Network and ISDN services, advanced mobile services with GSM European standard, and high-speed data applications by MANs.

The first introduction of digital switches began in 1980 and 1983 at the transit and local level, respectively. The industrial reconversion cycle lasted all through the 1980s, and since 1989, only digital switches are being purchased. At the end of 1992, over 13 million subscriber digital lines had been installed, i.e., 48.4% of the total lines; this percentage will exceed 75% by 1996, and the full digitalization of local switches is expected around the year 2000. Considerably higher is the digitalization rate of the transit switching nodes, and a full digital transit network has been available since 1994.

The digitalization of the trunk transmission network, started in the early 1970s, has at present reached a percentage of about 85%. The evolution of transmission facilities is dominated in all the network levels by the massive deployment of fiber optics; since 1986, no copper cable has been installed, except in the loop plant. At present, about 1,700,000 fiber·km are laid down at local and regional network levels, and about 2 million fiber·km will be reached by 1996, including the rapidly increasing deployment of fiber optics in the loop plant (Fig. 1). In addition, it is worth noting that over 300 000 fiber·km have been deployed in the long-distance trunk network. This large optical infrastructure is also complemented by existing copper and radio facilities, mostly equipped with digital transmission systems.

In parallel with the digitalization of switching and transmission facilities, in order to fully exploit the economical and operation benefits of digital technologies, a process of modification of the network structure is being carried out. Based on optimization studies accounting for cost profiles, modularities, and performances of digital systems, a new structure has been planned,

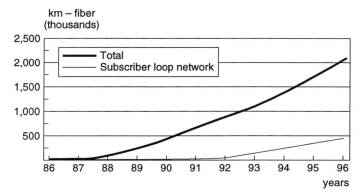

Fig. 1   Deployment of optical fiber in the local and regional networks.

and it is now in an advanced implementation stage. With respect to the analog network consisting of about 11 000 switch entities on five hierarchical levels, the digital network will exhibit a much simpler and flexible two-level structure based on about 600 local group stages (SGU) and 60 transit group stages (SGT) as in Fig. 2.

This network structure is based on a subdivision of the territory in "switching areas," each under the control of a single SGU with an average capacity ranging from 50 000 to 60 000 subscribers. Within the switching area, the subscribers are connected to line stages either remote or collocated to the SGU, and also by means of Digital Loop Carriers (DLC) having a capacity from 30 to 400 lines. These DLCs also can be installed in outdoor cabinets, and they can be effectively applied both into the loop plant and for replacing existing small electromechanical exchanges.

In the upper level of the digital network, based on the SGT transit switches, a dynamic nonhierarchical routing scheme will be adopted in order to assure maximum flexibility and grade of service even under unforeseeable peak traffic patterns or severe failure conditions. To fully achieve this ambitious goal, further experience and developments regarding real-time surveillance and control of the network will be essential. With this aim, a National Management Center located in Rome has been in operation since 1988.

Besides the traffic management features, the security and the protection of the upper layer transmission network are being significantly increased with the introduction of $1 + 1$ protection for the high-speed line systems, together with a distributed redundancy provided by a standby network through 140/155 Mb/s digital cross-connects, operated in quasi-real-time under centralized control.

Regarding the leased line services, since 1987, a digital transmission infrastructure (called the CDN network) is available, providing dedicated circuits in the speed range 2.4–64 kb/s. A remarkable evolution step has been the introduction of specialized digital cross-connect systems (DXC 1/0) to implement a "flexible network" for highest quality dedicated services up to 2 Mb/s. In particular, through the centralized management of the DXCs, improved availability and quality levels have been attained; moreover, new time-based services could be introduced

(on a preprogrammed or reservation basis), as well as "customer control" and "private virtual network" services. That allows a strategic positioning of the public infrastructures to comply with the needs of the most sophisticated business customers, and to cope with the proliferation and the possible competition of private networks.

3. SDH Network Architecture

The envisaged architecture of the SDH transmission network is basically founded on the economic and efficient utilization of ADM (Add/Drop Multiplexing) and DXC (Digital Cross-Connect) functions, particularly focusing on two main objectives:

- the implementation of a self-healing transmission network at the local level by the systematic deployment of ADM-based ring structures, at a higher level (where meshed structures are optimal) by rerouting 155 Mb/s flows through DXC 4/4 on a standby network; on the contrary, a poor level of protection is economically achievable in the PDH network, especially at the local level, as expensive redundancies and complex operations are required;
- the enhancement of the network flexibility by spreading on a larger number of nodes the transmission transit capabilities of ADMs and DXCs down to the 2 Mb/s level; on the contrary, the structure of PDH equipment now leads to economically allocating the transit functions at the highest multiplexing levels and in a limited set of nodes.

Looking at the transmission nodes, two main classes can be identified according to the relevant functionalities:

- end nodes, only with drop/insert and multi/demultiplex functions;
- hub nodes, providing cross-connection functions and service flexibility features.

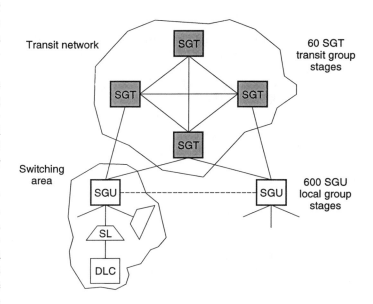

Fig. 2   Structure of the digital switched network.

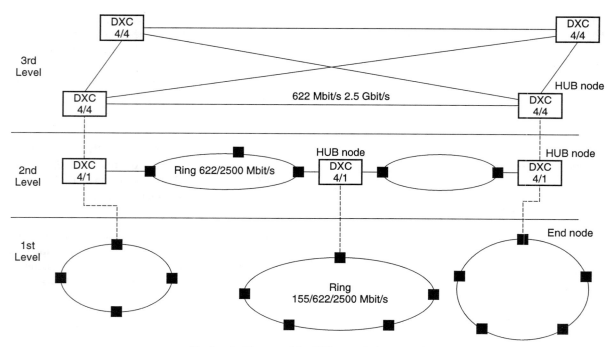

Fig. 3   Architecture of the SDH transmission network.

According to that, the transmission network architecture (see Fig. 3) basically consists of two layers: the lower layer linking end nodes to the relevant hub, and the higher one among the hub nodes. With regard to the physical implementation (and especially for large transmission networks) the hub functions can be properly split according to the hierarchical transit levels (2 and 34/45 Mb/s on DXC 4/3/1 and 155 Mb/s on DXC 4/4); thus, a three-level architecture generally results. In the lower level, ring structures are extensively used among end nodes, while the upper level is implemented by a quite meshed structure; the intermediate level can use both rings and direct line systems protected by 1 + 1 redundancy.

Referring to the structure of the digital trunk network, the lower level of this SDH architecture will include the urban junction and the regional trunk network, where DLCs and switching line stages are connected to the relevant SGUs and these latter are connected to the SGTs of the transit network. In addition to the efficient utilization of high-speed optical systems, the SDH rings allow there a fully automatic network protection, which on the contrary is generally expensive and difficult to operate with the existing PDH facilities.

The upper level of this architecture will roughly correspond to the long-distance SGT-SGT transmission network. The choice of a meshed SDH network offers the advantage of a smooth transition from the existing PDH structure, while an integrated protection of SDH+PDH facilities can be economically implemented through DXC 4/4 systems in the main transmission nodes.

Beside the trunk network, the SDH technology will gradually penetrate into the subscriber loop plant, in parallel with the copper-to-fiber modernization process. A fiber access architecture has been defined with the main strategic target to provide a multiservice and multivendor access network, and with future-proof capability to transport newer and wideband services. Referring to the present POTS/ISDN and 64 kb/s-based services, three reference configurations have been engineered, based on the transparent multiplexing and the extensive use of optical DLCs and SDH systems:

- "Fiber to the Office": It is the currently adopted solution tailored for large business customers; in the customer premises, a flexible multiplexer (called SAF) provides access for voice and data services, and it is directly linked to the central office via high-speed optical link with automatic switch protection and route diversity.
- "Fiber to the Curb": Multiservice high-capacity (up to 480 lines) DLCs (called H-MUX) are already available for both indoor and outdoor installation, to be linked to the digital exchange on a feeder optical cable and still exploiting the existing copper pairs of the terminal distribution plant; up to five DLCs can be connected to the exchange by a single 155 Mb/s SDH ring.
- "Fiber to the Premises": Compact small-capacity (30 lines) DLCs (called C-MUX) can be installed inside the customer buildings and star-connected via 2 Mb/s on fiber (or even copper) to hub SDH multiplexers, in turn connected to the central office on high-speed optical rings.

Figure 4 depicts these fiber access configurations, and it shows how they can easily coexist together; the deployment of fiber cables and equipment can be then properly scheduled according to specific territory/market needs and a graceful evolution from FTTC to FTTP can be planned. It is also worth noting that this access architecture will not produce negligible effects on the local switching plants, where the subscriber line interfaces will be gradually replaced by 2 Mb/s (or 155 Mb/s SDH) interfaces.

4. REQUIREMENTS OF SDH SYSTEMS

Starting from the above SDH architecture, the following equipment classes have been identified for the trunk transmission network:

- ADM with the ability to operate also as a terminal multiplexer, with tributary interfaces at mainly 2 Mb/s and aggregate interface at 155 or 622 Mb/s;
- DXC 4/3/1, with interfaces at 155 and 2 Mb/s, cross-connection capability at 2 and 34/45 Mb/s, and maximum size around 128 STM-1 ports;
- DXC 4/4, with interfaces and cross-connect function at both 155 and 140 Mb/s and maximum size around 256 ports;
- Line systems at 622 Mb/s and 2.5 Gb/s with integrated multiplexing functions and tributary interfaces at 155 and 140 Mb/s.

Considering the network requirements outlined in the previous paragraph, a number of equipment characteristics have been defined among, or in addition to, those specified in the relevant ITU-T or ETSI documents. Regarding, in particular, the STM-16 systems, the first planning results have shown a foreseen application range much wider than originally expected.

In fact, preliminary specifications for 2.5 Gb/s systems were essentially focused on the basic point-to-point terminal configuration, with optional section protection, suitable for long-distance high-capacity transmission. A first feedback to the above approach came from the regional planners, asking for the additional feature of a drop/insert configuration; moreover, in designing metropolitan structures, it turned out that STM-16

rings can be effectively used in many Italian cities for the deployment of high-speed rings. This result brought to the specification of the 2.5 Gb/s line system as an add/drop multiplexer, with tributary units at 622 Mb/s or 155 Mb/s. The same version of the 2.5 Gb/s system therefore will be used in terminal configuration for long distance trunk network (with 1+1 section protection) and in ring configuration for metropolitan and regional networks. First applications of these systems for the terminal configuration were started in early 1993.

In addition to the optical transmission systems, it also must be considered that the introduction of the SDH in the transmission network calls for the availability of radio-relay systems able to transmit the bit rates related to the new multiplexing levels, not only in the trunk network, but also in the subscriber loop plant.

Two basic applications are foreseen for STM-1 radio relays in the SDH network: i) to close up optical rings in specific environmental conditions; ii) as a back-up to optical fiber systems (multimedia protection). In all cases, an essential feature is the operating integration of the radio system within the SDH network, so that radio relays can be used in conjunction with other SDH equipment such as ADM or DLC (e.g., for outdoor applications), wherever economically justified. In fact, the SDH radio system will provide all the functions required for a fully integrated management of the SDH transmission network, as specified by ITU-T and ITU-R.

Finally, as mentioned in the previous section, the SDH technology will play a significant role in the modernization of the subscriber loop and the access network. In particular, a new high-capacity DLC has been developed, incorporating an SDH ADM multiplexing 63 channels at 2 Mb/s into a single 155 Mb/s opti-

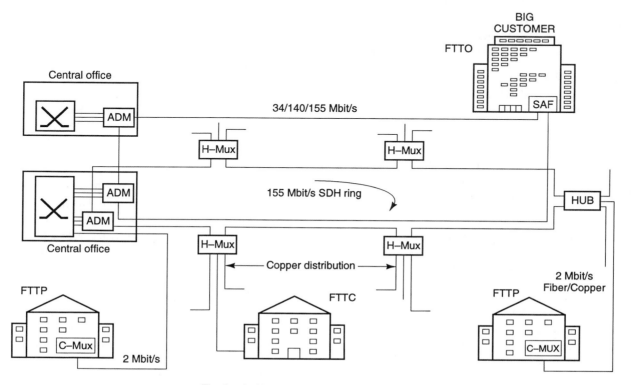

Fig. 4 Architecture of the fiber access network.

cal signal; two optical interfaces are provided allowing enhanced service availability through different protection schemes. This equipment can be housed in an outdoor cabinet, powered by ac mains and integrated batteries for 8-hour standby operation.

On the access side, up to 480 subscriber circuits can be terminated and mapped in up to 16 VC_{12} channels; a fully nonblocking duplicated interconnection mechanism provides all the customer interfaces with access to all the available time slots. Several customer interfaces have been developed for different switched and nonswitched services; at the exchange site, the 2 Mb/s channels are extracted from the STM-1 multiplex: the switched traffic is passed to the digital exchange, while the narrowband and wideband dedicated service are brought to the DXC 1/0 and 4/1 systems, respectively.

A massive deployment of these DLCs in the loop plant is foreseen in STM-1 ring configuration, allowing efficient utilization of optical cables and full protection even against a complete cable cut. Moreover, under the control of the centralized management system, the network capacity can be electronically reconfigurated to rehome the traffic, either in case of failure conditions or to better match the actual service demand.

5. Management Issues

The SDH management, as foreseen in Italy, will be implemented step by step in terms of covering area and performances; the interface between the equipment and the management center will be based on the ETSI standard as regards the OSI stack and the information model.

As a first step (second half of 1995), the SDH Management System (SDS) will consider simple network islands, as

- DLC ring in the access network
- ADM-1 ring in the local network
- ADM-4 ring in the metropolitan and regional networks

while STM-4 and STM-16 line systems will be operated as PDH equipment through the existing support system.

The main performances of this first release of SDS will include:

- provisioning management, i.e., network element installation and configuration, network configuration, date management
- alarm management, including alarm diplay, alarm file, alarm filtering, alarm aknowledge, and clearing
- software downloading

The second step of the SDS (first half of 1996) will allow the management of interconnected SDH rings and DXC 4/1 systems; quality measurements and test features also will be available.

The SDS hardware architecture will be subdivided in two levels: Element Manager (EM) at the regional level and Network Manager (NM) at the national level. The EMs are interconnected to the equipment and to the NM; they collect all the alarms, quality, and configuration data, and they send the commands for provisioning of the equipment. The EM database contains all information of the SDH managed equipment and subnet-works, while the NM database contains network configuration and services management information; a real time alignment of common data is needed.

The results of the present field trials on the first SDH systems will soon provide a better insight, especially from the operation and management perspective. It is, in fact, to be stressed that the deployment rate of SDH systems will be strongly affected by the availability of the management functionalities (alarms surveillance, performance monitoring, configuration management), in order to fully exploit in a multivendor environment all the expected operation benefits in terms of quality of service, flexibility, and OA&M cost reductions.

6. Deployment Strategy

The introduction of SDH systems in the Italian transmission network is currently seen as tightly correlated to the following main goals:

- to implement a generalized protection of the transmission network for all kinds of services and down to the local and peripheral level, through the extensive use of ADM-based ring structures;
- to maximize the operation cost reduction and the quality improvements achievable by the advanced management features of SDH systems;
- to promote the early availability of a managed transport infrastructure suitable for cross-connected wideband services, as a first viable step towards the B-ISDN, with particular attention to large business customers.

In accordance with these objectives, the highest priority will be given to the deployment of SDH systems within urban and metropolitan areas, both in the junction network and in the loop plant. The first step, already undertaken, mainly consists of the implementation of first SDH rings among the main local switches, together with the deployment of optical DLC linked on 155 Mb/s SDH rings. Note that an extensive use of these DLCs is being planned, both for modernizing the copper loop plant and for replacing existing small-size analog switches (thus sensibly reducing the total number of the central offices of the switched network).

In addition, deployment is underway of DXC 4/4 in the main nodes of the long-distance network, aimed at providing an enhanced protection even of the existing PDH facilities by rerouting 140 Mb/s flows. In this network level, 2.5 Gb/s SDH line systems will be also used, in particular where the exhaustion must be faced of existing optical cables operated by 565 Mb/s systems..

As a second step, in 1995–1996, extensive deployment will start of DXC 4/1 in all the main transmission nodes of metropolitan and regional networks: that allows the efficient interconnection of local rings among them and with the long-distance network, creating at the national level a first SDH network layer. At this stage, the SDH management system also will be fully available, thus allowing the additional utilization of this managed SDH infrastructure for wideband services from 2 to 155 Mb/s, as an extension of the existing flexible network for dedicated services.

Finally, around the end of this decade, the third evolution step will see the massive extension of the SDH technology in the whole transmission network by the gradual retirement of existing PDH systems. The replacement rate will be strongly affected by the actual evaluation of economic and operational benefits attainable through the full exploitation of the SDH systems capabilities and the relevent management features.

7. NETWORK PLANNING APPLICATIONS

According to the network architecture and the objectives previously described, several network planning applications are currently being carried out in order to derive the design rules and to define the territory of the SDH target network structures. These activities are very essential to assure a cost-effective and orderly evolution from the existing PDH infrastructure.

For this purpose, a computer-aided planning tool has been developed, specifically tailored to perform the sizing and the technical and economical evaluation of different structural alternatives of a given network. Basic inputs are the matrix of the node-to-node circuit (2 Mb/s) demand and the topology layout; to generate each network alternative, the configuration of the topology rings and the hub nodes locations are given as input, and they can be easily changed interactively.

This computer tool consists of three separate modules. The first module (RING) carries out the dimensioning of ring structures in terms of MUX and ADM equipment in each node and 2/155 Mb/s ports on DXC 4/1 of the Hub nodes. The second module (MESH) sizes the line systems of the upper level meshed network, including the standby network, and provides the needed 155 Mb/s ports on DXC 4/4 and 4/1. The third module (ROUT) performs the physical routing of rings and line systems on the given topology layout by a shortest path algorithm.

This computer procedure is provided with several interactive features and graphic aids to make easier and friendly the input/output data handling and the setting of different network alternatives to be evaluated. It has been written in C language and it runs on a standard PC workstation. The regional planners are currently using it to optimize the medium–long term structure of the SDH transmission networks at local junction and regional trunk level, accounting for the existing infrastructures and the forecast transport capacity.

As an application example to a metropolitan network, the main results are here summarized relevant to the junction network of Rome, consisting of 42 main local switches (SGU) and 3 toll transit switches (SGT); at a 10-year horizon, about 2 million subscriber lines are forecast with a total capacity of 14 000 2 Mb/s flows in the interoffice network including data services and dedicated lines.

Starting from the circuit matrix and the existing fiber cable layout, a first network level connecting all the SGUs has been optimized, based on 8 hub nodes (located in as many SGUs) and 8 topology rings, as depicted in Fig. 5; this part of the SDH transmission structure requires a total of 29 line systems in ring

configuration, 8 of which are at 622 Mb/s and 21 at 2.5 Gb/s. It is worth noting in Fig. 5 that each pair of rings is homed to two hub nodes: this configuration provides an enhanced network protection, even against a catastrophic failure of the hub node, as in this extreme condition, at least 50% of the traffic external to the ring survives (the internal traffic is fully protected).

Looking at the upper level network, which has to connect the 8 hubs among them and with the 3 hubs (SGTs) of the long-distance network, three different alternatives have been considered:

A) a meshed network among the resulting 11 hub nodes;
B) a single ring connecting the 11 hubs;
C) a second network level consisting of 3 rings (each homed to two hubs) and a third level simply connecting the 3 super-hubs collocated with the SGTs (this solution is shown in Fig. 5).

The main results of the technical and economical comparison of these three alternatives are summarized in the following table, in terms of number of ADM equipment, ports on DXCs, occupied fiber·km and relative total costs.

	ADM-4 #	ADM-16 #	DXC 4/1 ports #	DXC 4/4 ports #	Fiber km	Relative Cost
A)	50	234	723	1628	3830	100
B)	50	223	680	—	5300	82
C)	50	201	935	338	5620	88

Alternative C) has resulted in the most attractive solution, when jointly considering cost, implementation, and management aspects. Note that the extra cost with respect to alternative B) is mainly due to the presence in the three SGT nodes of DXC 4/4, whose cost should be actually shared with the long distance transmission network (which functionally belongs to the upper level mesh of Fig. 5); so, this apparent cost penalty of solution C) is offset when looking at the cost of the whole network.

8. CONCLUSIONS

By introducing the SDH technology, a very reliable, efficient, and flexible transmission network can be implemented, with integrated transport capabilities for all telecom services (including future broadband services) and with powerful management features.

Starting from the main characteristics of the Italian digital network, this paper has dealt with the expected impact of the new SDH systems, in both the loop and the trunk transmission network. Emphasis is placed on both the intrinsic advantages of enhanced flexibility and protection performances, and the expected benefits in terms of service quality and reduced operation cost.

The adopted SDH network architecture and the general requirements of the SDH systems have been outlined; the current development and deployment guidelines also have been presented. Moreover, the adopted network planning methodology

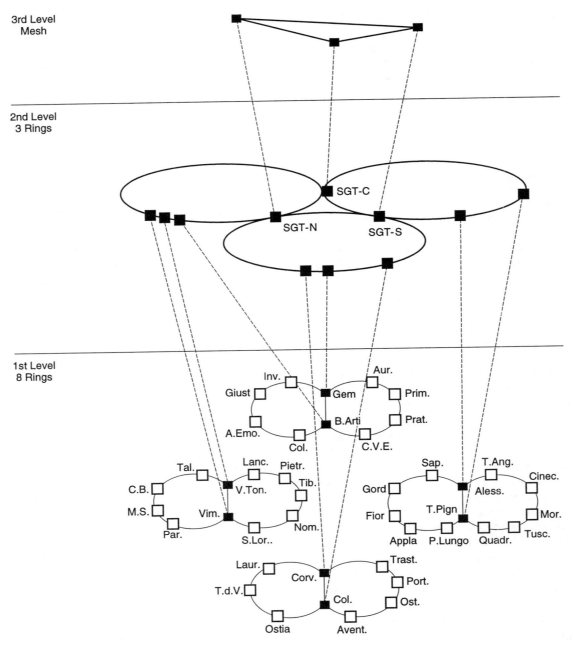

Fig. 5   SDH structure of the urban network of Rome (case C).

has been addressed as currently applied to metropolitan and regional networks, and an example result for the urban network of Rome has been shown.

References

[1] U. Mazzei, A. Palamidessi, P. Passeri, and F. Balena, "Evolution of the Italian Telecommunication Network towards SDH," *IEEE Commun. Mag.*, Aug. 1990.

[2] F. Parente, "SDH architecture in the Italian Access Network," 6th World Telecommun. Forum, Geneva, 10–15 Oct. 1991.

[3] A. Gambaro and R. Ledonne, "Planning SDH transmission networks: Computer tools and application results," 5th Int. Network Planning Symp., Kobe, Japan, May 1992.

[4] U. Mazzei and A. Palamidessi, "Planning of high-speed SDH systems in the Italian Trunk Network," GLOBECOM'92, Orlando, FL, 6–9 Dec. 1992.

[5] U. Mazzei and R. Pompili, "From copper to fiber in the Italian Loop Network: Architectures, economics, deployment strategies," ISSLS'93, Vancouver, Canada, 27 Sept.–1 Oct. 1993.

France Telecom's Deployment of SDH

G. BARS AND D. BOURDEAU

1. INTRODUCTION

The existing France Telecom network uses a synchronous multiplexing technique for rates up to the primary level of 2 Mb/s and a plesiochronous multiplexing technique based on justification for the upper levels. This scheme allows for flexible management of the 64 Kb/s channels in the switching equipment for telephony and ISDN services, and in cross-connect equipment used in data networks. Just a few years ago, the step-by-step multiplexing and demultiplexing structure of the Plesiochronous Digital Hierarchy (PDH) for the higher levels was not considered to restrain the evolution of telecommunication networks and services. However, these previously considered insignificant inherent limitations of the PDH are inhibiting the evolution of competitive communication networks and, consequently, the greatest revenue generating services by not allowing adequate responsiveness to new market demands.

New demands for higher bit rate services require transmission networks which are more flexible in terms of manageable bandwidth and swift provisioning times. The introduction of an optical fiber infrastructure satisfies these high bandwidth requirements by providing very high bit rate transmission capabilities. Network flexibility is accomplished by utilizing a synchronous multiplexing structure, which in turn makes the capability of accessing the tributaries easier and, as a result, reduces costs. Synchronous Digital Hierarchy (SDH) offers the advantages of synchronous multiplexing, in addition to new network capabilities, thanks to enhanced Operations, Administration, and Maintenance (OAM) features not provided by the existing PDH structure.

France Telecom has prepared a strategic program for the implementation of a single-mode fiber optic network infrastructure. This program coincides with the development of a new generation of digital transmission equipment based on SDH standards and the introduction of ATM technology. This paper begins with the introduction of the existing France Telecom network and outlines different aspects unique to the network. The target SDH network structure and deployment strategy, including the associated conception of Telecommunications Management Network (TMN), needed for taking full advantage of the potential management capabilities of SDH equipment, are also noted.

2. THE EXISTING TRANSMISSION NETWORK

The France Telecom network has some specific characteristics which have to be taken into account by network planning teams for accurately defining a successful evolution strategy.

The PSTN is organized in four switching levels: (figures reflect the state of the PSTN in 1992)

- Primary Transit..............(or Class 1) Exchanges :   5 PTE
- Secondary Transit..........(or Class 2) Exchanges : 70 STE
- Subscriber.................(or Class 3) Exchanges : 1,200 SE
- Remote Subscriber Unit (or Class 4) Exchanges : 10,000 RSU

The transmission network is organized into two main levels (Fig. 1):

- The long-distance network itself is subdivided into:

    ○ the main interconnection network, called the Réseau d'Inter-Connexion (RIC), which links the transit exchanges, and
    ○ the sectorial networks, which link the local exchange networks to the RIC.

These two networks, the main interconnection network and the sectorial networks, form the long-distance network.

- The local exchange networks which can be subdivided as follows (Fig. 2):

    ○ the Junction networks which:

        — link subscriber exchanges to the Long Distance Network Access Nodes and
        — link disparate subscriber exchanges together,

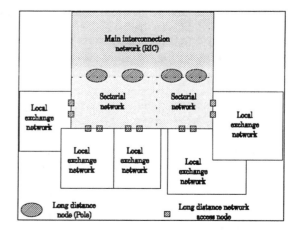

Fig. 1   Transmission network organization.

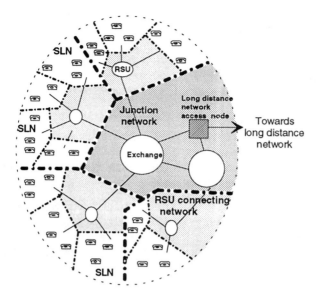

Fig. 2.   Local exchange network.

  ○ the Remote Subscribers Unit (RSU) that connects network links to the RSU or concentration units to the subscriber exchanges, and
  ○ the Subscriber Line Network (SLN) that links the subscribers to the RSU or to the subscriber exchanges.

2.1 The Long Distance Network

As noted, the existing long distance network is divided into two levels:

The Main Interconnection Network: Réseau d'Inter-Connexion (RIC).   The RIC is used for the interconnection of the transit switching nodes (Class 1 and 2 switches) and international nodes, i.e., approximately 50 nodes. It is presently based on 140 Mb/s systems, mainly coaxial cable or radio relay, and 565 Mb/s coaxial cable systems with new systems based on single-mode optical fiber. For protection against inevitable artery or node failures, the RIC is divided into two separate subnetworks interconnecting approximately 50 cable centers (*K points*) and 50 radio centers (*H points*). These K and H points are collocated in each of the 50 nodes that house the transit exchanges or international accesses [1,2].

At the 140 Mb/s level, transmission line systems between the two nodes (H and K subnetworks) are secured by switching on spare links between the same nodes or by rerouting via other nodes. This protection system, which is a subsystem of the SPARTE (Supervision et Protection des Artères de Transmission et des Equipements) project, is based on the use of switching equipment in the nodes controlled by a national operating system. Approximately 25% of the RIC capacity is assigned for spare links.

Protection functions were previously operated in this first stage using predetermined rerouting tables and implementing them by a manual step-by-step process. Restoration was then achieved within 15 minutes. Today, all necessary tasks required for complete network restoration are performed automatically. Mean achievement time is about 1 minute due to deliberately maintained human validation processes.

The Sectorial Networks.   Sectorial networks form the lower level of the long-distance network and link the RIC nodes to the junction networks (approximately 430), creating a star structure originating from the RIC nodes. Specifically, they utilize a double star structure so that each long-distance network access node (Noeud Interurbain de Desserte, or NID) is linked to its hierarchical H and K RIC centers. There are presently 37 sectorial networks. Sectorial networks are exclusively transmission networks.

The most frequently used transmission systems in the sectorial networks are 140 Mb/s coaxial systems, 140 and 565 Mb/s optical systems, and 2 × 34 Mb/s radio systems.

2.2 Local Exchange Networks

The existing 430 local exchange networks are linked to the long-distance network through 540 long-distance network access nodes.

As mentioned previously, these local exchange networks are subdivided into three levels:

  • Junction or subscriber exchange networks (430) are meshed networks that link the subscriber exchanges to the long-distance network access nodes,
  • Remote Subscriber Unit (RSU) connecting networks (1,200) are star networks which link the remote subscriber units to the switching exchanges, and
  • Subscriber Line Networks (SLNs) (10,000) are star networks linking subscribers to RSUs, or to the switching exchange.

Presently, all of these local networks are based primarily on copper and microwave technology, copper being generally dominant. During the 1980s, some multimode optical infrastructures were built in the largest urban networks. PDH systems, mostly from 2 to 34 Mb/s, are used in the local exchange networks with some 140 Mb/s monomode in urban networks (mainly in Paris).

Protection for these networks is achieved through diverse routing of all traffic between two points, utilizing two distinctly separate routes. In addition, specific 2 Mb/s equipment using 1 + 1 end-to-end protection has been developed for leased lines which have spare capacity left for the purpose of manual restoration.

The whole existing network is illustrated in Fig. 3.

3. SDH Evolutionary Strategy

3.1 Motivations for an Evolution

The motivations for the network structure evolution are being driven from two sides, technology and services:

  • The advantages of fiber optic transmission systems compared to other transmission media (coaxial and radio) are well known:

    • better transmission quality,

- reduction in investment costs, particularly in the long-distance network,
- reduction in operating costs, and
- greater flexibility in the event of an unforecasted increase in demand (increasing capacity of the systems and the rapid implementation of repeaterless links);

- SDH concepts make it easier to develop new functionality in transmission systems using add/drop facilities in a ring structure. These facilities are used by network planners to implement a more flexible and reliable network;
- The emergence of high-capacity switching exchanges (80,000 subscribers) warrants the reduction in the number of exchanges, which consequently leads to a restructuring and simplification of both the long-distance and local exchange networks; and
- The emerging demand for higher bit rate services requires operators to introduce networks which are more flexible and allow much faster provisioning of private line services, while improving critical network availability and providing efficient performance monitoring and OAM procedures.

3.2 The Long Distance Network

The Target Network. Taking into account the above motivations and driving forces, the target long-distance network will be based upon the following principles:

- RIC based upon an optical meshed network using 2.5 Gb/s systems (STM-16),
- as for the existing network, two separate centers (K1 and K2) in each node of the RIC for security precautions,
- RIC protection against link or node failure by rerouting based on the use of VC-4 cross-connect equipment,
- reduction in the number of sectorial networks (currently 37 to 19),
- use of STM-4 or STM-16 optical rings in the sectorial networks utilizing VC-4 add/drop muxes with 100% protection against link failures, and
- two long-distance network access nodes for each local exchange network connected by the ring to K1 and K2.

Evolutionary Steps. A new development plan for deploying optical fiber within the long-distance network was prepared in 1990. The first stage of 17,000 km of optical infrastructure will be available at the end of 1995, the final target consisting of an optical infrastructure of about 20,500 km including sectorial networks to be achieved in 1997.

A meshed SDH optical RIC, equipped with 2.5 Gb/s SDH

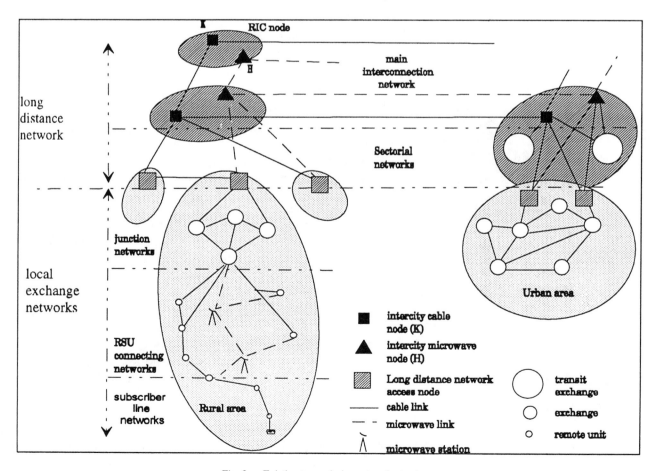

Fig. 3. Existing transmission network structure.

93

STM-16 line systems, began at the end of 1993, and VC-4 cross-connect (DXC 4/4) in the RIC nodes began in 1995. Initially this equipment was to be introduced with 140 Mb/s plesiochronous interfaces, the network management of the long-distance network remaining unchanged. In particular, VC-4 cross-connects are managed in the same way as existing protection switching equipment and are remotely controlled with the existing SPARTE system. The high capacity of STM-16 makes it difficult to protect them by existing PDH systems, necessitating a 1 + 1 automatic protection at deployment.

The second step in the evolution process is SDH connectivity. Plesiochronous interfaces appearing at 140 Mb/s are replaced by STM-1 interfaces connecting DXC 4/4 and STM-16 line systems, which provide full SDH connectivity at the VC-4 level. VC-4 management at this stage includes capabilities for network configuration control in relationship with network administration databases. The use of an object-oriented approach for information modeling and standardized Q interfaces is currently being studied.

In parallel, STM-4 and STM-16 optical rings with VC-4 add/drop muxes will be deployed in the sectorial networks providing the SDH interconnection between local exchange networks. The protection scheme in these rings will be based upon VC-4 path protection. For reliability and the maintenance of very high availability and quality of service, each ring will be connected to both K1 and K2 centers of the RIC node as depicted in Fig. 4. In the same way, each local exchange network will access the long-distance network through two long-distance network access nodes connected to the sectorial rings. This de-

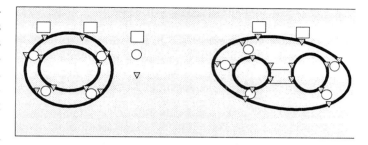

Fig. 5.   Target junction network.

ployment is scheduled to begin in 1996, the first step being network management based upon current proprietary solutions.

An experiment of DXC 4/1s with different suppliers began in 1994. The main function of the DXC 4/1s is to set up flexible VC-12 paths. These DXCs, interconnected through the VC-4 network, can form a meshed VC-12 network providing two types of transport services:

- interconnection between plesiochronous networks and quality control of the transfer and
- transparent VC-12 transfer in a synchronous environment.

At the present time a limited deployment of DXC 4/1s, in parallel with the introduction of terminal multiplexers, is scheduled for 1995 in the RIC nodes. For economical reasons, wider employment of this equipment will be decided if the context demonstrates the need of such services.

3.3 Local Exchange Networks

The common principles applied during the Local Exchange Networks evolution are:

- reduction in the number of local exchange networks from 430 today to 125 in 1998,
- local exchange access to the long-distance network through two long-distance network access nodes to meet service level and security objectives,
- implementation of high-capacity exchanges,
- construction of a monomode optical infrastructure, and
- deployment of an SDH ring structure using add/drop multiplexers (ADM) and VC-12 path protection.

This last point will ensure 100% protection against cable and system failures. Two suppliers of ADM were chosen in April 1994. Fifty rings have been ordered, half of which are STM-4 rings. The first rings were deployed at the end of 1994. A second equivalent order was passed at the beginning of 1995. Large-scale introduction of ADM is foreseen in 1995.

This optical infrastructure deployment plan for the local exchange networks gives priority for direct optical connectivity to professional subscribers.

Junction Networks.   In this section of the network, the optical infrastructure is scheduled for completion in 1996. The number of exchanges and long-distance network access nodes will remain limited (typically less than ten) in most cases.

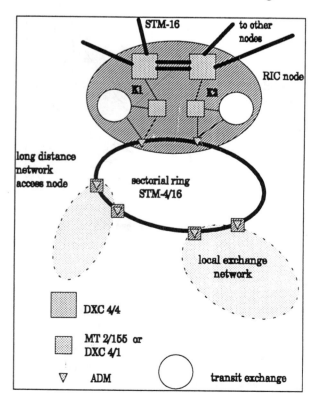

Fig. 4.   Target sectorial ring.

The target network will use an STM-4 ring structure and ADMs (Fig. 5).

Due to the size of the Paris network, its target structure will be similar to the long-distance network with the secondary level using ring structures (STM-4 and STM-16) and a meshed primary level using STM-16 line systems to interconnect DXC 4/4s. Protection within the primary level will follow the same principles as in the RIC.

RSU Connecting Network. These networks are subdivided into two levels: the primary level links the most important remote units to the hierarchical exchange and the secondary level links smaller remote units and remote concentrators to the primary level (Fig. 6).

The target primary level will use an STM-1 ring structure. For safety purposes, the optical ring will include two different exchange buildings.

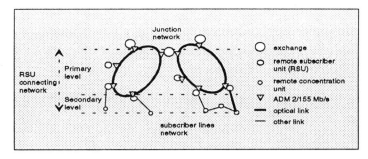

Fig. 6.   RSU connecting network.

The complete replacement of the existing copper and microwave infrastructure is a long-term objective, specifically in the primary level. In the first stage, ADM could be used in bus structures while awaiting the implementation of the optical rings.

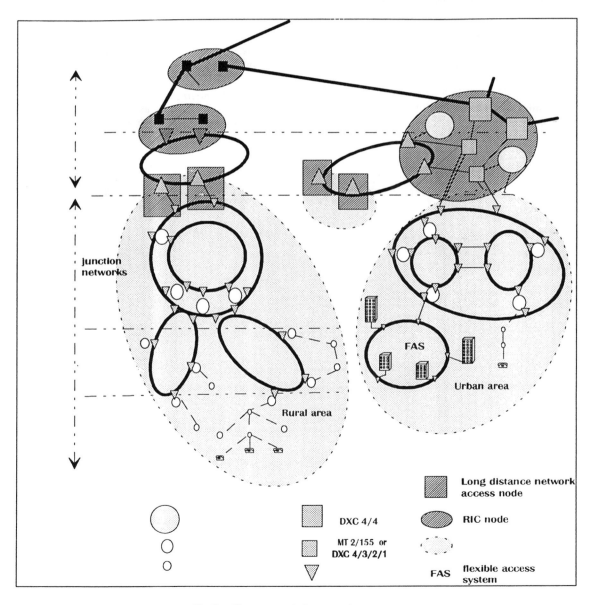

Fig. 7.   Target transmission network structure.

Subscriber Lines Networks. Subscriber line networks will evolve according to the fiber content in the local loop, but, taking into account economical restraints, it is a much longer termed plan. However, optical connection experiments for private subscribers is already underway.

After successful experiments with Flexible Access Systems (FASs) or Réseaux Optiques Flexibles (ROFs) in 1990, a large-scale program has been taken for optical connections of main business buildings, targeting professional subscribers.

Management Issues. During the first stage of the evolution process, add–drop multiplexers used in the SDH rings will not be able to provide standardized Q interfaces for management systems. Therefore, proprietary solutions for management interfaces and operating systems will have to be used for the control of all the groups of rings. For the longer term, the use of object-oriented and standardized Q interfaces to support an evolution towards SDH management systems, able to support a multi-vendor environment, are being studied.

4. Conclusions

The target transmission network structure, depicted in Fig. 6, reveals major architectural and technological maturation when compared with the existing network structure. Concurrently, critically important changes in the management system are planned in order to take full advantage of all the new possibilities offered by the SDH network architecture [3].

This evolution, however, must be realized under financial constraints and will therefore be conducted over many years; specifically, in the lower part of the network. This is where copper, microwave, optical fiber, PDH, and SDH will continue to coexist for the foreseeable future.

Evolutions such as this SDH deployment are part of France Telecom's global objective to improve continually the quality, availability, and flexibility of its transmission networks to better serve all its clients domestically and internationally.

References

[1] G. Bars, J. Legras, and X. Maitre, "Introduction of new technologies in the French transmission networks," *IEEE Communi. Mag.*, vol. 28, pp. 39–43, August 1990.
[2] P. Passeri, F. Balena, G. Bars, N. Vogt, and T. Wright, "Introduction strategies of SDH systems in Europe," *ICC 1991*, pp. 58–62
[3] X. Maitre, P. Incerti, J. Y. Serreault, and M. Le Gall, "Evolution of the France Telecom transmission network," Geneva, *Telecom 1991*, pp. 475–479.

Acknowledgement
The authors would like to thank X. Maitre, V. Pramil, and K. Robinson (France Telecom DRX) for their help in preparing this document.

SDH Network Evolution in Japan

HIDETOSHI MIURA, KAZUMITSU MAKI,
AND KAZUHIRO NISHIHATA

Abstract—Since 1989, NTT has been applying the Synchronous Digital Hierarchy to its network. The first generation of SDH-based network elements was developed to create a simplified and easy-to-operate transmission network. In 1993, NTT entered its second phase of the development to enhance the SDH network.

This paper describes the evolution of NTT's network from PDH to SDH, and overviews the enhanced network operation system developed for the management of the SDH network.

1. INTRODUCTION

The year 1989 marked the first commercial application of the Synchronous Digital Hierarchy (SDH) by Nippon Telephone and Telegraph (NTT) to its long-distance trunk line network. The successful deployment of the first generation SDH-based network elements, called SDH Phase 1, has achieved its original mission to create a simplified, easy-to-operate transmission network [1].

Now, NTT has entered the second phase of its deployment of SDH-based transmission systems [2]. The main focuses are to create a more reliable network with lower costs, and to improve network operation systems for these SDH-based network elements.

This paper describes the evolution of NTT's network from a Plesiochronous Digital Hierarchy (PDH)-based network to an SDH-based network. First, we see the evolution from a network element point of view. Then, we focus on the network operation system, which has been unified to achieve flow-through operation.

2. DEVELOPMENT OF NETWORK AND NETWORK ELEMENTS

Table 1 summarizes the evolution of network, network element, and operation system taking place at NTT. The first turning point came in 1989 when NTT succeeded in the first commercial application of SDH, which is now called SDH Phase 1. Now, NTT is in SDH Phase 2, seeking a more economical and reliable network with the enhanced network operation.

2.1 From PDH to SDH Phase 1

When NTT introduced the Plesiochronous Digital Hierarchy (PDH)-based network in the early 1980s, the multiplexing scheme was synchronous only up to the secondary level (6.3 Mb/s), leaving higher levels asynchronous. The PDH network required complex step-by-step multiplexer configurations in each transmission node, resulting in higher node costs. In addition, handling of various levels of paths was required to achieve high resource utilization of transmission lines, which were then expensive. Complicated network architecture made network design and operation difficult.

To solve these problems, the Synchronous Digital Hierarchy (SDH) was standardized in 1988 by CCITT (currently ITU-T), and in 1989 in Japan. Synchronous multiplexing enables the direct multiplexing of various kinds of signals into one transmission line and the cross-connection at higher levels. Thus, the introduction of SDH paved the way to a simpler network architecture, resulting in lower node costs and higher resource

TABLE 1. NETWORK, NE, AND OPS EVOLUTION AT NTT

	PDH	SDH Phase 1	SDH Phase 2
Network	Complicated	*Simple*	*Simple and Reliable*
Network Element (NE)	— Optical Transmission System — Multiplexer at each hierarchy	*Modularized* Module A Module B Module C	*Intelligent* Module AX Long Span Module A
Digital Cross-Connect System	M20: 384 kb/s cross-connect DF: Distribution Frame	*Module B: 1.5M/6.3M* *(VC-11/2)*	*Module AX: 52M/156M* *(VC-3/4)*
Operation System (OpS)	Primitive	*Remote*	*Flow-Through* Network Design and Operation with TMN Architecture
Introduction	Up to 1989	1989	1993

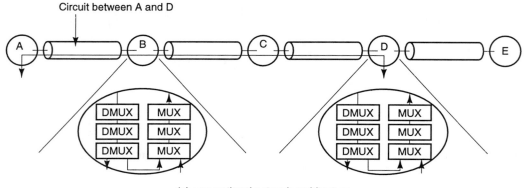

Circuit between A and D

(a) conventional network architecture

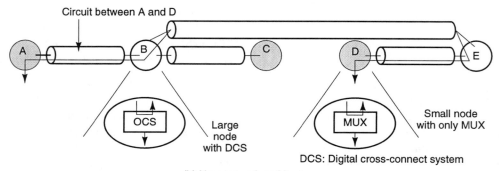

Circuit between A and D

Large node with DCS

Small node with only MUX

DCS: Digital cross-connect system

(b) New network architecture

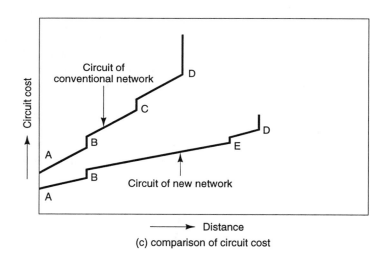

(c) comparison of circuit cost

Fig. 1.   An example of transmission network architecture.

utilization than were possible with the conventional PDH. Figure 1 illustrates major changes due to the introduction of SDH.

NTT has been aggressive in the development of the SDH. As early as 1989, the first generation of SDH equipment was introduced in NTT's network. It was the first commercial application of the technology in any telecommunications network.

In SDH Phase 1, network elements consist of a series of modularized equipment, called Modules A, B, and C, as shown in Fig. 2. Module A is an optical fiber line transmission system which consists of line terminating multiplexers and regenerators. Module A has interoffice optical interfaces of 156 Mb/s, 622 Mb/s, and 2.4 Gb/s. Module B is a VC-11 digital cross-connect system. Module C is a multiplexer transforming existing 2/8/1.5/6.3 Mb/s interfaces into SDH interfaces. Modules B and C have interoffice optical interfaces of 52 and 156 Mb/s. Modules A, B, and C can be connected to each other by intraoffice optical 52/156 Mb/s interfaces.

The Phase 1 SDH network was introduced as an overlay to the existing PDH network.

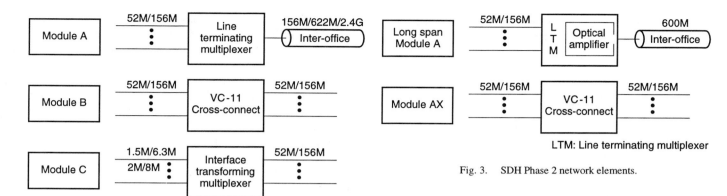

Fig. 2.   SDH Phase 1 network elements.

LTM: Line terminating multiplexer

Fig. 3.   SDH Phase 2 network elements.

2.2 SDH Phase 2 Network Elements

After the successful introduction of SDH Phase 1 network elements, NTT has now entered the second phase of its introduction of the SDH. In SDH Phase 2, the emphasis is placed on creating a more economical, reliable network and on achieving flow-through network operation. In order to create a more economical and reliable network, NTT has newly developed two SDH-based network elements: the 52/156 Mb/s Digital Cross-Connect System and Long Span Module A (Fig. 3).

(1) 52/156 Mb/s Digital Cross-Connect System. The 52/156 Mb/s Digital Cross-Connect System, called Module AX, has been developed to perform reconfigurations of the network at the higher order path level (52/156 Mb/s) [3]. It has the following major features.

Network restoration (Fig. 4): Network restoration in the event of a fiber cut will significantly improve the network survivability. Module AX, together with its operation system, executes network restoration in the event of a fiber cut by cross-connecting at the VC-3/4 levels. Upon receipt of a command from the operation system, Module AX changes cross-connection of VC-3/4 paths from their original routes to alternative routes. Before changing cross-connection, Module AX makes sure that the alternate path is logically the same as the original path by identifying the J1 byte, defined as a path trace on the VC-3/4 path overhead.

Hitless switching (Fig. 5): Module AX has the capability to change its cross-connection with no errors even if a phase difference exists between the original path and the alternate path. This function enables the changing of path routes without disrupting telecommunication services to customers. The changing of path routes is needed for several reasons, such as cable relocation for repair. In the PDH network, NTT also had the network restoration system. According to the data, however, only 25% of all cases used the system for its original purpose, that is, network restoration in the case of route failure. The remaining 75% of all cases proved to be using the system for cable relocation, affecting services to customers in each case. The data indicate the need of hitless switching capability, especially for cable relocation. The hitless switching is carried out by adjusting delays inserted

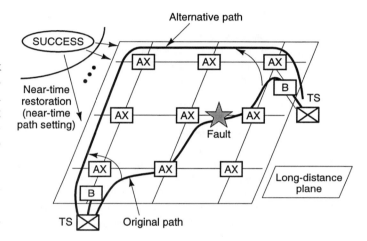

Fig. 4.   Near-time network restoration.

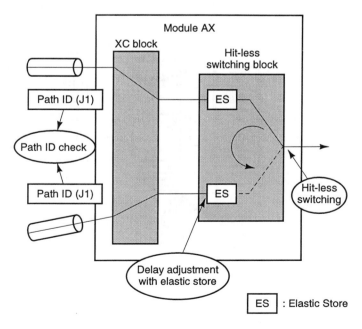

ES : Elastic Store

Fig. 5.   Hitless switching of module AX.

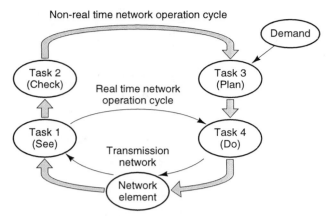

Task 1 (See)   : Network surveillance (performance monitoring, network configuration monitoring)
Task 2 (Check) : Network evaluation management (cost evaluation, efficiency evaluation)
Task 3 (Plan)   : Network design (efficient and reliable network design)
Task 4 (Do)   : Network provisioning and operations (path installation, path rearrangement, path restoration)

Fig. 6.   Network operation and management cycle.

in each path. Normally, there is a phase difference between the original path and the alternative path because the lengths of these paths are different. Normal switching of these paths results in data loss or redundancy. In Module AX, the J1 byte is used for the accurate recognition and phase adjustment of VC-3/4 paths. NTT uses the J1 byte as 64 multiframes, using a specific code in frames 63 and 64 for the mutiframe synchronization. The phase adjustment is achieved by controlling elastic stores inserted in each path.

(2) Long Span Module A. Long Span Module A has been developed to reduce the number of repeaters, resulting in a more economical trunk network [4]. Long Span Module A enabled 160 km repeater spacing using optical amplifier technology. An Erbium Doped Fiber Amplifier (EDFA) is used as an optical amplifier of Long Span Module A. The output optical power range has been increased from $-3 \sim +3$ dBm of conventional Module A to $+14 \sim +18$ dBm. The minimum received power level has been lowered from -35 dBm of conventional Module A to -41 dBm. The improved output–input level difference of 55 dB enabled 160 km repeater spacing. Another feature newly introduced in Long Span Module A is Automatic Laser Shutdown/Recover Control. This function is for safety purposes to lower the optical output level automatically when the optical fiber is disconnected. Whether the optical fiber is connected or disconnected is detected by observing that the reflection level occurred at connectors.

3.   DEVELOPMENT OF AN INTEGRATED OPERATION SYSTEM FOR SDH PHASE 2

With the advance of network and network elements aspects as described above, the operation system has also been reshaped to achieve flow-through operation. The network management architecture is designed to conform to the Telecommunication Network Management (TMN) architecture, recommended by ITU-T [5].

3.1 Flow-Through Operation

As a network operating company, NTT's competitiveness lies in the effectiveness of network operation, administration, maintenance, and provisioning (OAM&P). As shown in Fig. 6, NTT

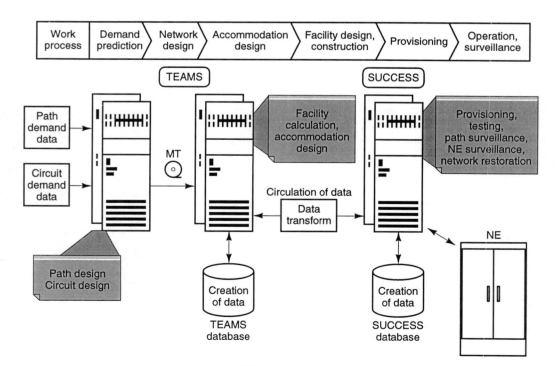

Fig. 7.   Flow-through SDH network design and operation system.

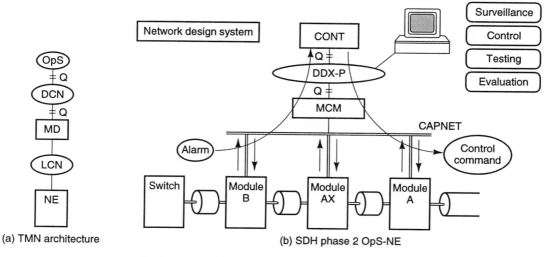

(a) TMN architecture   (b) SDH phase 2 OpS-NE

Fig. 8. TMN architecture versus SDH Phase 2 configuration.

has identified the four aspects of network management: 1) monitoring of the current status of the transmission network (= see), 2) evaluation of the collected data (= check), 3) planning of actions as a result of evaluation (= plan), and 4) execution of the action plan (= do).

The new network operation system, called SUCCESS, Surveillance Concentrated Control and Evaluation System for the SDH network, is developed and introduced with the SDH Phase 2 network elements. On the other hand, NTT had a network designing system, called TEAMS (Telecommunication Engineering And Management System). Thus, the introduction of SUCCESS, together with the use of TEAMS, enables a sophisticated flow-through process from network design to network operation (Fig. 7).

3.2 Network Management Architecture

NTT's SDH Phase 2 network is based on the TMN model, recommended by ITU-T, to conform its network to international standards. Figure 8 shows the network architecture of the SDH Phase 2 network in comparison with ITU-T's TMN model.

The standard Q interface is used between the operation system and network elements. The protocol stack is shown in Fig. 9. The lower protocol layers of intraoffice DCN (Data Communications Network) use NTT's existing proprietary protocol, CAPNET (Control and Access for Plant Network). The standard Q interface is used between SUCCESS and a mediation device, MCM (Message Communication Module), using CMISE (Common Management Information Service Entity) in the application layer. MCM acts as a protocol converter for lower layers.

3.3 Intelligent Network Elements

While SDH Phase 1 network elements could send limited information, mainly alarm information, SDH Phase 2 network elements are capable of sending more detailed information about network current status, thereby enabling more accurate assessment of the network (Fig. 10). Major features of intelligent network elements are as follows.

(1) Automatic Configuration Data Updating. Network elements are able to notify the operation system of any changes

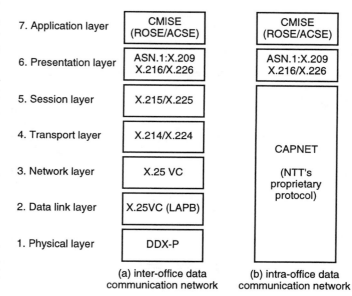

	(a) inter-office data communication network	(b) intra-office data communication network
7. Application layer	CMISE (ROSE/ACSE)	CMISE (ROSE/ACSE)
6. Presentation layer	ASN.1:X.209 X.216/X.226	ASN.1:X.209 X.216/X.226
5. Session layer	X.215/X.225	
4. Transport layer	X.214/X.224	CAPNET (NTT's proprietary protocol)
3. Network layer	X.25 VC	
2. Data link layer	X.25VC (LAPB)	
1. Physical layer	DDX-P	

Fig. 9. Protocol stack of two data communication networks.

in the status of component packages, thus enabling the operation system to recognize the configuration of network elements in real time. The automatic collection of configuration data reduces significantly the need for manually entering these data by maintenance personnel, and eliminates the impairment of data due to human error.

(2) Autonomous Alarm and Performance Monitoring Information. The management of alarm and performance monitoring information is essential for providing high-quality telecommunication services to customers. SDH Phase 2 provides alarm reporting and performance monitoring functions, conforming to the ITU-T standard. Alarms are categorized, indicating the actions to be taken by maintenance personnel. "Critical" needs immediate network restoration, "Major" requires maintenance, and "Warning or Minor" indicates no need for immediate action.

Performance monitoring information is used to assess the de-

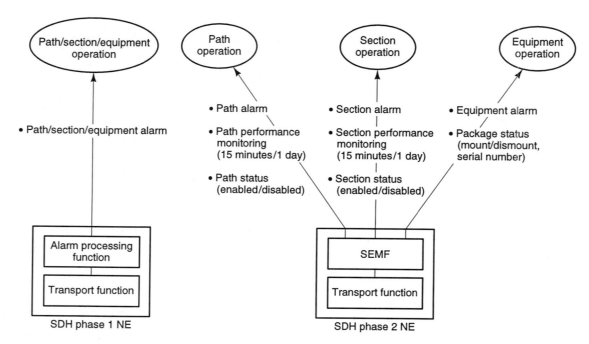

Fig. 10. Information from NE.

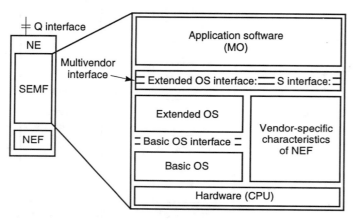

Fig. 11. Configuration of network element.

gree of service degradation, including Code Violation (CV), Errored Seconds (ES), Severely Errored Seconds (SES), and Unavailability Status (UAS). These performance monitoring data may be sent to the operation system on demand and/or on a scheduled basis. In addition, the Threshold Crossing Alarm will automatically be sent if the service quality is degraded lower than the predetermined level.

3.4 SEMF Configuration

To realize intelligent network elements, network elements are equipped with the Synchronous Equipment Management Function (SEMF). The configuration of SEMF is shown in Fig. 11. It employs a standard commercial real-time OS interface and S interface. The S interface is defined as the interface between the application software and the processing portion dependent on the hardware architecture. It compensates for differences in the way the hardware resources are treated by different vendors and provides a vendor-independent common interface.

These OS and S interfaces provide a multivendor interface environment for the application software. Application software uses object-oriented design for flexible modification and reusability, thus reducing the development period and costs.

In addition, the SDH Phase 2 network element has the standard Q interface, and uses the concept of managed objects. From a management aspect, the objects observed and manipulated through the Q interface are defined as managed objects. These include equipment, termination point, path, and so on. Managed objects are manipulated by the standard CMIS operation.

4. CONCLUSION

This paper described the evolution of NTT's network and network elements from PDH, SDH Phase 1 to SDH Phase 2, and the enhanced network operation system developed and introduced with SDH Phase 2 network elements. The flexible transport network and enhanced network management system together achieve the reliable and economical network, providing better service to customers at lower cost.

NTT continues to improve network technologies toward the realization of a broadband ISDN.

References

[1] H. Shirakawa, K. Maki, and H. Miura, "Japan's network evolution relies on SDH-based systems," *IEEE LTS*, vol. 2, pp. 14–18, Nov. 1991.
[2] H. Miura, "Special Feature: Construction & operation of the transmission line network," *NTT Rev.*, vol. 5, pp. 30–58, Mar. 1993.
[3] K. Nishihata, S. Umino, and T. Shinomiya, "SDH network in Japan and SDH 52/156 Mbps (VC-3/4) digital cross-connect systems with hit-less switching function," *ICC*, 1994.
[4] K. Nishihata, T. Kuwata, and H. Seki, "Introduction of SDH optical transmission system with 160km repeater spacing into SDH network," GLOBECOM'94, 1994.
[5] T. Kunieda, S. Sugimoto, and N. Sasaki, "A synchronous digital hierarchy network management system," *IEEE Commun. Mag.*, vol. 31, Nov. 1993.

Plans and Considerations for SONET Deployment[1]

N. B. SANDESARA, G. R. RITCHIE,

AND B. ENGEL-SMITH

1. INTRODUCTION

Optical transmission facilities have expanded rapidly in all segments of the telecommunications networks, especially in the Local Exchange Carrier (LEC) networks, because of the high service quality and low cost associated with the fiber transmission medium. SONET defines, for the first time, a family of standard optical interfaces for use in transport networks. This allows direct optical interfaces on Network Elements (NEs) and further integration of equipment and systems while offering opportunities for mix-and-match.

SONET introduces new network capabilities such as single-ended operations, integrated operations, network survivability, and flexible bandwidth allocation, all in a multivendor environment. Because of these advantages, SONET can facilitate introduction of new services, and reduces network equipment and network management and operations costs.

Completion of the fundamental aspects of the SONET standards, contained in the ANSI T1.105 [1] series of standards, and a family of equipment-specific requirements from Bellcore [2]–[10], have led to the availability of SONET-compatible systems. Initial SONET-compatible products appeared on the market in 1989. Globalization of SONET standards further accelerated development of products for different applications. A number of SONET equipment field trials occurred in Bellcore Client Companies' (BCCs) networks in 1990. Because of the BCC commitment to standards and the benefits of SONET described above, almost all newly purchased fiber transport systems were SONET-based, beginning in 1992. Likewise, there have been aggressive vendor plans for equipment development.

The intention of this paper is to give an LEC perspective on: (i) the advantages of deploying SONET, (ii) the rate of SONET deployment, (iii) some typical applications and architectures, and (iv) the role SONET will play in the evolution of the LEC network of the future.

There are three basic reasons why BCCs have sought to deploy SONET.

- Migration of existing services to fiber — BCCs have begun migrating existing services (DS1s and DS3s) to fiber because of quality of service and operations benefits obtained from such a migration. Most BCCs have capped (no

[1] This paper was originally published in the *IEEE Communications Magazine*, August 1990. It has been revised based on the information available as of June 1995.

[2] The dates on Figs. 1, 2, and 3 are revised based on the information available as of June 1995.

new assignment to) or retired (removal of) T1 carrier systems. Retirement of T1 carrier systems began as early as 1992 in some BCCs. Deployment of SONET systems is the preferred solution to fulfill these needs.

- New network capabilities — Sophisticated network capabilities are available using SONET. Some examples are: survivable and self-healing networks; DS1 and DS3 rapid provisioning, networking, grooming, and switching capabilities; network management and control capabilities that allow flexible bandwidth allocation.

- Deployment of new services — The network capabilities mentioned above facilitate development of new service offerings. Also, the broadband services will be deployed using SONET facilities.

1.1 SONET Deployment Timeline

Figure 1 shows a timeline for SONET deployment in the BCC networks. Corresponding network architectures are shown in Figs. 2 and 3.[2] Dates on these figures indicate the earliest dates for trials. General deployment typically begins six months after successful trials.

Initially, SONET transport systems and multiplexes were used in essentially the same configuration as the current proprietary fiber transport systems, namely, point-to-point. Trials and deployment have taken place in both the interoffice and access portions of the BCC networks providing capabilities similar to the current fiber optic systems. DS1s and DS3s are transported between point A and point B.

Trials of integrated access capability, i.e., direct multiplexing of customer access lines (e.g., POTS, DS1, etc.) to SONET systems using a remote digital terminal (RDT) with SONET interfaces, began in 1991, with general deployment in the 1992 timeframe. Initially, integrated access required a stand-alone Central Office Terminal (COT) for interfacing to switches, Central Office (CO) equipment, and interoffice facilities. Later on, integrated Local Digital Switch (LDS) access capability may be provided via proprietary interfaces. Generic integrated switch access capability as contained in Bellcore's Integrated Digital Loop Carrier (IDLC) proposal may not be deployed until 1996.

Multinode networks in simple linear and tree configurations were trialed in 1991 using Add-Drop Multiplexes (ADMs). Tree configurations are used where traffic tapers off as it moves from the hub node to the branches and on to end-nodes. Such a configuration is most suitable for distribution networks. Beginning in 1992, SONET self-healing ring systems using ADMs became available which facilitate building of survivable networks.

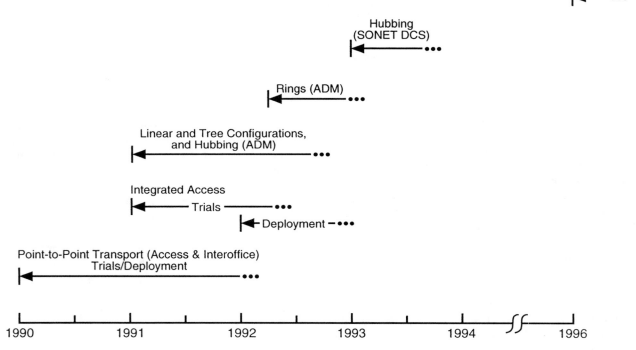

Fig. 1. SONET deployment timeline.

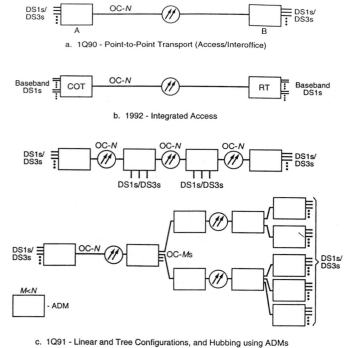

Fig. 2. SONET deployment configurations.

Full cross-connection capabilities at wideband (DS1, VT) and broadband (DS3, STS-1) levels in NEs such as Wideband and Broadband Digital Cross-connect Systems (W-DCS and B-DCS) with SONET interfaces are expected in the 1995/96 time-frame. These capabilities will make large-scale SONET facilities hubbing and grooming possible. W-DCS and B-DCS that have DS1 and DS3 (but no SONET) interfaces and provide DS1 and DS3 cross-connections, respectively, are already being deployed in the BCC networks. From a BCC perspective, these DCSs should be upgradable to provide VT and STS-1 cross-connection capabilities as the SONET networks grow. Because of the use of standard interfaces, it is expected that initially-deployed point-to-point SONET systems and SONET islands will be interconnected and smoothly evolve to provide multi-vendor multicarrier compatible end-to-end optical networks.

2. SONET CAPABILITIES

2.1 Rates and Mappings

OC-N and electrical STS-1 and STS-3 are suitable for a variety of applications as shown in Table 1. Use of different SONET signals in the network will be dictated by factors such as traffic capacity needed, network architecture, and technology breakpoints. Rates most likely to be deployed in the BCC networks include OC-3, OC-12, OC-24, and OC-48 and OC-192 (in the 1996 timeframe). Most loop applications will use OC-3 and OC-12, and interoffice applications will use OC-12, OC-24, and OC-48.

DS3s in the BCC networks will be carried primarily using the clear channel DS3 mapping. A BISDN ATM cell mapping into SONET payload has been specified for use of SONET to transport ATM cells. An STS-1 payload can be organized in a

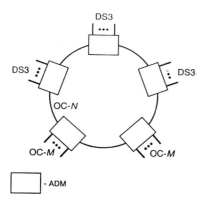

a. 2Q92 - SONET Rings using ADMs

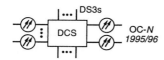

b. 2Q93 - Hubbing using DCS

Fig. 3. SONET deployment configurations.

VT structure that is suitable to carry sub-DS3 rate signals such as DS1s. This is expected to be the primary use of VTs in the initial deployment of SONET in the BCC networks.

VTs have two modes of operation: floating and locked. The floating VT mode is suitable for bulk transport of channelized or unchannelized DSn and for distributed VT grooming. The locked VT mode is optimized for bulk (i.e., an integral number of STS-1s) transport of DS0s and for distributed DS0 grooming between DS0 path terminating equipment as, for example, between DS0 switches. For most BCC applications, distributed DS0 grooming capability at an intermediate node is not needed. The increased administrative complexity and the need for DS0 level maintenance will outweigh potential benefits for all but a small number of applications needing this capability.

TABLE 1. SONET SIGNAL APPLICATIONS

Interface Signal	Applications
OC-1	Distribution networks
OC-3	Distribution networks, IDLC access, Thin interoffice routes, Broadband UNI
OC-12	Access rings, Loop feeder, Interoffice transport
OC-24 and OC-48	Interoffice backbone between major hubs, Interoffice rings, BISDN feeder
STS-1	Intraoffice interconnection
STS-3	Intraoffice interconnection, Broadband UNI

Table 2 compares network implications and applications of the two VT modes. Floating VT mode has been used quite extensively in the BCC networks. The majority of SONET applications involving VT structured payloads uses the floating VT1.5 asynchronous and byte synchronous mappings. Potentially, there are five different DS1-to-VT1.5 mappings that can be used, but the two mentioned above are optimum for most applications.

2.2 Compatibility

Compatibility of an NE with the standard does not necessarily guarantee easy networking. "Blind compatibility," allowing networking of NEs without any planning, is difficult to attain for a complex network involving many different types of NEs. To ensure that SONET NEs are cost-effective and meet BCC requirements, each BCC will attempt to limit the number of options available in the deployed equipment. Desire to achieve intervendor compatibility, operational complexity of attempting to administer several options, and potential interconnection problems as the network is rearranged also provide motivation for minimizing the set of options used. A small subset of the possible SONET rates and DSn mappings are expected to cover most BCC applications. Compatibility and cost-effective implementation also requires consistent usage of SONET overhead capabilities across a network. The SONET T1.105 standard [1] recognizes this fact and divides usage of various overhead functions into required and optional categories.

TABLE 2. VT MODES

Mode	Network Implications	Applications
Floating	• Minimal delay through the network and at VT cross-connects for DSn transport • VT path level maintenance and end-to-end performance monitoring possible	• Distributed VT cross-connection • Bulk transport and switching of channelized or unchannelized, synchronous or asynchronous DS1s (asynchronous mapping) • Unchannelized synchronous DS1 transport (bit synchronous mapping) • DS0 circuit switched traffic and IDLC (byte synchronous mapping)
Locked	• Large delay ($\geq 125 \ \mu s$) through the network and at VT cross-connects for DSn transport • VT path level maintenance and end-to-end performance monitoring not possible • DS0 level maintenance required at VT cross-connects • Cannot transport asynchronous DSn traffic	• Unchannelized synchronous DS1 transport (bit synchronous mapping) • Bulk transport of DS0s and distributed DS0 cross-connection (byte synchronous mapping)

3. SONET ARCHITECTURES

Some general observations can be made regarding BCC SONET deployment. Most existing interoffice networks were based on hub or star architectures, while the loop has been traditionally based on a tree architecture. SONET gives BCCs the ability to consider new, more flexible, and integrated architectures utilizing ADMs and DCSs rather than existing point-to-point transmission systems. Initial SONET deployment consisted of point-to-point transport systems in both the access and interoffice networks as shown in Figs. 1 and 2. Current systems in the loop operate at OC-1 and OC-3 rates, and those in the interoffice networks operate at OC-3, OC-12, and OC-48. In the loop, integrated access capability will allow integrated baseband and DS1 multiplexing to, and transport via, SONET between RDT and COT. Loop carrier systems will initially operate at OC-3 and subsequently at OC-12. More flexible network architectures using ADMs and further system integration due to integrated switch access will follow.

Terminal multiplex (ADM in the terminal mode) is like a conventional multiplex that generates a high-speed signal from a set of low-speed signals. The most common low-speed interfaces are DS1, DS3, electrical STS-1 and STS-3, and OC-3. The most common high-speed signal interfaces are at OC-N where N = 1, 3, 12, and 48. The add-drop multiplex (ADM in the add-drop mode) can drop and insert a variety of tributary signals (DS1, DS3, OC-3, etc.) from the OC-N signal passing through the multiplex. Such a multiplex is useful in applications where economical access to a small number of tributary signals from a high-speed signal is needed at some intermediate network node. ADMs are also useful for traffic grooming in a tree network.

We expect that a family of multiplexes designed with specific interfaces and optimized for specific applications will be deployed. Bellcore TR-NWT-000496[3] provides generic criteria for SONET ADM equipment to be used in a typical linear topology in BCC interoffice or local access networks.

Flexibility in the interoffice SONET networks is achieved first by using ADM capabilities. This will be supplemented by cross-connection capabilities of DCS, and bandwidth sharing and self-healing capabilities of rings.

3.1 Access Network

Technology has responded to the needs of local access by mainly offering feeder relief capabilities such as digital loop carrier systems. The equipment developed generally used proprietary interfaces, giving BCCs little opportunity to mix and match between suppliers. The concept of a generic DS1-based IDLC interface [9] applies to the RDT and the integrated digital terminal (IDT) at the LDS. An IDT-RDT combination defines an IDLC system.

Although DS1-based IDLC systems are adequate for most current applications, some cost savings can result from extending this concept to higher speed interfaces due to greater system integration. With increasing penetration of fiber in the loop, SONET is a prime candidate for such an interface. A generic Bellcore-proposed SONET IDLC [10] interface maintains compatibility above the physical layer with the generic DS1-based IDLC already defined. A SONET-based IDLC provides synergy between access and interoffice SONET networks, and allows greater integration of interoffice and access networks via use of common equipment. The generic SONET IDLC is the target SONET architecture for BCC access networks. It is expected to be trialed in 1996.

Generic DS1 and SONET IDLC interface concepts are illustrated in Fig. 4. Five RDTs are shown, all utilizing the add/drop capabilities of ADMs for transport, via an OC-N rate signal, to the OC-3 and DS1 interfaces of a switch. The SONET-IDLC interface operates at the OC-3 rate, and several IDLC systems (IDT-RDT system pairs) can be accommodated via each OC-3 termination at the LDS. OC-1, OC-3, and OC-12 are all good candidate rates for IDLC applications. However, OC-3 seems to be the best choice for most near-term applications. Power consumption is a major issue in RDT applications. Low-power CMOS technology can be economically used at the OC-3 rate. Also, OC-3 offers spare capacity at the RDT and consistency with BISDN transmission rate. It is difficult to envision an OC-12 being fully utilized at the switch interface for IDLC applications. Greater use of OC-12 is likely as new device technologies mature and broadband services are deployed.

The SONET option to directly map DS0s byte-synchronously into VT1.5 eliminates any intermediate DS1 framing, and is suitable for digitally-encoded voice services, and for DS0 clear-channel capability. Because of the advantages mentioned earlier, the floating mode is preferable to the locked mode for loop applications where VTs are add-dropped at multiple points along a feeder route and where intermediate DS0 cross-connects are not anticipated. It also appears that the floating mode will allow better switch utilization. Based on these considerations, the SONET-IDLC interface is required to support only the byte-synchronously mapped floating VT1.5.

3.2 Interoffice Transport Network

Prior to SONET, optical line terminating multiplexes used asynchronous multiplexing of DS3s to generate a nonstandard optical signal. Multiplexing of DS1s to DS3s was done with Digital Multiplexes (DM13). Fiber transport was point-to-point with equipment from the same vendor used at both ends. Different fiber systems were interconnected in a CO using DS1s and DS3s. For some large facilities hubs, DCS 3/1s are being

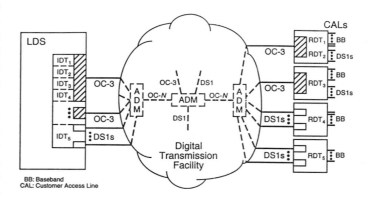

Fig. 4. A SONET IDLC application configuration.

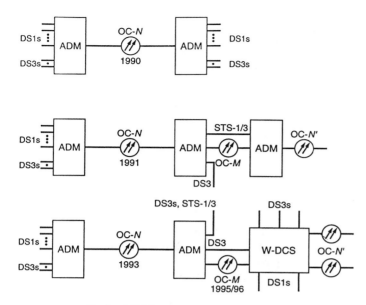

Fig. 5. SONET transport systems deployment.

deployed to provide efficient interconnection at the DS1 level. DCS 3/1s have DS3 and DS1 interfaces and provide automatic DS1 cross-connections. They provide grooming for better facility fill and eliminate back-to-back multiplex arrangements.

Initial SONET deployment, illustrated in Fig. 5,[3] provided point-to-point transport of signals of the existing digital hierarchy (DS1, DS3) with enhanced operations capabilities. ADMs provide efficient access to tributary signals and some grooming and hubbing capabilities. Because of the use of standard interfaces, it is possible to interconnect ADMs directly via SONET signals. In a large facilities hub, tributary signal rearrangements can also be effected using W-DCS and B-DCS rather than back-to-back multiplexes.

The W-DCS terminates asynchronous DS3s and SONET OC-Ns with floating VT1.5 payloads, and also provides clear channel DS1 interfaces. The W-DCS provides DS1 and VT1.5 cross-connect capabilities. A Broadband Digital Crossconnect System (B-DCS), like W-DCS, terminates OC-Ns and DS3s. However, the basic function of B-DCS is DS3 and STS-1 cross-connections. From the perspective of broadband services, STS-3c cross-connection capability will be quite desirable. W-DCS and B-DCS provide automated hubbing, grooming, and network interconnection. Bellcore TR-NWT-000233 [6] gives generic criteria for W-DCS and B-DCS to be deployed in the BCC networks.

As mentioned above, W-DCS and B-DCS have already appeared on the market and are being deployed. Current W-DCS products have DS1, DS3, and STS-1 interfaces, and B-DCS products have DS3 and STS-1 interfaces. It is expected that these products will be upgradable to provide SONET OC-N interfaces. SONET-based DCSs will offer advantages over DCS

[3]The dates in Fig. 5 are revised based on the information available as of June 1995.

[4]ANSI standards for bidirectional rings were completed in 1992. Prestandard implementations were trialed in 1992–93. Products conforming to the ANSI standard became available in 1993.

®TIRKS is a registered trademark of Bellcore.

3/Xs (DCS 3/1s and DCS 3/3s) that are designed for operation in an asynchronous network.

3.3 Ring Architectures

Increasing concentration of traffic in fiber optic transmission systems and large hubs and customer demand have stimulated BCC study of deployment of survivable architectures in both the loop and interoffice networks. Three methods of providing survivability in the event of a single link (e.g., facility) or node (e.g., central office) failure are being used: geographically diverse-routed fiber optic transmission systems, DCSs, and rings.

Diverse routing of fiber provides protection against cable cuts for point-to-point systems. This scheme provides fast restoration of service (< 50 ms restoration time) and can be more cost effective than other alternatives. The disadvantages are that capacity is dedicated on a point-to-point basis limiting bandwidth rearrangement, and a large number of fiber pairs are required compared to, say, a ring architecture.

Path rearrangement using DCSs is a second method of providing network survivability. In the event of failures, the DCSs rearrange paths according to prestored maps, craft actions, or dynamic algorithms which determine available alternate routes. Use of prestored maps to provide survivability is complicated by the administrative efforts required to maintain them. Restoration times using the path rearrangement capability of DCSs can range from minutes to hours, depending on whether human intervention is required.

Ring architectures are the third survivable network architecture. Advantages of rings include simplicity, flexibility to rearrange services, and fast restoration time. Disadvantages include cost as rings require significantly more bandwidth than the other alternatives, and concerns regarding planning and maintaining a ring-based network.

Interest in ring architectures to provide survivability is based on the following requirements, which are driven by customer demand for survivable services: (i) diverse routing of fiber facilities, (ii) flexibility to rearrange services to alternate serving nodes, and (iii) automatic restoration within less than 2 s in the event of a single link or intermediate node (central office) failure.

Both SONET ADMs and DCSs should provide the capability to be used in a physical SONET self-healing ring network. Bellcore GR-1400-CORE [4] and GR-1230-CORE [5] give generic criteria for SONET ADM with the capability to be used in unidirectional path-switched and bidirectional line-switched self-healing rings, respectively.[4]

3.4 SONET Operations and Control

SONET provides single-ended performance monitoring and proactive maintenance, and eliminates the need for many traditional testing procedures.

The SONET provisioning process differs from the current process due to the flexibility and options allowed by SONET. Two examples are the administration of time-slot assignments in ADMs, and the administration of different types and levels of tributary signals within one line signal. The provisioning and administration capabilities needed for simple SONET architectures have already been incorporated into TIRKS®[11].

SONET section and line Data Communications Channels (DCCs) can be configured to provide a control network embedded within the transport network for control and operations. It is desirable to initially limit control networks to simple tree architectures until more experience is gained in using them. This is illustrated in Fig. 6, where the sample transport network is overlaid with a SONET control network connected to the Telecommunications Management Network (TMN). This type of control network suffices for a string of regenerators within a line and for most loop applications, but perhaps not for ring architectures. With tree architectures, operations data from remote locations can be brought back to the CO via the section DCC and routed to the appropriate Operations System (OS). Similarly, the OS can send operations commands to remote NEs. In some cases, the OS may use the DCC for software downloads to the NEs.

The OS-NE communication may be direct, through a backbone data communications network or through a mediation device. Gateway NEs with mediation function, such as a DCS, may also be used later on, as shown in Fig. 6. In a large CO with many SONET NEs, a LAN may also be used for intrabuilding operations data communications between NEs, gateway NEs, mediation devices, and OSs.

4. NEW NETWORK CAPABILITIES AND SERVICES

SONET has been primarily designed as a transport concept, and has been deliberately structured to make the SONET technology as independent as possible of specific services or applications. SONET is intended to provide the transport infrastructure

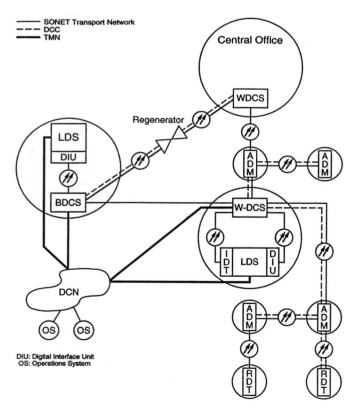

Fig. 6. Sample SONET transport and control architectures.

TABLE 3. SOME POSSIBLE SONET APPLICATIONS

Point-to-Point Transport	Survivability
Digital Loop Carrier	Rapid Provisioning
Digital Cross-connect Systems	Bandwidth on Demand
Switch Interface	Customer Control
Integrated DLC	Advanced Network Management
BISDN	DS3 CPE Interface
Supercomputer Networking	Intelligent CPE Multiplexes
FDDI Transport	Fiber to the Home

of the next three decades in much the same manner that T1 carrier technology and its descendants have provided the transmission infrastructure of the past three decades.

On the other hand, SONET has a number of attributes and capabilities that can be exploited to facilitate the development and delivery of new services, and it is definitely intended that SONET, in combination with new network switching and operations technology, will provide a powerful new service delivery platform.

For example, in the past, the ability to deliver large bandwidth services to the end user has been limited by the fact that the highest standardized module of bandwidth has been the 45 Mb/s DS3 signal; access to the full bandwidth of fiber transmission systems by the end user has been impractical. With the ability of SONET to concatenate payloads of multiple 51.84 Mb/s modules, it will, in principle, become possible to offer almost unlimited bandwidth transport services. This capability could be exploited to provide such services as high-speed dedicated interconnections between supercomputers.

As another example, the ability of SONET to transmit operational commands between NEs and OSs via its embedded DCCs, coupled with its ability to accommodate a very flexible mix of tributary signal types, could be exploited to build a new family of rapidly provisioned private line services. Conceivably, the capabilities and intelligence of the SONET NEs could cut the time required to provision such services from weeks or months down to a matter of hours or minutes.

Table 3 gives a list of some of the applications and services which have been mentioned for exploiting the capabilities of the SONET concept. It is still too early to understand which of these applications will be the most important, and how the use of SONET in these applications may either overlap or work synergistically with such concepts as the Broadband ISDN. It is clear, however, that in some of these areas, such as survivability, the LECs are planning major initiatives based on SONET.

5. EVOLUTION TO BROADBAND

An active international effort is currently underway to define the fundamental aspects of a forward-looking transport and switching architecture known as ATM broadband networking. The general concept of this architecture is that a wide range of customer data, voice, and video applications should be supported by a single network, using a limited set of user-to-network interfaces (UNIs) and internal network equipment configurations. The ATM Permanent Virtual Connection (PVC) services are available now. ATM Switched Virtual Connection (SVC) services are expected to be available in the 1995/1996 timeframe.

Because of the timing of SONET deployment (aggressive deployment beginning in the early 1990s), its maturity as a national and international standard, and its ability to carry a flexible mix of service payloads, SONET is supported in U.S. and international (ITU-T) standards bodies as the foundation for the physical layer of the ATM broadband network. Access to this network will be offered over a set of interfaces at 155 and 622 Mb/s, using a protocol known as the Asynchronous Transfer Mode (ATM), a form of high-speed, fixed-length packet switching. These rates correspond to the SONET STS-3 and STS-12 levels.

Because SONET's deployment timing in LEC networks will be coincident with a major emphasis on increasing deployment of fiber in the loop plant, SONET-based carrier systems will move very close to the end user during the 1990s. At first, the existence of SONET within the network will be invisible to the end user, who will continue to receive service via today's DS1 and DS3 service interfaces. But as more SONET technology is deployed, and as customer premises equipment (CPE) based on SONET becomes available, SONET-based interfaces to the network will be offered to the customer.

Initially, these SONET-based UNIs may support only conventional time-division multiplexed services. Major switching and transmission equipment manufacturers have indicated that the technology to support integrated SONET and ATM network access and switching is now available.

Because of the ability of SONET to mix virtual tributaries formed by different multiplexing techniques in a single bitstream, we can imagine an evolutionary strategy in which loop fiber systems operating, say, at 622 Mb/s may support a mix of 155 Mb/s tributaries, some of which are formed based on ATM and some of which are formed based on time-division multiplexing. The pace of evolution toward a purely SONET/ATM-based network will be determined by the growth in demand for high-capacity switched services.

6. SUMMARY

Deployment of SONET equipment has been taking place in BCC networks since 1990. In the initial SONET deployment, transport systems and multiplexes were used in both the interoffice and access networks in essentially the same configurations as the current proprietary fiber transport systems, namely, point-

to-point. Multinode networks in simple linear and tree configurations were trialed in 1991 using ADMs, which provide some grooming and hubbing capabilities. Integrated local access capability was available in about the same timeframe.

Aggressive SONET deployment began in 1992. Because of the use of standard interfaces, it is expected that initially-deployed point-to-point systems will be interconnected and smoothly evolve to provide multivendor multicarrier compatible end-to-end networks using ADMs and DCSs. SONET will provide cost savings due to increased network integration, improved bandwidth-management capabilities, enhanced operations capabilities, and the transport infrastructure for new services.

SONET offers flexibility and many options. Managing these effectively is crucial to meeting service requirements and maximizing operational efficiencies. BCC SONET deployment will implement a subset of the features to ensure cost-effectiveness and compatibility.

References

[1] ANSI T1.105, American National Standard for Telecommunications — Digital Hierarchy — Optical Interface Rates and Formats Specifications (SONET), and T1.105.01 to T1.105.09.

[2] GR-253-CORE, Issue 1, Dec. 1994, Synchronous Optical Network (SONET) Transport Systems: Common Generic Criteria (a module of TSGR, FR-440).

[3] TR-NWT-000496, Issue 3, May 1992, SONET Add-Drop Multiplex Equipment (SONET ADM) Generic Criteria (a module of TSGR, FR-440).

[4] GR-1400-CORE, Issue 1, Mar. 1994, SONET Dual-Fed Unidirectional Path Switched Ring (UPSR) Equipment Generic Criteria (a module of TSGR, FR-440).

[5] GR-1230-CORE, Issue 1, Dec. 1993, SONET Bidirectional Line-Switched Ring Equipment Generic Criteria (a module of TSGR, FR-440).

[6] TR-NWT-000233, Issue 3, Nov. 1993, Wideband and Broadband Digital Cross-Connect Systems Generic Criteria (a module of TSGR, FR-440).

[7] TR-NWT-000917, Issue 1, Dec. 1990, SONET Regenerator (SONET RGTR) Equipment Generic Criteria (a module of TSGR, FR-440).

[8] TA-NWT-000782, Issue 2, Oct. 1992, SONET Digital Switch Trunk Interface Criteria.

[9] R-NWT-000303, Issue 2, Dec. 1992, Integrated Digital Loop Carrier System Generic Requirements, Objectives, and Interface (a module of TSGR, FR-440).

[10] R-NWT-000303, Supplement 2, Issue 2, Dec. 1993. (Supplement 2 to TR-TSY-000303). IDLC System Generic Requirements, Objectives, and Interface: Feature Set C - SONET Interface (a module of TSGR, FR-440).

[11] SONET Add/Drop Multiplex Equipment Administration Using the TIRKS Provisioning System, *GLOBECOM '89 Conference Record*, Nov. 1989, vol. 3, pp. 1511–15.

SONET Implementation

Does the status of SONET deployment meet the original expectations of the system's developers?

Yau-Chau Ching and H. Sabit Say

In the early stages of standards development, it was recognized that the synchronous optical network (SONET) infrastructure would involve several complementary types of transport equipment introduced in an integrated but phased manner, and that a ubiquitous SONET deployment would lead to a restructured transport network, providing many innovative features, but requiring substantial new operations support. Several years have passed since the approval of the first set of SONET standards. Thus, in this article, we examine the status of SONET development and deployment to see whether or not they meet original expectations, we discuss issues arising from deployment, such as the use of new features and the necessary support from operations systems, and we also briefly discuss future directions of SONET.

Standards Perspective

When SONET was introduced in 1984, we proposed that fiber optic transport systems should be examined from a network perspective. This network approach to standardization was readily accepted by American National Standards Institute (ANSI) Committee T1 and later proposed to CCITT (renamed ITU-Telecommunications Sector). By 1988, CCITT had adopted a set of interface standards for synchronous digital hierarchy (SDH) as Recommendations G.707, G.708 and G.709, and ANSI had published standards for SONET rates and formats and for the optical interface, as T1.105 and T1.106, respectively. These standards provide specifications of interfaces common to all SONET equipment; in particular, they carefully provide operations overhead capacity to enable numerous new network features. At the same time, standards work was started on protocols for the operations communications network to control the SONET subnetwork, and on information models for the SONET subnetwork and its components. These activities resulted in a seven-layer OSI protocol stack for the data communications channel (DCC), and an information model for terminations, protection switching, and performance monitoring from an equipment view.

Once these standards were approved, Bellcore issued a similar set of proposed criteria in TR-NWT-000253, "SONET Transport Systems: Common Generic Criteria," a technical reference that includes transport equipment and transport interface specifications. Bellcore followed by a series of operations interfaces specifications. Table 1 lists interface specification documents relevant to SDH/SONET from ITU, ANSI, and Bellcore. A comprehensive list of references can be found in the above-mentioned technical reference.

As a global transmission standard, SDH needed to be backward compatible. Because existing networks around the world are based on a number of different digital hierarchies, intensive negotiations between T1X1 and CCITT took place in 1988. The result was that SDH in CCITT essentially became an umbrella standard that allowed local variations: SDH-Europe to accommodate CEPT hierarchy, SDH-SONET to accommodate North American hierarchy, and SDH-Japan to accommodate Japanese hierarchy. A T1X1 technical report enumerates these variations [1]. We should point out that, in addition to the commonalty of optical parameters and common rates that are universally accepted, local variations disappear for signals at and above 155 Mb/s that carry new payloads such as asynchronous transfer mode asynchronous transfer mode (ATM) cells.

Product Development

As SONET interface standards were being developed, products using this common interface started to emerge. Bellcore issued a series of equipment specific requirements to promote the development of SONET equipment. These documents (Table 2) cover terminal multiplexers, add drop multiplexers (ADMs), regenerators, digital cross connect systems (DCSs), and switch interfaces. All these were familiar products with proprietary counterparts — except for the ADMs, which represent a first application

Reprinted from *IEEE Communi. Maga.*, vol. 31, no. 9, pp. 34–40, Sept. 1993.

of the synchronous multiplexing techniques. Synchronous multiplexing could reduce the cost of equipment — since lower level tributaries can be directly derived from the high-speed line signal — but the equipment structure itself is a more compelling reason for the ADM development. One ADM can replace two back-to-back terminal multiplexers. Since a typical application requires only a fraction of the traffic to be dropped and added at each ADM node, only a fraction of the low speed termination line cards is needed, hence reducing the cost of ADM. It is estimated that a typical SONET ADM costs 25 to 40 percent less than a back-to-back terminal multiplexer arrangement.

During the early days of SONET, equipment suppliers took the lead in developing equipment in the United States. By 1989 Alcatel and Fujitsu had developed terminal multiplexers at 50 Mb/s and 150 Mb/s rates, and the first 2.4 Gb/s system was introduced by NTI in 1990. By 1991, there were at least 13 suppliers developing over 38 different network elements, together with a number of companies making test equipment and SONET IC chips. Table 3 provides a partial list of SONET transport products in the United States This list is from public sources and does not include ATM equipment or multiplexers for use in customer networks. (Bellcore does not recommend products and nothing contained in this article is intended as a recommendation of any product to anyone.)

In the long term, SONET is expected to become a universal fiber optic interface that will link not only equipment in the telephone network but also end users. In most cases, end users would still access the network through existing digital hierarchical signals but, with emerging applications, this will certainly change. In the standards bodies, mapping specifications have been completed for FDDI, DQDB, and ATM signals, and several ATM manufacturers have announced products using SONET as the physical transport for ATM cells. In 1991, Northern Telecom announced their "Fiber World" family of products, and AT&T followed with their "Service Net 2000" family of products. Both families of products envision a seamless network built by switching, transport, and access equipment; they highlight the backbone transport with SONET.

Operations Support

*B*y itself, SONET technology does not represent a dramatic departure from existing technology nor does it significantly alter the direction or pace of network evolution. With or without SONET, optical fiber has become the primary medium for high-speed telecommunications, and the broadband network elements being introduced into the fiber network have become more intelligent to provide more flexibility, maintainability, and manageability. SONET merely serves as a tool to facilitate this network evolution.

As intelligent SONET network elements are deployed, however, network operations systems have to be upgraded. These intelligent network elements break down the traditional boundaries between interoffice and loop networks, they impact the processes of network capacity provisioning, service activation, and service assurance, and they necessitate the establishment of communications links to the operations systems, so they can be controlled remotely. For SONET, the standards bodies decided that

Document	Subject	Year published
ITU-T (formerly CCITT) Recommendations on SDH		
G.707 to G.709	SDH rates and formats	1988
G.781 to G.784	Equipment functions	1990
G.957	Optical interfaces	1990
G.958	Line systems	1990
G.803	Network architecture	1993
G.831	Management capabilities	1993
G.774	Management information model	1992
ANSI Standards on SONET		
T1.105	SONET rates and formats	1991
T1.106	Optical parameters (superseded by G.957 and TR 253)	1988
T1.117	Optical parameters-short reach	1991
T1.118	OAM&P communications	1993
Bellcore Generic Requirements on SONET		
TR-NWT-000253	SONET common criteria	1991
TA-NWT-000253, Iss.7	Routing and LAN support	1992
TR-NWT-001042	SONET information model	1992
TA-NWT-001042, Iss.3	Ring information model	1992
TA-NWT-001250	File transfer	1992

■ **Table 1.** *SONET/SDH interface specifications and generic requirements documents.*

Document	Subject	Date published
TR-TSY-000496, Issue 3	Add-drop multiplexer	5/92
TA-NWT-000233, Issue 4	Digital cross-connect system	11/92
TR-TSY-000303, Suppl 2	Digital loop carrier system	10/89
TR-NWT-000917	Regenerator	12/90
TA-NWT-000782	Switch interface (trunk side)	10/92

■ **Table 2.** *Bellcore generic requirements documents on SONET equipment.*

two embedded overhead channels in the basic format, i.e., section DCC and line DCC, would be used for communication between a network element and operations systems and among network elements, to avoid an expensive overlay operations communications network. A seven-layer open systems interconnection (OSI) protocol stack would be used for this DCC network, gateway mediation devices would be allowed to interconnect with operations systems and, furthermore, the transaction oriented messages would be specified and communicated using the common management information protocol (CMIP).

Some upgrades of operations systems are relatively simple, and many have already been implemented to support the SONET network elements. Others require fundamental changes to existing operations systems or development of new operations systems, either at the network management layer or element manager layer. For example, only a few of the current operations systems for the trans-

Company	Product	Type
ADC Telecomm	DS3/OC-1	TM
	Soneplex	DLC
Alcatel	TM-50	TM (OC-1)
	ADM-50	ADM (OC-1)
	SM family, FTS family	ADM and RGTR (OC-3 to OC-48)
	AN family	DLC (OC-1, OC-3)
	1633 SX	B-DCS
	1631 SX	W-DCS
ANT	SLA 16	TM (STM-16)
Ascom Timeplex	SONET TX3/SuperHub	DCS, T3 Multiplexer
AT&T	DACS III-2000	B-DCS
	DACS IV-2000	W-DCS
	DACScan-2000	NM
	DDM-2000 family	ADM (OC-3, OC-12)
	FT-2000	ADM (OC-48)
	5 ESS-2000	Switch
	SLC-2000, BRT-2000	DLC
DSC Communications	DEX family	DCS
	Litespan-2000	DLC
Fujitsu	FLM 50/150	TM (OC-1, OC-3)
	FLM 150, 600, 2400 ADM	ADM (OC-3 to OC-48)
	FDLC	DLC
	FACTR	DLC (OC-3)
	FLEXR-Plus	EM
General DataComm	Megamux TMS/S family	TM (OC-3, OC-12)
NEC	IMT family	ADM (OC-1 to OC-12)
	ITS family	ADM, RGTR (OC-12, OC-48)
	ISC-303	DLC
Northern Telecom	S/DMS SuperNode	Switch, DCS
	S/DMS TransportNode family	
	OC-12 NE-TSS, TBM	TM
	OC-48 NE-TSS	ADM
	S/DMS AccessNode family	DLC (OC-3, OC-12)
	OPC	EM
Reliance Comm/Tec	SONET DISC*S	DLC (OC-3)
Seiscor	FiberTraq	DLC
Tadiran Telecomm.	T::DAX	W-DCS
Telco Systems	S-828	DLC
	HyperLynx 150, 600	TM
Tellabs	TITAN 5000, 5500	DCS

Key to Types
ADM — Add/drop multiplexer
B-DCS — Broadband digital cross-connect system
DCS — Digital cross-connect system
CPE — Customer premises equipment
DLC — Digital loop carrier
EM — Element manager
LTE — Line terminating equipment
NM — Network manager
RGTR — Regenerator
TM — Terminal multiplexer
W-DCS — Wideband digital cross-connect system

■ **Table 3.** *A partial list of SONET transport products in th U.S.A.*

mission network use CMIP; most communicate with transaction language 1 (TL1) or even with the primitive E2A telemetry. Bellcore operations systems initially support SONET equipment using TL1 interfaces. In the future, they will support SONET using OSI protocols with a common management service element (CMISE) at the application layer. Bellcore has issued an operations systems support plan for SONET in a special report [2].

SONET Deployment

Why should anyone deploy SONET? Network providers typically examine three areas of potential economical benefits derived from a new technology: first-cost savings, operations savings, and enhanced revenues. Because of its relatively new technology and new development, a SONET terminal multiplexer in 1990 typically cost 10 to 15 percent more than its mature, proprietary counterpart (asynchronous fiber optic terminal) and was not widely available. Early Bellcore generic requirements concentrated on SONET multiplexers for point-to-point and tree applications, reflecting the trend in the industry and needs of local operating companies. Likewise, Bellcore operations systems were enhanced only to support these simple architectures, i.e., network capacity provisioning systems were modified to recognize SONET equipment and facilities, and network monitoring systems were modified to recognize SONET alarms. For these simple applications, SONET systems perform much like proprietary systems, and operations savings and revenue enhancement are not realized.

SONET is now a stable standard, which should lead to longer product life, more competition, and reduced equipment prices. Also, SONET is conducive to more flexible, efficient, manageable architectures and new services, which should lead to reduced operations cost and enhanced revenues. These economic advantages are difficult to quantify before equipment is deployed. Telephone companies have been asked to deploy a technology whose impact on price competitiveness, network management, efficiency, flexibility, and new revenues could be years away.

Still, strategic decisions to deploy SONET were made by some network providers. These strategic decisions stemmed partly from the belief that SONET will eventually become the transport technology of choice for the future, and partly from the willingness to take a risk on the new technology. Deployment by these network providers could still, however, be considered prudent and cautious, as needs for new fiber systems developed and equipment prices dropped. Since 1989, therefore, network providers in the United States have deployed SONET steadily without much fanfare. By March 1992, about seven thousand SONET ADM terminals were deployed in a point-to-point architecture, and more than twenty thousand terminals were installed by the end of 1992. Along the way, the mid-span meet of terminals from different manufacturers was demonstrated in 1991 by Contel; for the first time in history, fiber optic terminals from two different companies (Alcatel and Fujitsu) were built to a common standard interface specification and were able to communicate with each other.

Two possible interpretations could explain the conservative level of deployment. On one hand, the numbers are small, activities are dominated by three operating companies that concentrate on point-to-point systems performing the same functions as earlier proprietary terminal multiplexers, and operations support is not much different from that for proprietary systems; therefore, there is no reason to expect SONET deployment to take off rapidly. From an optimistic view, however, the current deployment status can be seen as a major milestone for the industry and a significant first step in the building of a standardized fiber optic infrastructure; it represents a basic shift of mind set and, now that momentum has been reached, many other network providers are seriously considering the deployment of SONET equipment. Sales volume for SONET equipment in 1993 is projected to double from 1992. In addition, many suppliers have capped their proprietary systems and are building SONET equipment exclusively.

SONET Penetration

The SONET deployment schedules from 1989 to 1992 were mostly for growth traffic. Assuming SONET ADMs are primarily used for new growth traffic, it is estimated that only 40 percent of the interoffice network will be SONET by the year 2003; however, if existing proprietary systems are replaced upon retirement, SONET deployment would be much accelerated. Based on depreciation schedules, for example, if fiber optic systems are retired in 10 (or 15) years and replaced with SONET-compliant equipment then SONET penetration would reach 99 percent by 2003 (or 2008).

SONET deployment can be further accelerated by modernizing the existing telephone network. Significant network savings and other benefits can be realized through restructuring of the overall network. In fact, identification of SONET ring applications (and subsequent Bellcore generic requirements and industry standards work in 1991) has led to the possibility of modernizing the network with SONET rings. Even with this late start, it is estimated that more than 30 percent of SONET equipment will be for ring applications by 1997.

We are further encouraged by recent deployment announcements from MCI, Sprint, and several Bell regional companies. MCI has a contract with Northern Telecom for $250M, and Sprint has a procurement plan for about $500M. These new deployment scenarios would start a move from a simple point-to-point architecture to ring architectures, including loop applications, where the potential for expansion seems much higher.

SONET Rings

As a consequence of the standards process, SONET allows a plethora of extra capabilities in the network elements through its overhead structure. Bellcore requirement documents specify these capabilities but no one expects a super single transport equipment that implements all the potential capabilities. These features would be introduced into appropriate network elements as applications emerge.

What applications will drive SONET network infrastructure? Survivability and reliability will greatly impact the deployment of SONET. In 1992, we witnessed the first installation of a SONET ADM ring, which provides sharing of fiber capacity among multiple locations and added protection against fiber cable cut and equipment failures. SONET rings deliver the first of many promising new features offered by the SONET standards. Since a single ADM replaces two back-to-back terminal multiplexers, it saves capital expense. A ring shares fiber capacity and requires fewer terminals for the same traffic, while achieving protection against fiber cut without the more expensive one-for-one protection switching with diverse routing. Therefore, ring deployment results in better survivability and reliability of the network at reduced cost. Most network providers are expected to deploy SONET rings in 1993. Many of these rings will displace potential point-to-point applications.

The simplest ring architecture is the unidirectional path switched ring (UPSR) described in a Bellcore technical reference [3]. In this architecture, two copies of a signal traverse through the ring in opposite directions around the ring, and the destination selects the better one. Because of this route diversity, a good copy will reach the destination even when the fiber is cut. A UPSR is also called a dedicated protection ring because the two copies of the signal always occupy that particular bandwidth around the ring. The UPSR achieves its maximum efficiency when all traffic on the ring is between a hubbing node and other subtending nodes, as is the case for access applications.

A more complex, but bandwidth-efficient ring architecture is used in a bidirectional line switched ring (BLSR). In this architecture, signals spanning disjoint segments of the ring can occupy the same working and protection capacity, and therefore share that bandwidth. The ring becomes more efficient as more disjoint signals are accommodated. To coordinate the sharing, however, a set of messages has to be communicated among the nodes around the ring so that the protection capacity can be activated properly. This set of messages on the SONET overhead channel has been standardized in Committee T1 and is at the final approval stage [4].

To introduce a SONET ring into the network, we need equipment, planning tools, and other operations support. UPSR ADMs have been available since 1992 and BLSR ADMs will be available in 1993. Planning tools has been developed by Bellcore and ring ADM suppliers. To support rings, circuit provisioning features will be introduced into operations systems as well as element managers that control rings.

Digital Cross Connect Systems (DCSs)

What comes next? Manufacturers have been developing SONET DCSs, and products are expected to be available for deployment in early 1994. The basic functions of these SONET DCSs, including broadband DCS (B-DCS, cross-connecting at DS3 rate and above) and wideband DCS (W-DCS, cross-connecting at DS1 rate and above), are the same as the functions of existing proprietary B-DCSs and W-DCSs. They provide a centralized point for grooming and testing, and consolidation of traffic. SONET DCSs differ

SONET is now a stable standard, which should lead to longer product life, more competition, and reduced equipment prices.

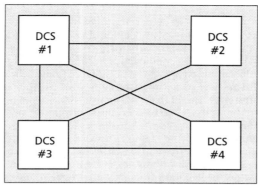

■ **Figure 1.** *DCS mesh architecture.*

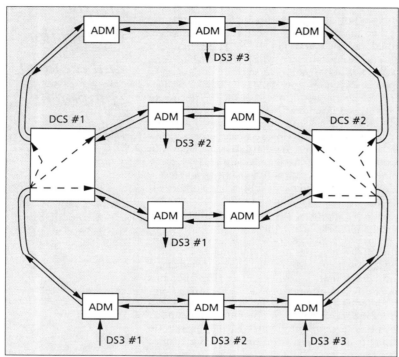

■ **Figure 2.** *Logical ring application with DCS.*

from proprietary DCSs as a result of two major additional features: standard operations interfaces and standard transport interfaces with direct optical termination, allowing direct optical interconnection of DCS with SONET terminal multiplexers, rings, and other DCSs. Thus, SONET DCSs can connect SONET islands (expanding end-to-end SONET continuity) and provide flexibility points for traffic management and restoration. Finally, the SONET DCS is a natural gateway point for operations communications. With a standardized DCC on all SONET links, operations messages from the operations network can be routed through a gateway DCS to all of its subtending network elements. In addition, SONET standards allow interconnection of equipment from different manufacturers, an important consideration in the deployment of hub equipment. Tellabs' Titan 5500 communicated with a Fujitsu terminal at SuperComm '92 and with a Northern Telecom terminal at SuperComm '93.

Three key areas that must be considered when deploying SONET DCS architectures are flexibility, survivability, and economics. Since the inherent reconfiguration capability of SONET DCSs provides greater flexibility than an ADM network, and the survivability of the architecture depends on the configurations in which the DCS is applied and the features available in the DCS, the question then becomes: how will SONET DCS-based configurations and features provide flexibility and survivability at a competitive cost?

The DCS mesh configuration is a possible SONET DCS architecture that provides capacity sharing, protection against fiber cable cut and equipment failures, and more flexibility than rings (Fig. 1). Traffic from DCS #1 could be routed to DCS #4 either directly or through DCS #2 and #3. This flexibility allows for more efficient use of the bandwidth as well as restoration of traffic during a cable failure. For example, spare capacity from DCS #1 to DCS #2 and from DCS #2 to DCS #4 can be reserved to protect the direct traffic between DCS #1 and DCS #4. This protection mechanism is similar to that of a ring but, unlike the ring, additional protection can be provided alternatively by the spare capacity from DCS #1 to DCS #3 and then from DCS #3 to DCS #4. Protection against DCS failures can also be provided by accessing the same user traffic from two DCSs, but this flexibility and protection against fiber and equipment failures with only DCSs carries a price tag. W-DCSs may be too big an equipment and — with sophisticated controllers — too expensive and unsuitable for sparse traffic applications. In addition, to provide automated survivability, a restoration mechanism for multiple DCSs in a meshed configuration must be developed, perhaps standardized, and coordinated amongst multiple suppliers and users.

Hybrid Network: Rings and DCSs

Since the SONET standards allow direct interconnection of different types of equipment, a more likely deployment scenario is a hybrid network that maximizes the simplicity of rings and the flexibility of DCSs. Many combinations can be envisioned. In one such example, called a "logical ring application," the DCSs serve as nodes on rings and provide interconnection between rings (Fig. 2). Each partial ring, which consists of a chain of ADMs connecting two DCSs, is designed to maximize the common community of interest. Most traffic from an ADM node therefore would terminate on other ADMs of the same ring or the two DCSs. For traffic terminating on a different ring, the DCS serves as a circuit tandem. The two-DCS configuration protects against DCS failures, a feature of growing importance as the traffic passing through the DCS becomes large. The bandwidth grooming functions of SONET rings allow for better filling of the optical facilities that terminate on DCSs and increase access efficiency. For a link that carries a high volume of traffic, of course, the DCS optical termination could still originate from a simple terminal multiplexer, possibly in a one-for-one protection switching arrangement with diverse routing. Similarly, a direct high capacity link can be deployed between the two DCSs, also using one-for-one protection with diverse routing, as a common connecting link for many partial rings terminating on these two DCSs.

New Services

*T*he most important driver for SONET deployment in the long run is the prospect for new services and new revenues. While the private line service can be a near term revenue source, SONET will prove itself over the long term as a physical platform for broadband services such as ATM and video.

In its simplest form, a private line is a pipeline of a specified capacity and format between two end-user premises, i.e., end-user-to-end-user service. It is more economical for an end user to lease lines in the public infrastructure from network providers than to build its own network. In addition, network providers can provide maintenance services and respond quickly to changes in a customer's traffic demand. Furthermore, an enhanced public network can provide better survivability and band-width management.

To implement a SONET private line service, a SONET infrastructure must connect the two end user locations. Furthermore, the SONET user-network interface (S-UNI) has to be standardized and customer premise equipment has to be developed. If the connections cross a local access and transport area (LATA) boundary, then a SONET inter-carrier interface (S-ICI) also has to be specified. None of these are major issues except for the building of the infrastructure. The interface specifications for S-UNI and S-ICI are subsets of the existing SONET interface specifications, and many existing network elements could easily be modified for use as customer premise equipment (CPE). Since SONET deployment is still on a small scale, however, the SONET infrastructure might not be available for all end users unless specific links are built at customer request. In addition, variations in customer needs may have to be met by the infrastructure; for example, to carry the concatenated signal at 155 Mb/s to provide ATM connections, the CPE should have provisions for an STS-3c line card and the SONET infrastructure would have to route these signals to the destination intact.

SONET private line services can also be provided as an end user network access service, in which case a SONET private line links the end user's SONET CPE to a network provider's SONET multiplexer. In one scenario, the SONET private line provides physical transport for accessing public network services such as frame relay, switched multimegabit data service (SMDS), integrated services digital network (ISDN), and ATM cell relay, as well as POTS (Fig. 3). Note that the network access service is a point-to-point connection at the SONET layer, but the underlying subrate services are multipoint or switched services.

Currently, several private line tariffs have been filed in the United States and many intercarrier interface discussions have also begun. In February 1993, Comdisco Disaster Recovery Services created a 622 Mb/s subnetwork within Bell Atlantic's 2.5 Gb/s SONET ring in northern New Jersey, and in April 1993 they began to test the same type of subnetwork within an Ameritech SONET ring in Chicago. Plans exist to do the same in other regions around the country. Trials are also going on for new services. Currently five national research and education network (NREN) projects (Casa, Blan-

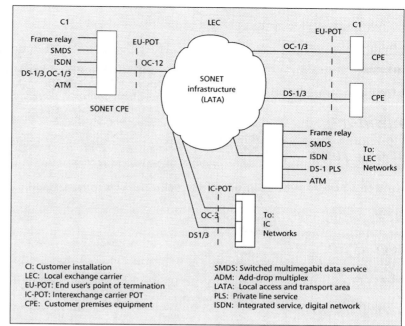

Figure 3. *Private line access scenario.*

ca, Nectar, Aurora, and Vista) have decided to use SONET as the physical transport layer. In addition, Pacific Bell has just announced a $35 million project called California Research and Education Network (CalREN) around San Francisco and Los Angeles. The basic transport platform for CalREN is also ATM over SONET.

Challenges

A number of challenging technical issues for further SONET deployment remain. In particular, planning tools and other operations support systems are needed if a hybrid SONET network is to operate in an integrated fashion. Published plans indicate possible availability in the 1994-95 time frame. These operations support systems would prove the capabilities of SONET and fuel the acceleration of the implementation of the SONET infrastructure. Bellcore has been working with local operating companies and equipment suppliers intensively to identify the industry trends and technical issues, and their special report on SONET network and operations should help suppliers, telephone companies, and Bellcore coordinate efforts in SONET deployment [2]. Also, in anticipation of the complex network issues, T1 Technical Subcommittee T1X1 has created a new working group T1X1.2 to study network architecture aspects.

The deployment of SONET can be further accelerated if new high bandwidth services are offered by network providers. Rapid growth of data communications traffic and ubiquitous video services would drastically alter the demand for a high capacity transport infrastructure. Also, if these services can be carried over SONET access networks, (e.g., SONET private line services) the demand for SONET equipment would again drastically increase. Technical issues for the SONET interfaces required to support new services have been resolved in T1E1 and the relevant standards are near completion. Many test beds and trials for gigabit computer networks and ATM switches with SONET terminations are in planning stages.

The introduction of new high speed services will accelerate SONET penetration.

SONET is already here. It will penetrate the entire transport infrastructure by deployment through normal growth today and by retirement and modernization in the near future. While we expect that the point-to-point applications will still be an important component of SONET architecture, a hybrid network that combines point-to-point, rings, and DCS equipment to maximize their respective benefits will eventually gain more acceptance. Moreover, the introduction of new high speed services will significantly accelerate SONET penetration.

References

[1] Draft T1 Technical Report on SONET-SDH Comparison contained in T1X1.2/93-024 RI.
[2] Bellcore Special Report SR-TSV-002387, "SONET Network and Operations Plan: Features, Functions and Support, Issue 1," August 1992.
[3] Bellcore Technical Reference TR-NWT-000496, Supplement 1, "SONET ADM Generic Criteria: A Unidirectional, Dual-Fed, Path Protection Switched, Self-Healing Ring Implementation," September 1991; also, Bellcore Technical Advisory TA-NWT-001400, April 1993, contains additional criteria on the same subject.
[4] Draft of proposed American National Standard SONET Automatic Protection Switching contained in T1X1.5/93-057.

The Role of SDH/SONET-Based Networks in British Telecom

S. WHITT, I. HAWKER, J. CALLAGHAN, G. L. BLAU, J. H. MACKENZIE, AND A. M. MANLEY

BT LABORATORIES

MARTLESHAM HEATH, IPSWICH, IP5 7RE, UK

Abstract—In order to satisfy ever-increasing customer demands for new services, and to further improve quality of service and operational efficiency, BT is at present introducing SDH/SONET-based systems and networks. These provide increased flexibility through easier realization of the multiplex and cross-connect capabilities within the network, and have enabled the introduction of integrated management and protection, resulting in the realization of the SDH/SONET Managed Transmission Network.

INTRODUCTION

The role of Synchronous Optical Networks (SONET), and its ITU-T twin Synchronous Digital Hierarchy (SDH), is to provide an international interface standard and frame format for digital transmission at the physical and management levels [1], [2]. This will result in easier interworking between telecommunication operators, reduced equipment cost, and integrated management and protection systems. SONET and SDH have been developed over the last few years, and detailed recommendations covering all aspects of the standards are complete or under study in various ANSI and ITU working parties. BT is currently deploying SDH-based systems in its terrestrial and submarine networks [3]–[7]. This paper describes the advantages, and possible disadvantages, of SONET-based networks that can be expected from the deployment of these networks, together with a network structure, control structure, and management philosophy that could be utilized within BT's networks.

SONET-Based Networks

The existing BT plesiochronous digital transmission network (conforming to ITU-T Recommendation G.702) supports advanced digital switching telephone exchanges and digital data services. However, in common with other digital telecommunications networks, there are limitations which impede the realization of the full potential of emerging technologies including:

- Absence of a worldwide standard for bit rates and frame formats, leading to international interworking problems.
- Complexity of the frame structure, such that a "multiplex mountain" is needed to assemble 2 Mb/s tributaries into 140 Mb/s, via 8 and 34 Mb/s, with bit interleaving, justification, and addition of overheads at each stage.
- Lack of user-accessible overheads so that additional systems are required for operations, maintenance, and control facilities.
- Lack of flexibility, since it is uneconomic to implement automatic cross-connects and "add/drop" facilities needed for flexible networking.

Most of these limitations are overcome by use of SDH/SONET equipment with the inherent advantages of:

- Direct access to tributary signals without the need to demultiplex the entire aggregate signal (leading to reduced hardware).
- Comprehensive and easily accessed overheads giving good control, operation, and maintenance capabilities.
- Planned growth to higher bit rate systems as technology evolves.
- Standard optical line interface signal structure and communication protocols to allow interworking of multivendor equipment.
- Standard management architectures and management communications proposals again, to support interworking of multivendor equipment within a single management domain.
- Utilization of the latest technology.

Use of SDH/SONET equipments in telecommunications makes flexible networking economic and eases the introduction of a Managed Transmission Network (MTN), with centralized or distributed control, and fully integrated OAM&P facilities.

However, there can also be risks due to the wholesale introduction of software control, and the simultaneous concentration of traffic in a few high-order links and cross-connects. For example, software can control every cross-connect and multiplexer in a network so that human error at the network control or element manager level, or the introduction of a software fault or virus, could lead to a network catastrophe. For these reasons, SDH-based networks should be introduced with care, using fully tested software, and topologies which have been modeled for robustness and reliability.

OBJECTIVES OF THE BT NETWORK

The next generation of the BT network must meet ever-increasing customer demands and be capable of future enhancements. The main objectives are:

(a) Flexibility: To respond quickly to the ever-changing customer demand and to adapt to growth of the network,

(b) Quality of Service: Path availability will be improved by use of less hardware in a digital path and automatic protection techniques [8], [9],

(c) Monitoring: Use of embedded operations channels (EOCs) to permit end-to-end monitoring of digital paths at all levels of the hierarchy,

(d) Reduce Costs: Transmission equipment costs should be reduced by purchasing equipment on a world market, and planning and operating costs reduced with enhanced flexibility and integrated management systems,

(e) Future Proofing: Use of optical systems, with almost unlimited bandwidth, and software-controlled switches, will ease future growth and readily support new initiatives and services such as ATM, from relay, etc.,

(f) Broadband: Advanced services such as High Definition Television (HDTV) will be supported on SDH/SONET using standard mappings.

In the medium term, these objectives will be met by the introduction of new SDH-based network architectures and management systems as described in the following sections. In the longer term, it is expected that advanced developments such as Broadband ISDN, optical access networks, and all-optical networks will play a major role [10]–[12].

Structures

Historically, the BT transmission network is divided into three layers, namely, trunk, junction, and local. The trunk network carries long-haul traffic over a highly meshed mixture of coaxial cable, microwave radio, and optical fiber links (Fig. 1) which is manually reconfigurable and monitored on a link basis. In contrast, the junction and local networks were based mainly on star topologies, but also manually configured.

The future MTN will be divided into hierarchical tiers, allowing autonomous control and protection of each tier and structured deployment and growth. In addition, it is likely that all long-haul Private and PSTN traffic will be transported on a common MTN to simplify control. Factors such as existing duct topologies and the location of BT buildings point towards a four-tier network (Fig. 2):

(a) Inner Core Tier 1. The inner core replaces the trunk network and has two tiers. Tier 1 consists of a small number (20–50) of large digital cross-connects linked by high bit rate optical systems. This will form the main long-haul transmission medium for the country with the cross-connects providing traffic flexibility, protection, and grooming facilities. Because of the volume of traffic, path availabilities and protection are of prime importance. In the future, coherent transmission and wave division multiplex (WDM) techniques may be used to augment TDM techniques at this level.

(b) Inner Core Tier 2. Each of the major Tier 1 nodes will have a catchment area of 20–60 lower level cross-connects. There will be approximately 400 Tier 2 nodes, acting as gateways between the inner core and outer core networks.

(c) Outer Core Network. The outer core replaces the junction (interoffice) network, and will consist of up to 6000 nodes interconnected by 155/622 Mb/s optical links as star, chain, or ring networks of add/drop multiplexers. These will distribute traffic between digital local exchanges (Central Offices) and their remote concentrator units within the outer core, and convey traffic to the inner core for wider geographical distribution via the gateways.

Fig. 1.   The BT trunk network.

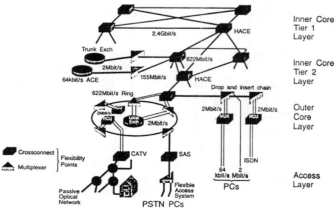

Fig. 2.   Possible future MTN/SDH network hierarchy.

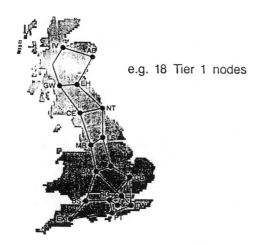

e.g. 18 Tier 1 nodes

Fig. 3.   Candidate inner core (Tier 1) network.

(d) Access Network.   This will provide local distribution to the customer. It is the largest part of the network, consisting of the links between customers and the 6000 outer core sites. The SDH network will carry existing and future services such as Megastream (an unstructured 2 Mb/s business service), Kilostream (31 × 64 kb/s data), PSTN, and vision services.

In addition to the four tiers, there is also a logical tier 0 which forms the international transmission layer from BT's international gateways/switching centers to corresponding international operators.

Topologies

The key network requirements influencing the choice of topology at each level are:

- High availability incorporating physical route diversity and automatic protection over digital paths,
- Avoidance of node or regional isolation under fault conditions (e.g., by use of dual parenting, where a node has diverse routes to independent nodes in the next hierarchical level),
- Rapid reconfiguration and provision of bandwidth,
- Structured network design for ease of introduction and growth,
- Use of existing ducts where possible to reduce costs,
- Use of nonblocking cross-connects in the inner core,
- Node spacings which give a near optimal balance of cross-connects and transmission costs.

i) Topology of the Inner Core Tier 1.   One possible network has 18 Tier 1 nodes (Fig. 3) located on a dual highway running north–south, supplemented by greater meshing in the regions of higher traffic density in the south. This network may be partially or fully interconnected, depending upon node sizes and grooming requirements.

ii) Topology of the Inner Core Tier 2.   A principal requirement here is for protection autonomy and containment of faults within each regional area. A possible topology is shown in Fig. 4 (although others are also being studied) where Tier 2 cross-connects are arranged into cells with dual access to each Tier 1 node.

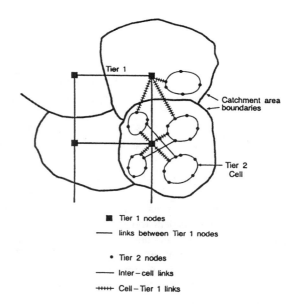

Fig. 4.   Tier 2 candidate network topology.

iii) Outer Core and Access Network Topology.   For the outer core, self-healing rings are favored, since they can be introduced as autonomous units and later linked into an MTN. Traffic can also be routed in both directions around the ring as a form of diverse 1 + 1 protection. SDH digital relay radio systems are also being considered for ring closure [13] for expediency or where the geographical landscape makes it economic. For the access network, star topologies are envisaged, following existing ducts, with 1 + 1 link protection for good availability. In addition, fibers can be shared using passive optical network (PON) architectures and time division multiple access (TDMA) transmission techniques.

Implementation

SDH/SONET-based networks in the U.K. core will be implemented in stages, to allow integration into the existing plesiochronous network and phased introduction of an MTN. For example, SDH multiplexer rings could be introduced into the outer core as autonomous units under local control. Similarly, SDH line systems could be introduced into the inner core, and later linked into SDH cross-connects to form a basic MTN. This would be gradually enhanced to full functionality over a number of stages.

PROTECTION AND RESTORATION

ITU-T Recommendation G.803 [14] presents a set of descriptive tools and concepts which can be used in the network design process to model the features of SDH/SONET-based transport networks [15]. G.803 uses the concept of a layered transport network as shown in Fig. 5, comprising a circuit layer, a path layer, and a transmission media layer with a client/server relationship between any two layers. Each layer has its own operations and maintenance capability, including protection capabilities, and this leads to a fully integrated protection strategy for networks which can provide high end-to-end circuit availabilities.

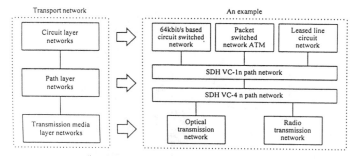

Fig. 5.   Layered model of the transport network.

Protection makes use of preassigned capacity between nodes, and once provisioned, the decision to use the protection capacity is normally autonomous upon detection of failures. Restoration makes use of spare capacity between nodes, and normally relies on the network management system identifying the spare capacity after a failure has been detected. Consequently, restoration is slower than protection.

The SDH/SONET standards [14], [16] allow the integration of transport, management, and protection facilities by comprehensive and easily accessed overhead bytes contained within the frame structure. These mechanisms can protect a portion of a connection across a transport layer (called "subnetwork connection protection") or it can provide protection of an end-to-end connection within a layer (called "trail protection"). These protection mechanisms can be implemented within each layer of the transport network as shown in Fig. 5. However, care must be taken to avoid interactions between various protection and restoration systems which may coexist within a managed network. Protection or restoration in a network layer should be delayed for a period sufficient to ensure satisfactory completion

of protection or restoration in any of its server layers [17]. For example, if an optical fiber were cut, then link protection (multiplex section protection) would be the most efficient protection mechanism, and should therefore operate faster than invoking protection at the path layer (e.g., 2 Mb/s) where a far greater number of protection switching actions would be required.

The principal features of a coherent protection strategy are:

a) Simple link and network protection methods. Note: dual-parenting will add to software complexity,

b) Fast response times using embedded signaling channels dedicated to protection, i.e., the K1 and K2 bytes [16] of the SDH frame overhead,

c) Preprogrammed protection and real time restoration plans downloaded from the network control level,

d) Flexible protection strategy downloaded according to network topology and path availability requirements,

e) Extensive equipment and link redundancy within each network layer leading to simpler protection software and fully duplicated switching nodes with MTBFs > 10 years.

Using the concepts outlined above, a typical protection and restoration strategy could use the following methods, listed in order of implementation:

Protection Method	Response Time	Element Protected	Control Scheme
Link (1 + 1, 1 : N)	Fastest	Links	Autonomous
Network (trail)	Fast	Links/Nodes/Ducts	Autonomous
Triangulation	Fast	Links/Nodes/Ducts	Preprogrammed
Network Restoration	Slow	Links/Nodes/Ducts	Real time rerouting
Manual	Slowest	Links/Nodes/Ducts	Human

(Note: Triangulation is a simple form of network restoration using preprogrammed make good routes.)

Computer simulators involving sensitivity studies have been used [8] to predict the gains in path availability using link protection and network restoration and to identify the best way to spend money on increasing availability. Generally, network restoration is more effective than simple link protection, but at the expense of more computer control software and a longer time to restore service. Another form of protection is whole network end-to-end (trail) protection with diverse routing of traffic paths through the network followed by terminal switching. This method should be cost effective for private circuits [6].

A comparison between whole network protection and simple link protection is shown in Fig. 6. Here, point-to-point availability is computed for a 1000 km path with nine intermediate nodes spaced at 100 km. It can be seen that whole network protection gives better availability than link protection, especially when link and node MTBFs are poor. For large node MTBF, total network protection is limited by link parameters and is relatively

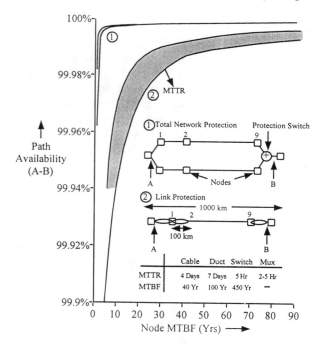

Fig. 6.   Comparison of link protection and total network protection for a 1000 km path.

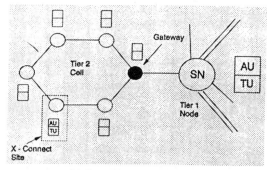

(a) Grooming at the inner core.

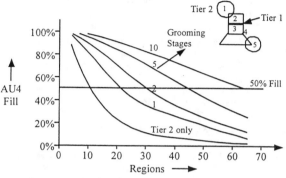

(b) Variation of fill factors with grooming strategy.

Fig. 7   Traffic grooming.

insensitive to node MTTR. Link protection, however, is limited by the parameters of the intermediate nodes and protection switches and is far less sensitive to link parameters.

TRAFFIC GROOMING

The primary aims of traffic grooming are to obtain acceptable fill factors, allowing optimal use of transmission plant and to allow diverse routing of traffic paths for protection purposes. SDH/SONET-based networks have the inherent flexibility to allow effective traffic grooming according to network requirements.

As an example, consider grooming within the inner core of an MTN in the UK. Traffic within each catchment area (Fig. 7(a)) is progressively groomed at each Tier 2 cross-connect according to final destination. The Tier 1 cross-connect further grooms the long-haul traffic from each Tier 2 cell within its catchment area, and combines it with long-haul traffic from other regions bound for the same destination. The fill factors obtained depend upon traffic per region and the number of grooming stages (Fig. 7(b)).

Clearly, extensive grooming requires large cross-connects and complex control, and the ideal grooming strategy would give acceptable fill factors using a minimal number of grooming stages.

MANAGEMENT AND CONTROL STRUCTURE

Work within BT to define a management structure for networks has resulted in the Open Network Architecture Management (CNA-M) structure [18], which is based on a five-layer Logical

Layered model consistent with the Telecommunications Management Network (TMN) Architecture [19] (Fig. 8). This management structure has enhanced the ability of BT to plan and control flexibility within a managed SDH/SONET network. Current work is focusing on the extension of this architecture to a distributed environment.

The management layers are made up as follows:

- Business Management is concerned with managing the complete undertaking in accordance with business objectives and customer requirements,
- Service Management allocates services to the network layer according to customer needs,
- Network Management provides the functionality to bind the individual elements, via their element managers, into a managed network,
- Element Management manages according to the requirements of the individual elements (e.g., cross-connects are managed differently from a passive duct system),
- Network Elements are the components which make up the network and include items such as multiplexers, cross-connect equipments, line systems, etc.

Each of the management layers is made up of seven functional blocks covering management of events, configurations, resource, performance, planning, finance, and security.

Extensive work has taken place in the standards arena, and in the ITU in particular, to define the interfaces which are needed to implement the management architecture. For the interface to the Network Elements, the requirements for each of the functional areas have been captured in ITU Recommendation G.784 [20]. Utilizing the principles of the functional architecture for SDH [14], [15], and the Functional Description of the Network Elements [16], a Managed Object model [21], [22] was developed for this interface. Current work is now addressing the Network Level interface.

A major issue is the mapping of the functional network model onto the hardware and software units, i.e., the practical distribution of intelligence between the Network, Element Manager, and Network Element layers. Failure to distribute the management

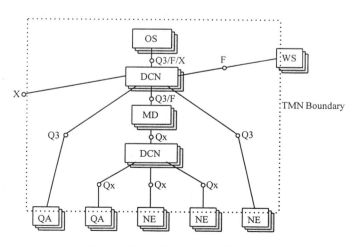

Fig. 8.   The TMN physical architecture.

121

functionality effectively will result in overloading of systems and communications links, and slower response times to network events. This problem is being addressed in the standards arena by using Open Distributed Processing techniques for the definition of the Network Level interface model. As broadband technologies and services are supported by SDH/SONET [10], the requirement for distributed management will increase.

In the U.K., network management functions will be duplicated on separate sites for security. A dedicated packet-switched network will provide communication from the management centers to the element managers, with proprietary communication from elements managers to network elements managers to network elements.

An important aspect of the control architecture at the element level is the use of the communication channel within the section overheads. This 768 kb/s channel provides communications for managing and controlling the SDH/SONET network and ANSI T1X1 and T1M1 committees, and the ITU has addressed the protocol stacks that will be used. The section overheads also contain dedicated communication channels for protection switching, alarm status, etc., which will be used for functions such as rapid service restoration and isolation of network faults.

CONCLUSION

SDH/SONET digital networks permit the introduction of higher levels of flexibility and quality into the BT transmission network. This flexibility will be used for functions such as traffic grooming (e.g., to maximize fill factors) and to implement an integrated protection strategy, whereby spare transmission capacity is managed by down-loading protection algorithms to meet the local requirements of each region or network layer. New network topologies will facilitate ease of growth, good path availability (including dual access), and resilience against disasters. These topologies will make maximum use of the existing duct routes. When combined with new network architectures and management philosophies, SDH/SONET-based networks will allow BT to provide improved quality of service and the ability to meet future customer requirements at a lower capital and current account cost than the plesiochronous digital network.

References

[1] K. R. Harrison, "The new CCITT synchronous digital hierarchy: Introduction and overview," *British Telecommun. Eng.*, vol. 10, part 2, pp. 104–107, July 1991.

[2] T. C. Wright, "SDH multiplexing concepts and methods," *British Telecommun. Eng.*, vol. 10, part 2, pp. 108–115, July 1991.

[3] W. R. Balcer, "Equipment for SDH networks," *British Telecommun. Eng.*, vol. 10, part 2, pp. 126–130, July 1991.

[4] I. Hawker, S. Whitt, and G. Bennett, "The future British Telecom core transmission network," 2nd IEE Conference on Telecommunications, York, England, 2–5 Apr. 1989.

[5] J. Marshall, R. Gallagher, and R. Cole, "Managing flexibility in an SDH network," 2nd IEE Conference on Telecommunications, York, England, 2–5 Apr. 1989.

[6] T. Wright, "The synchronous digital hierarchy standard," 2nd IEE Conference on Telecommunications, York, England, 2–5 Apr. 1989.

[7] M. Andrews, M. A. Anwar, R. Cole, and G. Reiffer, "BT Northern Ireland STAR SDH Network — NISTAR," *BTE*, vol. 12, part 3, Oct. 1993.

[8] J. Davidson, I. Hawker, and P. Cochrane, "The evolution of service protection in the BT networks," IEEE Global Telecommunications Conference & Exhibition, Dallas, TX, 27–30 Nov. 1989.

[9] R. Hall and S. Whitt, "Protection of SONET-based networks," IEEE Global Telecommunications Conference & Exhibition, Dallas, TX, 27–30 Nov. 1989.

[10] D. Clarke and T. Kanada, "Broadband: The last mile," *IEEE Commun. Mag.*, pp. 94–100, Mar. 1993.

[11] P. Cochrane and M. C. Brain, "Future optical fiber transmission technology and networks," *IEEE Commun. Mag.*, pp. 45–60, Nov. 1988.

[12] G. Hill, "A wavelength routing approach to optical communication network," *BTT J.*, vol. 16, no. 3, pp. 24–31, July 1988.

[13] G. D. Richman and P. C. Smith, "Transmission of synchronous digital hierarchy signals by radio," ICC/SUPERCOM 1990, Atlanta, GA, 16–19 Apr. 1990.

[14] ITU-T Recommendation G.803 (1992), Architectures of Transport Networks Based on the Synchronous Digital Hierarchy (SDH).

[15] A. B. D. Reid, "Defining network architecture for SDH," *British Telecommun. Eng.*, vol. 10, part 2, pp. 116–125, July 1991.

[16] ITU-T Recommendation G.783 (1993), Characteristics of Synchronous Digital Hierarchy (SDH) Multiplexing Equipment Functional Blocks.

[17] ITU-T Recommendation G.831 (1992), Management Capabilities of Transport Networks Based on the Synchronous Digital Hierarchy (SDH).

[18] K. Willetts, "A total architecture for communications management," Network 88 Conf., Wembley, England.

[19] ITU-T Recommendation M.3010, Principles for a Telecommunication Management Network (TMN).

[20] ITU-T Recommendation G.784 (1993), Synchronous Digital Hierarchy (SDH) Management.

[21] ITU-T Recommendation M.3100 (1992), Generic Network Information Model.

[22] ITU-T Recommendation G.774 (1992), SDH Management Information Model for the Network Element View.

Planning and Deploying a SONET-based Metro Network

SONET capabilities can change our concept of an optimum network architecture.

● ● ● ● ● ● ● ● ● ●

Michael To and James McEachern

Synchronous Optical Network (SONET) provides new features and capabilities not available in previous generations of asynchronous systems. The changes go well beyond a simple cost reduction due to standard interfaces, and have the potential to fundamentally alter the optimum network architecture. SONET can be deployed in a simple one-for-one replacement for asynchronous systems, yet the full potential of SONET cannot be realized in this way. To both fully understand the impact of SONET deployment strategies, and to identify the optimum architecture, network studies must consider factors such as growth, modernization and interworking with the installed base.

We will demonstrate how metropolitan networks can be analyzed to develop an architectural strategy that provides both cost-effective bandwidth management and survivability. The planning and deployment considerations, methodologies and planning tools are discussed. An example of a planned SONET metropolitan network, including the transition from an asynchronous network, is shown at the end of this article.

There are many advantages to the SONET format. SONET transport systems are now commercially available from a number of different vendors at several standard optical rates. Although one of the primary motivations for developing the SONET standard was to achieve multi-vendor optical interfaces, the value of SONET extends well beyond a "mid span meet" capability. The SONET format includes a coherent, structured OAM capability not found in any of the existing asynchronous formats that enables a host of OAM features in SONET transport products. For the first time, these features provide the operating company with the possibility of a uniform operations environment, allowing optimization of OAM operations, and significant OAM savings.

The synchronous nature of SONET also presents the telco with capabilities and options not previously available. SONET provides direct access to individual time slots in the optical signal, allowing the remaining bandwidth to be passed through untouched. This is a fundamental change from asynchronous systems which are forced to completely demux and re-mux a signal to gain access to even a single channel. This one subtle difference has the potential to completely change the economics

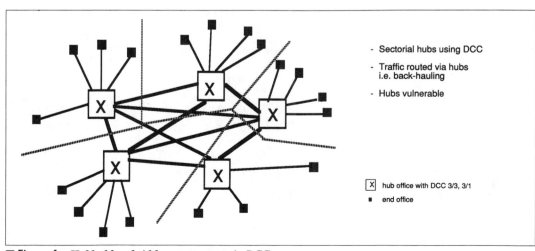

- Sectorial hubs using DCC

- Traffic routed via hubs
  i.e. back-hauling

- Hubs vulnerable

☒ hub office with DCC 3/3, 3/1
■ end office

■ Figure 1. *Hubbed bandwidth management via DCCs*

Reprinted from *IEEE Maga. Lightwave Telecommuni. Sys.*, vol. 2, no. 4, pp. 19–23, Nov. 1991.

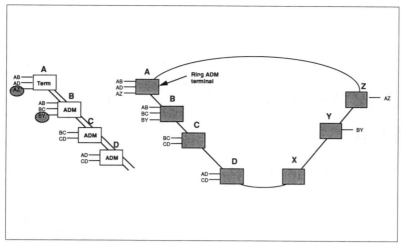

■ Figure 2. *SONET distributed bandwidth management via ADM chains or rings*

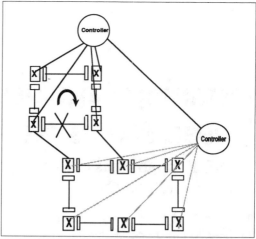

■ Figure 3. *Network survivability via DCCs and centralized intelligence*

of bandwidth management, and the corresponding optimum network architecture. In today's asynchronous environment, the cost of accessing a single channel is the same as the cost of accessing the complete signal. Thus, facility management is carried out in central locations, where the complete signal is demultiplexed. With SONET, because the cost depends on the amount of traffic being accessed, it becomes very beneficial to consider grooming traffic where it is needed, in a distributed fashion.

SONET metro objectives and architectural choices

S ONET provides telcos the potential to cost-effectively offer a range of new and existing services. To realize this potential, however, several issues must be addressed, namely, bandwidth allocation for various services, bandwidth management and network survivability.

Flexible bandwidth allocation is provided by the SONET format, which is organized into VT, nxVT, STS-1 and nxSTS-1 payloads. As a result, SONET systems can provide narrowband, wideband and broadband services.

Bandwidth management (routing, grooming, cross-connecting) is important in any network planning study. SONET has more bandwidth and a greater variety of services to satisfy growing customer demands for flexibility and control than other networks. For these reasons, it is essential that cost-effective bandwidth management be treated as

one of the key design objectives of the network.

Today, network survivability is a major concern to both end users and telcos. In the U.S., cable cuts are almost daily events now. The bandwidth available from fiber, and the growing volume of data traffic, make disruptions from link and node failures increasingly serious. Therefore, survivability must also be treated as a key design objective of the future SONET metro network.

Most network planning studies currently consider two alternatives for bandwidth management. One is the strategy used in asynchronous networks, where traffic from outlying offices is brought into hubs. In a hub office, traffic is completely demuxed, groomed or cross-connected for further routing. This scheme can be implemented with SONET cross-connects at the STS-1 or VT level (Fig. 1). The second approach takes advantage of the capability provided by synchronous multiplexing in SONET to eliminate the need to back haul traffic to central hubs. At each office, the SONET transport node directly accesses the required time slots in the bit stream without the need for total demuxing and remuxing. This approach, known as distributed SONET facility management, can be accomplished by SONET add-drop mux (ADM) chains or rings (Fig. 2).

Currently, network planning studies also consider two primary alternatives for survivability. One uses a network-wide scheme where network intelligence coordinates restoration for a link or node failure. In this scheme, either algorithms for the searching of restoration paths or predetermined maps are used. Usually this scheme is accomplished via a set of cross-connects under the control of a centralized computer system (Fig. 3). The second alternative is based on a sub-network approach where automated protection switching with diverse routing is used to provide restoration. As it is based on automated protection switching, there is no searching algorithm or predetermined maps needed. This approach is usually implemented via 1+1 systems or facility rings (Fig. 4).

The relative cost of different approaches to bandwidth management and survivability are usually analyzed by using simple network models. The real test of these concepts, however, is how they compare in real network-wide deployment. The economics of deployment can be accurately test-

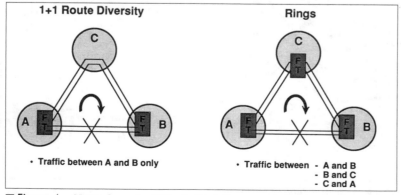

■ Figure 4. *Network survivability via 1+1 diverse routed system or ring*

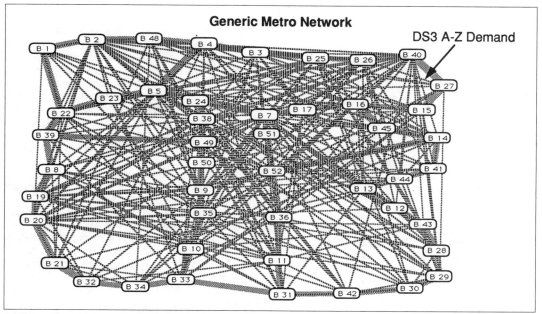

Generic Metro Network

DS3 A-Z Demand

■ Figure 5. *Office to office A-to-Z traffic demands in DS3s*

ed with network planning exercises, using actual traffic patterns and actual facility data as the base for study, rather than generic comparisons.

Target architecture based on eight metro analyses

*T*o test the feasibility of the various schemes for bandwidth management and survivability, individual North American metropolitan networks have been analyzed in detail, using data supplied by the operating companies. The goal was to identify network solutions that satisfied both the bandwidth management and survivability requirements synergistically, and that were applicable to a range of metropolitan networks.

In each metro study, total office-to-office demands were used as end-to-end traffic data, that is, without regard to current routing/hubbing constraints. Cable and conduit maps were then analyzed to understand the physical infrastructure that would be used to implement one of the networks. The key step in the analysis is to identify the traffic pattern or community of interest (Figs. 5, 6 and 7).

These studies have shown that SONET rings, based on traffic patterns or community of interest, are highly cost- effective, because the majority of traffic is confined within a ring as intra-ring traffic (Fig. 7). Although a ring design based on community of interest will achieve the optimum network cost, the ring layout is insensitive to variations in traffic patterns that result from growth and churn. The distributed bandwidth management capability of rings will be effectively utilized by the intra-ring traffic. This has the effect of reducing the need of stand-alone SONET cross-connects in many small sites. In these sites the inter-CO bandwidth management function will be achieved by SONET ring nodes. Inter-ring traffic management, via SONET cross-connects, will still be required at selected sites. They will, however, tend to be major COs or toll centers. In these locations, where multiple SONET rings meet, SONET cross-connects can be loaded more cost- effectively than they can in small offices.

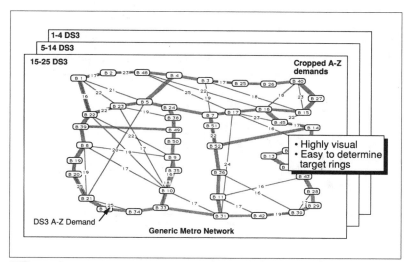

■ Figure 6. *A-to-Z demand cropped to identify potential rings*

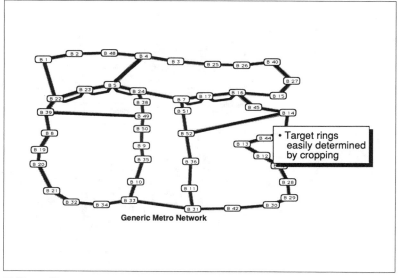

■ Figure 7. *Target ring well mapped onto community of interest by the cropping method*

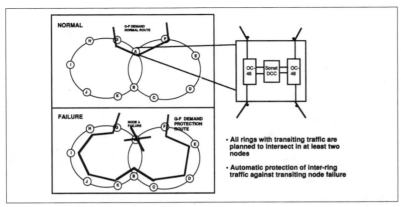

■ Figure 8. *Matched nodes A & B for Rings interconnect*

Objectives	SONET Architecture Hubbed	SONET Architecture Distributed
• Low cost network design	$36M	$26M
• Operations simplification	41 pt-pt systems	5 rings
• Survivability built-in	✓ sec-min	✓ ms
• Network read to support future services	✓	✓

■ Figure 9. *Sample hubbed and distributed SONET metro comparison*

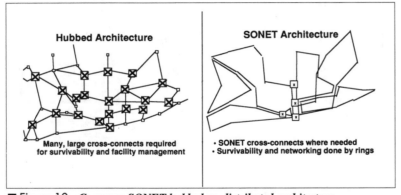

■ Figure 10. *Company SONET hubbed vs. distributed architecture*

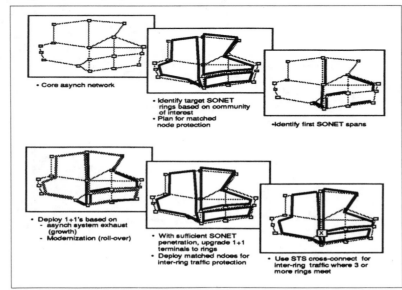

■ Figure 11. *Transition strategy to SONET metro*

Rings provide cost effective bandwidth management while they simultaneously provide link and node survivability. For traffic that has to traverse a number of rings, however, a scheme must be devised to provide end-to-end survivability. This is known in the industry as the matched node capability. The matched node scheme, also shown in Fig. 8, basically requires that two rings meet at two points. Inter-ring traffic through the first point will be duplicated through the second point and vice versa. This approach ensures that even inter-ring traffic can be protected against node failures.

Based on many metro analysis and sensitivity tests a cost-effective SONET metro target architecture has been identified. In this architecture SONET rings are used to provide survivability and the bulk of bandwidth management. SONET cross-connects are used at key locations where multiple rings meet and where they can be most cost-effectively loaded. A typical cost comparison of this target architecture to a hubbed architecture is shown in Fig. 9. These studies have consistently shown network-wide savings in the range of 26 percent to 36 percent depending on traffic patterns in each network. Fig. 10 shows the difference between the two architectures from an overall network point of view.

SONET metro deployment/ transition strategies

Although studies that identify the target SONET network are important, it is the migration from today's asynchronous network to the future SONET network that is of primary importance to operating companies. In addtion to providing guidance for year-to-year deployment plans, an effective migration strategy must also ensure a smooth evolution to the telco's future. There are many possible triggers for deploying SONET. A typical migration strategy might start with asynchronous system exhaust. Individual system exhaust, due to growth or new services, would be identified on a year-to-year basis. These exhausted links are the prime candidate for SONET 1+1 systems as an overlay (i.e., one working fiber pair and one protection). Since the target SONET rings have already been identified through traffic pattern analysis, the new SONET 1+1 system should be diversely routed following the path of the future target ring. As more asynchronous systems exhaust along the ring, the first SONET 1+1 system will be gradually upgraded to a full ring. As more target rings are established, inter-ring traffic will be managed by matched nodes and inter-ring SONET cross-connects.

The speed of transition can be varied by considering other triggers such as asynchronous system capping, network modernization and focused SONET deployment. Fig. 11 shows a series of transition steps directed toward a target SONET network. In Figs. 12, 13 and 14, these transition steps are translated into concrete deployment plans over a five-year period.

Planning tools and future

The planning methodology for identifying communities of interest, and the associated SONET rings in a target network, can be performed manually. This was done for the initial

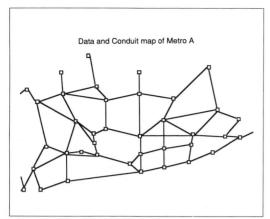

■ Figure 12 . *Case study of a SONET metro from transition to target*

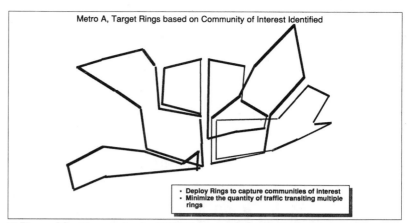

■ Figure 13 . *Case study of a SONET metro from transition to target*

studies used to develop and refine the methodology. The manual process, however, is slow and tedious, and requires a reasonable level of expertise to ensure sensible results. To alleviate this problem, the algorithms that identify target rings have been programmed into a MacIntosh-based planning tool. This planning tool manages end-to-end data, cable/conduit maps, searches for community of interest patterns, and provides automated traffic provisioning. The primary advantage of this kind of an automated tool is that since it is interactive, it immediately illustrates the result of each step. The planner can experiment with different alternatives, get a visual representation of how this changes the network, and in the process, develop an intuitive understanding of the architectures. This process reduces the time required to plan a network, gives the planner more confidence in the target design, and leads to a network that is more easily implemented.

This planning tool is now in use by more than six telcos. The next step is to provide planning tools that will assist with the transition to the target network. This tool would identify asynchronous systems nearing exhaust, propose SONET overlays, and show the optimum time to upgrade these SONET systems to rings. It would also allow the planner to input transition triggers such as modernization, capping or focused SONET deployment, in accordance with local directives. It is expected that this kind of a transition planning tool will be available by the end of this year.

SONET products are available today. Deployment has already started, and will soon be underway throughout North America. SONET, however, is more than an asynchronous fiber optic transmission system (FOTS) with mid-span meet capability. It is designed to provide the operating company with OAM and service capabilities previously not possible (this subject is beyond the scope of this article). The synchronous SONET mappings provide more efficient bandwidth management through direct access to individual payloads.

Although SONET inherently enables new features and capabilities in the network, the realization of this potential will not be automatic. Getting the maximum benefit from this new technology requires a target architecture and deployment strategies that are optimized for SONET.

Our studies have shown that the most effective strategy will be built around an infrastructure of survivable SONET rings, and automated facili-

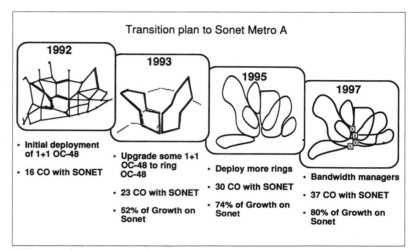

■ Figure 14. *Case study of a SONET metro from transition to target*

ty management based on a combination of SONET cross-connects and distributed bandwidth management in SONET ring terminals. This SONET infrastructure will provide the framework for the predicted expansion in communications services, and will meet expanding telecommunication needs into the next century.

References

[1] S. Yan, M. To, S. Oxner, "Bandwidth Management in a Sonet Transport Network," Globecom 90.

[2] T. Flanagan, S. Oxner, D. Elkaim, "Principles and Technologies for Planning Survivability-A Metropolitan Case Study," Globecome 89.

[3] T. Flanagan, "Planning a Sonet Network," Supercomm ICC 90.

Section 3
Survivability and Robust Architectures

RECENT failures in carrier networks, often affecting service across broad regional areas, coupled with trends toward higher speed services and facility management, have sensitized the communications community to the importance of robust, survivable networks. This topic is, however, especially challenging, insofar as failures can occur at transmission facility, nodal, or operational levels. Further, measures to ensure improved integrity are dependent upon: appropriately defining service objectives; assessing individual customer needs; evaluating OAM&P (operations, administration, maintenance, and provisioning) procedures; improving software quality and reliability; understanding the impact of standards; formulating strategies for physical-, datalink-, and network-level restoration; and planning network topology and growth with survivability as a forethought.

Survivable networking clearly encompasses almost every aspect of modern wide-area network design, implementation, and operation. The eight papers in this section focus, however, on physical-layer restoration and planning for survivable SDH/SONET networks; the remaining topics are discussed to a varying extent elsewhere in this book.

Presentation begins with self-healing rings, one of the two principal SDH/SONET survivable network architectures, the other being meshes with distributed-control digital cross-connect systems (DCSs).

"Self-Healing Rings in a Synchronous Environment," by I. Haque *et al.*, introduces the topic of survivable SDH/SONET networks with a comprehensive description of unidirectional path-switched and bidirectional line-switched rings, set in the larger context of other self-healing technologies, especially diverse routed protection and DCS-based systems. The alternative approaches and respective applications of rings are reviewed, along with a discussion of ring interworking for nodal and span failures, and a generic overview of ring interworking requirements.

The second paper, by T.-H. Wu and M. E. Burrowes, "Feasibility Study of a High-Speed SONET Self-Healing Ring Architecture in Future Interoffice Networks," complements the first in that Wu and Burrowes examine the economic feasibility of using self-healing rings in interoffice networks, offering case studies based on a metropolitan Local Access Transport Area (LATA) network. Additionally, this paper is particularly insightful in illustrating self- and dual-homing ring networks and hardware architectures.

Selection of either unidirectional or bidirectional self-healing rings depends upon application, network size, and demand pattern, points developed in "A Class of Self-Healing Ring Archi-

tectures for SONET Network Applications," by T.-H. Wu and R. C. Lau. The authors note that four-fiber bidirectional rings meld well with today's operations environment, while path-switched unidirectional rings could ease intervendor compatibility concerns. The above points are developed by way of a detailed architectural analysis, which considers cost, capacity, implications for standards, and multivendor compatibility.

The fourth paper transitions the discussion from rings to DCSs. J. Sosnosky, in "Service Applications for SONET DCS Distributed Restoration," provides a taxonomy of the network applications and services amenable for use with distributed-control cross-connects in mesh networks. The paper includes analysis of service outage impact and determination of how restoration time objectives affect the applicability of DCS-based network architectures. Based on a two-second restoration time as an appropriate target, and the need for fault tolerance and full survivability to support future services, he proposes integration of the distributed-control DCS mesh with self-healing ring architectures.

The following paper by R. D. Doverspike *et al.*, "Network Design Sensitivity Studies for Use of Digital Cross-Connect Systems in Survivable Network Architectures," examines the economic viability of DCSs in robust local exchange carrier networks. Three transmission technologies are considered (SONET self-healing ring, point-to-point system, and mesh with automatic DCS restoration) in various combinations to form six network alternatives for evaluation. Parameters considered include demand, network connectivity, and equipment cost sensitivities, with survivability of each option calculated on the basis of a major node failure. Special attention is given to DCSs with integrated optical terminations (i.e., OC-n). The authors conclude that hybrid networks, with combinations of point-to-point, mesh, and ring topologies, are most cost-effective.

"The Impact of SONET Digital Cross-Connect System Architecture on Distributed Restoration," by T.-H. Wu *et al.*, is the sixth paper of the section. Beginning with a two-second benchmark for restoration in large metropolitan local exchange networks, the authors study the applicability of distributed-control DCSs and find that they may not meet the stated restoration objective. To improve the restoration time, they suggest several options for DCS architecture enhancement, notably parallel processing and cross-connection, in contrast to serial processing/cross-connection.

The next paper, by S. Hasegawa *et al.*, "Control Algorithms of SONET Integrated Self-Healing Networks," differs from the previous papers in that it describes a control algorithm that could be used in a mesh network of DCSs, wherein the self-healing control is locally initiated and executed based on distributed intelligence in each network element. The authors first provide a valuable overview of route diversity, self-healing ring, and DCS-based restoration of STS-1/STS-3c channels, where the element-to-element control signal could be carried as either a 192-kb/s section Data Communications Channel (DCC) or dedicated channel in the SDH/SONET overhead. The authors then consider coherent integration of ring- and DCS-based network survivability.

Section 3 concludes with a discussion of *a priori* strategic planning for survivability. In "Survivable SONET Networks—Design Methodology," by O. J. Wasem *et al.*, a prototype software system is described which specifies SONET ring types (unidirectional and two- and four-fiber bidirectional fiber) and locations, based on inputs of multiyear traffic demands, route mileage, fiber material and splicing costs, multiplexers and regenerators, facility hierarchy, and nodes, links, and connectivity.

As previously noted, the papers in this section narrowly focus on SDH/SONET. For broader treatment of the topic, including the role of asynchronous transfer mode, software, standards, and user perspectives, readers are referred to a recent issue of *IEEE Journal on Selected Areas in Communications* (vol. 12, no. 1, January 1994), "Integrity of Public Telecommunication Networks," edited by T.-H. Wu, J. C. McDonald, T. P. Flanagan, and K.-I. Sato, from which several of the papers mentioned above were selected.

Self-Healing Rings in a Synchronous Environment

IZAZ HAQUE, WILHELM KREMER,
AND KAMAL RAYCHAUDHURI
AT&T BELL LABORATORIES

1. INTRODUCTION

Self-healing architectures are increasingly prominent in network provider plans to satisfy end user demand for network survivability. While self-healing architectures have been around for several years, increased equipment integration and add/drop functionality promoted by SONET standards have generated a range of architectures optimized for different applications.

A network designer in the 1990s will have several options for deploying self-healing networks. Recently, Self-Healing Rings (SHRs) have piqued the interest of the telecommunications world in a big way. In SHR networks, traffic between two points is carried in an add/drop network that is looped back on itself, forming a "ring." Rings protect against fiber cuts or node failures by providing an alternate path around the failure. The protection mechanisms vary depending on implementation. The two major architectures for ring deployment, unidirectional path switching and bidirectional line switching, are discussed in detail in this paper. The results of work in the area of interworking rings with nodal protection are also presented.

The paper also discusses other, nonring-based, self-healing technologies, with the intent of providing the reader a comprehensive view of self-healing approaches. These include Digital Cross-Connect System (DCS)-based self-healing, and path-based or line-based Diverse Routed Protection (DRP). These architectures, while accomplishing the same fundamental objective of providing protection against catastrophic failures in the network, are substantially different in implementation from ring-based architectures.

A typical self-healing network design for a Local Exchange Carrier (LEC) network may use all of these architectures.

The paper concludes with a discussion of network applications, and shows where the best fit for each approach may be in the evolving network.

2. SELF-HEALING RINGS

Many of the benefits of SONET stem from standardization of optical interfaces, simplified tributary visibility, and standard operations. For example, SONET provides an inherent potential for equipment integration by reducing the processing requirements for conducting simple networking functions such as multiplexing. This has led to the development of Network Element (NE) architectures that allow integrated add/drop (lin-ear or ring-based, as discussed in the following) functionality, replacing back-to-back multiplexing as the primary means of providing this capability.

In the linear add/drop application, shown in Fig. 1, traffic between two nodes is carried on a chain of connected nodes, where the service may originate or terminate on any node on the chain. The two end nodes on this network are called terminal nodes, and intermediate nodes are called add/drop nodes. A SONET add/drop node performs a function akin to back-to-back multiplexing, except that through circuits are more economically cross-connected electronically (instead of being electrically patched together at a DSX panel and requiring interface circuit packs on the back-to-back terminals). It also allows directional provisioning of traffic originating at a node (east or west).

The corresponding ring add/drop architecture, also shown in Fig. 1, is similar to the linear application, except that the add/drop chain connects back on itself. If the linear add/drop application can be likened to a "string of beads," the ring application can be thought of as a "necklace of beads." There are no terminal nodes on a ring, and each node has add/drop functionally.

SONET promotes the concept of layered maintenance through the use of path, line, and section boundaries. Overhead channels are provided at each of these layers to facilitate maintenance and inter-NE communications at the appropriate layer. Signal health criteria (for example, bit error rate indicators) are present in all three layers, and are used in the design of different ring architectures.

The directionality of service routing and the protection mechanism are two key attributes that distinguish different ring architectures. For example, a *unidirectional* ring carries service traffic

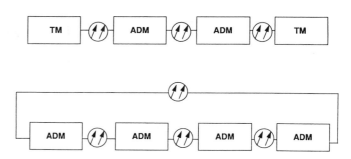

Fig. 1.   Linear add/drop network and ring add/drop network.

in only one direction of the ring, e.g., if the A to Z direction of a circuit is clockwise, the Z to A direction is also clockwise. Conversely, a *bidirectional* ring carries all service traffic in two directions, e.g., if the A to Z direction of a circuit is clockwise, the Z to A direction is counterclockwise.

Furthermore, a *path-switched* ring protects traffic based on the health of each individual path where it exits the ring, on a per-path basis. In a *line-switched* ring, on the other hand, switching is based on the health of the line between each pair of nodes. When a line is faulty, the entire line is switched out to a protection loop at the boundaries of the failure.

Based on the above, two architectures have gained prominence for deployment in SONET networks. These are, respectively, two-fiber unidirectional path-switched rings (alternately termed as Unidirectional Self-Healing Rings or USHRs, or simply path-switched rings), and two-fiber and four-fiber bidirectional line-switched rings (alternately termed as Bidirectional Self-Healing Rings or BSHRs, or simply bidirectional rings). The following discussion describes these architectures in detail.

2.1 Unidirectional Path Switched Rings

Figure 2 shows a two-fiber unidirectional path-switched ring. The basic routing mechanism on the path-switched ring is unidirectional in that transmit from A to C is clockwise on the Service (S) fiber and transmit from C to A continues around clockwise on the S fiber. Note also that the signal from the transmit side

is dual fed onto the protection fiber so there is a protection path on the Protection (P) fiber that flows counterclockwise from A to C and then back again from C to A. Each path is individually switched based on the signal health criteria mentioned earlier. In the case of a cable cut affecting the traffic flowing on the S fiber, as shown in Fig. 2 between nodes B and C, each individual path normally traversing the cut line is switched at its terminating node to the signal appearing off the protection fiber. Thus, traffic from A to C, being what would have normally flowed through the failed span, is switched to the protection fiber. Traffic flowing between C and A still flows on the S fiber, temporarily routing the traffic bidirectionally between the two nodes. All other paths unaffected by the cable cut stay where they are.

The two-fiber unidirectional path-switched ring has the following characteristics:

- Each node on the ring provides add/drop functionality; however, unlike the linear add/drop network, the path-switched ring actually does not require directional east–west provisioning since each tributary entering the ring is carried all the way around the ring regardless of where it enters or leaves the ring, and one only needs to know the two end nodes.

- The ring network service capacity equals the lightwave system capacity of the ADMs forming the ring. Network service capacity refers to the maximum number of tributaries

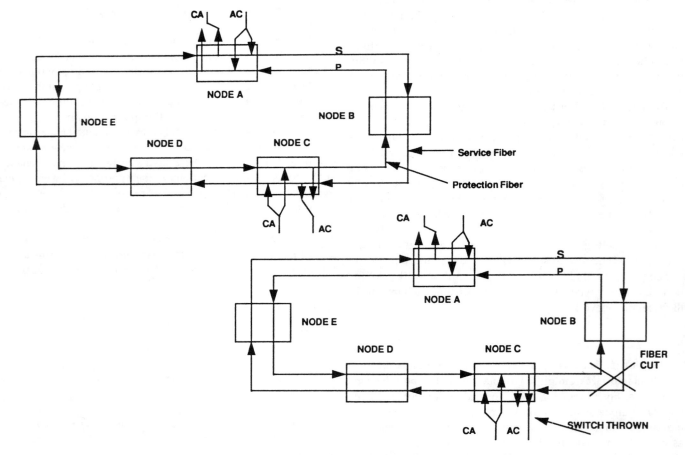

Fig. 2. Fiber unidirectional path-switched ring—architecture and response to cable cut.

the ring can carry. Since each tributary entering the ring is carried all the way around the ring, the amount of traffic on the ring is the aggregate of all the traffic entering the ring. Thus, for example, an OC-3 ring can carry upto 84 equivalent DS1 circuits. This characteristic of path-switched rings optimizes them for hubbing applications, since such applications require all traffic to aggregate to the hub.

- The equipment configuration looks like an unprotected OC-N add/drop chain. Thus, the start-up configuration can have half the number of receivers and transmitters of the linear add/drop configuration. This, along with the fiber count, represents a significant cost advantage of the path-switched ring over linear add/drop.
- Receiver, transmitter, fiber, or node failures are protected via a path switch on the NE terminating the path.

In general, path switching takes place at the point where signals leave the network, and hence is not constrained to a physical ring architecture. Such architectures are sometimes referred to as Virtual Path-Switched Rings. However, SONET technology has spawned more physically compact integrated Path-Switched Ring Architectures, which are typically referred to as Path-Switched Rings. The path-switched ring architecture requires no additional SONET standardization, and is an implementation within the existing standard. Its perceived simplicity and match to typical loop traffic patterns makes it a suitable architecture for NEs designed for loop deployment.

2.2 Bidirectional Line-Switched Rings

This section gives some intuitive views of a four-fiber and then a two-fiber bidirectional ring.[1]

The key to understanding the four-fiber bidirectional ring is that, for all intents and purposes, this ring looks indistinguishable from and acts just like an ordinary add/drop chain folded back on itself. The only exception is that, in the case of the complete fiber cut of a span or in the case of a node failure, the four-fiber bidirectional ring has ring self-healing and the add/drop chain does not. Figure 3 shows an OC-48 bidirectional ring.

In particular, the four-fiber bidirectional ring has the following characteristics:

- It has service and protection bidirectional fiber pairs on every span like a linear add/drop chain, and is provisioned as such. To get from A to Z (say, in the clockwise direction), the circuit is provisioned bidirectionally as an "add" at A, a "through" at intermediate nodes, and a "drop" at Z. Since traffic is only routed partway around the ring, the bidirectional ring has the property of service channel reuse which gives it a capacity advantage over a path-switched ring, where service capacity for a circuit needs to be dedicated the whole way around the ring.
- The span service capacity is equal to the lightwave system capacity like an add/drop chain (Fig. 3). Span service capacity is defined as the amount of service traffic that can be carried on a span between two terminals. The maximum

ring service capacity is the product of the number of spans times the span service capacity.

- The equipment configuration looks like a normally protected add/drop chain. Looking out at each span from a ring node are receiver/transmitter pairs for service and protection.
- Receiver, transmitter, or single fiber failures (i.e., single line failures) can be protected by normal span protection (see Fig. 4).

Only complete fiber cuts or node failures need be protected by a line switch, more specifically by a ring line loopback function (Fig. 5). Observe that the self-healing reaction to the cable cut shown in the figure amounts to "real-time" route diversity. That is, the protection loop acts like terminals next to the cut and like repeaters at intermediate sites, in the process forming a route-diverse protection line for the cut service (and span protection) lines. Thus, the only time the four-fiber bidirectional ring acts like a ring and not like an add/drop chain is precisely when

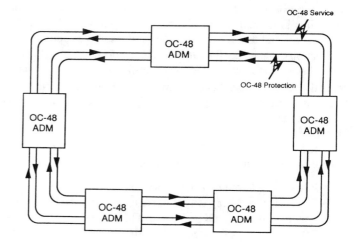

Fig. 3.   Four-fiber OC-48 bidirectional ring. It has service and protection bidirectional fiber pairs, such as a linear add/drop; circuits are provisioned, as on an add/drop chain; and the equipment configuration looks like a normally-protected add/drop chain.

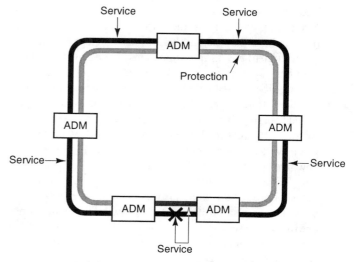

Fig. 4.   Four-fiber bidirectional ring span protection for electronics failures, typically of receiver or transmitter.

[1]W. Kremer, "On Ring APS Guidelines," T1X1.5/91-010, Feb. 6, 1991.

133

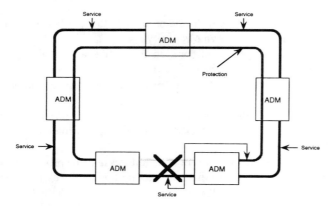

Fig. 5. For cable cuts (as shown) or node failures.

the ring functionality is needed — during a cable cut or a node failure.

It turns out that most of the intuitive and favorable properties of the four-fiber bidirectional ring can be retained in a two-fiber configuration by using four "logical" fibers. Note that in the four-fiber bidirectional ring, there are two directions of traffic on every span, and that for each direction, the service capacity and protection capacity are the same (see Fig. 3 once more). By using two fibers, one for each direction of traffic; by dividing capacity equally on each fiber between service and protection; and by performing ring switching with the "logical" fibers analogous to that in the real four-fiber case, the two-fiber bidirectional ring is born.[2]

The following lists the differences from the two-fiber bidirectional ring to the four-fiber bidirectional ring:

- It has a single bidirectional fiber pair (Fig. 6).
- The span service capacity is equal to one-half the lightwave system capacity ("OC-6" for an OC-12 system and "OC-24" for an OC-48 system). The other half of the lightwave is reserved for protection.
- It is provisioned like an OC-N/2 add/drop chain. Hence, the add/drop properties and the associated service channel reuse with its capacity advantages are retained.
- The equipment configuration looks like an unprotected OC-N add/drop chain. Thus, the start-up equipment configuration can have half the number of receivers and transmitters of the four-fiber configuration. This is the key, along with the fiber count, to the overall start-up cost advantage over the four-fiber (albeit with half the capacity of the four-fiber).
- Receiver, transmitter, fiber, and node failures are all protected by a ring line loopback function. The reliability of the two-fiber bidirectional ring is analogous to 1:M protection, where M is the number of nodes in the ring. Figure 7 shows how five spans each vie for the use of the shared protection loop. When one failure causes a ring loopback switch, no other failure (of the same priority) can now grab the loopback. This is just like 1:5 protection switching, where once one of the five service lines has grabbed protection, no other service line can grab it unless it has higher priority.

[2]T. Flanagan, J. Brule, and M. Betts, "Comparison of Bidirectional and Unidirectional SONET Rings," T1X1.5/90-123R1, Nov. 5, 1990.

No two-fiber ring, line-switched or path-switched, can protect electronics failures with a span switch. Rather, all two-fiber rings protect all failures with their ring protection switching mechanism, and trade off a fully acceptable increase in circuit outage due to electronics failures for protection against fiber/node failures. The four-fiber bidirectional ring does not increase outage due to electronics failures, while providing protection against fiber/node failures.

Bidirectional Line Switch Ring Architectures are an ANSI standard. At the time of writing, ITU-T has a recommendation (G.841) that is consistent with and embodies the ANSI standards (using SDH terminology and concepts).

2.3 Comparison of Ring Approaches

Apart from the obvious topological differences between the more familiar linear add/drop and ring networks, certain attributes of ring networks serve both to differentiate as well as

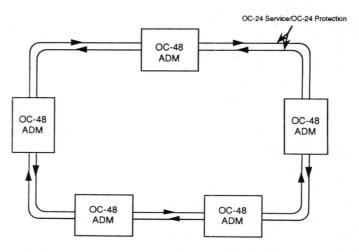

Fig. 6. Two-fiber OC-48 bidirectional ring. It has a single bidirectional fiber pair. The span capacity is equal to one-half the lightwave system capacity ("OC-6" for an OC-12 system and "OC-24" for an OC-48 system). It is provisional, like an OC-N/2 add/drop chain. The equipment configuration looks like an unprotected OC-N add/drop chain (folded back on itself), thus, the start-up equipment configuration can have half the number of transmitters and receivers of the four-fiber configuration.

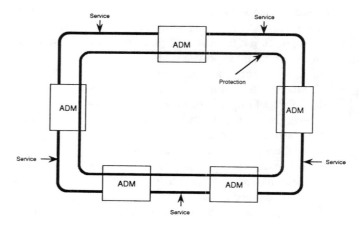

Fig. 7. Bidirectional ring-shared protection analogy to 1 : M protection. Five lines share single protection loop.

TABLE 1. COMPARISON OF ADD–DROP ARCHITECTURES

Attribute	USHR	Two-Fiber BSHR	Four-Fiber BSHR	Linear
# of Fiber Pairs Between Spans	1	1	2	2
# of R/T Pairs/Node	2	2	4	4
Span Service Capacity	n/a	$OC-N/2$	$OC-N$	$OC-N$
Network Service Capacity		$\geq M^*$ $(OC-N/2)$	$\leq M^*$ $OC-N$	$\geq(M-1)^*$ $OC-N$
Provisioning Mechanism	Add/drop	Add/drop	Add/drop	Add/drop
Rcvr/Trmt Protection	Path diversity	Line loopback	Span protection	Span protection
Complete Cable Cut	Path diversity	Line loopback	Line loopback	No protection or node failure
Span Reliability	Between $1 \times M$ and $1 \times 1^*$	Like $1 \times M$	Like 1×1 plus ring	1×1

*The reliability of the path-switched ring falls somewhere between a $1 \times M$ system (where M is the number of nodes on the ring) and a 1×1 system. Depending on where the failure occurs on the ring, the ring can tolerate multiple simultaneous failures.

identify these topologies with linear add/drop. These include equipment characteristics, service transport capacity characteristics, implementation and architectural differences, and reliability.

These comparisons are summarized in Table 1 (assuming an OC-N[3] lightwave system).

3. RING INTERWORKING

Although rings can independently handle both nodal and span failures, many applications call for multiple ring solutions with redundancy at the signal hand-off points. Figure 8 gives a high-level problem definition. The figure shows two rings having two adjacent nodes in common, and a circuit assigned between nodes A and Z of the two rings. The "X" over the interconnecting node for the A–Z circuit denotes the failure of that node. This section treats the issue of ring interworking, sometimes referred to as dual access or nodal diversity, where at least one of the rings is bidirectional. The nodal diversity problem is how to get a circuit with one termination in the first ring and the other termination in

[3]All references to OC-N rates in this paper necessarily relate to the North American SONET standard. SDH standards use STM-M terminology. To get equivalent STM rates, the OC-N rate should be divided by 3.

[4]B. Smith and S. Sigarto, "Dual Feed Ring Architectures," T1X1.5/90-184, Oct. 12, 1990.

[5]J. Sosnosky and T.-H. Wu, "SONET Ring Applications for Survivable Fiber Loop Networks," *IEEE Communications Magazine*, vol. 29, no. 6, June 1991.

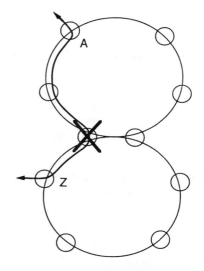

Fig. 8.   Ring interworking problem definition.

the second ring to survive the failure of the shared node currently carrying service for the circuit.

This problem of node failure is automatically handled in protection switching time frames for a circuit routed entirely within a single ring when an intermediate node fails, whether that ring employs path switching or line switching. The issue of automatic and speedy nodal diversity is for the two rings to respond to the node failure in such a way that the A–Z circuit is preserved by automatic alternate routing over the second shared node instead, again in protection-switching time frames.

3.1 Ring Interworking for Path-Switched Rings

The issue of ring interworking for path-switched rings has been treated in several references.[4,5] There, the problem of providing a protected connection between multiple path protection switched rings is treated in significant depth. As pointed out earlier, one application of ring interworking is an access ring coupled with an interoffice ring. The key example is illustrated in Fig. 9.

Please observe the following points in the architecture (this is symmetric, albeit mirror-image, for each of the two directions of transmission):

- A (unidirectional) signal is dual-fed from its origination point around both sides of the ring containing that origination point.
- When each of the dual-fed signals hits a shared node, it is dropped at that node and replicated to be continued to the other shared node. This is also called the "drop and continue" function.
- Thus, each shared node can and does select (on a unidirectional path basis) from two signals, each sent a different way around the ring.
- The selections are then each handed off at each of the shared nodes to the second ring.
- Each of the shared nodes in the second ring takes its respective hand-off and transmits it towards the signal termination node in a direction away from the other shared node.

- The final selection is made at the termination point for the signal.

Again, note that the symmetrical process takes place in the other direction of the circuit.

The path-switched ring interconnect example has some interesting and very desirable properties, which can be abstracted into generic ring interworking requirements, which can be used to extend this concept to bidirectional rings.

3.2 Abstracted Generic Ring Interworking Requirements

In this and the following section, we describe the results of some work that builds on the discussion in the previous section to include bidirectional rings. The generic rings interworking solution should satisfy the following requirements:

- The ring interworking should be such that it survives the failure of any one shared node.
- The rings have virtually complete maintenance independence.
- No inter-ring signaling should be required.

A bidirectional ring solution which observes all of these requirements can be used regardless of whether the second ring is path-switched or bidirectional.

Because of the dual inter-ring hand-off, it turns out that the bidirectional ring must have some path-switching mechanisms. Despite this requirement, a bidirectional ring retains its capacity advantage over path-switched rings, even in ring interworking scenarios when traffic patterns are more evenly distributed and not entirely hubbed.

3.3 An Architecture for Ring Interworking with a Bidirectional Ring

Figure 10 shows an inter-ring circuit in the nominal failure-

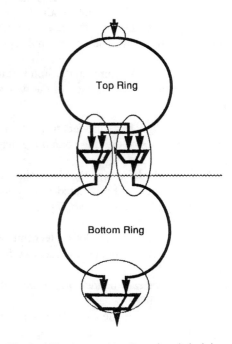

Fig. 9.   Ring interworking for path-switched rings.

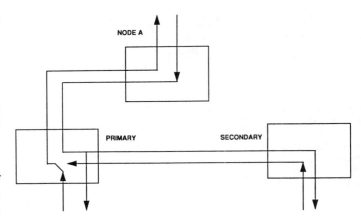

Fig. 10.   Nominal failure-free state of the A–Z circuit subpath in the bidirectional ring.

free state of the bidirectional ring, and illustrates a particular circuit, handed off at the primary and secondary nodes, that terminates at node A (the node which performs the path functions is referred to as the *primary* node, and the other shared node is referred to as the *secondary* node). Starting from the top in this figure, a bidirectional circuit from the (uppermost) termination node A is assigned counterclockwise towards the bottom on the ring. The primary node would normally be the one "closest" to the terminating node without respect to ring orientation, and the primary node performs the path functions described in the previous section. Primary and secondary nodes are defined on a per-circuit basis.

In particular, the required ring interworking functions are as follows:

- For the unidirectional signal transmitted from node A, the primary node dual feeds that signal both towards its own interface and towards the line to the secondary node.
- In the other direction, the primary node selects between the hand-offs to the primary and secondary nodes from the other ring, and transmits that selection to the upper terminating node. The signal handed off to the primary node is the one normally selected.
- It is assumed that the same channel assignment on the line is used between the secondary and primary nodes, and between the primary and terminating nodes.
- It is also assumed that in the case of a failure of the primary node, the connection of the signal between the termination A and the secondary node is maintained. This is called a *secondary* connection.

An example of how the ring responds in the case of a primary node failure is shown in Fig. 11. The failure of the primary node on the bidirectional ring results in a loopback switch which creates an unprotected ring just like the previous one, but without the primary node. The secondary connection of the circuit from A to the secondary node is automatically formed (as would any other secondary connections past the primary node), and undesired potentially crossed traffic is squelched. Observe that the signal selection and signal dual-feed functions used in the reactions to the failure scenarios of the previous section do not necessarily need to be integrated in the ring ADM. These interworking functions could also be integrated in a DCS, with or without the integration of the optics and the other ring functions. In these cases, the only further requirement on the bidirectional ring for ring interworking is that for node failures, circuits headed for the failed node can (when set up) be provisioned either as secondary connections to be maintained or as potentially crossed traffic to be squelched.

This proposal has been accepted as draft text for a later version of the standard. There is also an additional capability that allows use of some protection capacity to enhance overall service capacity capabilities of this ring interworking architecture.

Ring interworking is being worked into G.SHR-2 of the ITU standard. There has been little activity to date. However, an SDH version of the architecture described here has been proposed as an architecture for G-SHR-2.

It should be mentioned that the Bidirectional Line Switched Ring Architecture work has been a real success story. There are products in the field that are functioning to the specified standard, using configurations up to the full number of nodes specified in the standards, and meeting switch time requirements. Bidirectional Line Switched Rings have caught the imagination of the service provider community. There is a vendor/user consensus on the application of these architectures in the high density traffic network domains. The standards bodies can compliment themselves on a job well done.

4. OTHER SELF-HEALING ARCHITECTURES

In this section, we briefly discuss two other self-healing architectures. The first of these uses 1×1-protected lightwave systems with diverse-routed protection, and the second, digital cross-connect systems. Although sometimes posed as alternatives to, these architectures are more often complementary to the ring architectures described in the previous sections, and reflect the wide range of choices a network designer faces in the design of a self-healing network.

4.1 Diverse-Routed Protection

In this architecture, a pair of offices with traffic between them is served by a dedicated 1×1-protected point-to-point lightwave system (or more, if needed) with the service and protection lines routed along spatially diverse paths. Since the protection fiber does not physically accompany the service fiber that it is protecting, this arrangement provides immunity to service interruption from cable cuts.

The key point to note about DRP is that it is an express-routed architecture, in which each DRP system serves only the two end-offices, with no terminals (although there may be line repeaters) in between. Thus, the DRP is an excellent architectural choice when the demand between a pair of offices is very high.

[6]D. Doherty, W. D. Hutcheson, and K. Raychaudhuri, "High Capacity Digital Network Management and Control," IEEE Global Telecommunications Conference, San Diego, CA, Dec. 1990.

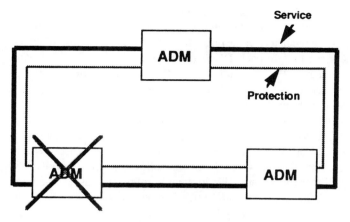

(a) Bidirectional ring loopback protection for node failure.

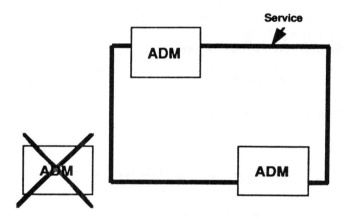

(b) Conceptual representation of loopback protection response to node failure.

Fig. 11.   Response to node failure.

4.2 DCS-Based Self-Healing

This architecture employs DCSs at each node, with lightwave spans forming the connections between the nodes. When a span fails, causing a break in the service path between two nodes, a restoration path is formed on an arc complementary to the service path by cross-connecting spare channels on appropriate spans, via the DCSs. The DCS architecture can be more generally applied to a mesh and not be constrained by a ring formation. It relies on finding a diverse path for every path on the network, and can be likened to "real-time" path diversity for the paths on a failed span or paths heading through a failed node.

Current instances of this methodology use a central controller[6] to conduct the restoration process. Automated restoration with such a controller typically calls for the controller to maintain a view of the network in its database, which includes awareness of the live facility routings as well as of the spare channels, with the associated port assignments. On receipt of failure information (either from auxiliary transport monitoring systems or from the DCSs themselves), the controller identifies the affected facilities, algorithmically determines alternate paths skirting the failure for each of these, and commands the DCSs to implement these alternate paths, thus restoring service.

Another way to restore the network is through the implementation of precomputed alternate routes for each possible major failure (such as a cable cut), stored as "alternate maps" of cross-connections in each DCS, and invoked by a single command to each DCS to institute the particular map desired. Both approaches require maintenance of network maps, which is an administrative overhead.

In addition to the centralized method, there are several proposals to evolve the DCS-based restoration methodology to distribute processing to the DCS nodes.[7,8,9] These methods rely on signaling between DCSs (or collocated processors) to establish restoration paths; in what is referred to as a "flooding" mechanism, a path-search attempt emanates from a source node, branching in all available directions at each node it encounters, until it finally intersects the target node, whereupon a path is synthesized. The obvious advantage of distributed restoration methods is that processor load is insensitive to network size, since processing activity at any node is not concerned with the far reaches of the network, being strictly confined to exchanges with the immediate vicinity. This leads to a significant improvement in restoration time, which can be reduced to the order of 2.5 s or less. Further, because of tight coupling with the DCSs, the problem of "database synchronization" (the need for a central controller to maintain an accurate view of the DCS cross-connect fabric at all times) is eliminated.

5. APPLICATIONS PERSPECTIVE

A given network design will most likely be a hybrid design, incorporating most, if not all, of the alternatives discussed in previous sections. This section offers some insight into how the characteristics of these architectures make a particular architecture attractive for a given application.

From the standpoint of applications, the network is traditionally compartmentalized into two sectors, the loop and the interoffice. The loop sector is the last leg in the communications path, the connection between the service provider and the end-customer. It is typically characterized by a hubbed demand pattern, in which traffic from all end-customers within a cluster is aimed at a central office. Because the loop is at the boundary of the provider's network, its connectivity is low, as are the traffic cross-sections; SHRs seem made-to-order for the loop. Because of the hubbed demand pattern, as well as for its simplicity and independence from additional SONET standardization, the path-switched ring is favored as the architecture of choice for loop applications. For applications where a large amount of traffic exists between two points, point-to-point route diversity could also be used.

The interoffice portion of the network is typically mesh-like in

structure. This portion of the network is characterized by a more homogeneous demand pattern than the loop, and is therefore suitable for deployment of bidirectional rings or DCSs. (Office-pairs with significant demand between them can be satisfied with DRP systems, as noted earlier.)

Small to medium-sized networks could, in the near term, profitably use bidirectional rings, appropriately segmented into osculating rings. These networks may contain some DCSs, placed for reasons such as multiplexer replacement, flexibility, or faster provisioning. For these networks, any embedded DCSs could be used for nodal restoration of inter-ring circuits for which 60 ms restoration time is not crucial for node failures.

Further, the DCS could become an integral part of the SHRs. DCS manufacturers are looking at providing both path-switching and line-switching functionality in their high capacity products. This would be a most desirable alternative, since it provides, in effect, a monolithic ring junction.

Finally, the interoffice sector of large networks can be modeled as a highly connected core surrounded by low-connectivity peripheral subnetworks serving clusters of end-offices. Ideally, all other things being the same, a DCS-based solution is the most cost effective in these scenarios. However, while capital cost exerts strong influence in choice of restoration architecture, speed and reliability are also crucial to self-healing, and represent major considerations. In the near term and for both of these major considerations, the pure-lightwave (i.e., non-DCS) methods, with their almost "reflex" actions (within 60 ms), have a distinct advantage over the contemporary, controller-driven, DCS-based method. The latter can take minutes to restore a cable cut, and needs constant vigilance to maintain controller database accuracy, or its reliability will be compromised. Hence, the most likely near-term architecture for the core of a large network will use some combination of bidirectional rings and DRP for their speed and reliability with respect to cable cuts. However, since the core will tend to be large and highly-connected, it cannot be served by any single ring, nor even less by any single DRP. Also, the core will undoubtedly have embedded DCSs for various reasons already discussed, and for inter-ring grooming and management. These DCSs can therefore profitably be used for nodal restoration in the core — even for some cable cut (or individual circuit) restoration — where a 60 ms switch time for the end-to-end circuits involved is not critical.

Over the longer term, assuming the existence of the DCS-based distributed restoration techniques with restoration speeds under 2.5 s as mentioned earlier, the most plausible with regard to the speed criterion, it is not clear that a 60 ms restoration speed across the entire network is critical, and 2.5 s may be entirely adequate for the bulk of the traffic, especially if it brings with it other advantages, not the least of which is lower cost.

6. CONCLUSIONS

To summarize, this paper has provided an overview of different self-healing architectures, with a particular focus on ring architectures.

While SHRs offer an elegant and simple restoration solution

[7]W. D. Grover, "The Selfhealing Network," IEEE Global Telecommunications Conference, Tokyo, Japan, Nov. 1987.

[8]C. Yang and S. Hasegawa, "FITNESS: Failure Immunization Technology for Network Service Survivability," IEEE Global Telecommunications Conference, Ft. Lauderdale, FL, Dec. 1988.

[9]J. D. Bobeck, S. P. Lee, and J. E. Waninski, Jr., "DPAS Network Control System, A Real-Time Distributed Self-Healing Network Capability," MILCOM '91, Nov. 1991.

for ring network islands, the complex topology of practical networks makes it necessary to have multiple interconnected rings, which poses an interworking problem. For path-switched rings, this problem has a relatively straightforward solution; however, for bidirectional rings, it is necessary to employ some auxiliary, path-level switching at the transit nodes to achieve the desired function. Alternatively, this interworking problem can be solved through the use of DCSs as transit managers; it likely possible that DCSs will acquire SHR functionality, in which case an integrated solution results.

The paper also presented two other important self-healing architectures — 1×1 point-to-point lightwave systems with diverse-routed protection, and digital cross-connect systems. The express-routed DRP is ideal for heavy traffic situations in short-haul networks. DCS-based restoration has the best network capacity utilization in mesh networks, and is therefore very popular in long-haul networks, as well as highly connected short-haul networks. It does, however, have the slowest restoration speed of the self-healing architectures, although the speeds are expected to increase significantly with the advent of distributed restoration techniques.

Although each network has its own idiosyncrasies and individual network providers have different sets of constraints which makes it risky to generalize, the consensus at the present time indicates the following trends:

- Path-switched rings will dominate the loop.
- Small- to medium-sized networks are likely to see bidirectional rings being deployed in the interoffice network.
- Large networks will probably be best served by DCSs in the highly-connected core of the interoffice network, with bidirectional rings towards its boundaries.

Feasibility Study of a High-Speed SONET Self-Healing Ring Architecture in Future Interoffice Networks

Tsong-Ho Wu
Maurice E. Burrowes

Optical fiber systems play an essential role in telecommunication transmission systems due to low cost, high capacity, and good service quality. To economically utilize the fiber's high capacity, the network architecture often used is organized as a hubbing structure that is based on a Single-Homing (SH) concept. SH is a centralized demand routing concept that aggregates demands from any office to their destinations through an associated home hub [1]. However, such a hubbing network architecture is inherently vulnerable due to fiber cable cuts and major hub failures. Figure 1 depicts an example of the facility hubbing architecture.

To provide fiber cable protection capabilities for the hubbing networks, the Diverse Protection (DP) architecture, which places the protection fiber in a physically diverse route, has been commonly used or planned in current and future optical fiber networks [2]. The DP architecture protects only against fiber cable cuts and multiplex equipment failures, not major hub failures. In order to provide protection against hub failures, the DP concept can be extended to a Dual-Homing (DH) concept, which connects two hubs from each Central Office (CO) requiring hub protection by duplicating fiber spans to both hubs. Figure 2 depicts such a DH architecture. In case of home hub failure, that office can still access other offices via the foreign hub to continue service. However, this DH architecture using DP has been shown to be expensive for interoffice networks [2]. Since network survivability is a required overhead for Operating Telephone Companies (OTCs), continuously reducing network cost while maintaining an acceptable survivability level is a challenging and necessary task for OTC strategic network planners.

The recently standardized SONET technology [3] and associated high-speed add/drop multiplexing technology [4] make a Self-Healing Ring (SHR) architecture [5] [6] practical due to the SONET simple control scheme, ease of adding and dropping subchannels, and high-speed add/drop multiplexing capability (e.g., 2.4 Gb/s), which may meet the intra-Local Access and Transport Area (intraLATA) interoffice demand requirement. (For convenience, throughout this article, unless specified otherwise, SHR represents the SH SHR.) Figure 3a depicts a generic SONET self-healing arrangement for the interoffice application, which provides STS-3c (155 Mb/s) connections between three COs and one hub. In Figure 3a, each CO is equipped with an Add/Drop Multiplexer (ADM) that terminates two fibers for transmit and two fibers for receive. The two transmitters in each ADM are arranged to continuously transmit signals into the ring in opposing directions. Similarly, receivers in each ADM receive transmitted signals from opposing directions. Selectors at each ADM receive both signals and usually select the signal from the primary ring, but select the signal from the secondary ring to recover from a network failure. The depicted SHR would be capable of surviving one or more fiber system failures affecting a single ring, one ADM failure, or a cable cut severing both rings. The SHR architecture can be arranged to include one or two hubs to provide functions corresponding to the SH or DH architecture in the fiber-hubbed networks [7]. Figures 3a and 3b depict examples of the SH SHR and the DH SHR, respectively. The SHR architecture can also provide a "pseudo" point-to-point link between any two COs in the SHR that may down-size the hub, which may result in an increased survivability in case of major hub failures. The depicted SONET SHR architecture could be used in interoffice networks to reduce survivable network cost (compared to its hubbing counterpart discussed above) by sharing facilities and ADMs. However, the penalty associated with using SHRs may be an expensive system upgrade when the ring capacity is exhausted.

In this article, we study the economic feasibility of using SONET SHR architecture in survivable interoffice fiber networks. We first discuss a model for the SHR feasibility study. We then discuss results of case studies based on a metropolitan LATA network. Finally, conclusions are given.

Feasibility Study Model

The goal of this SHR feasibility study is to investigate if there are any subareas of the interoffice fiber-hubbed networks that may show a promising role for the SHR architecture. This feasibility study compares the SHR architecture with its hubbing counterpart based on the same survivability level for fiber cable cuts. The counterpart of the SHR in this feasibility study is the hubbing architecture with 1:1/DP. The 1:1/DP architecture is selected as the counterpart architecture here, since it has the same survivability for cable cuts as the SHR architecture (i.e., 100% protection for cable cuts). The following de-

Reprinted from *IEEE Communi. Maga.*, vol. 28, no. 11, pp. 33–42, 51, Nov. 1990.

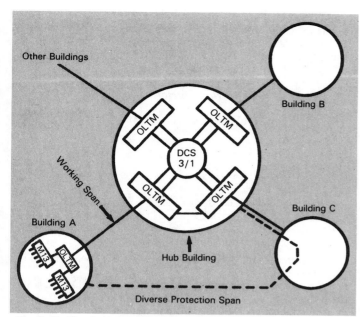

Fig. 1. Facility hubbing with diverse protection.

scribes a general model used for our feasibility study. This feasibility study mode includes a factor due to network growth, since the comparison results could be less meaningful unless all factors showing advantages and disadvantages of the SHR architecture have been considered.

The feasibility study model is done as follows:

- Manually find areas in a given fiber-hubbed network that can form physical rings and match the given selection criteria.
- For each year unit during a planning period, perform the following capital cost comparison for each candidate area until all candidate areas have been processed: Convert the network demand requirement to the demand requirement of the area being considered, compute the area's cost and survivability for using SHRs, and compute the area's cost and survivability for using the 1:1/DP architecture.

Note that the demand requirement used in this study is the DS3 demand requirement, since this is the basic signal rate for current interoffice transport systems. Thus, a demand bundle algorithm is needed to bundle the given circuit demand requirement to the DS3 demand requirement before conducting the feasibility study. Different demand bundle algorithms could result in different results for feasibility studies. The demand bundle algorithm used in this feasibility study model, which is designed primarily for fiber-hubbed networks, is from [8]. The basic idea of this bundle algorithm is to build direct or indirect DS3s between each pair of COs based on a demand threshold (e.g., if there are 16 DS1s required between CO 1 and CO 2, and if the demand threshold is 14 DS1s, then the algorithm builds a direct DS3 between CO 1 and CO 2). A direct DS3 is a point-to-point DS3 between two COs, while an indirect DS3 is a DS3 including DS1s originating or terminating at different COs. An indirect DS3 must terminate at the Digital Crossconnect System (DCS) of the hub for grooming. More details of this demand bundle algorithm can be found in [8]. The major reason for using this hubbing-oriented demand bundle algorithm in this feasibility study is that the SHR architecture will be considered as a good alternative architecture only if it can first be demonstrated to have cost advantages compared to its hubbing counterpart.

Application Areas in Interoffice Fiber-Hubbed Networks

The selection criterion for the candidate area in this study depends upon the homing network architecture in demand (e.g., SH or DH networks). For SH networks, a criterion based on its facility hierarchy is a natural choice, since it minimizes the operational impact when introducing the SHR architecture into fiber-hubbed networks (i.e., the hub serves as a point of contact between the SHR and the remainder of the network). Thus, a possible application area is within each cluster of COs or the interhub subnetwork, which can form a physical ring topology.

For DH networks, a possible candidate area is within a group of COs with two near hubs that can be the home and foreign hub of each CO in this group.

SHR Cost Model

The SHRs depicted in Figures 3a and 3b can be engineered by using SONET ADMs in the add/drop mode. Figure 4 depicts a hardware configuration of an ADM's add/drop mode in the SHR to support the generic requirement as described in [5] [6]. In Figure 4,[1] an incoming optical OC-48 demultiplexed into a maximum of 16 STS-3 (155.52 Mb/s) channels. Some of the STS-3 channels may directly pass through and terminate at an outgoing STS-48 (2.488 Gb/s) multiplexer if the STS-3 channels need not drop to this CO. Those "through" STS-3 channels, together with other STS-3 channels added from this CO, are then multiplexed and converted to an OC-48 optical signal and sent to the next CO in the SHR over an outgoing fiber. If an STS-3 channel must drop at this CO, an STS-3 multiplexer/demultiplexer (mux/demux) is used to demultiplex an STS-3 signal to a maximum of three STS-1 (51.84 Mb/s) channels and then convert each STS-1 to the DS3s (44.7 Mb/s) via STS-1 interface cards. The assumption made in using the DS3 rate in our cost model here is that we will interface with existing equipment; the cost model can easily be modified to incorporate new signal rates if appropriate. The 1:2 selector is used to accept the STS-1 signals from the primary ring in the normal situation, and is changed to accept the STS-1 signals from the secondary ring should network components fail. The 1:2 generator is used to generate duplicate STS-1 signals from an added STS-1 to both the primary and secondary rings. Thus, the protection level (i.e., redundancy level) in our SHR ADM configuration model here is the STE-1 and higher signals. The minimum SONET electrical signal considered in our ADM mode here is the STS-1.

In the following, we discuss a cost model for SHRs using OC-48. The SHRs using OC-N rates other than OC-48 follow a similar model. To compute the cost for the SHR mode, we first convert the network pair-wise DS3 demand requirement to the pair-wise DS3 demand for the area being studied. Given the

---

[1] In Figure 4, we assume that the optical line rate of the SHR is OC-48. However, any OC-N with $N \geq 3$ can be applied to our model.

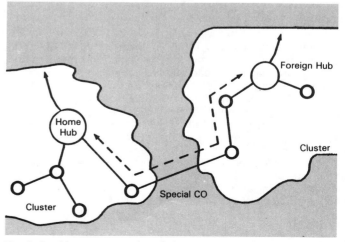

Fig. 2. Dual-homing network architecture.

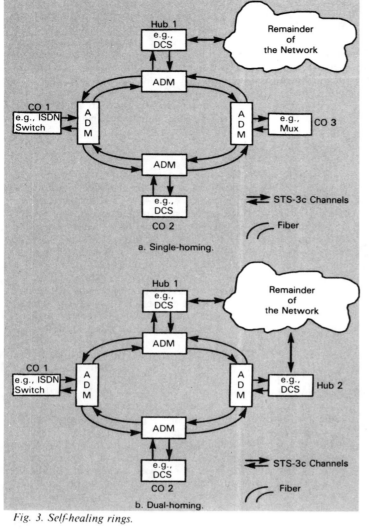

Fig. 3. Self-healing rings.

DS3 demand requirement for CO pair *(j,k)* (denoted by $d_{j<k}$) in the SHR, the number of DS3s adding from and dropping at CO *i* (denoted by D_i) is

$$D_i = \sum_{j=i \text{ or } k=i} d_{jk}.$$

The total number of DS3s carried on each ring of the SHR is

$$\sum_{jk} d_{jk}.$$

Note that the above D_i is only applicable to SH SHRs. It requires a minor modification for DH SHRs. This modification is to ensure that the second hub has enough spare capacity to restore demand if the first hub fails and vice versa. Let CO $k1$ and CO $k2$ be the two hubs in the DH SHRs and DD_i be the number of DS3s adding from and dropping to CO *i*. $DD_i = D_i$ for CO *i*, which is not the hub. If CO *i* is one of two hubs, $DD_i = (D_i + \text{DS3s that pass through but do not terminate at the hub other than CO } i)$.

Let $nSTS3_i$ be the number of STS-3 mux/demuxes (multiplexing STS-1 channels to an STS-3 channel and demultiplexing an STS-3 channel to STS-1 channels) required to support D_i at CO *i* in the SHR. $nSTS3_i =$

$$\left\lceil \frac{D_i}{3} \right\rceil$$

where [*x*] is the smallest integer greater than *x*. The number of STS1 interface cards (mapping between STS-1 and DS3) required at CO *i* is D_i. Thus, the cost of the ADM at CO *i* (denoted by C_ADM_i) is computed as follows:

$$C\_ADM_i = (market\_factor) \times (C\_start\_ADM + C\_hardware\_ADM_i) \quad (1)$$

and

$$C\_hardware\_ADM_i = 2 \times C\_o/e + 2 \times C\_M48 + 2 \times nSTS3_i \times C\_M3 + C\_M1 \times D_i \quad (2)$$

where:

$C\_hardware\ ADM_i$ = cost due to optoelectronic hardware in the ADM of CO $_i$,
C_start_ADM = getting-started cost for each ADM in add/drop mode,
C_o/e - cost of each (O/E,E/O) pair for OC-48,
C_M48 = cost of each mux/demux (STS48 <-> STS3s),
C_M3 = cost of each mux/demux (STS3 <-> STS1s), and
C_M1 = cost of each STS1 interface card.

The getting-started cost for the ADM is the nonoptoelectronic hardware cost, which includes costs due to protection switching, frames, power, microprocessors (control, interface, and monitor), and control software. The sum of C_start_ADM and $C\_hardware\_ADM$ is the manufactured cost of ADM components. The value of *market_factor* is used to recover the overhead of design and development of ADMs by vendors. We estimate that the *market_factor* could range from 1.5 to 2.0 in the SONET ADM. The factor of 2 in Equation 2 represents the required component redundancy based on our SHR ADM model (in add/drop mode) depicted in Figure 4. The cost of the SHR is the sum of costs for ADMs, fiber material/splicing and placement, and regenerators.

Hubbing Network Cost Model

Hubbing networks can be designed by using either SONET Optical Line Terminal Multiplexers (OLTMs) or SONET ADMs in the terminal mode [4]. The hubbing network design using SONET OLTMs is similar to that using today's OLTMs [2]. A back-to-back OLTM configuration is used to provide protection for the working system. Figure 5 depicts a model using one ADM in the terminal mode to equip working and protection terminals for the fiber-hubbed networks. Unlike the add/drop mode depicted in Figure 4, all STS-3 channels must be dropped to the CO and demultiplexed into STS-1 channels that are eventually converted into DS3s. The added DS3 signals from this CO follow a reverse path to the one described above. Also, unlike the operations of the add/drop mode used in the SHR cost model, the working transmitting and receiving signals travel over the same path, and the protective bidirectional signals travel over a diverse path. The 1:2 selector or protection switch is used for each STS-1 channel to ensure that the signals are processed through the working portion in the normal situation and through the protection portion should a network component fail. Note that the 1:2 generator required in Figure 4 is no longer needed for the ADM in the terminal mode (see Figure 5), since there is no need to send duplicate signals here. The difference between OLTMs and ADMs in the terminal mode from our cost model point of view is that the getting-started cost (see "SHR Cost Model" above) of the OLTM is less than that of the ADM in the terminal mode, which is the same as the ADM in the add/drop mode. This assumes that the ADM is sold as an integrated equipment package that can be configured in either the add/drop mode (Figure

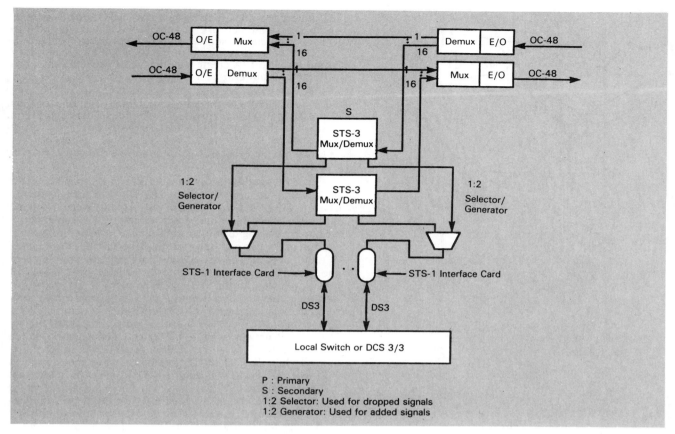

Fig. 4. SONET SHR's ADM hardware configuration in add/drop mode.

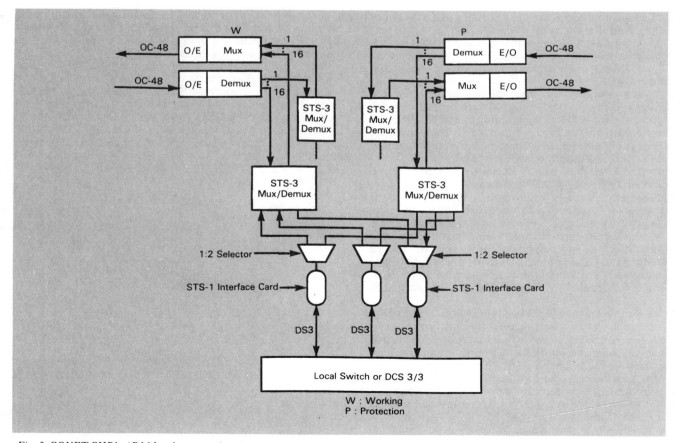

Fig. 5. SONET SHR's ADM hardware configuration in terminal mode for point-to-point span with DP.

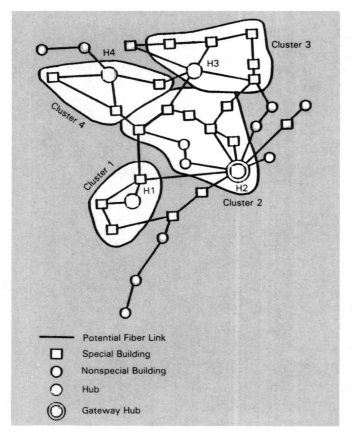

Fig. 6. LATA network model and study areas for single-homing SHRs.

- Create a dedicated span layout for each CO and hub pair for the SH/1:1/DP network, i.e., create a working span and a diverse protection span for each (CO, hub) pair; for the DH/1:1/DP network, create a similar span pair for the hub pair; and for each CO, create a similar span pair not only to its home hub but also to its foreign hub.
- Calculate DS3s on each span.
- For each candidate line rate, calculate the cost of each span for the SH/1:1/DP network or the DH/1:1/DP network, and select the line rate with the least span cost as the line rate of the span.
- Calculate the network cost by summing all span costs plus the route placement cost.

Given the candidate topology, the dedicated span layout for the hubbing subnetwork is created as follows. For each nonhub CO, say CO i, find two link-disjoint paths between CO i and the hub by using algorithms described in [8]. A working span is created by placing terminal equipment at the two ends of the shortest of the two paths and splicing fibers at the intermediate COs in that path. The protection span uses the remaining path and a similar span creation procedure. Thus, for a cluster of COs with n COs for the SH/1:1/DP network, there are $(n - 1)$ working spans and corresponding $(n - 1)$ diverse protection spans. However, it has $(2(n - 2) + 1)$ working spans and corresponding $(2(n - 2) + 1)$ diverse protection spans for the DH/1:1/DP network. Thus, the SH/1:1/DP network with n nodes requires $2(n - 1)$ OLTMs or n ADMs (in terminal mode), while the DH/1:1/DP network with n nodes requires $2(2(n - 1) + 1)$ OLTMs or $(2(n - 1) + 1)$ ADMs in terminal mode. Therefore, terminal costs of two overlaid SHRs with n nodes are higher than their counterpart, since these two overlaid SHRs require $2n$ ADMs (in add/drop mode), and the SH/1:1/DP network requires only $2(n - 1)$ ADMs in terminal mode or the DH/1:1/DP network requires $(2(n - 1) + 1)$ ADMs. Two SHRs are defined as overlaid if these two SHRs have the same set of nodes and fiber routes. So, as will be discussed in "Case Studies" below, adding one more overlaid SHR may not be an economical strategy for growth when the first SHR's capacity is exhausted.

After creating the span layout, we convert the network pairwise DS3 demand requirement to the area's pair-wise DS3 demand requirement, as in the SHR cost model. Given the DS3 demand requirement for CO pair (j,k) (say d_{jk}) in the candidate area, the number of DS3s carried over span i (denoted by D_SP$_i$) is computed as follows. Let $s\_span(i)$ and $d\_span(i)$ be the two ends of span i where $d\_span(i)$ is assumed to be to the hub:

$$D\_SP_i = \sum_{j = s\_span(i)\ or\ k = s\_span(i)} d_{jk}.$$

Given D_SP_i, the cost of span i, denoted by C_SP_i, (either working span i or protection span i), can be computed as follows:

$$C\_SP_i = (market\_factor) \times (2 \times C\_start\_TM + C\_hardware\_TM_i) + SPF \tag{3}$$

and

$$C\_hardware\_TM_i = 2 \times C\_ove + 2 \times C\_M48 + 2 \times nSTS3 \times C\_M3 + Cost(M1) \tag{4}$$

$$Cost(M1) = \begin{cases} 2 \times C\_M1 \times D_i & for\ the\ working\ span \\ 0 & for\ the\ protection\ span \end{cases}$$

4) or the terminal mode (Figure 5), depending upon the applications. The getting-started cost of the OLTM is less than that of the ADM, since the alarm and control circuitry for the SHR's ADM involves the entire ring network and, therefore, the SHR's ADM is more complex than the OLTM used in the hubbing network.

To provide working and protection terminals for the hubbing network, two separate OLTMs are needed (one for working and one for protection); but only one ADM, with the configuration depicted in Figure 5, is needed. This ADM terminal mode configuration can be obtained from reconfiguring the ADM in the add/drop mode (Figure 4).

Unlike the SHR, which usually must use high-speed terminal equipment (e.g., OC-48 ADMs) to carry all ring demand, the hubbing network can be engineered by using lower-speed terminal equipment (e.g. OC-12), since each fiber span in the hubbing network usually carries a much lower demand than the SHR, which must carry the total ring demand. There is a significant difference in component costs using the OC-12 (622.08) Mb/s and OC-48 or OC-24 (1.244 Gb/s or 2.488 Gb/s) in our cost model due to different requirements of power, density, number of channels, and probably different technologies.

The basic design principle of the SH networks is that all demands from each CO must be routed via the hub. For example, in Figure 1, connections between Building A and Building C must use two fiber spans: one between Building A and the hub, the other between the hub and Building C. Each fiber span may or may not pass through intermediate COs. A diverse protection span is needed to protect the network from service interruptions in case of a working span failure (see Figure 1). Given a set of candidate line rates (e.g., OC-12, OC-48) and the pairwise DS3 demand in the area being studied, the following model is used to calculate costs of the hubbing network. In the following, the SH network and the DH network are denoted by SH and DH only, respectively.

Table I. ADM Component Costs Not Including Market_Factor

Component	Unit Cost ($)
(O/E,E/O) Pair for OC-12	1,000
(O/E,E/O) Pair for OC-24 or OC-48	3,000
Mux/Demux for STS-12 to STS-3	500
Mux/Demux for STS-24 to STS-3	600
Mux/Demux for STS-48 to STS-3	800
Mux/Demux for STS-3 to STS-1	50
STS-1 interface card	200
Getting-started cost for a SONET OC-12 ADM	15,000
Getting-started cost for a SONET OC-24 ADM	25,000
Getting-started cost for a SONET OC-48 ADM	32,500

and SPF = span costs due to fiber material, splicing, and re-generators where C_start_TM = getting-started cost for each ADM in the terminal mode or each OLTM = (TM_ADM ratio) $\times$ C_start_ADM, TM_ADM_ratio = ratio of OLTM getting-started cost to ADM getting-started cost, and $C\_hardware\_TM_i$ = cost due to optoelectronics hardware at CO i.

An estimation of TM_ADM_ratio ranges from 0.5 to 0.99, since it is estimated that one OLTM costs less than one ADM, but two OLTMs are expensive than one ADM. Note that if the TM_ADM_ratio equals 0.5, then ADMs in the terminal mode are used for the hubbing network. The total cost for the hubbing subnetwork is the sum of all span costs plus the total route mileage cost (i.e., fiber placement cost).

Network Survivability Measure

The area survivability is defined as the percentage of total demand that is still intact when a network component fails. Since the survivability of fiber cable cuts for two competitive architectures (the SHR and hubbing with 1:1 diverse protection) is the same (100% protection for cable cuts), we compare network suvivability here for both architectures only for the case of major hub failures. "Hub survivability" is defined as the percentage of the total demand that is still intact when that hub building fails. Note that if a DS3 or DS1 is terminated at the hub building, this DS3 or DS1 may be terminated at either the hub DCS or other equipment within the hub.

For SH SHRs, only demands not terminating at the hub will survive when that hub fails. For DH SHRs, the DS3s or DS1s that pass through the hub can be restored via the other hub when that hub fails. For fiber-hubbed networks, a model used to calculate the hub survivability for SH and DH cases can be found in [2].

Network Growth Model

Evaluation of network growth impact in this study is based on the following heuristic model. Two SHRs are defined as "overlaid" if they have the same set of nodes and fiber routes. Two SHRs are defined as "overlapped" if they are not overlaid and have a nonempty common set of nodes and fiber routes.

For 1:1/DP networks, the OC-N rate for each span in a planning period is determined from one of the following two procedures, depending upon which one results in the lowest total network cost for a given planning period of N years.

Procedure 1 is as follows:

- a. Assign each candidate OC-N rate as the line rate for the entire planning period, and compute the span cost associated with that line rate.
- b. A least-cost OC-N rate is selected as the line rate for the entire planning period.

Procedure 2 is:

- a. Initially, set remaining_planning_period = a given period of N years, and set current_year = 0.
- b. Select a least-cost OC-N rate that can carry the cumulative demand throughout the remaining_planning_period if possible and stop; otherwise, go to the next step.
- c. Select the maximum available OC-N rate as the OC-N rate for years between the current year and year k where the cumulative demand at year $k + 1$ exceeds this maximum OC-N rate; set current year = k, set remaining_planning_period = {year k, ..., year N}; and repeat step b.

For the SHR architecture, the OC-N rate selection is based on the following principles:

- Select the minimum OC-N rate that can carry the cumulative demand of the planning year as the SHR capacity throughout the planning period; otherwise, place the available maximum OC-N rate as the initial SHR capacity and go to the next step.
- If the growth demand does not exhaust the SHR capacity, then put all of this demand into the SHR.
- If the growth demand exhausts the current SHR capacity, then partition the current SHR into two smaller overlapped SHRs if possible; otherwise, add one more overlaid SHR to accommodate growth demand. Under this step, the first SHR should carry demand as fully as possible, and the second SHR the remainder of the demand.

The procedure for the SHR growth model described above is based on observations that the SHR continues to have a cost

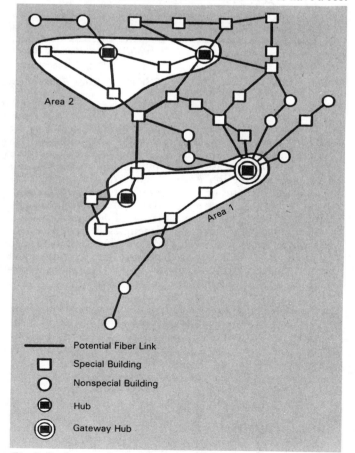

Fig. 7. Study areas for dual-homing SHRs.

145

Study Areas	DS3 Demand	Number of Nodes	Hubbing Architecture			SHR		
			Number of Span Pairs*			Number	OC-N	N (CO)
			OC-12	OC-24	OC-48			
Cluster 1	24	4	3	0	0	1	24	4
Cluster 2	71	7	5	1	0	2	48	(4,4)**
Cluster 3	45	6	4	1	0	1	48	6
Cluster 4	40	4	2	1	0	1	48	4
Interhub Subnetwork	91	4	0	0	3	2	48	(3,3)

N(CO): Number of COs in the SHR (also in the hubbing subnetwork)

*Each span pair includes a working span and a diverse protection span.

**Both first and second SHRs include four COs.

Table III. Cost/Survivability Comparisons Between SHR and Single-Hubbing Structures

Study Areas	Relative Cost		Hub Survivability (%)	
	SHR/Hub (ADM)*	SHR/Hub (OLTM)**	SHR	Hub
Cluster 1	0.98	0.85	8.3	0.0
Cluster 2	1.00	0.84	9.8	0.0
Cluster 3	0.94	0.80	15.6	0.0
Cluster 4	0.95	0.83	5.0	0.0
Interhub Subnetwork	0.78	0.69	7.7	0.0

SH/1:1/DP: Single-homing fiber-hubbed network with 1:1 diverse protection

*SHR/Hub (ADM): Relative cost of SHR compared to SH/1:1/DP using ADMs in the terminal mode

**SHR/Hub (OLTM): Relative cost of SHR compared to SH/1:1/DP using OLTMs (TM_ADM_ratio = 0.7)

advantage compared to its hubbing counterpart as long as the SHR capacity does not exhaust, and that two overlapped SHRs cost less than two overlaid SHRs since the latter requires more ADMs. We will discuss those observations in more detail below. Note that the above growth model does not include the possible use of point-on-point systems to support demand growth, since the purpose here is to evaluate how well the SHR achitecture can deal with the situation of demand growth.

Under a growth scenario, it is sometimes desirable to evaluate architecture alternatives based on a representative single cost measure. A commonly used measure in the growth model is called "present worth" (denoted by PW). It takes into account the interest rate in that planning period.

Let C_0 = the network cost at year 0, I_i = incremental network cost at year i, N = number of years in a planning period, and r = yearly interest rate:

$$PW = C_0 + \sum_{i=1}^{N} \left(\frac{I_i}{(1+r)^i} \right)$$

Case Studies

The purpose of the case studies in this section is to answer the following questions: what should the SONET SHR capaci-

ty be in the interoffice application; what are the relative capital cost economics of the SHR compared to its hubbing couterpart; and what are the general conditions that can best utilize the SHR characteristics to reduce survivable network cost. We follow the model and philosophy developed above and use a LATA network as our model network here to answer the above questions. The considered LATA network is a metropolitan area LATA network including 36 nodes and 64 links. Each link is a candidate location for realizing the fiber system. Of these 36 nodes, 4 nodes are hubs (one hub is designed as the gateway), 20 nodes are special COs, and 12 nodes are nonspecial COs, where only special COs have DP routing provision in the hubbing network. The newtwork topology depicted in Figure 4 has been optimized for the hubbing case as taken from the model network in terms of costs of fiber material, placement, and regenerators by using a network design tool called Fiber Options [8]. The circuit demand data for the considered LATA network are from OTC personnel. This is total demand in the network, including all demand now carried by T1s and Voice-Frequency (VF) copper cable. The given circuit demand requirement was bundled into the DS3 demand requirement by using a module of Fiber Options whose design is primarily for fiber-hubbed networks. The reason for using this hubbing-oriented demand bundle algorithm here is that the SHR architecture will be considered a good alternative architecture only if it can be demonstrated to have cost advantages compared to its hubbing counterpart.

Table IV. Candidate Study Areas, Layouts for SHRs, and Dual-Hubbing Networks

| Study Areas | DS3 Demand | Dual-Hubbing Architecture | | SHR | |
		Number of Nodes	Number of OC-12 Span Pairs *	Number	OC-N
Area 1	46	6	9	1	48
Area 2	28	5	7	1	48

*Each span pair includes a working span and a diverse protection span.

Table V. Cost/Survivability Comparisons Between SHR and Dual-Hubbing Structures

| Study Areas | Relative Cost | | Worst Hub Survivability (%) | |
	SHR/Hub (ADM)*	SHR/Hub (OLTM)**	SHR	DH/1:1/DP
Area 1	0.69	0.57	63.6	21.7
Area 2	0.75	0.63	72.7	17.8

*SHR/Hub (ADM): Relative cost of SHR compared to DH/1:1/DP using ADMs in the terminal mode

**SHR/Hub (OLTM): Relative cost of SHR compared to DH/1:1/DP using OLTMs (TM_ADM_ratio = 0.7)

There are three line rates considered in this study: OC-12, OC-24, and OC-48. Table I shows ADM component costs used in this case study. The optoelectronic component costs and the regenerator costs are estimates from Bellcore subject experts and do not include costs due to design and development (i.e., not including market_factor). Assume that the market_factor equals 2 in this case study. Note that the ADM costs do not include equipment installation costs. This ADM cost is based on an ADM equipment model using SONET path-level protection switching for the ring application [6]. The distance threshold beyond which a regenerator is required is assumed to be 30 miles for OC-12, 25 miles for OC-24, and 23 miles for OC-48. The fiber material and splicing cost per mile per fiber pair is assumed to be $470, which reflects future fiber technology. The route mileage cost (installation cost) is assumed to be $5,000 per mile.

Single Homing Interoffice Networks

Using the model discussed in "Feasibility Study Model" above, we first study the case of SH networks. In this case, we

select four clusters and the interhub subnetwork as candidate areas to study the SHR application feasibility problem. Figure 6 depicts these five study areas. Note that in the study, only special COs are considered as candidate nodes that can be included in SHRs.

Table II describes study areas with corresponding layouts of SHRs and their hubbing counterparts. The hubbing counterpart of the SHR considered here is the SH/1:1/DP network, since we would like to compare their relative capital costs under the same survivability level for fiber cable cuts (i.e., 100% protection for fiber cable cuts for both cases). Table III reports feasibility results associated with Table II and compares cost and suvivability between two competitive architectures. The market_factor and TM_ADM_ratio used in Table III are assumed to be 2.0 and 0.7 (for hubbing networks using OLTMs), respectively. Each span pair in Table II represents a working span and its corresponding diverse protection span.

First let us consider the case of using SONET ADMs in terminal mode to implement the SH/1:1/DP networks. As shown in Table III, the SHR architecture plays a promising role (compared to SH/1:1/DP) in all areas except Cluster 2. The demand requirement of Cluster 2 requires two SHRs with four nodes each. In Cluster 2, the terminal cost for using SHRs approximately equals that for using SH/1:1/DP, since the former requires eight OC-48 ADMs and the latter requires ten lower-cost OC-12 ADMs and two OC-24 ADMs. The case for Cluster 1 is similar to that for Cluster 2. For Cluster 3 and Cluster 4, the cost saving from using the SHR compared to SH/1:1/DP is due to lower termination costs for the SHR. For the interhub net-

Table VI. Sensitivity Analysis of SHR Carrying Low Demand in Cluster 4

| DS3 Demand | SHR OC-N | Relative Cost | |
		SHR/Hub (ADM)*	SHR/Hub (OLTM)**
5	12	0.75	0.65
9	12	0.75	0.65
16	24	0.98	0.85
23	24	0.87	0.75
30	48	0.97	0.84
40	48	0.95	0.83

*SHR/Hub (ADM): Relative cost of SHR compared to DH/1:1/DP using ADMs in the terminal mode

**SHR/Hub (OLTM): Relative cost of SHR compared to DH1:1/DP using OLTMs (TM_ADM_ratio = 0.7)

Table VII. Sensitivity Analysis Due to Area Size (Width)

| Area | Relative Cost | |
	SHR/Hub (OLTM)	SHR/Hub (ADM)
Cluster 1	0.85	0.98
3E-Cluster 1*	0.78	0.84
Cluster 3	0.80	0.94
3E-Cluster 3**	0.69	0.77

*3E-Cluster 1: Cluster 1 enlarged by a factor of 3 to link distances between COs

**3E-Cluster 3: Cluster 3 enlarged by a factor of 3 to link distances between COs

Table VIIIa. Network Capacity Arrangement for Cluster 1 Network Growth

Year	Cluster 1						
	NR	DS3s(R)	NO(R)	OC-N(R)	NS(H0)	DS3s(H)	OC-N(H)
0	1	(24)	(4)	(48)	3	(11,4,11)	(48,12,48)
2	1	(30)	(4)	(48)	3	(14,6,14)	(48,12,48)
4	1	(36)	(4)	(48)	3	(17,8,17)	(48,12,48)
6	1	(43)	(4)	(48)	3	(21,10,20)	(48,12,48)
8	2	(26,24)	(3,3)	(48,48)	3	(25,12,23)	(48,12,48)
10	2	(30,27)	(3,3)	(48,48)	3	(29,14,26)	(48,12,48)

NR: Number of SHRs

DS3s(R): Number of DS3s in each SHR

DS3s(H): Number of DS3s in hubbing span layout

NO(R): Number of nodes in each SHR

NS(H): Number of spans in the SH/1:1/DP network

OC-N(R): OC-N line rate for each SHR

OC-N(H): OC-N line rate for each span in the SH/1:1/DP network

Table VIIIb. Network Capacity Arrangement for Interhub Subnetwork Network Growth

Year	Interhub Subnetwork						
	DS3s(R)	NO(R)	OC-N(R)	NS(H)	DS3s(H)	OC-N(H)	
0	2	(47,44)	(3,3)	(48,48)	3	(32,29,27)	(48,48,48)
2	3	(48,48,7)	(3,3,3)	(48,48,48)	3	(36,33,43)	(48,48,48)
4	3	(48,48,20)	(3,3,3)	(48,48,48)	3	(41,37,49)	(48,48,48)
6	3	(48,48,34)	(3,3,3)	(48,48,48)	3	(46,42,55)	(48,48,48)
8	4	(48,26,48,24)	(3,3,3,3)	(48,48,48,48)	3	(52,47,62)	(48,48,48)
10	4	(48,34,48,32)	(3,3,3,3)	(48,48,48,48)	3	(58,52,69)	(48,48,48)

work with four hubs, its high demand requires two overlapped OC-48 SHRs with three hubs each. A significant cost difference between SHRs and SH/1:1/DP in the interhub subnetwork is due to three expensive OC-48 regenerators needed for diverse protection systems of SH/1:1/DP. In this area, both SHRs and SH/1:1/DP require six OC-48 ADMs. Note that there is another option here to use two overlaid SHRs with four hubs each for the interhub subnetwork, but the result has shown that two overlaid SHRs are more costly than two overlapped SHRs, since the former requires more ADMs in terminal mode than the latter. Thus, it implies that whenever the area demand requires multiple SHRs, use of multiple overlapped SHRs should be preferable to use of multiple overlaid SHRs. This observation is useful for planners in developing a cost effective plan for growth.

It may be intersting to see the result for high-demand areas (e.g., the interhub area) with relatively short distances among COs. To do that, we reduce link distances in the interhub subnetwork by half so that no regenerator is needed for SH/1:1/DP. The result shows that the relative cost has been increased from 0.78 to 0.88 for SH/1:1/DP using ADMs in the terminal mode, and 0.69 to 0.75 for SH/1:1/DP using OLTMs. However, this new result still shows that a SHR architecture is more attractive in the interhub area than in other areas. This seems to suggest that the SHR architecture will be attractive if its hubbing counterparts have to use very-high-speed systems.

In the case of using SONET OLTMs to implement the SH/1:1/DP networks, the relative cost savings of SHRs are generally better than SHRs using ADMs in the terminal mode, since the number of OLTMs needed for SH/1:1/DP networks is much greater than the number of ADMs in the terminal mode needed (see "Hubbing Network Cost Model"). As shown in Table III, the SHR architecture is attractive in all areas, and the maximum cost savings of SHRs may reach 31%.

We also see a better hub suvivability for SHRs than for the hubbing subnetwork, should the hub building fail. For each candidate area (an intracluster subnetwork or an interhub subnetwork) in the SH/1:1/DP network, the hub survivability is always zero because all demands from COs or hubs in the area pass through or terminate at the hub or gateway [2] in the SH architecture. Note that it is always very expensive to increase survivability in major hub failures (i.e., hub survivability) by using the hubbing architecture. The only architecture option based on today's automatic protection switching technology that can increase hub survivability is the DH architecture, which requires hundreds of thousands of dollars to increase gateway survivability a percentage point under the same study model network [2]. Such a hub survivability figure for using

Table IX. Relative Present Worth Comparison for Network Growth*

Planning Years	Cluster 1		Interhub Subnetwork	
	RPW(O)	RPW(A)	RPW(O)	RPW(A)
6	0.75	0.91	0.78	0.88
8	0.81	0.98	0.81	0.94
10	0.74	0.95	0.76	0.87

*5% yearly growth and 10% yearly interest rate

RPW(O): Ratio of present worth of the SHR to present worth of SH/1:1/DP using OLTMs (assume TM_ADM_ratio = 0.7)

RPW(A): Ratio of present worth of the SHR to present worth of SH/1:1/DP using ADMs in the terminal mode

SHRs could be increased when the demand bundle algorithm designed for point-to-point connections is used. Also, using a DH SHR arrangement, depicted in Figure 3b, can also increase hub survivability, as will be discussed later.

Dual Homing Interoffice Networks

DH is an architectural option for fiber-hubbed networks to increase hub survivability. Table IV describes the area layouts for using DH SHRs and DH hubbing networks, and Table V reports cost/survivability comparisons between two competitive architectures. Two areas depicted in Figure 7 are studied here (see Table IV). Area 1 covers six nodes, i.e., two clusters with three nodes in each (served by two hubs). The DS3 demands in this area can be supported by DH OC-48 SHR. Area 2 covers five nodes, i.e., two clusters with four nodes in one cluster and the remaining node in the second cluster (this node is also the hub of the second cluster). Again, the demand in Area 2 can be carried by DH SHR.

Comparing Table V with Table III, we see that the cost savings for using DH SHRs (compared to DH/1:1/DP networks) is significantly higher than the cost savings for using SH SHRs (compared to SH/1:1/DP networks). The reason for this is that the design cost for DH hubbing networks is much higher than for SH hubbing networks, since the number of span pairs required for the DH network architecture is almost twice that for SH architecture needs [2]. The gain of worst hub survivability for using DH SHRs is much higher than that for SH SHRs (63.6% to 72.7% here compared to 5% to 15% in Table III). Thus, the DH SHR architecture is definitely the best-choice architecture to maximize the potential cost and survivability advantages inherently associated with the SHR architecture. Note here that we consider all demand to each hub, but do not include the cost of completing the path from the second hub to the destination, i.e., we consider costs for the study area only.

Sensitivity Analysis

The examples shown in Table III are SHRs, all with median or high demand, since the studied area is metropolitan downtown. It is interesting to see the case for the SHR carrying low demand, which is likely to be the case in suburban areas. To do that, we select Cluster 4 as a test case and reduce its demand requirement, keeping the rest of the conditions unchanged. Table VI describes the sensitivity analysis results in terms of DS3 demand requirements. As shown in Table VI, the SHRs can be attractive if demands can be carried by a single OC-12 ring. There is a significant difference between the 16 DS3 and 23 DS cases, since two OC-24 ADMs and four OC-12 ADMs are required for the 23 DSs, but six OC-12 ADMs are required for the 16 DS3 in the SH/1:1/DP networks. That makes SHRs attractive for the 23 DSs.

Also, the above analysis focuses on a metropolitan area that has a relatively short distance between two consecutive COs. For some suburban areas, regenerators may be needed to con-

nect COs to the hub. To simulate this situation and see the impact of the regenerator factor on the SHR application, we study two cases here. The first and second cases use Cluster 3 and Cluster 1, respectively, and their associated demand requirements as the network models, except that the link distances between two COs are triple those of the original distances of the model reported in Table II. In the enlarged Cluster 3, no regenerator is needed for the SHR, but five regenerators are needed for the SH/1:1/DP network (four for DP spans, and one for the third working span). For the area of enlarged Cluster 1, one OC-48 regenerator is needed for the SHR and two OC-12 regenerators are needed for the SH/1:1/DP network. The results shown in Table VII suggest that the wider the area covered, the more cost savings we can expect from SHR compared to its hubbing counterparts, as long as the demand in this area can be supported by a single SHR.

The sensitivity analysis was conducted to see the impact of two parameters: market_factor and TM_ADM_ratio. The numerical results indicate that the cost savings of using SHRs is not sensitive to the parameter of market_factor, and there is a stable rate of increase as the value of TM_ADM_ratio increases for the hubbing network design using OLTMs.

Network Growth Impacts

The analysis in the previous two subsections deals with the problem of the SHR initial placement, which clearly shows the advantages of SHR due to its characteristics of sharing facilities and equipment. In this subsection, we discuss impact analysis due to one of the major disadvantages of the SHR, which is capacity exhaustion in a growth environment. Tables VIIIa and VIIIb describe network capacity arrangements for Cluster 1 and the interhub subnetwork, respectively, in a growth scenario. These capacity arrangements are obtained from a heuristic model discussed under "Network Growth Model" above. Table IX shows a relative present worth comparison due to network growth in Cluster 1 and the interhub subnetwork selected to represent relatively low and high demand areas. *PW*, commonly used in the growth model, takes into account the interest rate in that planning period. In Table IX, we assume a 5% yearly growth rate and a 10% yearly interest rate.

For Cluster 1, as shown in Table VIIIa, the OC-48 capacity is used to carry demand from year 0 to year 6, and then this OC-48 SHR is "rearranged" to two smaller OC-48 SHRs to carry growth demand from year 8 to the end of the planning period. For example, the OC-48 SHR for Cluster 1 carries 43 DS3s at year 6, and the DS3 demand requirement with a total of 50 DS3s at year 8 is the following: $d(1,2) = 15$, $d(1-3) = 7$, $d(1,4) = 18$, $d(2,3) = 5$, $d(2,4) = 5$ where $d(a,b) = c$ means the demand requirement between CO a and CO b is c DS3s. So at year 8, OC-48 SHR including four modes is "rearranged" into the following two smaller, overlapped SHRs. The first SHR includes COs 1, 2, and 3 with associated demand requirement $\{d(1,2) = 14, d(1,3) = 7, d(2,3) = 5\}$, while the second SHR in-

149

cludes COs 1, 2, and 4 with associated demand requirement $\{d(1,2) = 1, d(1,4) = 18, d(2,4) = 5\}$. For the hubbing approach in Cluster 1, there are three spans with one OC-48 line rate for spans 1 and 3 carrying demand throughout the planning period, one OC-12 system for span 2 carrying demand through year 8, and an additional OC-12 system carrying the extra two DS3s in year 10. For the interhub subnetwork, the capacity selection problem is much simpler, because the demand requires the use of the highest available OC-N rate.

In the following discussion, we discuss only the case of SH/1:1/DP networks using ADMs in the terminal mode (see columns RPW(A) in Table IX). For Cluster 1 (a relatively low demand area), the SHR is more attractive than SH/1:1/DP in terms of PW for a planning period ranging from six to ten years. This result indicates that the SHR architecture has higher cost saving than its hubbing counterparts as long as the growth demand can be carried by the maximum available OC-N SHR capacity. Compared with Cluster 1, the SHR shows a much better result in terms of present worth for the interhub subnetwork. This is partly because many expensive OC-48 regenerators are needed for the SH/1:1/DP networks in the interhub subarea. For those areas where SHRs show only insignificant cost savings (e.g., 2%) over their hubbing counterparts, the decision to use SHRs may depend not only on capital costs but also on operations costs.

Conclusions

We have discussed a study model to analyze the feasibility of applying a SONET selfhealing ring architecture in the survivable interoffice fiber networks. A metropolitan LATA network is used as a model network. Case study results have suggested that a SHR architecture, in general, is attractive in terms of costs and survivability in metropolitan areas. Particularly, the SHR architecture offers a cost effective solution to applications requiring dual hub protection that cannot be implemented economically by using today's DP methods.

Acknowledgments

The authors would like to thank D. Kong for providing ADM cost estimates, and R. H. Cardwell, J. E. Berthold, and D. J. Kolar for their valuable comments.

References

[1]   M. Kerner, H. L. Lemberg, and D. M. Simmons, "An Analysis of Alternative Architecture for the Interoffice Network," *IEEE J. on Sel. Areas in Commun.*, vol. JSAC-9, no. 9, pp. 1,404–1,413, Dec. 1986.

[2]   T. H. Wu, D. J. Kolar, and R. H. Cardwell, "Survivable Network Architectures for Broadband Fiber Optic Networks: Model and Performance Comparisons," *IEEE J. of Lightwave Tech.*, vol. 6, no. 11, pp. 1,698–1,709, Nov. 1988.

[3]   R. Ballart and Y.-C. Ching, "SONET: Now It's the Standard Optical Network," *IEEE Commun. Mag.*, pp. 8–15, Mar. 1989.

[4]   "SONET Add-Drop Multiplex Equipment (SONET ADM) Generic Requirements and Objectives," Tech. Ref. TR-TSY-000496, Bellcore, issue 2, Sept. 1989.

[5]   R. Lau, "An Architecture for a SONET Self-Healing Ring," Bellcore Contribution to T1x1.5 Standards Project, May 10, 1989.

[6]   "SONET Add-Drop Multiplex Equipment (SONET ADM) Generic Requirements and Objectives for Ring Applications," Tech. Adv. TA-TSY-000496, Bellcore, issue 2, Nov. 1989.

[7]   T.-H. Wu, D. J. Kolar, and R. H. Cardwell, "High-Speed Self-Healing Ring Architectures for Future Interoffice Networks," *IEEE GLOBECOM '89*, pp. 23.1.1–23.1.7, Nov. 1989.

[8]   R. H. Cardwell, C. L. Monma, and T.-H. Wu, "Computer-Aided Design Procedures for Survivable Fiber Optic Telephone Networks," *IEEE J. on Sel. Areas in Commun.*, vol. 7, no. 8, pp. 1,188–1,197, Oct. 1989.

A Class of Self-Healing Ring Architectures for SONET Network Applications

Tsong-Ho Wu, *Senior Member, IEEE,* and Richard C. Lau, *Member, IEEE,*

Abstract—SONET technology has made high-speed self-healing ring (SHR) architectures practical and economical for use in intra-LATA telecommunication networks. This paper reviews a class of SONET SHR architectures and their control schemes. The SHR architectures can be divided into two basic classes: bidirectional SHR's (B-SHR's) and unidirectional SHR's (U-SHR's). Bidirectional SHR's using four fibers have evolved from a majority of today's point-to-point automatic protection switching systems (APS's), while unidirectional SHR's use two fibers and may evolve from today's APS or a new SONET path-level protection switching system made possible by SONET. The cost and capacity tradeoffs between B-SHR's and U-SHR's depend strongly upon the application, the network size and the demand pattern. B-SHR's with four fibers can be operated under today's operations environment without a significant change. However, U-SHR's using a path protection switching method may ease intervendor compatibility problems and can be implemented on a timely basis without changing the well-defined SONET APS protocol. The selection of appropriate SONET SHR architectures will depend upon the operating telephone companies' economic analysis, emphasis on multivendor environment, SHR implementation time frame, and standards progress on making change to support a bidirectional ring architecture.

I. INTRODUCTION

NETWORK survivability continues to be required for high-speed fiber optical networks, and investigating cost-effective survivable network architectures continues to be important. The ring architecture has been considered as a cost-effective survivable network architecture due to bandwidth sharing and improved survivability [1]. A recently adopted SONET standard [2], [3] along with very high-speed add/drop multiplexer (ADM) technology [4], [5] (e.g., OC-48 with a bit rate of 2.488 Gb/s) has resulted in an opportunity for designing cost-effective future survivable SONET interoffice networks [1], [6], [7] by incorporating self-healing ring architectures.

A self-healing ring is a network architecture that connects a set of offices in a physical ring topology with bandwidth sharing and a self-healing capability to mitigate network component failures. The ADM in the SONET ring supports bidirectional transmissions [4], [5]. This paper discusses SONET-based SHR architectures, their control schemes, their impacts on network planning, and their impacts on the SONET standard. We discuss a class of SHR architectures in Section II. Section III discusses differences among those SONET SHR

Paper approved by the Editor for Communication Switching of the IEEE Communications Society. Manuscript received October 22, 1990; revised February 19, 1991. This paper was presented in part at IEEE GLOBECOM '90, San Francisco, CA, December 1990.

The authors are with Bell Communications Research, Red Bank, NJ 07701.

IEEE Log Number 9203573.

architectures in terms of cost/capacity tradeoffs, impacts on the SONET standards, and intervendor compatibility. A summary and remarks are given in Section IV.

II. A CLASS OF SHR ARCHITECTURES

Several SONET SHR architectures have been proposed in [5], [8]–[12] for telecommunications applications. The self-healing ring (SHR) architectures can generally be divided into two categories: bidirectional SHR's (B-SHR's) and unidirectional SHR's (U-SHR's). The type of ring depends upon the path traveled by a duplex communication channel between each office pairs. The SHR is called a *bidirectional SHR (B-SHR)* if both directions of a duplex channel travel over the same path; a *unidirectional SHR (U-SHR)* if the directions of a duplex channel travel over opposite paths. Fig. 1(a) and (b) depict examples of B-SHR's, and Fig. 1(c) depicts an example of a U-SHR. For example in Fig. 1(a) and (b), both directions of a duplex channel between offices 2 and 4 use the same path $(2 \leftrightarrow 3 \leftrightarrow 4)$ which travels through office 3. For the case of U-SHR [see Fig. 1(c)], a duplex channel between offices 2 and 4 travels over two opposite paths: path1: $2 \rightarrow 3 \rightarrow 4$, and path2: $4 \rightarrow 1 \rightarrow 2$. Thus, a B-SHR requires two working fibers to carry a duplex channel, and a U-SHR requires only one working fiber to carry a duplex channel. In order to provide a protection capability for fiber system failures and fiber cable cuts, a B-SHR may use four fibers (i.e., one working fiber pair and one protection fiber pair) or two fibers (i.e., all working fibers with the spare capacity for protection), and a U-SHR requires only two fibers (i.e., one working fiber and one protection fiber).

For each type of rings, two possible SONET self-healing control schemes may be used: line protection switching and path protection switching. The line protection switching scheme uses SONET line overhead for protection switching and restores line demand from a failed facility; while the path protection switching scheme uses SONET path overhead, and restores individual end-to-end service channel such as STS-1 or VT.

A. Bidirectional SHR's (B-SHR's)

A B-SHR may use four fibers or two fibers depending upon the spare capacity arrangement. For convenience, a B-SHR with 4 fibers and a B-SHR with two fibers are denoted by B-SHR/4 and B-SHR/2, respectively.

1) Bidirectional Ring with Four Fibers (B-SHR/4): The B-SHR/4 architecture has essentially evolved from a majority

Reprinted from *IEEE Trans. Communi.,* vol. 40, no. 11, pp. 1746–1756, Nov. 1992.

151

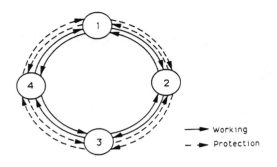

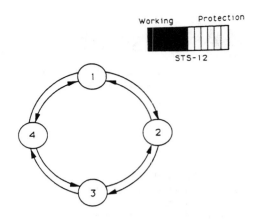

- Point-to-point traffic arrangement

- Protection/Restoration uses separate fibers

- Use SONET line protection switching (e.g. SONET APS)

- Point-to-point traffic arrangement

- Working and protection use the same fiber (reserve half bandwidth for protection)

- Use SONET line protection switching with time slot interchange

- Split Traffic for bandwidth efficiency

(a)

(b)

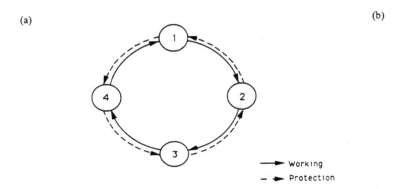

- Counter-rotating ring

- Protection uses separate fiber

- Duplicate and route signals to both rings in opposite directions

- Ring capacity is determined by sum of demands

(c)

Fig. 1. SONET self-healing rings. (a) Bidirectional SHR with 4 fibers (B-SHR/4). (b) Bidirectional SHR with 2 fibers (B-SHR/2). (c) Unidirectional SHR (U-SHR).

of today's asynchronous point-to-point systems (or hubbed protection systems). The protection capability of this B-SHR/4 architecture is achieved by using automatic protection switching systems (APS's) to perform a loop-back function in case of cable cuts or node failures. For example, in Fig. 1(a), working traffic in the B-SHR/4 travels bidirectionally on separate fibers of the same path and dual protection fibers serve as back-up. In the case of a cable cut, traffic is intercepted at the next CO and re-routed back to its destination on the protection fibers. The B-SHR/4 architecture requires a protection ADM for each working ADM and a 1:1 nonrevertive lower-speed electronic protection switch at each office. The *nonrevertive* 1:1 switch is a protection switch such that the signals need not be switched

back when the failed line is repaired. Another alternative implementation for this architecture is to use regenerators as the protection components, rather than duplicated ADM's. However, using regenerators as protection components requires a relatively complex control scheme since regenerators have no intelligence on determining whether signals should be dropped or not. Such a control scheme for this alternative implementation requires a further study. Thus, in this paper, we assume that the B-SHR/4 architecture uses duplicated ADM's as protection components.

2) Bidirectional Ring with Two Fibers (B-SHR/2): This type of architecture was proposed in [10], [11]. For this scheme traffic is routed on a two-fiber ring with one ADM at each office. In the normal situation, traffic is evenly split into the outer ring (ring B) and the inner ring (ring A) by filling even and odd numbers of time slots, respectively [10] (or filling the first half of time slots, as depicted in Fig. 1(b) [11]). In the event of a fiber break or equipment failure, traffic is automatically switched into vacant time slots in the opposite direction to avoid the fault. The simplest implementation of this can be achieved using STS-n where n is an even number. In this case, traffic would only be loaded onto $n/2$ of the STS-n's. Each ADM would then incorporate a modified loop back time slot interchange (TSI) switch which can loop back loaded STS-1's into the unused STS-1 time slots. In this event of a fiber break, local action by the ADM's adjacent to the break is sufficient to restore service.

For the case of STS-n where n is odd, the same principle can be applied but the design of the loop-back switch is more complicated. If a traffic utilization greater than 50% is required then more complex algorithms must be used to control the ring. Individual tributaries must be assigned priority according to traffic and in the event of a fiber break all ADM's on the ring must cooperate to restore priority services. Based on the analysis in [10], for simplicity of control, loadings of ≤50% should be used. Although this may seem wasteful of system bandwidth, the cost of increasing the system capacity should be traded against the cost of more complex control and administration.

B. U-SHR Architectures

Unlike the B-SHR's, the U-SHR architecture uses only two fibers with one for working and the other for standby and an ADM at each office. For each U-SHR, the directions of a duplex channel travel on different routes between two offices in the SHR. The self-healing capability is achieved by either using APS loop-back systems (as discussed in Section 2) or path selection. A U-SHR is called a *folded U-SHR* if the self-healing capability is achieved by using APS loop-back; or a *path protected SHR* if its self-healing capability is achieved by using low-speed path selection.

1) Folded U-SHR Architecture (U-SHR/APS): The folded SHR is defined based on the method used to perform loop-back protection switching using the APS system. The architecture and implementation are similar to the one described in Section II-A.

2) Path Protected SHR Architecture (U-SHR/PP) A SONET U-SHR/PP architecture discussed in this section was proposed in [5], [8] and is based on a concept of signal dual-feed (i.e., $1 + 1$ protection). The dual-feed concept has also been used in other similar ring architectures [13]–[15]. Fig. 2 shows a generic architecture of the U-SHR/PP ring of OC-N rate, which consists of one ADM at each office and a pair of fibers with traffic going in opposite directions. In the normal state of the ring, one or more STS-M ($M < N$) signals or different services that can be carried inside the STS path layer signals are transmitted onto both the clockwise and the counter-clockwise directions of the ring. These two identical signals will propagate along the ring and finally be dropped at one of the offices. Thus, at the receiving node, two identical

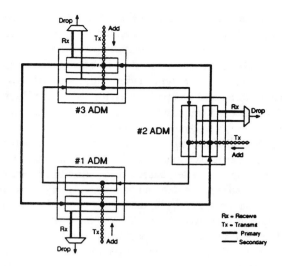

Fig. 2. A Generic architecture for U-SHR/PP.

tributary signals with different delays are observed. Suppose these two signals are designated as the primary and secondary signals. During normal operation, only the primary signal is used, although both signals are monitored for alarms and maintenance signals. If the ring is broken due to certain catastrophic failure (e.g., fiber cut or hardware failure), it is possible to resume the service by performing proper tributary protection switching to select the secondary signal. A control scheme for this tributary protection switching will be discussed in Section III–B.

III. ARCHITECTURAL ANALYSIS

Comparisons of several SHR architectures are discussed in this section based on the following criteria:
1. Cost and capacity tradeoff,
2. SONET standard impacts in terms of SONET APS operations, and
3. Multivendor compatibility.

A. SHR Cost and Capacity Tradeoff

In the following economic analysis for different types of rings, we only consider the equipment (i.e., ADM) cost since the equipment cost is a dominant factor of the total fiber transport cost[1] for intra-LATA networks [1].

1) B-SHR/4 Versus U-SHR: Both the B-SHR/4 and the U-SHR use the dedicated spare capacity for protection that can makes the control system simpler. In other words, the ADM cost for a B-SHR/4 and a U-SHR can be assumed to be approximately the same. Thus, comparing the B-SHR/4 with the U-SHR in terms of cost and capacity can be fair. In general, the B-SHR/4s have higher capacity than the U-SHR's, but at the penalty of additional components and facilities such as more fibers, regenerators (if any) and ADM's. The cost of a B-SHR/4 is approximately twice the cost of a U-SHR at the same rate since the amount of equipment and facilities for a B-SHR/4 is twice the amount for a U-SHR.

[1]The fiber transport costs include costs for terminating equipment, fiber material, and fiber placement.

The capacity comparison between a U-SHR and a B-SHR/4 highly depends upon the demand requirement pattern and the network design method. In this section, we will not investigate all of the above factors since that is beyond the scope of this paper. Instead, we will review three special types of demand patterns and compare costs for using U-SHR's and B-SHR/4s to carry required demand. Most of the practical demand patterns are somewhere between those special demand patterns. Note that the *capacity requirement* here is defined to be the largest cross-sections (e.g., DS3s) in the ring. The line rate of the ring is selected based on its capacity requirement.

Three special types of demand patterns are *centralized, mesh* and *cyclic* patterns. Let d_{ij} be the duplex demand requirement for office pair (i,j). Also let n be the number of nodes in the ring. The centralized demand pattern is represented by $\{d_{1j}\}$ where $j \neq 1$ and node 1 is the central node. The mesh pattern is $\{d_{ij}\}$ where $d_{ij} \neq 0$ for any i and j $(1 \leq i, j \leq n)$. The cyclic demand pattern is represented by $\{d_{i,i+1}\}$ where node $n + 1$ is node 1.

The main difference between a U-SHR and a B-SHR/4 from a traffic point of view is the way the capacity is computed. The total demand carried by a U-SHR is $\sum d_{ij}$ which is the capacity requirement on a U-SHR (denoted by C_U), i.e., $\sum d_{ij} = C_U$ where $i < j$.

For the case of B-SHR/4s, let m be the number of links (cross-sections) in the ring where cross-section is defined as the demand carried on the link. Assume $\{L_k, k = 1, .., m\}$ and L_C to be a set of cross-sections for links and the largest cross-section in the ring, respectively. L_k can be computed as follows, and $\max_k\{L_k\} = C_B$ where C_B is the capacity requirement on a B-SHR/4.

$$L_k U = \sum_{i,j} d_{i,j} \; where \; the \; working \; path \; of \; pair \; (i,j)$$

$$passes \; through \; link \; k.$$

Therefore, the worst case of B-SHR/4 engineering in terms of capacity is when demand for every demand pair is routed through the same link.

Assume the given demand requirement requires h_B B-SHR/4s with rates $\{OC-N_1, \ldots, OC-N_{h_B}\}$ or h_U U-SHR's with rates $\{OC-M_1, \ldots, OC-M_{h_U}\}$. Let $Cost_B(N)$ and $Cost_U(M)$ be capital costs of a B-SHR/4 with OC-N rate and a U-SHR with OC-M rate, respectively. The terminal cost ratio, CR, is defined as follows.

$$CR = \frac{\sum_i^{h_B} Cost_B(N_i)}{\sum_i^{h_U} Cost_U(M_i)} \qquad (1)$$

Let F be the ratios of $Cost_B(N)$ to $Cost_U(N)$ for the same OC-N rate. The value of F equals 2 in our analysis here since, as explained early in this section, the cost of an OC-N B-SHR/4 is twice the cost of an OC-N U-SHR.

a) Centralized Demand Pattern: Let n be the number of nodes in the ring. To make the analysis feasible, we assume that all demand pairs carry the same demand requirement, i.e., $d_{1j} = d$ where $j \neq 1$ and "d" is a constant. The total demand is $d \times (n-1)$. For a U-SHR, the capacity requirement (denoted by C_U) is $C_U = d \times (n-1)$.

For a B-SHR/4, the optimum routing for demand pairs in terms of minimization of capacity is when demand for half the demand pairs is routed to the central node in a clockwise direction, while demand for the other pairs is routed to the central node in the counter-clockwise direction. Based on this routing principle, the capacity requirement, C_B, can be computed as follows.

$$C_B = \begin{cases} d \times \frac{n-1}{2} & if \; n \; is \; odd \\ d \times \frac{n}{2} & if \; n \; is \; even \end{cases} \qquad (2)$$

Fig. 3(a) shows examples for Equation (2). There are 5 nodes in Fig. 3(a) with each demand pair having 4 DS3s. In Fig. 3(a), demand for pair (1,2) and (1,3) is routed to node 1 (hub) in a clockwise direction, while demand for pairs (1,4) and (1,5) is routed to node 1 in the counter-clockwise direction. The capacity requirement C_B is 8 DS3s. Note that the routing principle discussed here and examples depicted in Fig. 3(a) assume that all demand for a node pair is routed in the same path. This assumption is referred to as a *demand nonsplitting* assumption which has been used in Bellcore TIRKS provisioning system. This demand nonsplitting assumption will also apply to discussions of the next two subsections.

Table I shows a cost/capacity comparison between U-SHR's and B-SHR/4s for $d = 5$ DS3s and the centralized demand pattern. In Table I, we assume $F = 2$, three line rates OC-12, OC-24, and OC-48 and their relative costs are assumed to be $Cost_U(12) = 1.0$, $Cost_U(24) = 2.0$, $Cost_U(48) = 3.0$. Results for up to 10 nodes are shown in Table I, which is a practical upper bound for interoffice applications. For the particular example shown in Table I, the use of U-SHR's is more economical than or equivalent to the use of B-SHR/4s in terms of costs. This observation remains true when parameter "d" varies from 1 to 10.

b) Mesh Demand Pattern For a mesh demand pattern, there exists demand between any two nodes, i.e., $d_{ij} = d (\neq 0)$ for each i and j. The total number of demand pairs for the mesh pattern is $n(n-1)/2$ where n is the number of nodes. Thus, for a U-SHR with the mesh demand pattern, the capacity $C_U = d(n(n-1)/2)$.

For a B-SHR/4, it is difficult to formulate an equation for the required minimal capacity even when the demand for each demand pair is a constant. Fig. 3(b) shows an example of capacity arrangement for B-SHR/4s with 5 nodes. Each demand pair is assumed to have 4 DS3s. The capacity requirement for this example is 12 DS3s.

Table II shows a cost/capacity comparison between U-SHR's and B-SHR/4s for $d = 5$ DS3s and the mesh demand pattern. The minimum capacity requirement for a B-SHR/4 is obtained by manual capacity assignments. Results shown in Table II uses the same assumptions used in Table I. Note that in Table II, OC-N line rate of 60 (i.e., $L_U = 60$) means that the area implements two U-SHRs: one with OC-48 and another with OC-12.

For particular examples shown in Table II, the use of B-SHR/4s is not more economical compared to U-SHR's until the number of nodes in the ring increases to 7. The sensitivity analysis (in terms of the value of "d") indicated that the B-

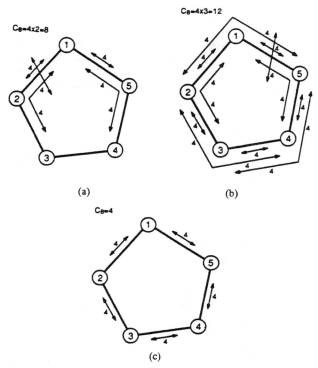

Fig. 3. Capacity arrangement for B-SHR's. (a) Centralized demand pattern. (b) Mesh demand pattern. (c) Cyclic demand pattern.

TABLE I
COST/CAPACITY COMPARISON FOR CENTRALIZED DEMAND PATTERN (FOR $d = 5$ DS3s)

| # of Nodes | U-SHR | | B-SHR/4 | | $R_{B/U}$ | CR |
	C_U	L_U (OC-N)	C_B	L_B (OC-N)		$(d = 5)$
3	10	12	5	12	0.5	2.0
4	15	24	10	12	0.7	1.0
5	20	24	10	12	0.5	1.0
6	25	48	15	24	0.6	1.3
7	30	48	15	24	0.5	1.3
8	35	48	20	24	0.6	1.3
9	40	48	20	24	0.5	1.3
10	45	48	25	48	0.6	2.0

CR = ratio of B-SHR cost to U-SHR cost
$C_{U \text{ (or } B)}$ = the capacity requirement of a U-SHR (or B-SHR/4)
$L_{U \text{ (or } B)}$ = the OC-N line rate of a U-SHR (or B-SHR/4)
$R_{B/U} = C_B/C_U$

TABLE II
COST/CAPACITY COMPARISON FOR MESH DEMAND PATTERN (FOR $d = 5$ DS3s)

| # of Nodes | U-SHR | | B-SHR/4 | | $R_{B/U}$ | CR |
	C_U	L_U (OC-N)	C_B	L_B (OC-N)		$(d = 5)$
3	15	24	5	12	0.3	1.0
4	30	48	15	24	0.5	1.3
5	50	60	15	24	0.3	1.0
6	75	96	25	48	0.3	1.0
7	105	108	30	48	0.3	0.9
8	140	144	40	48	0.3	0.7
9	180	192	50	60	0.3	0.7
10	225	240	65	72	0.3	0.7

TABLE III

COST/CAPACITY COMPARISON FOR CYCLIC DEMAND PATTERN (FOR $d = 5$ DS3s)

# of Nodes	U-SHR		B-SHR/4		$R_{B/U}$	CR ($d = 5$)
	C_U	L_U (OC-N)	C_B	L_B (OC-N)		
3	15	24	5	12	0.3	1.0
4	20	24	5	12	0.3	1.0
5	25	48	5	12	0.3	0.7
6	30	48	5	12	0.2	0.7
7	35	48	5	12	0.1	0.7
8	40	48	5	12	0.1	0.7
9	45	48	5	12	0.1	0.7
10	50	60	65	12	0.1	0.5

SHR/4 may be more economical than its counterpart when the demand requirement (i.e., the value of "d") increases.

c) Cyclic Demand Pattern Mathematically, the cyclic demand pattern is represented by $\{d_{i,i+1}\}$ where node $n + 1$ is node 1. For a U-SHR, the capacity $C_U = n \times d$. For a B-SHR/4, the capacity $C_B = d$ is for the single failure case. Fig. 3(c) shows an example of a 5 node-network. Therefore, the capacity ratio of a B-SHR/4 to a U-SHR, denoted by $R_{B/U}$, is $1/n$.

Table III shows a cost/capacity comparison between U-SHR's and B-SHR/4s for $d = 5$ DS3s and the cyclic demand pattern. Note that results shown in Table III uses the same assumptions used in Table I.

As shown from Table III B-SHR/4s appear to be more economical than U-SHR's for the cyclic demand pattern. This is obvious due to characteristics of B-SHR/4s as explained in Section II. A sensitivity analysis in terms of the parameter "d" showed a similar result as described in Table III. Note that simple point-to-point links with diverse protection may be used for this demand pattern since it may have a lower operations cost than the B-SHR/4s. However, the use of B-SHR/4s in this case may save regenerators and equipment for the protection systems.

2) B-SHR/2 Versus U-SHR: Unlike the B-SHR/4, working and protection channels for B-SHR/2s are routed on the same fiber. In case of network system failures, protection switching is accomplished by using the time-slot interchange method. In order to simplify the ADM system design complexity and minimize impact on network operations, the fiber system dedicates half of bandwidth to protection [10], [11]. For example, for an OC-12 B-SHR/2, STS-1 channel 1 to channel 6 may be assigned to working STS-1s and channel 7 to channel 12 are dedicated to protection. Since the B-SHR/2 only uses half of bandwidth, the load balancing (demand splitting) traffic arrangement is usually used to increase ring utilization. Note that the load balancing arrangement used for B-SHR/2s is not supported by the present Bellcore TIRKS provisioning system.

When comparing B-SHR/2s with U-SHR's under the condition that can be supported by current TIRKS systems (i.e., no demand splitting is allowed, results are similar to the ones described in Section IV-A1) (i.e., B-SHR/4 versus U-SHR). Even if we take into account the demand splitting factor for the B-SHR/2, we still find that the relative cost comparison depends strongly on the demand pattern. In this comparison,

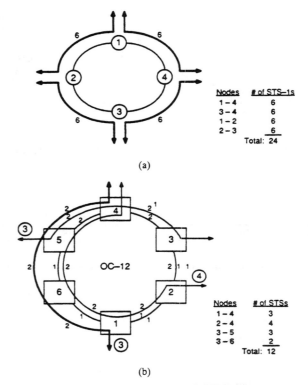

(a)

(b)

Fig. 4. Comparison between U-SHR's and B-SHR/2 with demand splitting. (a) B-SHR/2 traffic arrangement. (b) Comparison between U-SHR's and B-SHR/2 with demand splitting.

we assume that the ADM cost for the B-SHR/2 is higher than that for the U-SHR since the former requires the time slot interchange capability to perform protection switching. Fig. 4(a) and (b) depict examples that B-SHRs/2 works better and worse than U-SHR's, respectively. For the example shown in Fig. 4(a), the demand requirement can be carried by an OC-12 B-SHR/2, but requires an OC-24 U-SHR. For the example depicted in Fig. 4(b), it requires an OC-24 B-SHR/2 to carry demands, but only require an OC-12 U-SHR to carry the same demands.

B. Impacts on SONET Standard

In order to evaluate SONET impacts in terms of K1 and K2 operations, it would be helpful first to review the SONET standard K1 and K2 operations. Two types of automatic protection switching (APS) architectures are defined in the

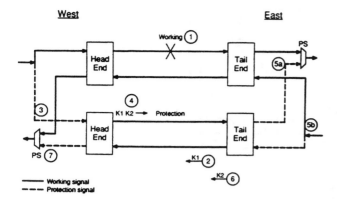

Fig. 5.   SONET automatic protection switching.

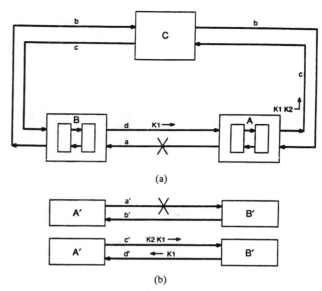

Fig. 6.   The first proposal of healing method for USHR/APS. (a) Proposed protocol. (b) SONET reference.

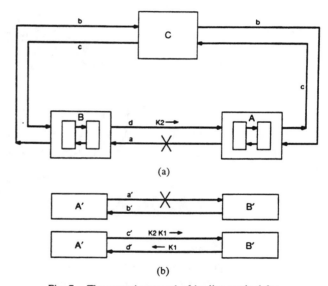

Fig. 7.   The second proposal of healing method for USHE/APS. (a) Proposed APS protocol. (b) SONET reference.

SONET standard [2]. The 1 : N APS architectures allows any one of the N (permissible values for N are from 1 to 14) working channels to be bridged to a single protection channel. The APS protocol communication is via the K1, K2 bytes that are located in the SONET Line Overhead. The operation of the protocol can be summarized as follows: referring to Fig. 5, when a failure is detected, the tail end sends out the K1 byte, which contains the number of the troubled channel. The head end, after receiving the K1 byte, bridges the west-east channel and sends out both the K1 and K2 bytes. The K1 byte is for reverse request (for bidirectional switching) and the K2 byte is for confirmation. At the tail-end node, the received K2 byte confirms the channel number and the west-east channel protection switching is completed. At the same time, the east-west channel is bridged as requested by the K1 byte. To complete the bidirectional switching, the K2 byte is sent out from the tail end. When this K2 byte is received by the head end, the east-west channel is switched and the APS process is completed.

Another type of APS that is also defined in the standard is the 1 + 1 protection switching, which is a form of 1 for 1 APS with the head end permanently bridged. Thus, a decision to switch is made solely by the tail end. For bidirectional switching, the K1 byte is used to convey the signal condition to the other side and the actual switching is decided by the tail end.

1) The Folded Ring Using APS (U-SHR/APS) Two proposals of healing methods for the folded ring were submitted to T1X1.5 [9]. The first one is briefly described in Fig. 6. This approach attempts to follow a similar protocol as that defined in the standard for the 1 : N APS with $N = 1$. However, the definition of the K2 byte needs to be modified. This can be explained by comparing the protocol proposed in [9] and the standard. Referring to Fig. 6(a), when the failure is detected in node B, a K1 byte is transmitted to node A via link d. After receiving the K1 byte, node A bridges its input to both links a and c and transmits the K1 and K2 bytes on link c. When node B receives the K2 byte, it completes the protection switching by acquiring its data from link c. For comparison, the APS described in the SONET standard is repeated in Fig. 6(b). The working lines are labeled a', b' and the protection links by c', d', which correspond to the a, b, c, d paths of Fig. 6(a). By comparing the protocols for SONET APS and that of

the folded ring, we notice that the folded ring exercises a unidirection APS protocol similar to that of the SONET APS. The difference, however is in the use of the K2 byte. In the SONET standard, the K2 byte is transmitted directly to the head end node, whereas in the folded ring, it has to travel through all the intermediate nodes on the protection path. This means that the definition of the K2 byte needs to be modified to accommodate this ring application.

The second approach [9] for the folded ring attempts to use the K2 byte in the standardized way for the protection switching. The comparison between this second folded ring APS and that of the SONET standard is shown in Fig. 7. In the folded ring APS, node B switches *autonomously* and sends the K2 byte to node A via the protection link d. At node A, the detection of the K2 byte triggers the bridging function so that the input data at A is also transmitted to B

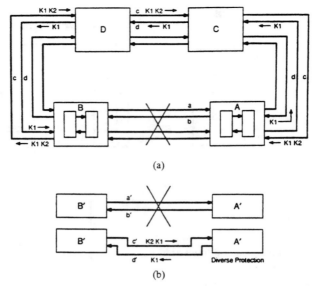

(a)

(b)

Fig. 8.   A proposal of healing method for B-SHR/4. (a) Proposed protocol. (b) SONET reference.

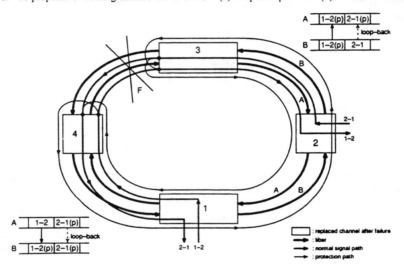

Fig. 9.   B-SHR/2 (slot ring) failure scenario.

via link c. This approach avoids the need for the change of the definition of the K2 byte. However, the protocol itself is very different from that of the APS standard: the tail end (B) switches autonomously even before the bridging of the head end. As a result, the K1 byte is never used in the switching.

In summary, the first folded ring approach follows the spirit of the SONET APS standard but requires some modification of the definition of the K2 byte. In the second folded ring approach, no change of K2 byte is needed, however, the actual APS protocol is different from that of the SONET standard.

The other proposal for the folded ring using APS which was described in Reference [12] has a similar impact on the present SONET standard.

2) Bidirectional Ring with Four Fibers (B-SHR/4) The recovery procedure for the bidirectional ring is similar to that of the folded U-SHR/APS [9] (see Fig. 6). One important difference between the bidirectional ring and the unidirectional slotted ring is that only the nodes adjacent to the failure are required to perform recovery functions. Therefore, APS procedures, similar to that described in Section III–B1) can be

used for the signaling process between the affected nodes. As a result of the similarity with the folded U-SHR, the APS required for the bidirectional ring suffered from the same disadvantages as that of the folded ring, i.e., requires modification of the SONET APS standard.

The foregoing discussions only focus on a single fiber failure case. As to the case of cable cuts, Fig. 8 depicts a possible APS protocol for the B-SHR/4. In Fig. 8, intermediate nodes C and D must have knowledge of destinations of K1 and K2 bytes which pass to them from node A or node B in order to perform the correct protection switching. This addressing capability is not within the current SONET APS standard. Thus, it requires changes of the current SONET APS standard for the ring application.

3) Bidirectional Ring with Two Fibers (B-SHR/2) To understand the impact of the slotted ring (B-SHR/2) on the SONET standard, we have to take a closer look at its control mechanism. This is best illustrated with an example. Fig. 9 shows a slotted ring consisting of four nodes. Assume that each node is equipped with an STS-1 level ADM, and that

158

communication channels among the nodes are in the unit of an STS-1. To illustrate the healing mechanism, we focus on one particular duplex STS-1 channel from node 1 to node 2. In normal operation, both loops of the ring are used in such a way that only alternate slots (STS-1 channels) are occupied so that only half of the available bandwidth is used. In the following, we describe the healing mechanism for the case of a cable failure (i.e., both loops are broken).

In this cable cut scenario (see Fig. 9), the connection between nodes 3 and 4 are broken (failure F) in both loops A and B. The healing mechanism involves the loop-back at both nodes 3 and 4. After the detection of the failure (F) at node 4, the clockwise traffic (loop A, channel 1-2) is routed onto the corresponding spare slot in the counterclockwise loop (loop B). The protected channel 1-2(p) is then routed through nodes 1 and 2. At node 3, 1-2(p) is inserted onto the corresponding slot on loop A, thus replacing the original 1-2 channel. The healing mechanism is exactly the same in the opposite direction. At node 3, after the detection of the failure (F), the 2-1 channel on loop B is routed onto the corresponding spare slot on loop A. This protected channel 2-1(p) is then routed through nodes 2 and 1. At node 4, this 2-1(p) channel is inserted onto loop B, thus replacing the original 2-1 channel.

From this example, we observe a few features of the slotted ring control mechanism. First a signaling scheme is needed for the healing control. After a node has detected a failure, it is required to signal the node at the other end of the loop. This signal is essential since the failure may occur on one of the fibers and both the nodes adjacent to the failure are required to perform protection switching. Second, only the nodes that are adjacent to the failure are involved in the healing mechanism; all the intermediate nodes are not required to perform protection switching. Third, all the nodes of the slotted ring are required to have the time slot interchange capability that is used to perform the loop-back of protected channels.

It is evident from the above discussion that a signaling procedure is needed for the control of the healing mechanism. One possible solution is to use the SONET data communication channel (DCC) for carrying the signaling messages. In such a case, the control protocol and DCC messages will need to be standardized.

Another solution is to use automatic protection switching (APS). However, a simple APS scheme, as described in Section III-B1) [9] and which is standardized for point-to-point systems, is not applicable for the slotted ring (B-SHR/2). Thus, modification of the SONET APS procedure is necessary for the slotted ring application.

4) The Path protected Ring (U-SHR/PP) A simple control mechanism of the U-SHR/PP architecture described in Section II-B2) for recovery from failure is given below [5], [8]:

1. Detection of a loss of signal (LOS) or a line AIS (alarm indication signal) triggers the insertion of path AIS (STS or VT level, depending on ADM application) onto all the downstream tributary paths (STS or VT level).
2. Detection of a path AIS on one of the two tributaries initiates protection switching to the other tributary.

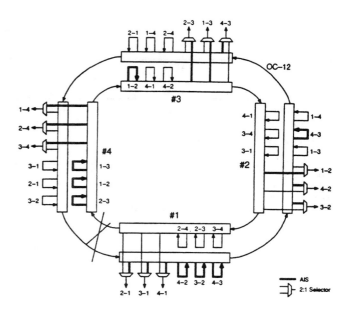

Fig. 10. USHR/PP using STS-1 ADM's (only drop side is shown).

3. Detection of path AIS on both of the two identical tributaries signifies a multiple failure situation and will trigger the generation of an AIS in the dropped signal.

Note that with the above control mechanism, the protection switching can be nonrevertive (i.e. the system does not switch back to the original state even after the failure has been removed) with no performance penalty. After the ring is brought back to normal working condition when the failure is removed, the identification of the primary and the secondary loop is irrelevant. Each low speed interface of the ADM's just monitors the Path AIS and chooses the valid signal to perform protection switching.

Fig. 10 shows an example of how the self-healing ring recovers from a fiber cut [8]. For simplicity, the capacity of the ring is assumed to be OC-12 and there are four ADM's in the ring, which add/drop STS-1 signals. Furthermore, we shall assume that each ADM communicates with all the other ADM's via a single STS-1 channel (the numbers in Fig. 10 such as 2-1 indicates communication from node #2 to node #1). All of these assumptions are given so that the description of the operation of the ring is simplied and have no effect on the generality of the scheme.

Referring to Fig. 10, during normal operation, STS-1 signals are inserted onto both the clockwise and the counterclockwise loops with the latter designated as the primary loop. When the fiber pair between nodes #1 and #4 fails, the receiver at node #1 detects the failure and declares a red alarm state. Since the control mechanism applies to both the clockwise and the counterclockwise directions, the following discussions refer to the counterclockwise loop but is applicable to both loops. After ADM #1 declares LOS or detection of line AIS, it inserts STS path AIS onto all the tributary channels. At the STS-1 interfaces of node #1, the status of the tributary signal is monitored. The detection of STS path AIS in one of the two tributary STS-1 signals triggers tributary protection switching, which allows the 2:1 selector to select the valid tributary signal. Since node #1 is adjacent to the fiber cut, all

TABLE IV
A RELATIVE COMPARISON AMONG A CLASS OF SHR ARCHITECTURES

SHR Architectures	ADM Type	Carry Demand	Node Component Costs	SONET K1, K2 Changes	System Complexity
B-SHR/4	basic ADM[a]	more	higher	yes	simple[d]
B-SHR/2	ADM/TSI[b]	more/less[c]	medium	yes	complex
U-SHR/APS	basic ADM	less	lower	yes	simple
U-SHR/PP	basic ADM	less	lower	no	simple

[a]Basic ADM: ADM without TSI capability
[b]ADM/TSI: ADM with TSI capability
[c]Depends on the demand pattern
[d]Simple because no cross-connect capability in the ADM is needed

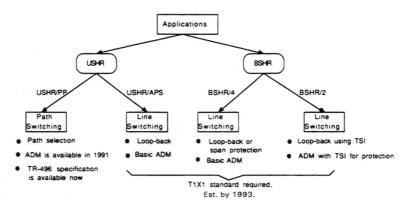

Fig. 11. SONET ring architectures and control schemes.

the STS-1 interfaces of node #1 perform protection switching. The output of node #1 on the counterclockwise loop thus has STS Path AIS on all of its STS-1 channels except those that are added at node #1.

At node #2, the high speed demux will not detect any alarm since SONET framing is never interrupted. After extracting the dropped channels, the STS-1 interfaces of node #2 (#2a,#2b) detect the STS-1 Path AIS and performs protection switching. Interface #2c corresponds to communication between nodes #1 and #2 and is not affected by the failure, thus no protection switching is required. For those channels that propagate through node #2, only the 4-3 (communication from node #4 to node #3) channel carries a STS-1 path AIS. Both the 1-3 and 1-4 channels originate from node #1, thus both are carrying valid data.

If the same algorithm is applied to node #3, we notice that only the 4-3 channel is required to perform protection switching. Moreover, all the three loop-through channels (2-1,1-4,2-4) are carrying valid data that has been inserted onto the ring from an ADM that is downstream from the failure location. Using the same reasoning, none of the STS-1 interfaces in node #4 is required to performed protection switching. The ADM's in the clockwise loop operates in exactly the same way as those in the counterclockwise loop. Since the U-SHR/PP ring is a form of channel switching and the indication of switching is by using the path level AIS signal, no APS protocol is required. As a result, this architecture is totally independent of the development or standardization process of the APS protocol. Note that the operation of inserting path AIS is simple since it is indicated by STS pointers with all 1's.

C. Multivendor Compatibility

Since there exists no universal K1/K2 protocol for the ring application now, all SHR architectures using the APS control scheme may not meet the requirement of intervendor compatibility. For example, if the company's SONET transition plan is to deploy SONET point-to-point systems first, and then connect those offices to form a B-SHR later on providing there exists connectivity there. For this scenario, the company may install those point-to-point systems using multivendor SONET ADM (in terminal mode) and those ADM's have features to support the B-SHR. However, we may find that it is not possible to connect those offices to form a B-SHR since different vendor's ADM's at different offices on the ring are not compatible since each vendor may implement a propriety APS protocol for the ring application. In other words, it is not possible to implement the B-SHR architecture under this scenario unless there exists a common APS protocol for the ring application.

The U-SHR/PP architecture is easy for intervendor compatibility since it does not use SONET K1 and K2 bytes, but instead it uses path AIS which has already been defined in the SONET standard. Also the U-SHR/PP requires minimal changes to the requirements of the ADM described in [5], e.g., the use of the AIS signals is exactly the same as in the case where ADM's are connected in a chain.

D. Summary

The following Table IV summarizes discussions in Sections II and III. Fig. 11 depicts a diagram for SONET ring architecture classification. As shown in Fig. 11, the line protection switching scheme which does not require the time slot interchange capability may be applied to both the USHR/APS and BSHR/4 architectures.

IV. CONCLUSION

We have reviewed a class of SONET-based self-healing ring (SHR) architectures and associated control schemes. The cost and capacity tradeoffs between bidirectional SHR's (B-SHR's) and unidirectional SHR's (U-SHR's) depend strongly upon the application, the network size and the demand pattern. An analysis of the impact on the SONET standard suggested that using automatic protection switching (APS) schemes with SONET SHR architectures requires a change of currently standardized SONET K1 and K2 overhead byte definitions and functionality. Alternatively, a simple distributed control scheme using path AIS avoids a change of the SONET standard in terms of K1 and K2 operations, and thus can be deployed on a timely basis and may ease intervendor compatibility problems. The selection of appropriate SONET SHR architectures will depend upon the operating telephone companies' economic analysis, emphasis on multivendor environment, SHR implementation time frame, and standards progress on making change to support a bidirectional ring architecture.

ACKNOWLEDGMENT

The authors would like to thank R. H. Cardwell, R. E. Clapp, J. E. Berthold, and S. D. Personick for their valuable comments to this paper.

REFERENCES

[1] T.-H. Wu, D. J. Kolar, and R. H. Cardwell, "Survivable network architectures for broadband fiber optic networks: model and performance comparisons," *IEEE J. Lightwave Technol.*, vol. 6, pp. 1698–1709, Nov. 1988.

[2] American Standard for telecommunications "Digital hierarchy optical interface rates and formats specification," ANSI T1.105, 1988.

[3] R. Ballart and Y.-C. Ching, "SONET: Now it's the standard optical network," *IEEE Commun. Mag.*, Mar. 1989, pp. 8–15.

[4] Bellcore Technical Reference, "SONET add-drop multiplex equipment (sonet adm) generic requirements and objectives," TR-TSY-000 496, Issue 2, Sept. 1989.

[5] Bellcore Technical Advisory, "SONET add-drop multiplex equipment (sonet adm) generic criteria for a unidirectional, path protection switched, self-healing ring implementation," TA-TSY-000 496, Issue 3, Aug. 1990.

[6] T.-H. Wu, D. J. Kolar, and R. H. Cardwell, "High-speed self-healing ring architectures for future interoffice networks," *IEEE GLOBECOM'89*, Nov. 1989, pp. 23.1.1–23.1.7.

[7] T.-H. Wu, and M. Burrowes, "Feasibility study of a high-speed SONET self-healing ring architecture in future interoffice fiber networks," *IEEE Commun. Mag.*, vol. 28, pp. 33–42, Nov. 1990.

[8] R. Lau, "An architecture for a SONET self-healing Ring," Bellcore Contribution to T1X1.5, T1X1.5/89-088, May 10, 1989.

[9] S. Hasegawa et al., "Protection switching in a sonet ring architecture," NEC Contribution to T1X1.5/89-053, Apr. 12/July 20, 1989.

[10] I. Hawker et al., "Self-healing fiber optic rings for SONET networks," British Telecom Contribution to T1M1/T1X1 Ad Hoc Committee, Oct. 1988.

[11] G. Copley and B. Malcolm, "SONET rings: Proposal for K byte definition," Northern Telecom Contribution to T1X1.5/90-124(R2), Nov. 5, 1990.

[12] R. Boehm et al., "Overhead usage for protection switching in a self-healing SONET ring," Fujitsu Contribution T1X1.5/90-066, Apr. 20, 1990.

[13] J. K. Conlisk, "How fragile is your network ?" *Telephony*, pp. 27–35, Oct. 31, 1988.

[14] A. Alexander, "The critical umbilical," *Telephony*, pp. 62–70, May 1989.

[15] D. L. Howells, "High capacity light wave technology comes to age," *AT&T Technol.*, vol. 3, no. 4, 1988.

Service Applications for SONET DCS Distributed Restoration

Joseph Sosnosky

Abstract— This paper determines the scope of network applications and services that could be offered using a SONET DCS-based self-recovering mesh architecture with distributed control. The study includes an outage impact analysis on network services and a determination of how network restoration time objectives will affect the applicability for the distributed controlled DCS network architecture. It is concluded that using SONET DCS distributed control architectures to provide more complete survivability of a network would support numerous applications. Future services will demand a fault-tolerant network with complete survivability; this may only be reached through integration of SONET DCS distributed control architectures with other survivable architectures such as cell relay networks (e.g., supporting SMDS) and self-healing rings.

I. Introduction

FIBER optic transport networks (e.g., SONET) are being designed to have fault tolerance against the impact from service outages that occur due to events such as cable cuts, central office failures, and hardware failures. Survivability issues are significant because of the fibers' high traffic carrying capacity. New broadband network service concepts are also demanding high reliability and high throughput.

Two current approaches to enhance the survivability of networks are diverse protection (DP) architectures and SONET-based self-healing rings. These approaches have been shown to be cost-effective self-healing network architectures [1], [2]. Such architectures, which are based on physical redundancy, have recovery times of less than 60 ms (10 ms to detect, and 50 ms to complete the switch) and meet current network facility protection switching requirements [3]. In general, protection switching times lasting less than 50 ms is considered to be "transparent" to most customers.

There are two types of DCS mesh restorable network architectures: centralized and distributed. With the centralized architecture, as shown in Fig. 1, DCS reconfiguration is primarily performed by an Operations System (OS) or centralized "controller." However, this scheme may take upwards of 10 min for network restoral and may cause severe degradation for most customers and services that pass through the failed link.[1] The results of this study show that most existing services are impacted by a 2–10 s service outage. This is the main motivation for considering DCS reconfiguration schemes with

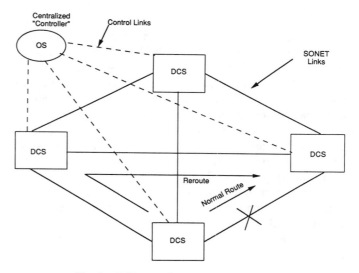

Fig. 1. DCS centralized control architecture.

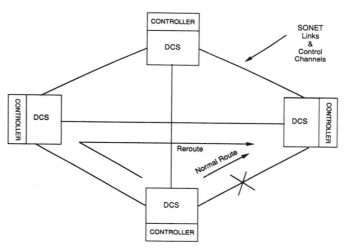

Fig. 2. DCS distributed control architecture.

distributed control, which exhibit much faster restoral times than centralized control.

Digital Cross-Connect System (DCS)—based mesh restorable networks with distributed control, as shown in Fig. 2, have been studied by industry over the past few years, and preliminary results from these studies [4], [5] indicate that they may be a potential self-recovering network-wide alternative (trading reduced spare capacity for longer restoration time) which complements overall survivable network planning.

Distributed control involves communications among the DCS's for rerouting failed traffic. Intelligence resides locally

Manuscript received July 15, 1992; revised February 15, 1993.
The author is with Bellcore, Redbank, NJ 07701-7030.
IEEE Log Number 9212470.

[1] Note that, on the other hand, it is much faster than manual DCS reconfiguration or cable jumping in DSX frames (as it is done today in the event of failure) which may disrupt communications for hours or more.

Reprinted from *IEEE J. Selected Areas Communi.*, vol. 12, no. 1, pp. 59–68, Jan. 1994.

162

at each DCS, and therefore the DCS controllers including central processing units (CPU's) act like parallel processors computing alternate routes in real time. The signal restoration technique in a given network can be either link (or line) restoration (i.e., routing failed link(s) over alternate links where all traffic is restored intact) or path Synchronous Transport Signal/Virtual Tributary (STS/VT) restoration (i.e., paths are restored individually on an end-to-end basis and may travel over different links). A mix of line and path restoration is not allowed under the current proposals.

Actual restoral times for these DCS distributed architectures are still under investigation, and may range from, e.g., 150 ms up to tens of seconds depending, among other factors, on network traffic, failure severity, and size. Restoral times are also strongly dependent on parameters such as DCS cross-connection time, which is discussed later.

This study determines the type of network applications and services that would benefit from a SONET DCS-based self-covering mesh architecture with distributed control. The study includes an outage impact analysis on network services. One of the most crucial parameters for determining the impact of an outage is the outage duration. Various network restoration time ranges (each having a different severity impact to customers) are established from the analysis, as targets for SONET DCS distributed control architectures. Key network topologies and services, both present and future planned, are considered in deriving these restoration time ranges. The goal of the study is to understand the technical implications and determine the usefulness of SONET DCS distributed control architectures in meeting these network service restoration time targets. Note that this study does not address the restoration procedure of returning to normal after failures have been repaired. It is assumed that the customer will only experience a "hit" of service in this process. Conclusions are drawn on the possible applications of SONET DCS distributed control architectures as a result of this study. The study summarizes the potential applications satisfied by the target restoration times using SONET DCS distributed control architectures to provide more complete survivability of a network.

The remainder of this paper is organized as follows. Section II discusses the network model, the key networks and services considered in the study, and their sensitivity to service outages. Section III establishes objectives for network restoration times and describes their technical implications on SONET DCS distributed control architectures. Some application guidelines are also given in this section. Section IV provides conclusions and a summary.

II. Network Model

Wideband and broadband SONET DCSs [6] are considered as intelligent network elements in the SONET transport network. They serve as a convenient way to groom traffic and provide network facility management functions to SONET or the Synchronous Digital Hierarchy (SDH). The SONET Optical Carrier (OC-N) facilitites that will be managed by DCS's will carry a multitude of services, and these facilities most likely will act as backbone links in the evolving future

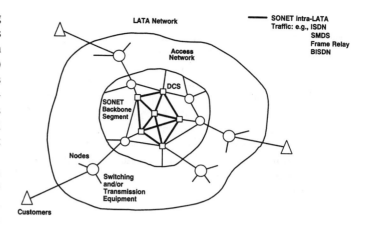

Fig. 3. LATA network backbone architecture example.

digital high-speed networks. The new Asynchronous Transfer Mode (ATM)/DCS technology utilizing the virtual path concept will also play a role with DCS mesh networks. This new ATM/DCS network technology shows promise for enhancing DCS distributed control architectures with greater flexibility and responsiveness [7].

This section identifies the key services that transport network architectures should support along with their required restoration time against failures, and the major type of network architecture in which SONET DCS distributed control architectures may be used providing restoration for these services. Service sensitivity to network outages will also be discussed.

A. Network Architecture

The network architecture assumed for this study is shown in Fig. 3. It is assumed that DCS's are deployed in nodes which process large traffic volumes (e.g., hub nodes or core nodes in metropolitan networks).

The example in Fig. 3 shows a LATA network consisting of access networks and a SONET backbone mesh network with SONET DCS's. The SONET DCS backbone mesh network is assumed to provide distributed control restoration. The vision here is one of a SONET backbone network providing reliable high-speed transport between switches. The backbone segment in this model has no intermediate nodes (e.g., Add-Drop Multiplex [ADM] type nodes) between the DCS nodes. The model is a high-level vision of the application for SONET DCS distributed control architectures. Application of SONET DCS distributed control architectures interworking with other NE's is considered later in a discussion of integrated architectures for achieving survivability in more realistic transport networks. The backbone network is part of a larger public digital network supporting services such as ISDN, SMDS, frame relay, and BISDN. The access network will provide customers access to these end-to-end digital services. Support for private line (PL) services (analog or digital) could be provided as an overlay network, with the vision of eventually migrating PL customers over to the public digital network.

Under a network failure condition such as a break in a SONET span between nodes in the backbone network, the scenario is that any interrupted traffic would be automatically rerouted within a reasonable amount of time before there is any serious impact to the customer. The objective would be to allow the recovery of lost data through retransmissions for high-speed data services. During the restoration process, SONET would apply Path Alarm Indication Signal (Path AIS) to all affected services. AIS is removed after the services are restored. AIS [8] is an indication to downstream equipment of an upstream alarm, but it also acts as a "keep alive" signal (i.e., no loss of signal) to downstream access equipment (e.g., digital terminals) and end customer equipment (e.g., digital CPE's). Thus, the insertion of AIS is important in a total digital network.

B. Network Service Groupings

The networks and services listed below are categorized into network service groupings. The majority of network traffic that might be carried by the SONET backbone network should fall under these basic groupings. Note that signaling networks are included in this list because they provide the call processing necessary to support some of the other services listed.

- Public-Switched Services
  - Voice   ○ Data   ○ Public Switched Digital Service (PSDS)
- Private Line Services
  - Voice   ○ Data
- Packet-Switched Services
  - X.25   ○ X.75   ○ Services supported by frame relay technology
- ISDN Services
  - Basic rate   ○ Primary rate   ○ Channel switched
- MAN Services
  - Public (SMDS)   ○ Private (FDDI)
- Broadband Services
  - Services supported by SONET/ATM (BISDN) technology   ○ Services supported by SONET/STM technology
- Signaling Networks
  - In-band   ○ CCSN/SS7
- Intelligent Services
  - AIN   ○ INA
- Wireless Services
  - PCS

In the following subsections, the impact of outages on each of these group of services is discussed.

1) Public-Switched Services: Public-switched voice services such as Plain Old Telephone Service (POTS) and public-switched data services (e.g., FAX) over voiceband (DS0) channels will be dropped if an outage lasts for 2 to 3 s. A digital terminal system (e.g., channel bank, DCS) or a digital switch reports a major red alarm and begins Carrier Group Alarm (CGA) processing (release call, stop billing) within 2.5 ± 0.5 s of initial detection of a hard failure.[2] Customers must then reestablish interrupted calls. Note that once the CGA is initiated, CGA deactivation requires 10–20 s before clearing the outage. Trunks (DS0 circuits) will not be created by the switch or channel bank until the CGA is removed.

The Public-Switched Digital Service (PSDS) is a high-speed digital data service that requires 56 kb/s bandwidth over the public-switched network, and is also subject to the standard 2–3 s CGA procedure.

2) Private Line Services: Private line voice circuits (e.g., PBX access trunks) and private line data circuits [e.g., voiceband or Digital Data Service (DDS)] carried on DS0 trunks are also subject to the same 2–3 s call dropping threshold as switched circuits, but automatically come back up after the CGA is removed.[3] Note that private line data circuits set up as "full transparent" (CGA mode) will clear as soon as the outage is cleared since these circuits use no signaling.

In the private digital data communications networks, a connection (e.g., user to end host) is declared "dead" after exceeding a predetermined outage that is set by the software application (e.g., 30 s), and the disrupted communications session must then be reinitialized. Setting up a session involves numerous control messages and is considered a complicated procedure. A session is the capability to transmit error-free data using higher layer protocols, independent of the physical makeup of the connection. Today, session-dependent applications such as file transfer commonly use SNA and TCP/IP protocol architectures. With Systems Network Architecture (SNA), in particular, a session-dependent application has a software programmable session timeout of from 1.1 to 225 s [9] and can be specified by users based on file size, for example. Lost data due to shorter interruptions are retransmitted, a process which in some applications is triggered by receiver timeouts. Receivers may begin to timeout based on twice the roundtrip delay.

3) Packet-Switched Services: Packet-switched traffic in X.25 networks, e.g., Public Packet-Switched Network (PPSN), are carried on DS0 trunks. X.25 services have throughputs up to 56 kb/s. X.25 is a connection-oriented service that provides virtual call services and permanent virtual circuit services. The PPSN provides a transport capability for digital data, suitable for nonreal-time services. Typical applications for the PPSN are bursty in nature, such as interactive communications between terminals, point-of-sale, and information (database) exchange between computers. The X.75 protocol is for interconnection between X.25 subnetworks at speeds up to 56 kb/s and also supports virtual call services and permanent virtual services.

Packet networks incorporate idle channel state condition timers that are settable from 1 to 30 s, in 1-s increments (suggested time is 5 s) [10]. When physical links are lost between switches, these timers may expire; and if they do, they will trigger the disconnection of all virtual calls that were up on those links. The customers/computers must then restart their sessions. It is common (but not required) in these networks to feature a virtual circuit reconnect capability, through alternate routing around the failed links, but only after the call is reestablished. Connection reestablishment procedures have not been standardized in X.25 networks (not a priority) that could allow for automatic virtual circuit reconnect through alternate routing.

---

[2]Old channel bank types (e.g., D1A, D1D, D2, D3) exhibit CGA times of 300–500 ms.

[3]Old data multiplexers (T1DM) exhibit CGA times of 300–500 ms.

Frame relay is an upgrade of X.25 packet switching, allowing speeds of 56 kb/s to DS1. Frame relay networks may incorporate algorithms in their network switches/routers (although not yet standardized) that permit alternate routing of frames around failed links. These rerouting algorithms are proprietary, and the speed of the recovery process can vary greatly between vendors. Note that this type of recovery is associated with the service layer rather than the transport layer, which may incorporate physical protection (e.g., rings, APS) and logical protection layers (e.g., DCS restoration).

4) ISDN Services: ISDN provides for both circuit and packet-switched call services on the public switched network. A basic rate ISDN line access provides a customer with two "B" channels (each 64 kb/s) and one "D" channel (16 kb/s) for call control. Each of these channels is independent of one another, meaning that there may be multiple paths involved with an ISDN call. Thus, a network outage may interrupt one or more of these channels.

The "B" channels can support circuit-switched voice or data calls, packet-switched data calls, or channel-switched services. Channel switching is a DS0 rearrangement or nail-up capability integrated into the ISDN switch. The "D"-channel supports only packet data. Packet traffic can either be handled by the ISDN switch directly (as an integral part) or the packet traffic can be routed to a separate packet switch (i.e., PPSN). ISDN uses out-of-band signaling.

Primary rate ISDN provides 23 "B" channels (64 kb/s each) plus a "D" channel (64 kb/s) on a DS1 access line.

Circuit-switched calls and channel-switched services on ISDN will rely on the standard CGA (2–3 s) procedure established for handling service outages in the trunk network. During trunk failure (outage), packet-switched calls on ISDN will rely on the packet disconnection (1–30 s) procedure established in the trunk network for PPSN.

SWF-DS1 service (switched DS1/switched fractional DS1) [11] will be a primary rate ISDN service capability, also subject to the standard 2–3 s CGA procedure.

5) MAN Services: Metropolitan Area Networks (MAN's) are high-speed communications networks for the delivery of voice, video, and image services. Speeds are in the 100 Mb/s and higher range. MAN's offer high bandwidth, low delay, and high reliability. A primary application for MAN's is LAN interconnection, with LAN-like performance. Another application is high-speed image transfer. Two key MAN standards are IEEE 802.6 (DQDB) and FDDI.

Switched Multimegabit Data Service (SMDS) is a connectionless, cell relay public service using no call setup or teardown. It can be provided over a metropolitan network designed to work with DS1 and DS3 facilities and switching over the public network.

SMDS is designed to be a highly reliable service based on DQDB (IEEE 802.6) MAN standard. SMDS utilizes its own routing protocol, which means packets can take different routes during a customer's session. This lends itself to a robust network.

SMDS supporting networks provide redundancy for rerouting packets around failures, including node or link failures. A single failure should not adversely affect performance. A break in the links will not cause a loss of service, only a slight disruption of service. The supporting network should be able to eventually recover completely (through cell rerouting). Note that the packet loss probability is not zero with alternate routing because of network congestion and buffer limitations, but it is significantly better than without alternate routing. Having physical protection and/or logical layer protection would still be beneficial to prevent lengthy periods of congestion.

Normal cell routing processes are interrupted and rerouting processes may be started after a certain interval. This interval can be estimated at 200 ms based on SMDS objectives. The objective in SMDS supporting networks for a switch/router to update its routing tables upon notification of a link set change through lower-level Open Systems Interconnection (OSI) protocols is less than 100 ms [12]. It is reasonable to expect that within 200 ms from the onset of the link outage, the rerouting process should have been started, allowing time for topology updating information to propagate to all switches (50 ms), and delay (50 ms) for waiting on physical level protection, if any (e.g., point-to-point systems using APS). However, in large or more complex networks, this interval may be longer than 200 ms. During the recovery process, the network may discard packets. The SMDS network will not retransmit lost packets at the lower layers (Inter-Switching System Interface Protocol [ISSIP] levels 1 through 3)—this falls to the end system. SMDS provides a packet transport service to higher layer end-to-end (OSI) data protocols. The end-to-end higher layer has windows (typically set at 1–2 s) for error correction on packets (i.e., receiver requests retransmission of lost packets).

Fiber Distributed Data Interface (FDDI) is an ANSI, private, high-speed MAN interface standard for use in private networks. Fault tolerance is built in to FDDI networks through using dual-ring technology for its backbone network. This allows the system to recover automatically when a link between nodes fails. FDDI traffic may be transported over metropolitan facilities to provide interfaces between multiple LAN networks. To maintain the FDDI restoration time, these facilities would also need to provide high reliability (i.e., built-in automatic protection switching).

6) Broadband Services: The Broadband Integrated Services Digital Network (B-ISDN) promises to provide a common platform for offering all types of network voice and data services, including real-time applications in video, image, multimedia, plus other new services that are as yet defined. B-ISDN could also support nationwide gigabit data communications networks for applications requiring gigabit speeds (e.g., high-speed supercomputer interaction).

The B-ISDN concept is based on SONET and ATM, where SONET is the physical layer of the ATM transport architecture. ATM is a cell relay technology with a fixed cell size and is bit rate and service independent. The ATM network is connection oriented, but supports both the connection-oriented and connectionless services, either of which can be continuous or bursty.

There are no firm requirements yet for an alternate routing capability at the ATM layer. Note that the ATM network will not retransmit lost cells at the ATM layer based on current

CCITT standards. Work in standards is just beginning on introducing restoration techniques in B-ISDN's. One proposal [13] calls for the development of a fault-tolerant network for BISDN, introducing a restoration hierarchy in which there are restoration options available to the network provider. It basically introduces restoration schemes at each level, i.e., ATM layer (virtual channel, virtual path, rerouting) and physical layer (protection switching). The contribution also points out that it would be possible for the ATM layer to detect fault conditions faster than the physical layer allowing for fast total protection switching times as link speeds are increased. Faster switching times may be needed in the future gigabit transport systems (e.g., 10 Gb/s and beyond), simply because of the larger amount of customer data that can be lost in 50 ms, as compared to present transport speeds which are up to 2.4 Gb/s. Also, faster switching times may be needed for high-speed data services, since they may be required to be retransmitted at the application layer (e.g., TCP/IP associated protocol). As noted earlier, cell relay may include such services as real-time interactive applications (e.g., image), multimedia, and high-speed switched video. The protocols to recover lost data for these types of services are not yet developed.

At this time, it can only be assumed that the ATM network will be made fault-tolerant. If experts are correct in predicting that huge bursts of data (i.e., large amounts of data transferred in such a short time) will account for a major portion of the traffic on the public network in the future, this scenario will demand highly fault-tolerant networks. It is important to note that a fault-tolerant network will still include the recovery of lost data.

There are also concepts being explored on using SONET Synchronous Transfer Mode (STM) technology (i.e., new ways of payload concatenation called virtual concatenation) for providing greater bandwidth flexibility for data traffic than is currently standardized in SONET. These concepts could also be extended to circuit switching by a high-speed broadband circuit switching architecture.

It is important to mention that the red alarm timing threshold in SONET is settable from 0 to 10 s (default 2 ± 0.5 for DS0-based services) [14]. Broadband service concepts based on SONET/ATM or SONET/STM could have higher thresholds associated with them before a red alarm is declared (e.g., 5–10 s). Having higher thresholds would give more time for the broadband network to recover from service outages before initiating the outage (i.e., CGA) and dropping broadband calls (e.g., DS3 calls).

7) Signaling Networks: Signaling networks provide the underlying connection/call control in the public network.

Traditionally, signaling has been done in-band. The in-band signaling network has incorporated signaling freezing to protect against network outages. When loss of frame occurs for a DS1 signal, the signaling channel information is maintained in the state that existed before detection of the out-of-frame condition. This capability provides a 95% assurance of freezing in the correct signaling state (i.e., no change in the signaling state) [15]. Thus, there is less than a 5% probability of a call in progress being dropped by a switch, prior to a CGA declaration (see Section III-A).

The common channel signaling network (CCSN) will provide better performance by virtue of being out-of-band and separate from the path of the call. The CCSN uses 56 kb/s links (DS0 channels) bunched together on a DS1 trunk (carrier). With CCSN, calls will not be dropped prior to a CGA declaration by a DS0 processing network element (e.g., switch) in the message trunk.

The CCSN backbone network segment uses a mesh architecture with full redundancy (i.e., mated Signaling Transfer Point [STP] pairs and alternate links). This network is robust against single failures and does not require additional physical level protection (e.g., APS). Calls only in the process of being set up may be lost with an outage on the links between STP's. Note that according to signaling system number 7 (SS7) specifications [16], a 56 kb/s signaling link is considered failed and taken out of service when a loss of alignment (i.e., no signaling flags) lasts for approximately 146 ms at the STP. A changeover to the alternate signaling link is then initiated. This is an event (typically lasting 500 ms) marked by large network management activities (messages) to accommodate the changeover.

CCSN/SS7 will provide the high quality and highly reliable call processing necessary for the new telecommunications services offered under ISDN, B-ISDN, and intelligent networks.

8) Intelligent Services: An Intelligent Network Service (e.g., AIN) introduces control to the customer and offers the ability to rapidly create telephone "features" to match customer's applications. AIN features will initially be tied to voice services and later to broadband and video services.

Customers will have more interaction with the network through a software control architecture or platform, and will assume more responsibility in network management for maintaining databases, reconfiguring their networks, and managing switching resources. AIN is reliant on SS7, and so is dependent on the performance of the CCSN (see the previous subsection).

Information Networking Architecture (INA) is a service and operations concept that will enable the evolving broadband network to support new advanced data networking applications by end users. The functional architecture includes software control and management for the rapid creation of new service capabilities (sessions) such as multimedia calls, full-motion video transport and switching, and other data networking applications. INA, like AIN, is reliant on CCSN/SS7 for call setup. Delivery of these network information services will be based on SONET/ATM and SONET/STM technologies.

9) Wireless Services: Personal communications networks will offer wireless access to services that will permit Personal Communications Services—PCS (voice and/or data)—to take place (with a single number access) wherever and whenever the subscriber desires. PCS is considered as the next generation of mobile communications.

Service capabilities planned will be voice and voiceband data and, eventually, speeds up to the basic ISDN line data rate of $2B + D$ (144 kb/s) [17]. Thus, PCS will use resources of the public-switched network, as discussed in subsection 1, and ISDN, as discussed in subsection 4. Correspondingly, PCS should have the same sensitivity to network failures as other services provided by the public-switched network and ISDN.

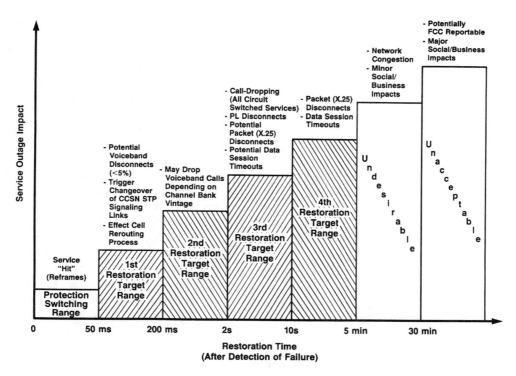

Fig. 4. Restoration time impact on customers.

III. Network Restoration Characterization

In Section II-B, various networks and services were analyzed for their sensitivity to network outages. This section establishes objectives for restoration times, based on that analysis. Included are technical implications to SONET DCS's with distributed control and application guidelines.

A. Network Restoration Time Objectives

Fig. 4 shows what the impact on various network services are as a function of restoration times for SONET-based architectures due to a network failure. Restoration time in Fig. 4 is the time taken to restore traffic after the initial detection of the failure. Outage durations include restoration times at the SONET level. As indicated, the impact on network services (and customers) associated with a service outage depends largely on the outage duration (the outage duration here is defined to be the time interval between first loss of a particular unit of service until that unit of service is fully restored). Note that actual service outage time will exceed facility (SONET) restoration time. Various restoration time targets (50 ms–5 min) for SONET DCS distributed control architectures are accordingly defined in this figure. Restoration times within a given interval will have roughly the equivalent impact on services. Restoration times under 50 ms will be transparent to most services. These time targets are tabulated in Table I. It is important to point out that DCS-based mesh restorable networks cannot restore services in less than 50 ms mainly because a DCS network is not based on a physical layer network design, such as DP and self-healing rings, that use separate or dedicated spare capacity for protection. In

TABLE I
RESTORATION TIME TARGETS

Time Target	Response Time
1	50 ms to <200 ms
2	200 ms to <2 s
3	2 s to <10
4	10 s to <5 min

essence, the protection resource acts as a "hot" standby. On the contrary, DCS networks are logical layer networks that share capacity and routing decisions must be made at the time of failure.

Restoration times below 50 ms will meet network protection switching time requirements at the SONET level [14]. Recovery within this time frame will be without interruption (only "hit") of service. A "hit" is a temporary interruption of serivce that causes only a reframing of a distant terminal (e.g., digital channel bank) off of the SONET backbone. A "hit" does not cause a CGA at the distant terminal.

For the public-switched network, the minimum disconnect timing interval for a possible false disconnect to occur at a downstream switch trunk interface is 150 ms [18] (i.e., 150 ms on-hook supervisory signal state). A failure in the interoffice trunk network may cause the supervisory signal state to appear as on-hook, rather than off-hook, to the switch. The network allocation of this 150 ms is as follows:

+ 10 ms	failure detection time	
+ 40 ms	reframing of lower-rate multiplexes	
+ 50 ms	reframing of distant terminal	
Subtotal	100 ms	
+ 50 ms	switching time (SONET level)	
Total	150 ms	

Restorations taking longer than 50 ms are out of the range of protection switching times and are defined as restoration times (i.e., the time taken beyond 50 ms to restore services associated with a network failure).

The restoration range from 50 ms to under 200 ms is considered to be the first target range for SONET DCS-based network restoration if protection switching was not present. This time frame will have minimal impact on services. Any affected voiceband switched calls (voice or data) will have less than a 5% probability of being dropped (related to signaling freezing limitations with in-band signaling). A restoration time in the range from 50 ms to under 200 ms will trigger a changeover of CCSN STP signaling links. However, calls only in the process of being setup may be lost; existing calls will not be affected. In this same time frame, the normal cell routing process in cell relay networks (e.g., supporting SMDS) should have been interrupted, and the cell rerouting process started. Again, this estimated 200 ms interval for cell relay networks may be longer in large or more complex networks. Keeping the restoration time to under 200 ms will prevent any affected voiceband circuit-switched calls that were on trunks associated with older channel banks from being dropped. Note that a 200 ms restoration time appears as a 300 ms outage to a channel bank (when including detection time and reframe times).

If the first target range cannot be achieved, the next best target range for network restoration is the second time frame from 200 ms to under 2 s. Voice service is not significantly degraded in this range because only affected voiceband circuit-switched calls that are on trunks associated with older channel banks will be dropped. Actual voice performance is only slightly degraded in this range. Video services can become degraded in just 1/30 s (loss of 1 frame). Furthermore, outages resulting in a loss of frame, and generally greater than 100 ms, become a customer-reportable impairment [19]. Nevertheless, an actual outage of less than 2 s is not critical for entertainment/educational-type video services (1–2 s is close to human reaction time). This observation applies to video services with no protocol for recovery of lost frames. When keeping the restoration time to under 2 s any affected voiceband circuit-switched calls that were on trunks associated with older channel banks will be dropped. Network data from one exchange carrier indicate that there is still a percentage of DS0 channel bank circuits (roughly 12%) carried on older-type channel banks (including DS0's that carry switched circuits as well as nonswitched circuits). Thus, the total number of trunks in the public-switched network that are carried on older-type

channel banks is less than 12% because significant numbers of DS0 trunks do not route over channel banks (e.g., DS0 trunks that directly interface into digital switches at the DS1 rate).

The next best time frame to meet, if the second restoration target range is not practical, is the third time frame from 2 s to under 10 s. The 10 s upper bound is consistent with the SONET CGA requirements [14]. When outages exceed 2 s (minimum CGA threshold) DS0, NxDS0, and DS1 call-dropping will occur. Private line disconnects will also occur. Voiceband data modems will timeout (typically 2–3 s) after detecting a loss of incoming data carrier. The modems must then be resynchronized (usually a 15-s procedure) after the data carrier is restored. Broadband calls in the future may have CGA thresholds set higher than 2 s (e.g., 5 s), but B-ISDN calls would still be dropped in this time range, assuming the outage lasts longer than 5 s. In this time range, there is a potential for packet (X.25) virtual calls to be dropped depending on the idle link timer system setting. Disruption of data communications sessions (connection oriented) may occur in this range depending on the session termination time. High-speed, interactive applications (e.g., video image) may be impacted in this time range since these applications are sensitive to delay. This time range is also unattractive for high-priority data applications, e.g., transfer of funds using automated teller machines. Automated teller machines use private data communications networks based mostly on SNA and TCP/IP protocol architectures. Services that are less sensitive to delay such as electronic mail (e-mail) should not be greatly impacted. Frame relay data services may be sensitive to service outages because of their high speed and lack of ability of the frame relay network layer protocol (roughly equivalent to the third layer of the OSI stack) to recover lost data.

The last time frame to meet, if the third restoration target range cannot be realized, is the fourth time frame from 10 s to under 5 min. In this fourth target range, packet (X.25) calls and data communications sessions will be disrupted and customers/computers would attempt to reinitialize their connections. The 5-min upper bound is not meant to be a firm number based on particular software or hardware specifications, but rather an estimate based on general planning goals for various exchange carriers. For example, the 5-min restoration time is the goal for completing service restoration by some network management systems and is considered responsive to most customers. Also, the 5-min upper bound covers the maximum SNA session timeout of 255 s.

For purposes of this study, the time interval beyond 5 min but less than 30 min is considered to be undesirable to most customers. In this time frame, digital switches in switched networks may experience a buildup of network congestion. There may be minor social and business impacts with this outage event. The 30-min level is an FCC requirement of phone companies that major outages (affecting at least 30 K customers and exceeding 30 min) be reportable. Note that even through the 30-min FCC requirement for notification applies to very large outages, it was still used as a benchmark in this study. Accordingly, outages exceeding 30 min in duration are considered to be unacceptable to most customers, causing major social and business impacts.

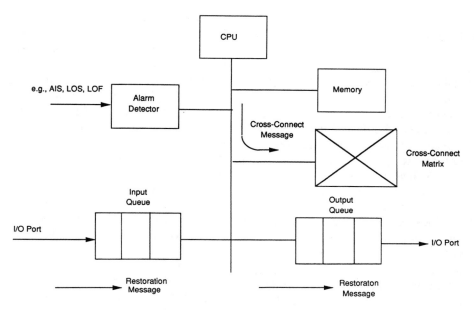

Fig. 5. DCS model for restoration time.

Note that the characterization of these ranges could also be done by their cost. A customer that wants fast restoration (or protection switching time requirements) could be charged more. Some customers may be willing to accept 20-s restoration times if the cost of service is much lower than that for 200-ms restoration. Another variable is the cause of failures. Customers may be more understanding about 10-s outages because of major catastrophies, etc.

B. Implications to SONET DCS's with Distributed Protection Control

In a DCS with distributed control, various parameters that can affect total restoration times are as follows:

- CPU (message) processing time
- message transfer time
- alarm detection time
- cross-connection time.

These parameters relate to DCS functionality and are illustrated in Fig. 5. CPU processing time (processing of restoration messages) depends on the CPU implemented in the DCS, the DCS architecture, and the size of the restoration algorithm. Bellcore simulation conditions have assigned 1–10 ms (per restoration message) for CPU processing time. The 1–10 ms represents the low end for this parameter. Message transfer time through the DCS depends on the type of queue, and the time from input queue to CPU memory or from CPU memory to output queue. This time is assumed to be the same as the CPU processing time. Alarm detection time is the duration between alarm occurrence and recognition by the CPU; 10 ms is generally assumed for this parameter. Current DCS cross-connect time requirements only require the time to be under 1 s per path [6]. Some vendors have claimed that cross-connect times in next generation DCS's can be achieved in 10 ms per path, but times under 100 ms per path may be more common. This parameter will mainly dictate which network

restoration time target is met with SONET DCS distributed control architectures.

The cross-connect time is the time associated with a single cross-connection. This is the time to establish a single cross-connection after acceptance of the cross-connect message (command), and includes time for the command acceptance procedure. Under an actual network failure condition, there may be large numbers of SONET paths (e.g., STS-1's) to be restored, and a corresponding amount of cross-connect commands generated at the busiest node. In this case, the method of cross-connect becomes important. Unless all the required cross-connections at a given DCS node (in the restoration path) are performed in parallel, rather than sequentially the total restoration time (to restore all paths) could fall into one of the higher restoration time ranges (e.g., 4th target range). This can be simply demonstrated as follows. If the DCS architecture was based on sequential cross-connections and a total of 35 STS-1 paths needed restoring, the total cross-connect time at the sender or chooser node (or at busiest node) in the restoration path is 35 times longer, and if 10 ms cross-connect times per STS-1 are assumed, this increases the total restoration time to 350 ms, at a minimum (2nd restoration target range). If the cross-connection time is 100 ms, this increases the total restoration time to 3.5 s (3rd restoration target range). If the cross-connection time is 500 ms, the total restoration time is increased to 17.5 s (4th restoration target range).

A final point to make is that in a practical mesh network application of DCS's from different suppliers, cross-connection times and architectures may vary among suppliers. This may make it difficult to further quantify total restoral times. In view of this, it may be necessary to significantly reduce the cross-connection times for SONET DCS distributed restoration, in order to quantify restoration time for next generation DCS's.

In a SONET DCS distributed control architecture (mesh network), other factors that can affect the total restoration time are:

- line versus path level restoration (longer restoration times with path level)
- abnormal conditions in DCS (e.g., alarmed)
- failure severity (single failure versus multiple failure versus node failure)
- network size (# nodes, # links)
- high restoration ratio (restored circuits/failed circuits)
- hop limits (e.g., depends on network connectivity)
- message (network) propagation time [this depends on message length (e.g., 64 bytes), line speed (e.g., 64 kb/s), distances between nodes, and the type of message channel (e.g., bit oriented channel versus DCC)]
- restoration path verification to prevent misconnections (e.g., using path trace mechanism)
- delay time for waiting on APS (50 ms).

These additional factors to the cross-connection time issue could make it very difficult for SONET DCS distributed control architectures to achieve total restoration times that fall within the 1st or 2nd restoration target range.

C. Application Guidelines

Different services have different outage-duration tolerances. For example, voiceband call tolerances (i.e., the interval in which calls in progress are abandoned) can vary anywhere from 150 ms to 2 s, while data (packet) session timeout can vary from 2 to 300 s. Most existing services are impacted by a 2–10 s service outage. As was discussed in the previous section, the 3rd restoration target range (2 s to <10 s) is considered the critical time range for which circuit-switched services could be dropped. Disconnects could also occur for private line services and packet-switched (X.25) services. It should be emphasized that dropped calls is the principle (or critical) customer-perceived impairment, based on human factor studies. Data communications sessions (connection-oriented) potentially could timeout as a result of outages within this time range. Hence, all of these services may not benefit from the standpoint of service impact with SONET DCS distributed control restoration. As noted earlier, customers may be willing to accept this level of performance if the service is priced accordingly. For example, if the cost to provide 10-s restoration is much lower than 2 s, then many customers may be willing to accept dropped calls and retrials.

Delay sensitive, high-speed-type services, such as interactive image applications, interactive information services (e.g., interactive databases), FDDI traffic, and other highly reliable and high-throughput services, are also not candidates for SONET DCS distributed control architectures. In general, high-speed applications that need large amounts of data transferred in a short time will be severely degraded if their restoration times take seconds. A large amount of data will be lost in 2 or more seconds, resulting in excessive retransmissions and low throughput.

Cell relay networks (e.g., supporting SMDS, ATM, including frame relay[4]), with their cell rerouting capabilities, on the other hand, may relax the restoration time requirements for

[4] Note that frame relay is not a cell relay service.

SONET DCS distributed control architectures. These types of networks (and their services), as they evolve and become fault-tolerant, are possible candidates for SONET DCS distributed control architectures.

Clearly, not all networks and services may benefit with SONET DCS distributed control architectures, as with other survivable architectures (DP, rings). However, SONET DCS distributed control architectures should not be restricted to less impacted services, such as voice, because of the longer term trend to more data traffic on the public network; also, it would be an inefficient use of DCS's as facility managers. Therefore, some application guidelines for SONET DCS distributed control architectures, other than a primary survivable architecture, are necessary based on this scenario.

It is recommended that SONET DCS distributed control architectures be included in the public network in conjunction with other faster forms of survivability (e.g., DP, rings). SONET DCS distributed control architectures could add another layer of survivability to proctect against larger failure events (multiple failures, node failures, or other failure types), thereby increasing overall network survivability. This type of application is referred to as providing "background" survivability. Under this concept, restoration times are not as critical (3rd time range and higher) and can be scaled based on the severity of the failure. For example, node failures put a strain on the network and will take longer to restore services due to more links needing restoring and more messages to be processed. There is also a loss of a computing node in the network. This outage event may not require 100% survivability (i.e., 100% restoration ratio). It has been shown [20] that SONET DCS distributed control architectures can increase network survivability through integration with other survivable architectures (e.g., interworking with rings).

It is recommended that distributed algorithms have the capability to prioritize traffic, i.e., in terms of faster restoration time requirements. This could be in conjunction with other survivable architectures as noted above, or as a primary architecture. In principle, DCS networks should be able to identify which services (or paths) need to be restored first, before other lower priority services (or paths). Note that this requires path level restoration, and it would also require bundling of priority services into individual STS-1's. In concept, the algorithm would attempt to restore most priority circuits within 2 s, such that lower target ranges (1st and 2nd time range) are met first for priority services, and all other services with less priority could be restored within the higher target ranges. The goal would be to have all circuits restored by 10 s (at worst, by 5 min). The algorithm should be rerun as many times as necessary to achieve this goal. It should be noted that a goal of 100% survivability would require sufficient spare capacity in the network. In principle, DCS distributed control architectures should be able to invoke their restoration algorithm a multiple number of times.

IV. Summary

A model of a public digital network consisting of an access network portion and a SONET DCS backbone network segment was used in a study to determine the type of network applications and services that would benefit by a SONET DCS distributed control architecture. This network model assumed support of digital services such as ISDN, SMDS, frame relay, and B-ISDN. A total digital network is the preferred network application for SONET DCS distributed control architectures. These services and other network applications were analyzed for their sensitivity to service outages, and from this analysis four restoration target ranges (each having a different severity impact to customers) were derived for SONET DCS distributed control architectures. The critical time range was 2 s to under 10 s because this is where most existing services become impacted. Cross-connection time was highlighted as a major parameter which could make it very difficult for SONET DCS distributed control architectures to achieve faster times. In view of this, it may be necessary to significantly reduce the cross-connection times for SONET DCS distributed restoration, in order to quantify restoration time for next generation SONET DCS's.

It was also shown that not all networks and services may benefit from the standpoint of service impact with SONET DCS distributed control restoration. Service pricing was mentioned as another way that customers may benefit. Networks that may benefit were fault-tolerant networks, such as SMDS supporting networks, which may relax the restoration time requirements for SONET DCS distributed control architectures.

There appear to be potential applications for SONET DCS distributed control architectures to provide more complete survivability in the public network. Future services will demand a fault-tolerant network with complete survivability; this may only be reached through integration of SONET DCS distributed control architectures with other survivable architectures such as cell relay networks and self-healing rings. It was recommended that distributed algorithms have the capability to prioritize traffic as part of the network integration process, and affected channels are restored using the path restoration technique.

ACKNOWLEDGMENT

The author benefited greatly from private conservations with the following individuals: C.-M. Chiang, A. L. Clark, W. Cruz, S. C. Farkouh, R. C. Keinath, T. F. La Porta, and P. L. Patrick on data communications issues; E. J. Anderson on CGA issues; J. F. Ingle on FCC requirement for notification; A. R. Jacob, A. Leung, and W. W. Wauford on CCSN/SS7 architecture and protocol issues; and M. Azuma on DCS model for restoration time. The author would also like to thank R. H. Cardwell, R. E. Clapp, R. D. Doverspike, H. Kobrinski, N. A. Marlow, A. Reidy, and T.-H. Wu for their comments on this paper.

REFERENCES

[1] T.-H. Wu, D. J. Kolar, and R. H. Cardwell, "Survivable network architectures for broadband fiber optic networks: model and performance comparisons," *IEEE J. Lightwave Technol.*, vol. 6, pp. 1698–1709, Nov. 1988.

[2] T.-H. Wu, and M. Burrowes, "Feasibility study of a high-speed SONET self-healing ring architecture in future interoffice fiber networks," *IEEE Commun. Mag.*, pp. 33–51, Nov. 1990.

[3] Bellcore Tech. Ref., "Transport systems generic requirements (TSGR): Common requirements," TR-NWT-000499, issue 4, Nov. 1991.

[4] W. D. Grover, "The selfhealing network: A fast distributed restoration technique for networks using digital cross-connect machines," presented at the IEEE GLOBECOM Commun. Conf., 1987.

[5] C. Han Yang and S. Hasegawa, "FITNESS: A failure immunization technology for network service survivability," presented at the IEEE GLOBECOM Conf., 1988.

[6] Bellcore Tech. Ref., "Wideband and broadband digital cross-connect systems generic requirements and objectives," TR-TSY-000233, issue 2, Sept. 1989.

[7] K.-I. Sato, H. Veda, and N. Yoshikai, "The role of virtual path cross-connection," *IEEE Lightwave Commun. Syst. Mag.*, Aug. 1991.

[8] Bellcore Tech. Ref., "Alarm indication signal requirements and objectives," TR-TSY-000191, issue 1, May 1986.

[9] F. Ellefson, "Migration of fault tolerant networks," presented at the IEEE GLOBECOM Conf., 1990.

[10] Bellcore Tech. Ref., "Public packet switched network generic requirements (PPSNGR)," TR-TSY-000301, issue 2, Dec. 1988.

[11] Bellcore Tech. Ref., "Generic requirements for the public switched DS1/switched fractional DS1 service capability," TR-NWT-001068, issue 1, Nov. 1991.

[12] Bellcore Tech. Advisory, "Inter-switching system interface generic requirements in support of SMDS service," TA-TSV-001059, issue 1, Dec. 1990.

[13] J. Anderson, "OAM aspects for fault-tolerant B-ISDN's," T1S1.5/91–199 contrib., May 6–9, 1991.

[14] Bellcore Tech. Ref., "Synchronous optical network (SONET) transport systems: Common generic criteria," TR-NWT-000253, issue 2, Dec. 1991.

[15] Bellcore Tech. Ref., "Digital cross-connect system requirements and objectives," TR-TSY-000170, issue 1, Nov. 1985.

[16] Bellcore Tech. Ref., "Bell Communication Research specification of signaling system number 7," TR-NWT-000246, issue 2, June 1991.

[17] Bellcore Framework Tech. Advisory, "Generic framework criteria for universal digital personal communications systems (PCS)," FA-NWT-001013, issue 2, Dec. 1990.

[18] Bellcore Spec. Rep., "BOC Notes on the LEC Networks-1990," SR-TSV-002275, Issue 1, March 1991.

[19] D. Cormier, "Rationale for unidirectional protection switching," T1X1.5/92-087 contrib., Aug. 8, 1992.

[20] Y. Okanoue, H. Sakauchi, and S. Hasegawa, "Design and control issues of integrated self-healing networks in SONET," presented at the IEEE GLOBECOM Conf., 1991.

Network Design Sensitivity Studies for Use of Digital Cross-Connect Systems in Survivable Network Architectures

Robert D. Doverspike, Jonathan A. Morgan, and Will Leland

Abstract—This paper provides the results of an economic study on the use of SONET Digital Cross-connect Systems (DCS's) to provide survivable transmission network architectures in local exchange networks. Three fundamental survivable transmission technologies are considered: 1) a SONET self-healing ring, 2) a SONET point-to-point fiber system with 1:1 automatic protection switching and diverse routing of protection facilities, and 3) a DCS mesh with automatic DCS restoration (rerouting) protection. These three technologies are used in various combinations to form six survivable network alternatives for evaluation. Two Local Exchange Carrier (LEC) networks are used (a 15 node network and a 53 node network) and demand, network connectivity, and unit equipment cost sensitivities are evaluated on these alternatives. In addition, the survivability of each alternative in the event of a major node failure is calculated. The motivation for the study is to determine the viability of DCS-based survivable network architectures and, in particular, the viability of SONET DCS's with integrated optical terminations. The study has two objectives: 1) given a specific survivable network technology, underwhat conditions is it economical to place a Broadband DCS (B-DCS) in a central office as opposed Add-Drop Multiplexers (ADM's); and 2) which survivable technologies with B-DCS's are economical, and under what conditions. We conclude that the most cost-effective networks consist of "hybrids" of SONET point-to-point, ring, and mesh technologies, and that the B-DCS is economically viable for interconnection between these technologies.

I. INTRODUCTION

A BROADBAND Digital Cross-connnect System (B-DCS)[1] is a type of SONET transmission equipment used in telecommunications networks [1]. Telecommunications network providers must determine under what conditions it is economical to deploy B-DCS's as opposed to other SONET transmission equipment, e.g., an Add/Drop Multiplexer (ADM), and under what conditions are different DCS-based technologies/configurations economical. Reference [2] provides results of a related SONET fiber optic network suvivability/cost study that primarily focused on ADM-based architectures (although two of the thirteen architectures studied were DCS-based architectures). This paper provides the results of an extensive economic study on the use of SONET Broadband DCS's to provide survivable transmission network architectures in local exchange fiber networks.

Manuscript received July 15, 1992; revised February 18, 1993.
The authors are with Bellcore, Red Bank, NJ 07701.
IEEE Log Number 9212471.

[1] As defined here, a *Broadband DCS* cross-connects at the DS3 rate and/or the Synchronous Transport Signal Level 1 (STS-1) rate of 51.84 Mb/s.

A protection architecture in a transmission network is defined as an autonomously working subnetwork that carries working digital signals (traffic) as well as protects the traffic against equipment and line failures within that subnetwork. There are three general categories of topologies for protection architectures. In each of these architectures, the nodes of the subnetwork represent transmission equipment assemblies that have the ability to switch signals to alternate paths.

1) Point-to-Point: The subnetwork consists of a pair of nodes that are connected by two physically diverse transmission paths. In case of a failure of the working path, the working traffic of the single point-to-point is switched to the protection path, i.e., a simple two-point switch. Many autonomous point-to-point topologies must be placed to protect a network using point-to-point protection methods.

2) Ring: The subnetwork is a ring topology. In case of a failure, the demands of any affected traffic pair within the ring subnetwork can be rerouted in the opposite direction. Thus, the spare protection capacity of the ring is shared by all of the working traffic on the ring. Note that, theoretically, the point-to-point topology is a two-node ring.

3) Mesh: A mesh subnetwork is one where the degree of connectivity (i.e., the number of links coincident on that node) of at least one node exceeds two. In case of a failure, the affected traffic can reroute over a variety of diverse paths, in contrast to a single alternate path for the point-to-point and ring architectures. Generally, in the mesh architectures, as network connectivity increases, required protection capacity decreases because of the sharing of capacity among many potential pairs of nodes.

There are a variety of specific protection technologies or implementations of each of these three topologies. The three topologies considered here are: 1) a point-to-point transmission system with 1 : 1 Automatic Protection Switching (APS) and diverse routing; 2) a Unidirectional Self-Healing Ring with Path protection switching (USHR/P); and 3) a mesh with DCS restoration protection. The first architecture can be implemented with stand-alone Terminal Multiplexers (TM's) or with optical terminations integrated into B-DCS's. The second technology can be implemented with ADM's or DCS's, while the third technology can be implemented only with DCS's. The details of the three technologies are discussed below.

Point-to-Point with 1 : 1 APS and Diverse Routing: Here, the APS performs the role of a two-point switch to automatically reroute traffic to a dedicated, diversely routed protection path.

Reprinted from *IEEE J. Selected Areas Communi.*, vol. 12, no. 1, pp. 69–78, Jan. 1994.

A single point-to-point architecture is usually implemented in a facility hubbing (routing) arrangement. In a hubbing arrangement, traffic from multiple locations is aggregated and sent to a hub central office (CO) for cross-connection. At the hub CO, the traffic is segregated and distributed to other destinations attached to the hub CO. In a hubbing arrangement, either a B-DCS or back-to-back terminal multiplexers interconnected by a manual Digital Signal Cross-connect Frame (DSX-3) are used in the hub CO to provide cross-connection at the STS-1 (or DS3) level. In this paper, single homing arrangements are used, i.e., all traffic in a subtending CO is sent to a single hub CO. In hubbing arrangements, direct links between two nonhub CO's can be deployed when the traffic is sufficient to economically justify bypassing the hub CO.

USHR/P: A USHR/P consists of two fibers: one is designated the working fiber, the other the protection fiber. In a USHR/P architecture, the two directions of a duplex channel travel over different routes. The sending node transmits simultaneously on both the working and protection fibers. During normal operation, only the working signal is used, although both are monitored for alarms and maintenance. If the working signal is lost or disrupted because of a network failure, a 2 : 1 selector at the receiving node switches the traffic from the working to the protection signal. Details can be found in [3].

Mesh with DCS Restoration Protection: This technology consists of DCS's placed at each of the nodes of a mesh subnetwork. Capacity consists of fiber systems with optical terminations integrated into DCS's at both ends. This capacity is used for both working traffic and potential reroutes for traffic lost due to a network failure. When a network failure occurs, the DCS's cross-connect the affected demands onto alternate paths that bypass the failure. There are many potential DCS rerouting schemes, e.g., using centralized control (see [4] and [5]), distributed control (see [6]–[12]), dynamic routing or precalculated routes, and "point-to-point" or "link" rerouting algorithms.[2] See [13] for a general description of DCS restoration methods.

In the study, six network alternatives that are combinations of the above three protection technologies are considered. Figs. 1 and 2 use a hypothetical 9 node network to illustrate the six alternatives. Fig. 1 illustrates the available fiber links in the network. Fig. 2(a)–(f) shows the six alternatives, also described below. The links shown in Fig. 2(a)–(f) denote logical OC-N fiber systems (fiber systems used in each alternative to connect node pairs) that route over the fiber links shown in Fig. 1.

A) ADM APS & Ring: A combination of SONET USHR/P (technology 2) and point-to-point technologies with 1 : 1 APS (technology 1). The example shown in Fig. 2(a) consists of two rings and four point-to-point systems with 1 : 1 APS.

B) DCS/ADM APS & Ring: Same as A except using B-DCS's with integrated optical terminations as replacements for back-to-back USHR/P ADM's in key CO's, e.g., hub CO's. Also, the B-DCS will contain functionality to terminate

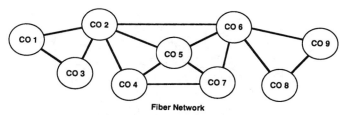

Fig. 1.   Example nine node network.

1 : 1 APS optical terminations directly. In Fig. 2(b), offices that contain back-to-back ADM's and/or Terminal Multiplexers in 2(a) are replaced with B-DCS's.

C) ADM Ring: SONET unidirectional self-healing rings only, i.e., technology 2. The example shown in Fig. 2(c) consists of three rings.

D) DCS/ADM Ring: Same as C except B-DCS's are placed in key offices (e.g., hub CO's) instead of ADM's. In Fig. 2(d), offices that contain back-to-back ADM's in Fig. 2(c) are replaced with B-DCS's.

E) DCS Mesh: A "pure" mesh solution, i.e., B-DCS's with integrated optical terminations in all offices with an automatic DCS restoration method for protection [as illustrated in Fig. 2(e)].

F) DCS Core Mesh: A combination of B and E, where a combination of SONET USHR/P and point-to-point technologies with 1 : 1 APS are deployed in the perimeter of the network and a DCS mesh in the "core" network [as illustrated in Fig. 2(f)].

All study alternatives assume the existence of SONET DCS technology that is expected to be available in the 1994–1995 time frame, in particular: 1) a SONET B-DCS with integrated optical terminations; and 2) a SONET B-DCS with integrated self-healing ring functionality. The advantage of using DCS's to provide these capabilities is that multiple ADM's (associated with multiple rings), stand-alone point-to-point terminal multiplexers, and DSX-3 functionality within a single central office can be integrated into a SONET DCS with integrated optical terminations (these terminations essentially consist of optical-to-electrical conversion, multiplexing/demultiplexing functionality, and possibly ring functionality).

The alternatives are compared using two Local Exchange Carrier (LEC) networks (a 15–node network and a 53–node network) and by performing demand, network connectivity, and equipment cost sensitivities. In addition, worst-case network survivability ratios of the alternatives are evaluated. Section II presents the study methodology; Section III gives a general analysis of the sensitivity of the three basic technologies to various network characteristics; Section IV describes the study networks; Section V presents the results of the study and the correlation with the analysis presented in Section III; and Section VI provides the conclusions to the study. Appendix A lists the unit costs used in the study.

II. STUDY METHODOLOGY

To aid in the evaluation of the study alternatives, two study tools developed by Bellcore are used. Strategic Options [14] is a Bellcore research prototype software system that economically places self-healing rings and point-to-point

---
[2] *Point-to-point* routing methods reroute the demand from endpoint node to endpoint node, rather than simply rerouting or "patching" the segment of those demands that route over the failed link, as in "link"-type methods. Point-to-point rerouting methods are also referred to as *Path* rerouting methods.

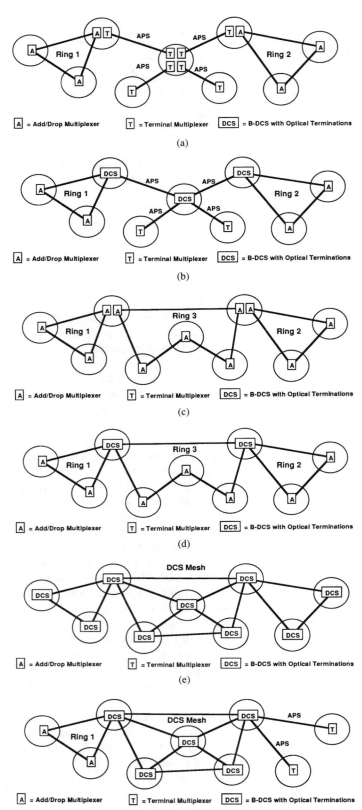

A = Add/Drop Multiplexer   T = Terminal Multiplexer   DCS = B-DCS with Optical Terminations

(a)

A = Add/Drop Multiplexer   T = Terminal Multiplexer   DCS = B-DCS with Optical Terminations

(b)

A = Add/Drop Multiplexer   T = Terminal Multiplexer   DCS = B-DCS with Optical Terminations

(c)

A = Add/Drop Multiplexer   T = Terminal Multiplexer   DCS = B-DCS with Optical Terminations

(d)

A = Add/Drop Multiplexer   T = Terminal Multiplexer   DCS = B-DCS with Optical Terminations

(e)

A = Add/Drop Multiplexer   T = Terminal Multiplexer   DCS = B-DCS with Optical Terminations

(f)

Fig. 2.   (a) Alternative A. (b) Alternative B. (c) Alternative C. (d) Alternative D. (e) Alternative E. (f) Alternative F.

fiber systems with 1 : 1 APS and diverse routing. Strategic Options uses various optimization methods for substeps of the design problem. Surcap [15] is a Bellcore research tool that economically sizes links between a given network of DCS's to study restoration algorithms for DCS mesh networks.

In the studies, both study tools are set to design networks that will survive any single link failure. The inputs to the design problem are the nodes, potential fiber links, and multiyear point-to-point DS0 traffic (demand) forecast. The DS0 level demands used in the study are for a single time period, both switched and nonswitched, corresponding to the cumulative, forecasted demand at the end of a 10-year horizon. To evaluate all the study alternatives equally, each of the study alternatives uses a common DS3 forecast generated by Strategic Options. Strategic Options lays out the network based on a hierarchical structure where subtending CO's are organized into clusters that home on a hub CO. Hub CO's can be further clustered into sectors that home in on a gateway CO. Strategic Options first bundles the DS0's into DS1's and then uses a "hubbing" method for routing and bundling of DS1's into DS3's (see [14] for more details). After the DS3 forecast is generated, a combination of Strategic Options, Surcap, and manual methods are used to design the "best" network for each of the study alternatives.

The goal of Surcap is to place fiber systems between the DCS's to minimize network costs. The constraints of the solution are that Surcap places sufficient capacity into the network to route the working demand plus satisfy the network restoration requirements. To satisfy the network restoration requirements, for each potential link failure, Surcap computes and stores the alternative routes for each point-to-point rerouting used. Surcap assumes that, given a network failure, the stored routes (reroutes) for the traffic will be implemented. Traffic between a given point-to-point is allowed to be rerouted over several different routes.

This is an intractable problem to solve exactly; therefore, Surcap uses a heuristic approach called "simulated annealing" to solve the problem to the extent possible. Generally, the longer the algorithm is allowed to run, the better the solution generated.

Note that to achieve 100% restoration for any single link failure in a DCS network designed by Surcap, one would have to implement a DCS restoration method that implements the exact precomputed reroutes determined by Surcap. Since these routes are computed by a complex optimization algorithm, other DCS restoration methods that compute routes dynamically at the time of failure (either on a centralized or distributed basis) will likely achieve a lower degree of restoration on the same network or, equivalently, require a network with more capacity to achieve the same level of restoration. Thus, the costs for networks with DCS restoration given in this study should be regarded as lower bounds.

The costs included in the study include per pair-mile fiber costs and electronics costs. Electronics costs include cost of B-DCS's with integrated optical terminations (OC-3, OC-12, and OC-48 systems are considered), ADM's, and terminal multiplexers. The electronics costs are separated into getting started and termination costs. The details of installed first costs (material plus installation costs) used in the study are given in the Appendix where, to protect proprietary costs, the costs are listed as multiples of the cost of an OC-3 stand-alone terminal multiplexer. Offices without B-DCS's will generally have manual DSX-3's; therefore, this cost is also included in the Appendix.

When DCS restoration protection is placed in the network, costs for the implementation of the restoration method are incurred, in addition to the DCS costs above. These costs are estimated based on a centralized control architecture. Thus, the cost of implementing DCS restoration methods is incremental to existing centralized systems (or "Operations Systems") that control DCS's, and this cost is pro-rated per central office. The cost of a data communications network between the DCS's and the centralized controller is not included because it will be shared by communications to other transmission equipment as well (including monitoring and provisioning of ADM's).

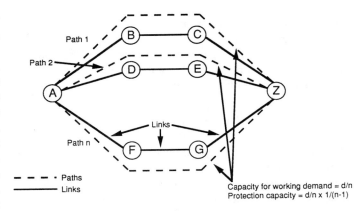

Fig. 3. Simple connectivity analysis.

III. GENERAL NETWORK COST CHARACTERISTICS

Before discussing the study results, this section provides some insight into the correlation between the protection technology deployed and some general network characteristics. The cost of providing protection technologies in a network is directly related to the amount of capacity required by that technology. The amount of capacity needed to achieve a given level of network survivability is a function of three broad network factors: 1) size of network, i.e., number of nodes and links; 2) quantity of traffic; and 3) network connectivity. Network connectivity measures the number of diverse paths between pairs of nodes in a network. The amount of network capacity required by all protection technologies is sensitive to the first two factors. However, the mesh technology with DCS restoration is highly sensitive to the third factor (network connectivity), while the other technologies are relatively insensitive to network connectivity (as the results in Section V-B indicate).

Thus, in networks with low connectivity, network designs with self-healing ring and/or point-to-point technologies tend to have lowest cost. As the network connectivity increases (with other parameters held constant), the cost curve of network designs with self-healing ring and point-to-point technologies tends to be flat while the cost curve of mesh networks with DCS restoration decreases. If the network size and quantity of traffic are large enough to overcome the DCS getting-started cost modularities, then as network connectivity increases, there is a cross-over point where DCS restoration costs less. The costs of DCS integrated optical terminations (as opposed to fiber costs and getting-started costs) in the DCS mesh are the primary cost components responsible for the correlation of decreasing protection costs with increased network connectivity.

These correlations are experimentally verified through the network sensitivity studies in Section V-B. However, to gain some insight into this phenomenon, the remainder of this section gives some simple measures that illustrate the correlation of the costs of DCS integrated optical terminations with network connectivity in DCS restoration technologies. Note that while these measures can be helpful in evaluating the effect of connectivity on placement costs, optimizing network capacity for the mesh network with DCS restoration is an intractable problem[3] and, therefore, these simple formulas will not accurately measure the economic tradeoffs under a full variety of network characteristics.

To understand the relationship between protection capacity for DCS restoration and network connectivity, consider the following. Express the amount of extra capacity needed to be placed in the network to achieve a given level of survivability as a ratio = *protection capacity/working capacity*, called *network redundancy*. With DCS restoration, as the average network connectivity increases, the needed network redundancy decreases. A simple analysis illustrates this relationship.

Fig. 3 shows a pair of network nodes, {A, Z}, connected by n link-disjoint paths, and d is the amount of traffic between the node pair. A "link" here is a "logical" link in the sense that it represents the collection of OC-n capacity that connects a pair of DCS's.[4] Such a link may physically route over fiber cable through intermediate offices, but not through intervening DCS's, and for simplicity it is assumed that these logical links are physically diverse from one another, that is, they do not intersect the same fiber cable or conduit.

Assume that we wish to protect against any single link failure. Given such a failure, one of the n paths may fail in Fig. 3 (note that a similar analysis can be done for a node (DCS) failure when the paths are node-disjoint). The DCS's at nodes A and Z, as well as the DCS's at intermediate nodes along the paths, will reroute traffic from the lost path onto the other $n - 1$ diverse paths using a point-to-point DCS rerouting method. The minimum capacity configuration is to spread the d working traffic equally among the n paths (ignoring equipment modularity constraints). For the failure of any single path, spare protection capacity must be provided among the remaining $n - 1$ paths to protect the d/n working traffic on the failed path. Thus, any subset of $n - 1$ of the n paths (there are n such sets) must have total protection capacity $\geq d/n$, and hence a total quantity of $d/n \cdot 1/(n - 1)$ protection capacity per path or redundancy = $1/(n - 1)$ over all paths between A

---

[3] This problem is called "NP hard" in complexity theory terminology.

[4] In this context, the term *span* is sometimes used (see [10]) rather than link. Here we prefer the term "link" because span is usually interpreted by telecommunications operations personnel to represent physical fiber or copper cable cross-sections, rather than a family of co-terminus (logical) digital signals.

and Z. Thus, as the connectivity, n, increases, the protection capacity decreases.

One can approximate the network redundancy, R_1, needed for the whole network by weighting the protection capacity by the proportion of traffic per pair, k,

$$R_1 = \sum_k \frac{d_k}{D} \cdot \frac{1}{n_k - 1} \qquad (1)$$

where d_k is the traffic between node pair k, n_k is the connectivity between node pair k, and $D = \Sigma_k d_k$.

In some cases, the above formula may overestimate the protection capacity required. This is because the protection capacity is computed separately per traffic pair, d_k, and does not consider possible sharing of protection capacity by different traffic pairs for potential nonsimultaneous network failures. Therefore, a general network connectivity measure, the *weighted link-connectivity, W*, can be used to formulate another approximation for network redundancy for DCS restoration. W is defined to be the sum of the link connectivity between each pair of nodes times the fraction of traffic originating and terminating between that pair, i.e., $W = \Sigma_k n_k d_k / D$ (weighted *node-connectivity* can also be defined where the number of node-disjoint paths between node pair k is used rather than n_k). Thus, a second approximation of needed network redundancy, R_2, for DCS restoration is

$$R_2 = \frac{1}{W - 1}. \qquad (2)$$

In a given network, we can obtain a rough rule-of-thumb for whether the placement costs of DCS integrated optical terminations in a design with DCS restoration are less than placement costs for a design with the point-to-point 1 : 1 APS technology. The DCS integrated optical termination without APS is used by the mesh with DCS restoration. Protection capacity with the DCS mesh is provided by placing additional integrated optical terminations without APS (the amount of which is reflected by the network redundancy), while protection capacity with the point-to-point technology is built into the integrated optical termination with APS. First, estimate the network redundancy, $R = R_1$ or R_2, for example, by computing one of the two approximations (1) or (2) above. Next, compare $1 + R$ to the ratio of cost of a single DCS integrated optical termination with APS, C_1, to one without APS, C_2. If $C_1/C_2 > 1 + R$, then it is more likely that DCS restoration will cost less from the perspective of placement of DCS integrated optical terminations.

IV. STUDY NETWORKS

In the study, two LEC networks are used in comparing the costs of different network architectures and NE's.[5] Two different networks are used to help determine how sensitive the results of the study are to network size and logistics, and to see if results of one network can be extended to

[5] The two networks are typical representations of metropolitan LEC networks. Another study methodology that could have been used in to perform the study by generating random networks, however, a random network will not necessarily represent an LEC network. Also, the tradeoffs are not necessarily sensitive to traffic patterns, but to connectivity.

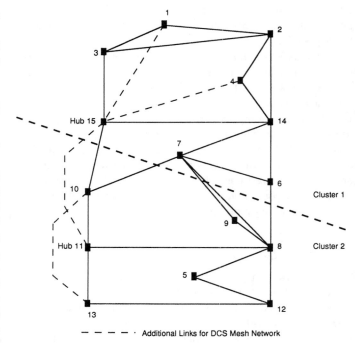

Fig. 4. 15-node network.

TABLE I
DEMAND DISTRIBUTION—15-NODE NETWORK

CO[a] to Hub 1	38%
CO to Hub 2	16%
Hub 1 to Hub 2	13%
CO to CO	33%

^a CO represents a nonhub central office.

other networks in general. The first network analyzed is a 15-node network serving a small metropolitan area. The second network analyzed is a 53-node network serving a large metropolitan area. The demand for each network consists of the number of point-to-point DS0's between two offices. As described in [14], Strategic Options determines the number of DS3's needed between any two offices based on its hubbing algorithm and the quantity of routed DS0's.

A. 15-Node Network

The 15 node study network, shown in Fig. 4, consists of 15 Central Offices (CO's) in a two level hierarchy, of which two are hub CO's and the remaining are subtending CO's. The network has 24 links representing potential interoffice fiber optic routes between central offices, although all 24 may not be used in a given scenario. For the DCS mesh alternative, the average connectivity for the 24-link network is less than three which will generally give poor protection capacity ratios for DCS mesh architectures. Therefore, four links are added for the DCS mesh network to increase the pairwise connectivity. These links are generated by routing through offices that are outside of the 15-node network, therefore preserving physical diversity of the "links" within the study network.

The total (baseline) number of point-to-point DS0 paths in the 15 node network is approximately $107,000$. Table I shows the distribution of the DS0 demand within the network. The figure shows that 67% of the demand either originates

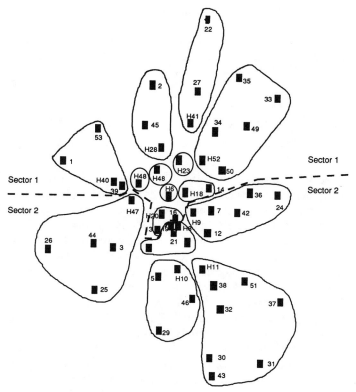

Fig. 5. 53-node network.

TABLE II
DEMAND DISTRIBUTION—53-NODE NETWORK

CO^a to Hub	56%
Hub to Hub	34%
CO to CO	10%

^a CO represents a nonhub central office.

Demand Multiple	Alternatives	Electronics Cost ($M)	Electronics + Fiber Cost ($M)
1x	A (ADM APS & Ring)	3.45	3.75
	B (DCS/ADM APS & Ring)	3.20	3.50
	C (ADM Ring)	3.50	3.77
	D (DCS/ADM Ring)	3.30	3.57
	E (DCS Mesh)	3.82	4.08
2x	A (ADM APS & Ring)	5.68	6.17
	B (DCS/ADM APS & Ring)	4.70	5.17
	C (ADM Ring)	6.32	6.76
	D (DCS/ADM Ring)	5.69	6.12
	E (DCS Mesh)	5.92	6.19
3x	A (ADM APS & Ring)	8.01	8.71
	B (DCS/ADM APS & Ring)	6.45	7.15
	C (ADM Ring)	9.81	10.43
	D (DCS/ADM Ring)	9.08	9.70
	E (DCS Mesh)	7.50	7.82
4x	A (ADM APS & Ring)	10.14	11.00
	B (DCS/ADM APS & Ring)	8.36	9.22
	C (ADM Ring)	11.67	12.44
	D (DCS/ADM Ring)	10.77	11.54
	E (DCS Mesh)	9.16	9.56
	F (DCS Core Mesh)	8.61	9.28

or terminates in a hub CO. Strategic Options multiplexes and routes the DS0 demand between 67 pairs of nodes with a total required capacity of 207 DS3's. The DS3 traffic is the baseline point-to-point traffic for comparing the different survivable alternatives.

V. STUDY RESULTS

B. 53-Node Network

The 53-node study network, shown in Fig. 5, consists of 53 CO's, of which 15 offices are designated hub CO's. The hub offices are numbered 6, 8, 9, 10, 11, 16, 17, 18, 23, 28, 40, 41, 47, 48, and 52, and the clusters that correspond to each hub are outlined (circled) in Fig. 5. The hub CO's serve as the interconnection point between subtending CO's. The network is further separated into two sectors and two of the hub CO's, 6 and 8, are designated gateway CO's (or "super hubs") for each sector. In some of the study alternatives, traffic between the two sectors routes through the gateway CO's. Table II shows the distribution of demand within the network. Since the 53-node network contains 15 hubs, Table II only shows the overall level of demand flowing between different types of offices. The table shows that 90% of the DS0 demand either originates or terminates in a hub CO. Strategic Options multiplexes and

routes the DS0 demand between 519 pairs of nodes with a total required capacity of 1324 DS3's. The DS3 traffic becomes the baseline point-to-point traffic for comparing the different survivability

This section provides the results of three sensitivity studies, including demand, cost, and network connectivity sensitivites. Also, a worst-case survivability analysis is presented.

A. Demand Sensitivity

The impact of changes in demand on the results are analyzed by performing demand sensitivities. This consists of: 1) varying the demand in the smaller 15-node network and running the results, and 2) analyzing the 53-node network to see if the results can be extended to larger networks.

1) 15-Node Network: The demand sensitivity on the 15-node network is performed by multiplying all of the baseline DS0 demand in the network by factors of 1, 2, 3, and 4, and running the results for five of the six study alternatives. (Alternative F is only analyzed for the "4x" value.) The resulting total number of DS3's is 207, 368, 530, and 693 for each of the 4 demand factors, respectively. The numerical cost results for the demand sensitivity are presented in Table III.

The results show that Alternative B (DCS/ADM APS & ring) has the lowest cost for all levels of demand. Alternative

177

A (ADM APS & ring) has lower cost than the DCS mesh alternative (Alternative E) for levels of demand between 1x and 2x. Alternative E has lower costs than Alternative A (ADM APS and rings) for levels of demand slightly greater than 2x and up to 4x. The difference between Alternative B (DCS/ADM APS & ring) and Alternative E (DCS mesh) decreases as the demand increases, however, the DCS mesh alternative never has a lower cost than Alternative B in this sensitivity.

For the 1x demand case, Alternatives C and D (ADM Ring and DCS/ADM Ring) have lower costs than the DCS mesh alternative (Alternative E). Also, Alternative D (DCS/ADM Ring) has a lower cost than Alternative C (ADM Ring). However, for 3x and 4x cases, the DCS mesh (Alternative E) has a lower cost than both Alternatives C and D (ADM Ring and DCS/ADM Ring). As the demand increases, the capital cost advantage of the DCS mesh alternative increases.

At a minimum, the results from this study show that B-DCS's are less expensive than back-to-back ADM's interconnected with manual digital cross-connect frames (DSX-3) in large CO's (e.g., hub CO's). In every scenario within the demand sensitivity (e.g., the 2x demand multiple scenario), an alternative with B-DCS's is less expensive than alternatives without any B-DCS's. Another conclusion from the study is that as the demand increases, the economic value of the DCS tends to increase as well.

Because of the lack of automated tools for planning hybrid networks consisting of DCS mesh and other survivable technologies, Alternative F (DCS Core Mesh) is calculated on the 4x network only, i.e., the network where it would be most competitive. For this alternative, four CO's are redesignated hub CO's (as opposed to the original two hubs) and interconnected in a mesh arrangement. This alternative had a cost of $9.28 million. This is only slightly higher—$60,000—than the least expensive solution ($9.22 million), Alternative B (DCS/ADM APS & Ring), shown in Table III. Further gains with Alternative F are not expected in this network, even as demand is increased because of connectivity limitations, and because the demand pattern is focused toward only two hub CO's.

2) 53-Node Network: For the 53-node network, the results are given for one demand level. The alternatives presented for the 53-node network are A (ADM APS & Ring), B (DCS/ADM APS & Ring), and F (DCS Core Mesh). Other alternatives were either not competitive or practical. For all three alternatives, the resulting networks consisted of rings connecting the subtending CO's to the 15 hub CO's (a total of 10 rings are placed). In alternatives A and B, the 15 hub CO's are interconnected in a hubbing arrangement (2 of the 15 hub CO's are gateway CO's) via point-to-point with 1 : 1 APS.

Alternative F (DCS Core Mesh) is generated by interconnecting the 15 hub CO's in a core mesh network. The logical "links" (i.e., direct fiber systems placed between the B-DCS's) of the core network are chosen by using shortest path routing between all pairs of hub offices while retaining physical diversity of the links. This results in 27 links generated for the core mesh network. However, for the resulting core B-DCS mesh network, some demand pairs had only connectivity of two. Connectivity of less than three gives generally poor

TABLE IV
COST COMPARISONS OF ALTERNATIVES (53-NODE NETWORK)

Alternatives	Electronics IFC ($M)	Electronics + Fiber IFC ($M)
A (ADM APS & Ring)	27.64	30.59
B (DCS/ADM APS & Ring)	20.88	23.85
F (DSC Core Mesh) (27 link mesh)	23.34	25.63
F (DCS Core Mesh) (30 link mesh)	22.11	24.65

protection capacity ratios for DCS meshes. Therefore, for study purposes, we generated another network by adding three key links to raise the minimum demand pair connectivity to 3.

Table IV shows the results of the 53-node network for four alternatives. The results show that Alternative B (DCS/ADM APS & Ring) has the least cost. Alternative F (DCS Core Mesh) is substantially less expensive than Alternative A (which costs 19.3% more), but exceeds Alternative B by 7.4%. One reason that Alternative F (DCS Core Mesh) is more expensive than Alternative B (DCS/ADM APS & Ring) is that the network connectivity is on a borderline for savings of the DCS mesh over the point-to-point APS technology (see Section III). Also, the connectivity and demand within the 15 hubs are similar to the 4x baseline demand in the 15-node network. Therefore, just as in the 15-node network, Alternative B (DCS/ADM APS & Ring) has lower cost than a mesh network to interconnect the 15 hub CO's. However, adding the three links (the 30 link mesh network) reduces the cost of Alternative F by 3.8% and is very close to Alternative B (costs 3.4% more).

B. Sensitivity to Network Connectivity

To experimentally validate the network connectivity discussion of Section III, we increased the connectivity of the 15-node network by adding six key links (increased from 28 links to 34 links). The DCS mesh alternative is then designed for the original 28-link network and the new 34-link network, using 4x the baseline DS0 traffic, by running Surcap using only DCS integrated optical termination costs, i.e., no fiber pair costs and no other DCS costs. The results shown in Table V compare the 28-link network results to the 34-link network results. The second column of the table is the weighted link connectivity (W), the third column is $1 + R_1$, where R_1 is the network redundancy determined from equation (1) in Section III, the fourth column is $1 + R_2$, where R_2 is the network redundancy determined from equation (2) in Section III, and the fifth column provides the results from Surcap showing only the DCS integrated optical termination costs.[6]

As Table V illustrates, the proportional change in cost is close to the proportional change of the second estimator, $1 + R_2$, that is, the weighted connectivity formula (equation 2 from Section III).

To experimentally validate the assertion in Section III that the ring and point-to-point technologies are relatively insensitive to network connectivity, we also ran Alternatives

---

[6]DCS costs besides the integrated optical terminations are generally unaffected by the network connectivity.

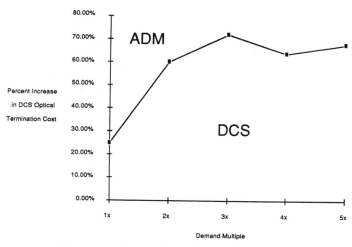

Fig. 6.   Cost sensitivity—region of applicability for Alternative A versus B.

TABLE V
CONNECTIVITY TRADEOFFS IN DCS MESH ALTERNATIVE (15 -NODE NETWORK)

Network	Weighted Link-Connectivity (W)	First Redundancy Estimate $(1 + R_1)$	Second Redundancy Estimate $(1 + R_2)$	Int. Opt. Term. Cost
28 Links	3.21068	1.53331	1.45235	$5.311M
34 Links	4.26840	1.31631 (-14.15%)	1.30596 (-10.08%)	$4.741M (-10.7%)

TABLE VI
CONNECTIVITY TRADEOFFS IN DCS MESH ALTERNATIVE
(15-NODE SUBNETWORK OF 53-NODE NETWORK)

Network	Weighted Link-Connectivity (W)	First Redundancy Estimate $(1 + R_1)$	Second Redundancy Estimate $(1 + R_2)$	Int. Opt. Term. Cost
27 Links	3.23623	1.52015	1.44718	$6.877M
30 Links	3.73109	1.38430 (−8.94%)	1.36615 (−5.60%)	$6.249M (−9.13%)

A and B (ADM APS & Ring and DCS/ADM APS & Ring) on the 34 link network with 4x traffic. The Electronics + Fiber cost decreases $99,000. Almost all of this decline is attributed to fiber savings because of shorter protection paths. The total decrease in the DCS mesh is approximately $580,000 or almost a factor of 6 increase over Alternative B (DCS/ADM APS & Ring); therefore, it appears that the DCS mesh is highly sensitive to connectivity, whereas the nonmesh technologies are insensitive.

We also compare the connectivity parameters and network costs of the 15-node hub CO (mesh) subnetwork of the 53-node network. We compare the 27-link network to the 30-link network as shown. The results are shown in Table VI.

C. Cost Sensitivity

Since SONET DCS's with integrated optical terminations are not expected to be available until the 1993–1994 time frame, a critical sensitivity is the cost of the B-DCS integrated optical terminations. A sensitivity is performed on the 15-node network to show the impact of variations in the cost of SONET integrated optical terminations. Fig. 6 shows a curve representing the region of applicability for a SONET DCS as a function of demand and cost. The graph shows a comparison of Alternatives A (APS & ADM Ring) and B (DCS/ADM APS & Ring). In Fig. 6, the x-axis represents the baseline demand multiple, and the y-axis represents the percentage increase in B-DCS optical termination costs over the estimated values given in the Appendix. The curve represents the break-even point for using a B-DCS in the hub CO's (Alternative B) versus using all ADM's (Alternative A). Points below the curve represent points where it is economical to deploy B-DCS's in the hub CO's. For example, given the baseline demand (1x on the graph), B-DCS's are economical if the cost of integrated optical terminations is not more than 25% of the estimated value in the Appendix. Points above the curve represent points where it is economical to deploy all ADM's in the hub CO's.

The results show that the cost of the integrated optical termination would have to be considerably underestimated to make the B-DCS with integrated optical terminations uneconomical.

D. Worst-Case Survivability (WCS)

Worst-Case Survivability (WCS) is defined as the lowest percentage of circuits (DS0's) surviving all possible single failures, e.g., node or link failures. In general, the worst case survivability will be based on a major node (for our study, a hub) failure. The WCS for the 15-node network (1x demand) was calculated in a related study [2]. Because of the demand distribution in the 15-node network, the WCS cannot exceed the maximum value of 49%, i.e., none of the study alternatives will have WCS greater than 49%. For alternatives A and B (ADM APS & Ring and DCS/ADM APS & Ring), the WCS is 29%. For the DCS Mesh alternative (Alternative E), the WCS is 48%. Therefore, Alternative E (DCS mesh) will provide greater survivability than Alternative B for major disasters such as node failures. (Note: there are several other survivability metrics besides WCS that are described in [2]).

We calculate the WCS for the 53-node network by determining the lowest percentage of DS3 circuits surviving all possible single failures.[7] Because of the demand distribution in the 53-node network, the WCS cannot exceed the maximum value of 76%. For Alternative B, the WCS is 45%. For Alternative F (DCS Core Mesh with 27 links), the WCS is 58%. Therefore, Alternative F will provide greater survivability than the other alternatives for major disasters such as a node failure (i.e., the mesh network provides better WCS than the point-to-point with 1 : 1 APS network in the 15 hub CO's).

VI. CONCLUSIONS

The results of the study show that local exchange transmission networks may consist of "hybrids" of ring, mesh, and point-to-point survivable transmission technologies. The results also indicate that the B-DCS is economically viable for interconnection between these technologies. The cost of providing a protection technology in a network is directly related to the amount of capacity required by that technology. The amount of capacity needed to achieve a given level of network survivability is found to be a function of three broad network factors: 1) size of network, i.e., number of nodes and links; 2) quantity of traffic; and 3) network connectivity. The amount of network capacity required by all protection technologies is sensitive to the first two factors. The DCS restoration (mesh) transmission technology is highly sensitive to the third factor (network connectivity), while the other technologies are relatively insensitive to network connectivity. Networks with higher connectivity have lower terminal costs for DCS mesh protection because of more extensive sharing of protection capacity among different traffic node pairs.

The results of the study show at a minimum that study alternatives that use B-DCS's with integrated optical terminations are less expensive then back-to-back ADM's interconnected with manual digital cross-connect frames in large CO's; and, as network demand increases, survivable technologies with B-DCS's become relatively less expensive. These conclusions for B-DCS integrated optical terminations are still valid under large variations in the cost estimates.

The least expensive alternatives studied consisted of self-healing rings and point-to-point technology using ADM's in the smaller perimeter offices with a core hub network consisting of B-DCS's. The economic viability of using B-DCS's with integrated optical terminations for providing ring transport and interconnection, i.e., between the perimter CO's and the hub CO's, is demonstrated.

The general results show that for interconnecting B-DCS's within hub CO's (or any large CO's), the point-to-point 1 : 1 APS technology had a slight cost edge over the DCS mesh. This tradeoff is found to be particularly sensitive to both network demand and connectivity. The networks analyzed in this study have an average weighted connectivity around

3.2—this tends to be a lower bound (but not uncommon) value for LEC fiber networks. Higher connectivity will tip the balance more in favor of the DCS mesh. Also, higher levels of demand within the network favor the DCS mesh.

The justification for placement of DCS restoration methods in a hybrid network hinges on cost/survivability objectives. The DCS restoration alternatives are not always the minimum cost alternative. Generally, DCS restoration methods give higher levels of survivability by providing restoration over a wider family of network failures (e.g., node and multiple link failures). If survivability is the primary concern, the DCS restoration alternatives are attractive. Therefore, the justification on the deployment of DCS restoration technologies should be based on both survivability and cost objectives. Given these conclusions, future work should concentrate on development of automated planning/optimization methods for design of networks with hybrid DCS mesh, self-healing ring, and point-to-point APS technologies.

APPENDIX

EQUIPMENT AND FIBER COSTS

Equipment	Component	Cost	Comments
TM OC-3	Getting-Started	X	
TM OC-12	Getting-Started	1.77X	
TM OC-48	Getting-Started	4.97X	
USHR/P ADM OC-3	Getting-Started	X	
USHR/P ADM OC-12	Getting-Started	1.77X	
USHR/P ADM OC-48	Getting-Started	4.97X	
STS-1/DS3 Termination	For all TM and USHR/P	0.097X	
DSX-3 Termination	DS3/STS-1 Termination	0.0186X	Per DS3/STS-1
B-DCS	Getting-Started	5.94X	960 Equivalent DS3 Ports
	Cost per Bay	1.93X	240 Equivalent DS3s
	Terminations:		
	DS3/STS-1	0.0371X	Per DS3/STS-1
	DS3/STS-1 Ring	0.097X	Per DS3/STS-1
	OC-3 (1 : 0 APS)	0.429X	
	OC-3 (1 : 1 APS)	0.726X	
	OC-12 (1 : 0 APS)	1.04X	
	OC-12 (1 : 1 APS)	1.54X	
	OC-48 (1 : 0 APS)	3.60X	
	OC-48 (1 : 1 APS)	4.97X	
Fiber	Material	0.060X	Per Pair-Mile
DCS Restoration Capability	Software	0.171X	Per central office

[7] The WCS calculation for the 53-node network is in units of DS3 (rather than DS0) because a calculation at the DS0 level for the hybrid DCS alternative (Alternative F) is manually intractable for a network of this size (recall that we use two disparate design tools and combine the solution by manual techniques for this alternative).

ACKNOWLEDGMENT

The authors would like to thank O. Wasem for helping and answering questions regarding Strategic Options.

REFERENCES

[1] Bellcore Technical Advisory, "Wideband and broadband digital cross-connect systems generic requirements and objectives," TA-NWT-000233, Issue 4, Nov. 1992.

[2] Y. Kane-Esrig, G. Babler, R. Clapp, R. Doverspike, et al., "Survivability risk analysis and cost comparison of SONET architectures," in Proc. GLOBECOM '92, Orlando, FL, Dec., pp. 841–846.

[3] Bellcore Technical Reference TR-NWT-000496, "SONET add-drop multiplex equipment (SONET ADM) generic criteria: A unidirectional dual-fed, path protection switched, self-healing ring implementation," Issue 3 (Bellcore, May 1992), plus Supplement 1, Sept. 1991.

[4] D. Doherty, W. Hutcheson, and K. Raychaudhuri, "High capacity digital network management and control," in Proc. GLOBECOM '90, San Diego, CA, pp. 60–64.

[5] B. J. Wilson and R. D. Doverspike, "A network control architecture for bandwidth management in Proc. ICC'92, Chicago, IL, June 1992.

[6] W. Grover, "The selfhealing network: A fast distributed restoration technique for networks using digital crossconnect machines," in Proc. GLOBECOM '87.

[7] C. Yang and S. Hasegawa, "FITNESS: A failure immunization technology for network service survivability," in Proc. GLOBECOM '88, pp. 1549–1554.

[8] T. Chujo, H. Komine, K. Miyazaki, T. Ogura, and T. Soejima, "The design and simulation of an intelligent transport network with distributed control," presented at the Network Operat. Manag. Symp., San Diego, CA, Feb. 1990.

[9] H. Sakauchi, Y. Nishimura, and S. Hasegawa, "A self-healing network with an economical spare-channel assignment," in Proc. GLOBECOM '90, San Diego, CA, pp. 438–443.

[10] W. Grover, B. Venables, J. Sandham, and A. Milne, "Performance studies of a selfhealing network protocol in Telecom Canada long haul networks," in Proc. GLOBECOM '90, San Diego, CA, Paper 403.3.

[11] R. Pekarske, "Restoration in a flash using DS3 cross-connects," Telephony, pp. 35–40, Sept. 10, 1990.

[12] B. Coan, M. Vecchi, and L. Wu, "A distributed protocol to improve the survivability of trunk networks," in Proc. Int. Switching Symp. (ISS) 1990, vol. 4, Stockholm, Sweden, June 1990, pp. 173–179.

[13] R. D. Doverspike, "A multi-layered model for survivability in intra-LATA transport networks," in Proc. GLOBECOM '90, Phoenix, AZ, Dec., pp. 2025–2031.

[14] O. Wasem, T.-H. Wu, and R. Cardwell, "Survivable SONET networks: Design methodology," see this issue, pp. 205–212.

[15] B. A. Coan, M. P. Vecchi, and L. T. Wu, "A distributed protocol to improve the survivability of trunk networks," in Proc. Int. Switching Symp. ISS'90, vol. 4, Stockholm, June 1990, pp. 173–179.

The Impact of SONET Digital Cross-Connect System Architecture on Distributed Restoration

Tsong-Ho Wu, *Senior Member, IEEE*, Haim Kobrinski, *Member, IEEE*,
Dipak Ghosal, *Member, IEEE*, and T.V. Lakshman, *Member, IEEE*

Abstract—The viability of distributed control restoration using Digital Cross-Connect Systems (DC's) depends on its capability for restoring services within specified time requirements, and its economics for providing restoration compared to other alternatives. In this paper, we report a Bellcore study for the impact of the DCS architecture on distributed restoration. This study concludes that currently proposed distributed control DCS self-healing schemes may not meet the 2-s restoration objective for large metropolitan Local Exchange Carrier's networks, regardless of the distributed algorithm used, if the present DCS system architecture which uses serial message processing and serial path cross-connection remains unchanged. This paper also discusses several DCS architecture enhancement options, including a parallel processing/cross-connect DCS architecture, which may improve the service restoration time.

I. INTRODUCTION

NETWORK survivability is a major service concern for Local Exchange Carriers (LEC's) using SONET networks due to significant impact on customers, revenues, and society. Several survivable SONET network architectures have been proposed and studied in order to find the most appropriate architecture for specific network applications. These survivable SONET network architectures include point-to-point systems with diverse routing and Automatic Protection Switching (APS) [1], SONET self-healing rings [2], and SONET mesh-type self-healing network architectures based on reconfiguration of Digital Cross-Connect Systems (DCS's) [3]–[10]. Among them, the DCS mesh survivable network architecture may be economically attractive in "core" networks with high demand requirements and high embedded fiber connectivity [11]–[13]. Examples of "core" network are the interexchange network and the interhub subnetwork in intra-LATA networks.

Comparing to APS and ring architectures, the DCS self-healing network architecture may require less spare capacity or may provide higher degree of survivability particularly for major failures. However, it has a slower restoration capability due to its relatively complex control scheme. The APS and ring architectures may restore services within 50 ms [14]–[16]. Such outage duration will not generally affect service integrity following a network failure. For the DCS mesh network architecture, services may be restored completely within 5–15 min following a network failure affecting 100 DS3's using

today's centralized control DCS network architecture [17]. This restoration time can be significantly reduced using the distributed control approach. The objective for distributed control DCS networks is to reduce the restoration time from the current 5–15 min to less than 2 s, since most existing services will not see adverse impacts when the service outage lasts less than 2 s [11], [18].[1] The viability of using the distributed control DCS network (to meet the 2-s service restoration objective) has been extensively studied by Bellcore in 1992, and the study results have been summarized in a Bellcore Special Report [11].[2] One of the key issues of this Bellcore study is to analyze the impact of the DCS system architecture on DCS distributed network restoration, which is the core issue in this paper. It is noted that this paper only considers distributed control, dynamic DCS restoration schemes. Preplanned DCS restoration schemes may result in faster algorithm execution and, more specifically, faster cross-connection since the cross-connection time may be reduced if the internal route for the cross-connect switching matrix is known *a priori*. However, the preplanned maps in each DCS must be synchronized, and updating the preplanned maps may add a large overhead on the network administration and maintenance system.

The purpose of this paper is to study the viability of using the distributed control DCS network to meet the 2-s service restoration objective. Section II reviews the distributed control restoration operation using present DCS systems. Section III discusses a feasibility study that shows the 2-s restoration objective may not be met for a metropolitan LEC LATA network, using existing DCS system architecture and cross-connect technology. Several options for improving the restoration time are discussed in Section IV to meet the 2-s restoration objective. Section V discusses design impacts due to these enhanced restoration options. A summary and remarks are given in Section VI.

II. RESTORATION ALGORITHM AND OPERATIONS PERFORMED WITHIN PRESENT DCS

A. Concept of Distributed DCS Restoration

For the purpose of the analysis in this paper, we assume a mesh network topology with DCS's in the nodes of the

---

Manuscript received August 10, 1992; revised February 15, 1993.
The authors are with Bellcore, Redbank, NJ 07701-7040.
IEEE Log Number 9212472.

[1] Although several papers have suggested that the distributed DCS self-healing algorithms may restore services within 2 s, they have not included the effects of DCS cross-connection delay.

[2] This Bellcore Special Report is a nonproprietary report.

Reprinted from *IEEE J. Selected Areas Communi.*, vol. 12, no. 1, pp. 79–87, Jan. 1994.

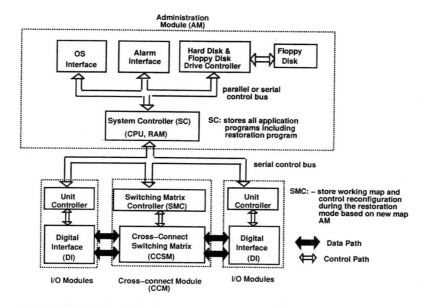

TODAY SYSTEM: Serial Processing, Serial Cross-Connect

Fig. 1. An existing DCS architecture model.

mesh, and we consider dynamic restoration as opposed to the preplanned approach which requires large databases and more complicated database updating mechanisms. Further, we concentrate here on *path level* restoration.[3]

Several algorithms have been proposed for the distributed control self-healing mesh architecture [5]–[10]. For the purpose of this paper, this section describes a generic algorithm for distributed control self-healing mesh networks. The actual implementation could vary depending on the specific algorithm considered.

For the distributed self-healing control architecture, each DCS stores local information that includes working and spare capacity associated with each link terminating at that DCS. When a failure is detected, one of two ends of the affected STS path on the failed facility is designated as the Sender and the other is designated as the Chooser. All other nodes that participate in the restoration process are called tandem nodes. In the restoration process, the Sender first broadcasts (floods) restoration messages to all adjacent nodes. To restrict the number of restoration messages and constrain the algorithm execution time, selective message flooding is implemented, for example, by the hop count limit. The tandem node updates received restoration messages and rebroadcasts them to other adjacent nodes based on the particular flooding algorithm used. When the message reaches the Chooser, it implies that one or more rerouting paths[4] for restoration are identified, and acknowledgment (ACK) messages are conveyed back to the Sender to reserve the spare capacity for the selected restoration route. Note that reservation of DCS ports for the alternate

routes are made either in the first or second phases in each of the DCS nodes involved, whereas cross-connections are made in the last phase.

When the Sender node receives the ACK and verifies the restoration path, it sends a confirmation message back to the Chooser node via the selected restoration path. When the tandem node receives the confirmation message, it reconfigures its DCS switching matrix according to instructions stored in the confirmation message. After the Chooser receives the confirmation message, it changes its DCS switching matrix and cross-connects the affected STS path from the failed facility to newly identified alternate routes. The process described here applies separately for each of the affected STS paths. The restoration process is completed when all the affected paths are restored.

B. Restoration Algorithm Considered

The Bellcore study for DCS distributed restoration analyzed five distributed algorithms currently proposed by the industry [11]. However, the impact of the DCS system architecture on distributed restoration in terms of the total restoration time for these five algorithms is similar. Thus, for the purpose of this paper, we only discuss the study result for a restoration algorithm. The distributed restoration algorithm considered here is "Algorithm B" which appeared in [11]. Descriptions and performance analysis for Algorithm "B" and the other four restoration algorithms can be found in [11].

C. Restoration Operations Performed in Present DCS Systems

In the distributed control DCS self-healing network architecture, broadband SONET DCS's are used to perform STS path restoration. A generic broadband DCS system, as shown in Fig. 1, is divided into three major modules: Input/Output interface module (I/O). Administration Module (AM). and Cross-Connect Module (CCM). The I/O module

---

[3] Another restoration technique, called *line restoration*, uses the line layer information to trigger the restoration process. The path restoration method restores end-to-end STS paths, while the line restoration method restores all channels on a failed line by rerouting demands around the failed facility.

[4] It is possible that demands through the failed facility are restored via multiple alternate paths due to the limited spare capacity available in each link.

includes interface ports, which terminate DS3/STS-1 signals and pass them to the CCM module, and a microprocessor-based controller which monitors the I/O connections.[5] The AM module controls the entire cross-connect system and interfaces with external operations systems or local operators. The AM module includes main memory and nonvolatile memory (e.g., hard disk), a CPU-based system controller with some secondary memory, alarm interface, and an OS interface. The system controller within the AM stores all application programs (in ROM) including the restoration program and controls and coordinates I/O and CCM modules via a LAN-based bus or star medium. Modules within the AM are communicating with each other via a serial or parallel bus depending upon the system design. The main memory within the AM stores all configuration maps[6] and other key information. The CCM module includes a Switching Matrix Controller (SMC) and a Cross-Connect Switching Matrix (CCSM) which is typically protected on a 1:1 basis. The SMC stores the present cross-connect matrix configuration and monitors the accuracy of the cross-connections. The CCSM is used for data (e.g., STS-1) transport and is typically composed of several (3 or 5) switching stages.

The basic operations for the DCS system during the normal and restoration modes are as follows. In the normal mode, the DS3/STS-1 channel enters the input module (i.e., I/O), then is forwarded to the CCSM of the CCM module for transport through based on the residing cross-connection map.

When a restoration process is in progress, the restoration messages are terminated at the I/O module and then forwarded to the system controller of the AM module for message processing. The purpose of the restoration message processing is to execute the restoration algorithm and eventually to obtain a new DCS configuration either from the preplanned database stored in the hard disk or the main memory of the AM module, or by dynamically computing the new configuration for each received message.

Once the system controller has the new configuration map, it transfers the new map to the SMC which then updates the current configuration map. The primary SMC verifies and tests whether the new ports specified in the new configuration map are correct and working. Once the verification and testing are complete, the SMC initiates the change of the CCSM matrix configuration. The physical path rearrangement process within the CCSM usually takes less then 2 ms, as specified in Bellcore Technical Advisory TA-TSY-000241 [19]. Note that STS-1 restoration messages are processed and STS-1 paths are cross-connected on a serial basis under the present DCS system architecture.

Based on the above restoration process within the DCS, the DCS reconfiguration time per STS-1 can be divided into two parts: message processing time and cross-connect time, as shown in Table I. The message processing time, which is part of the restoration algorithm execution time, is the time period from the restoration message entering the I/O to the time when

[5] Note that some commercially available DCS's have a CPU on each I/O port.

[6] Some existing DCS systems use the preplanned method for network reconfiguration. Others calculate new paths on line.

TABLE I
COMPONENTS OF INTRAMODEL NODAL DCS RECONFIGURATION TIMES USING PRESENT DCS SYSTEMS

DCS Reconfiguration Time Components		Times
Message[a] Processing ime	CPU Processing Time	1–5 ms
Cross-Connect Time	New Configuration Map Computation within AM	10 ms
	New Map Transfer Time, Spare Port Testing and Verification, and Physical Path Rearrangement[b] within CCM	100 ms

[a] The message processing time is part of the restoration algorithm execution time.

[b] The time requirement for physical path rearrangement per path is 2 ms [19].

the system controller of the AM modules starts to compute the new configuration map. The cross-connect time includes two phases: 1) time for computing new configuration map within the AM module; and 2) time for new map transfer to the CCM, spare port testing and verification, and path rearrangement (cross-connect) within the CCM module. Note that the values shown in Table I are given per STS-1 (or DS3) that needs to be reconfigured. They represent approximate and partial values under the restoration procedure (i.e., values not included here are the time components representing message parsing, database updating, and message verification). Note that these components currently consume more than 100 ms, but could be significantly reduced under moderate modifications for restoration purposes.

III. DCS NETWORK RESTORATION TIME USING PRESENT DCS SYSTEMS

To see whether the present DCS system architecture and cross-connect technology can be used to support the 2-s service restoration objective for distributed control DCS self-healing networks, we simulated network performance under failures which includes algorithm execution and cross-connections. In this study, a metropolitan LATA network, shown in Fig. 2(a), is used as the model network, and its traffic demand corresponds to the core of a major metropolitan area.

Two network simulators are used in this study: one for simulating the protocol execution time, and the other for simulating the cross-connect delay within the DCS. The latter simulator allows not only for the present serial processing/cross-connect DCS model, but also for a future parallel processing/cross-connect DCS model (as will be discussed in Section IV). The working and spare capacity assignments depicted in Fig. 2(a) are optimized for 100% restoration for all single link failures.

Fig. 2(b) shows the restoration values as a result of all single link failures based on the model network and STS-1 capacity design shown in Fig. 2(a). In Fig. 2(b), we assume that the

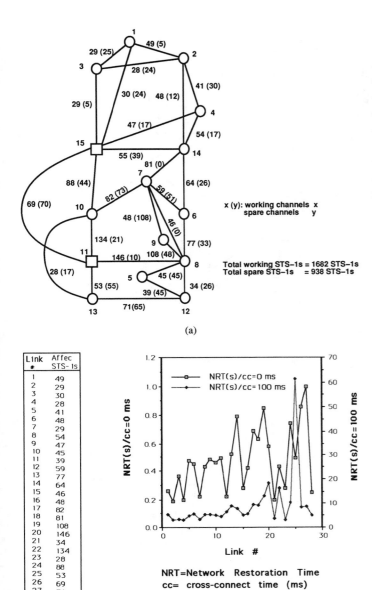

x (y): working channels x
spare channels y

Total working STS-1s = 1682 STS-1s
Total spare STS-1s = 938 STS-1s

(a)

Link #	Affec STS-1s
1	49
2	29
3	30
4	28
5	41
6	48
7	29
8	54
9	47
10	45
11	39
12	59
13	77
14	64
15	46
16	48
17	82
18	81
19	108
20	146
21	34
22	134
23	28
24	88
25	53
26	69
27	71
28	55

NRT = Network Restoration Time
cc = cross-connect time (ms)

(b)

Fig. 2. (a) The model network simulating metropolitan area demand. (b) Simulation results for STS-1 restoration using present DCS system.

hop[7] limit is 9, the CPU message processing time plus the internal bus contention time is 5 ms per STS-1, and the data throughput for restoration message is 64 Kb/s. The condition of cross-connection [i.e., "cc" in Fig. 2(b)] time of 100 ms per STS-1 path represents today's DCS cross-connect technology with moderate software modifications for restoration purposes. The restoration time shown in Fig. 2(b) is the time from when the restoration algorithm is invoked to the time when the last affected STS-1 path is restored. The restoration time for the case of cc = 0 in Fig. 2(b) represents the restoration algorithm execution time only.

As shown in Fig. 2(b) (see the curve with cc = 100 ms), it is not possible to meet the 2-s objective across the study network using the present serial processing/cross-connect DCS system. Results from Fig. 2(b) also indicate that the cross-connect time is a dominant factor for the total restoration time.

[7] The hop count of the path is defined as the number of links on that path.

IV. DCS RESTORATION TIME ENHANCEMENT OPTIONS

This section discusses several approaches that may improve the DCS restoration time to meet the 2-s restoration objective. These alternatives include: 1) reducing the number of restoration messages or paths needed to be processed, 2) reducing the DCS cross-connect time, or 3) a combination of above two approaches.

In the first approach, there are two possible ways to reduce the number of paths or restoration messages needed to be processed: *bundle restoration* or *priority restoration*. The bundle restoration system restores affected STS-1 paths in a group manner, i.e., the algorithm handle a bundle of affected STS-1's, all with the same originating and terminating nodes as an STS-Nc, where N may equal 3, 12, or 48. It is important to note that since today's DCS's do not cross-connect channels in bundles, this option will not reduce the contribution of the cross-connection delay to the total restoration time.[8] Priority restoration only restores STS-1 paths with higher priority in the first run of the self-healing algorithm execution. Therefore, it reduces both the algorithm and DCS execution delay and may provide 2-s restoration to high-priority services. However, both the bundle and priority restoration methods would affect the provisioning system design. In this section, we only discuss the restoration time effect using the parallel DCS system architecture. Issues related to bundle and priority restoration (e.g., administration) will be discussed in the next section.

For the approach of reducing the DCS cross-connect time, two possible methods may be used: 1) a parallel processing architecture in the Administration Module (i.e., the AM module, see Fig. 1), and a parallel path cross-connect architecture within the CCM module; or 2) the fast cross-connect (reconfiguration) system within the CCM module by modifying existing DCS hardware technology.

A. Parallel DCS Processing/Cross-Connect Architecture

The parallel DCS architecture allows for parallel message processing (including computing) for new configuration map generation within the AM module and/or parallel path cross-connections within the CCM module. Fig. 3 depicts a possible parallel DCS system configuration for the network restoration application. As shown in Fig. 3, a parallel DCS system includes I/O modules, an AM module which has two or more CPU-based system controllers, a control bus or star communication medium, a primary memory, and a parallel CCM module. The functions performed in these components are essentially similar to those described for the existing DCS system architecture (with the serial processing and cross-connect architecture) as described in Section II. However, the parallel AM modules compute the new configuration map in a parallel manner. Each CPU-based system controller processes restoration messages independently from the other and has a local memory which stores a subset of spare ports for restoration. In this system, a set of available spare output ports is partitioned disjointly, and each subset of spare ports is placed in the RAM of each system controller. In this case,

[8] It is important to note, however, that SDH DCS's do provide cross-connections of STS-3c paths.

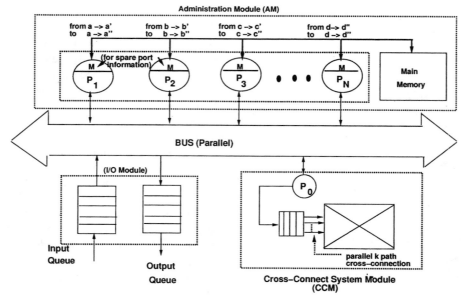

Fig. 3. A general DCS architecture with parallel processing and cross-connects.

x : input port,   x': failed outport,   x": spare output port

no spare output port contention is possible due to parallel processing. However, since the incoming messages are randomly assigned to available system controllers, it is possible that some system controllers' spare ports could be exhausted more quickly than other system controllers. Thus, spare output port sharing is needed among system controllers to maximize the probability of successfully accessing available spare output port. Other designs may also be implemented to accomplish distributed processing and parallel multiple task execution (e.g., hierarchical CPU structure, a CPU in each I/O port).

The primary memory stores the current cross-connect matrix configuration and information on the use of spare ports for restoration. The communication medium connecting different system modules can be a bus or star architecture.

In the parallel CCM module, the path cross-connect function can be performed in a parallel manner. The number of paths (denoted by the batch size in our DCS simulation model) that can be cross-connected simultaneously depends on the restoration size (in terms of the number of affected paths) and processing capacities for other system modules. In this case, the procedure of testing and verification for spare output ports associated with "k" paths in the queue of the CCM can be processed simultaneously, where "k" > 1. It is also possible that the physical "k" reconfigurations within the cross-connect switching matrix of the CCM can be performed in parallel.

B. Queuing Model for DCS Restoration Time Simulation

This section describes a simulation model for the DCS restoration time using the parallel DCS system architecture. This DCS restoration time simulation program has also been used to study the feasibility of meeting the 2-s restoration objective in Section III. A block diagram of the parallel DCS system architecture is shown in Fig. 3. It is a single shared-bus shared memory multiprocessor consisting of $N+1$ processors, $P_0, P_1, \cdots, P_N$. Processor P_0 in the CCM module interfaces

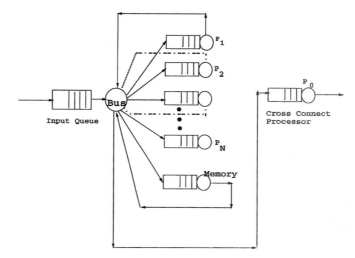

Fig. 4. Queuing model for parallel architecture.

the multiprocessor within the AM module and performs the function of transferring path reconfiguration code into the controller of the CCM module. Processors $P_1, \cdots, P_N$ in the AM module are identical and perform the function required to generate the new configuration map based on the received restoration messages.

Restoration messages arrive to the DCS from the network and are queued up at the input queue within the I/O module. The output queue of the I/O module buffers restoration messages which are sent out to the other nodes in the network. In the following simulation study, we do not consider the output queue since it is part of the restoration algorithm execution.

The primary function of the memory is to maintain a table for the current state of the cross-connect matrix including the information about the spare paths which are currently available. The current state matrix is also replicated in the local memory in each processor.

186

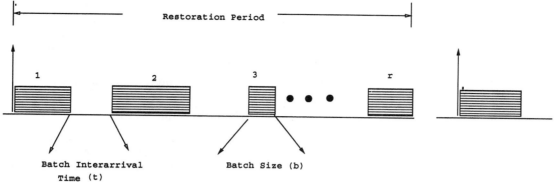

Fig. 5. Source arrival process in simulation model.

The queueing model of the system is shown in Fig. 4. Each processor is modeled as a server with its own associated queue. The input queue buffers the restoration messages which arrive from the network and distributes them randomly among the processors.

Processors $P_1, \cdots, P_N$ serve jobs from the queue in an FIFO (first-in-first-out) policy. The service cycle of each job consists of four phases in series: two computation phases interleaved with two memory access phases.

- Phase 1 consists of local computation for processing restoration messages.
- Phase 2 involves updating the cross-connect matrix in the shared memory.
- Phase 3 requires generating the new configuration map.
- Phase 4 involves transferring the new configuration map to processor P_0 in the CCM module.

The time for each phase, $s_i, i = 1, \cdots, 4$, is assumed to be constant.[9] The memory access (Phases 2 and 4) is performed while controlling access to the bus. The bus is assumed to have a centralized bus access arbitrator which enforces the FIFO service policy on the requests generated by the processors.

Processor P_0 performs the batch service with the batch size (b) and the service time (s_0) being fixed. In this study, the simulation program allows for a serial path cross-connection (i.e., $b = 1$) or parallel cross-connections of b paths. The service time (s_0) is the path cross-connect time in the cross-connect switching matrix for each path. In this study, the cross-connect times for each path are assumed to be 100 and 10 ms, which simulate present and future DCS cross-connect technologies, respectively.

The source arrival process is shown in Fig. 5. The time between two restoration periods is sufficiently long so that there are no interrestoration message conflicts. Each restoration period consists of a random number of bulk arrivals with a given distribution and mean r where the value of r is the number of affected STS-1's required to be restored at that DCS. Both the bulk size and the batch interarrival time are random variables with mean b and t, respectively. Note that the batch size of "b" represents the number of STS-1 paths that will be restored via one of k restoration paths, where $k \geq 1$.

[9] The times for Phases 1 and 3 are assumed to be 1 ms (see Fig. 8) or 5 ms (see Figs. 6 and 7) for each phase. The times for Phases 2 and 4 in our study are assumed to be 1 μs and 0.5 ms, respectively.

C. Network Simulation Results

As discussed in Section II, the considered algorithm provides inherent message bundling. Therefore, when combining with the parallel DCS architecture, it could meet the 2-s restoration objective. Table II shows the conditions (level of parallelism) under which the restoration algorithm can meet the 2-s restoration objective using a parallel DCS system architecture with the present cross-connect technology (i.e., 100 ms). As shown in Table II, the network cannot restore affected services completely within 2 s for any link failure, if the present DCS system architecture and cross-connect technology (i.e., 100 ms cross-connect time) remain unchanged. However, using the present DCS cross-connect technology (i.e., 100 ms cross-connect time per path), but adding up to 2 CPU-based controllers in the AM module and allowing the CCM module to cross-connect up to 14 paths simultaneously, would enable the DCS network to restore affected services completely within 2 s for all single link failure scenarios.

V. Design Impacts for DCS Restoration Enhancement Options

Section IV discussed the restoration time effects using several restoration time enhancement options. In this section, we discuss impacts of these options on the present DCS system design and the present LEC operations environment. Bundle and priority restoration options primarily impact present LEC operation systems (i.e., provisioning, administration, bandwidth management), while the DCS system enhancement option will primarily affect the DCS vendor community.

A. Bundle Restoration

The basic idea of bundle restoration is to restore a bundle of STS-1's that have the same path termination points with a single restoration message, instead of a restoration message for each individual STS-1 path. The purpose of bundle restoration is to reduce the number of restoration messages and paths needing to be processed within DCS's. A similar bundle restoration concept has been proposed in the ATM Virtual Path (VP) self-healing networks (i.e., VP group restoration) [20]. The basic idea of bundle restoration is to create an STS-Nc frame at the end nodes (for path restoration) to accommodate a maximum of N affected STS-1's. Bundle restoration can

TABLE II

TABLE II
CONDITIONS OF MEETING 2-S OBJECTIVE USING PARALLEL DCS

link #	aff. ch.	NRR(%)	NRT(ms) (cc=0)	NRT(ms) (s-s,100)*	Cross-connect time of 100 ms		
					#CPUs	"k"**	NRT(ms)
1	49	100	256	5463	1	6	1306
2	29	100	194	3214	1	6	902
3	30	100	359	3478	1	6	1078
4	28	100	200	3108	1	6	895
5	41	100	466	4796	1	6	1387
6	48	100	449	5540	1	6	1484
7	29	100	220	3240	1	6	928
8	54	100	428	5451	1	6	1574
9	47	100	483	5444	1	6	1505
10	45	100	463	5190	1	6	1470
11	39	100	488	4572	1	6	1387
12	59	100	220	6588	1	6	1467
13	77	100	516	8949	1	8	1848
14	64	100	792	7752	1	8	1918
15	46	100	282	5144	1	6	1298
16	48	100	419	5487	1	6	1463
17	82	100	686	9694	1	10	1970
18	81	100	633	9410	1	10	1898
19	108	100	845	13351	2	12	1995
20	146	100	580	18160	2	14	1940
21	34	100	198	3737	1	6	999
22	134	100	431	16255	2	12	1828
23	28	100	276	3184	1	6	771
24	88	100	740	10495	2	10	1832
25	53	100	488	4997	1	6	1624
26	69	100	860	8383	1	6	1432
27	71	100	1003	8767	2	10	1900
28	55	100	250	5180	1	6	1167

be achieved either through provisioning in the network engineering stage (called *external bundle restoration*) or by the restoration algorithm (called *inherent bundle restoration*).

For the service provisioning approach, recognizing which STS-1's should be bundled together during the network restoration process requires these STS-1's to be provisioned in an OS during the service provisioning phase. There are two possible impacts on the current provisioning system and the DCS system design due to the use of bundle restoration. First, in the provisioning system, information regarding which STS-1's should be bundled to which STS-Nc for each link failure scenario should be created during the service provisioning phase, and should be downloaded and stored in DCS's for appropriate cross-connection establishment, whenever needed. Thus, a database (table) is needed in each DCS to store this bundle restoration information, and this bundle restoration table should be continuously updated to ensure the correct bundling during the network restoration mode. The correct table update is important, since an STS-1 assigned to one user could be dynamically changed to another user due to the customer network management. Thus, keeping track of these dynamic STS-1 assignments to correctly update the bundle restoration table is a challenge for engineering and operations system personnel.

If bundle restoration is achieved through the restoration algorithm (i.e., inherent bundle restoration), no additional requirements are imposed on the DCS. However, it is also important to recognize that inherent bundling may require a somewhat larger restoration message size and longer message processing than that needed for a single channel message. The DCS must have a capability to recognize the new demand bundle and reconfigure its switching fabric allowing those bundled STS-Nc's to be processed correctly. This DCS feature does not exist in present DCS systems.

B. Priority Restoration

The concept of priority restoration is to first attempt to restore the STS-1 paths with higher priority within the 2-s objective. The other STS-1 paths with lower priority will be restored after these higher priority paths have been completely restored. This concept may be easy to implement in a vendor-specific subnetwork controller, but may be more complex in a large OS (like TIRKS) in the LEC environment. Thus, from the vendor point of view, the concept is workable. However, this concept may add a significant complexity to existing operations systems and would require bandwidth management capabilities in LEC networks. Further studies are needed to determine operations impacts for using the priority restoration concept.

C. DCS System Architecture Enhancement

This subsection discusses the delay performance for a single DCS system and the effects of parallel processing and cross-connections on the DCS's delay performance. The enhancement of the present DCS system for fast restoration (2-s objective) includes the parallel processing architecture within the AM module, and the parallel path cross-connect system or a fast reconfiguration system (i.e., 10 ms) within the CCM module. The fast DCS cross-connect system design relies on both the hardware and software improvements. In this subsection, we study several DCS system designs using various system parameters and their effects on the DCS restoration time. This system design effect analysis may identify the performance bottleneck of the DCS components that then may help design a cost-effective DCS system which supports the distributed restoration application. The results observed in this subsection are from the DCS resolution time simulation program described in Section IV-B. Note, in this subsection, that parallel processing is related to the processing capacity of multiple CPU's in the AM module, and parallel cross-connection (represented by the batch size) is related to the cross-connection capability in the CCM module.

Fig. 6 depicts the restoration time gain due to parallel processing and/or cross-connect effects in the DCS, using 100 ms for the DCS cross-connect time and 10 ms for new map computation (generation). As depicted in Fig. 6, with current technology, i.e., cross-connect time of 100 ms and the CPU processing time of 5 ms per computing phase (see Section IV-B), there are only small gains due to parallel processing. The cross-connect processor is the bottleneck of the system. As a result, increasing the batch service size in the cross-connect processor almost results in linear speedup for small batch service size. Fig. 6 also shows that the system may meet the 2-s objective if the batch size is greater than 6, which is consistent with results shown in Table II.

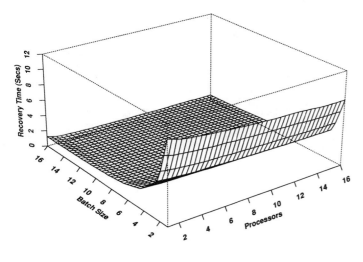

100 ms Cross-Connect Time;
New configuration map computing time (2 phases) = 10 ms

Fig. 6. Restoration time gain due to parallel processing/cross-connect (100 ms cross-connect time and 10 ms for path computation).

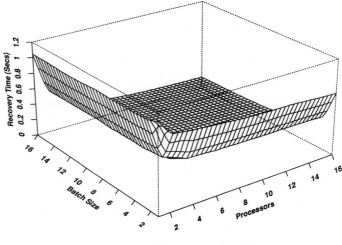

10 ms Cross-Connect Time;
New configuration map computing time (2 phases) = 10 ms

Fig. 7. Restoration time gain due to parallel processing/cross-connect (10 ms cross-connect time and 10 ms for path computation).

Fig. 7 depicts the restoration time gain due to parallel processing and/or cross-connect effects, using the next generation DCS reconfiguration technology of 10 ms switching time and 10 ms for new map computation (generation). As depicted in Fig. 7, the speedup achieved with increasing batch service size is restricted due to several factors. First, as the batch service size is increased, there is a point beyond which there will be no significant further gains. This is the point at which the processors have become the bottleneck. These observations are also evident from Fig. 6. To improve recovery time further, more processors have to be added or the processing time reduced. Second, the manner in which the cross-connect module services a batch is important. The results shown are for the case where the cross-connect (whenever it has finished servicing a previous batch) picks all the waiting messages in its queue up to the batch service size limit and processes them simultaneously. This batch of messages finishes the cross-connect processing at the same time that the cross-connect once again chooses another batch. If the average queue length at the cross-connect is lower than the batch service size limit, then further increasing the batch size does not improve the recovery time. However, this manner of cross-connect processing is not as efficient as the case where the cross-connect acts as a system of parallel processors (the number of processors equal to the batch service size limit) where a new message is chosen for processing as soon as a previous message is processed. For this latter case, if the cross-connect is the bottleneck, the recovery time for a given batch service size will be lower for the considered case than that for the case where the cross-connect takes the next batch of messages only after it has finished processing all messages of the previous batch. In summary, decreasing the cross-connect time to 10 ms may yield higher recovery time gains due to parallel processing or the larger batch service size.

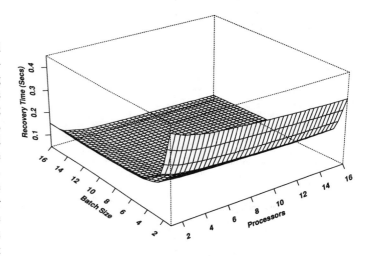

10 ms Cross-Connect Time;
New configuration map computing time (2 phases) = 2 ms

Fig. 8. Restoration time gain due to parallel processing/cross-connect (10 ms cross-connect time and 2 ms for path computation).

Fig. 8 depicts the restoration time gain due to parallel processing and/or cross-connect effects, using the next generation DCS reconfiguration technology of 10 ms and a next generation computing power of 2 ms for new map computation (generation). As depicted in Fig. 8, the bottleneck moves from the cross-connect to the processing phase resulting in higher speedup gains with larger batch service size.

VI. SUMMARY AND REMARKS

We have studied the viability of distributed control restoration schemes to meet a 2-s restoration objective in DCS mesh networks. Our study uses an LEC metropolitan-like LATA network and network simulators for different distributed control DCS restoration schemes. This study concludes that all currently proposed distributed control DCS self-healing schemes may not meet the 2-s restoration objective for metropolitan LEC LATA networks, as long as the present DCS system architecture (i.e., serial processing and serial cross-connection) and its switching hardware technology remain unchanged. In order to meet the 2-s restoration objective, two basic requirements have been identified. The first requirement is to design a DCS self-healing algorithm with a minimum set of restoration messages; and the second is to enhance the DCS performance by including a parallel CPU-based processing architecture and a parallel path cross-connection capability. Alternately, the 2-s service restoration objective may be met by implementing priority service restoration, or this objective may be relaxed based on hybrid network restoration architectures. These hybrid restoration architectures deploy fast restoration mechanisms (APS or rings) to meet specific customer needs on top of DCS mesh networks with distributed control that provide high survivability, enhanced protection against node failures, and the adequate restoration time for most other customers. Which approach should be used depends on the cost for these system enhancements, and on the revenue expected from services supported by the distributed control DCS network restoration system.

Acknowledgment

The authors would like to thank R. H. Cardwell, R. Clapp, J. Sosnosky, and J. E. Berthold for their valuable comments on this paper.

References

[1] T.-H. Wu, D. J. Kolar, and R. H. Cardwell, "Survivable network architectures for broadband fiber optic networks: Model and performance comparisons," *IEEE J. Lightwave Technol.*, vol. 6, pp. 1698–1709, Nov. 1988.

[2] T.-H. Wu and R. C. Lau, "A class of self-healing ring architectures for SONET network applications," in *Proc. IEEE GLOBECOM'90*, Dec. 1990.

[3] D. Doherty, W. Hutcheson, and K. Raychaudhuri, "High capacity digital network management and control," in *Proc. IEEE GLOBECOM'90*, San Diego, CA, Dec. 1990, pp. 301.3.1–301.3.5.

[4] J. Yamada and A. Inoue, "Intelligent path assignment control for network survivability and fairness," in *Proc. IEEE ICC'91*, Denver, CO, June 1991, pp. 22.3.1–22.3.5.

[5] W. D. Grover, B. D. Venables, M. H. MacGregor, and J. H. Sandham, "Development and performance assessment of a distributed asynchronous protocol for real-time network restoration," *IEEE J. Select. Areas Commun.*, vol. 9, pp. 112–125, Jan. 1991.

[6] C. H. Yang and S. Hasegawa, "FITNESS: A failure immunization technology for network service survivability," in *Conf. Rec. IEEE GLOBECOM'88*, Ft. Lauderdale, FL, Dec. 1988, pp. 1549–1554.

[7] B. A. Coan, M. P. Vecchi, and L. T. Wu, "A distributed protocol to improve the survivability of trunk networks," in *Proc. XIII Int. Switching Symp.*, May 1990, pp. 173–179.

[8] H. Komine, *et al.*, "A distributed restoration algorithm for multi-link and node failures of transport networks," in *Proc. IEEE GLOBECOM'90*, San Diego, CA, Dec. 1990, pp. 403.4.1–403.4.5.

[9] H. Sakauchi, *et al.*, "A self-healing network with an economical spare-channel assignment," in *Proc. IEEE GLOBECOM'90*, San Diego, CA, Dec. 1990, pp. 403.1.1–403.1.6.

[10] H. Fujii, T. Hara, and N. Yoshikai, "Characteristics of double search self-healing algorithm for SDH networks," IEICE CS91-48, 1991.

[11] Bellcore Special Rep., "The role of digital cross-connect systems in transport network survivability," SR-NWT-002514, issue 1, Jan. 1993.

[12] T.-H. Wu, *Fiber Network Service Survivability.* Norwood, MA: Artech House, May 1992.

[13] R. D. Doverspike, J. A. Morgan, and W. E. Leland, "Network design sensitivity studies for use of digital cross-connect systems in survivable architectures," *IEEE J. Select. Areas Commun.*, this issue, pp. 69–78.

[14] TA-NWT-000253, *Synchronous Opt. Networks (SONET) Fiber Opt. Transmission Syst. Require. Object.*, issue 6, Sept. 1990.

[15] Bellcore Technical Advisory, "SONET add-drop multiplex equipment (SONET ADM) generic criteria for a unidirectional, path protection switched, self-healing ring implementation," TR-TSY-000496, Sept. 1991.

[16] "SONET bidirectional line switched rings standard working document," T1X1.5/92-004, Feb. 3, 1992.

[17] "AT&T submits report on network reliability," *Commun. Week*, Mar. 23, 1992.

[18] J. Sosnosky, "Service applications for SONET DCS distributed restoration," *IEEE J. Select. Areas Commun.*, this issue, pp. 59–68.

[19] TA-TSY-000241, *Electron. Digital Signal Cross-Connect EDSX Syst. Generic Require. Object.*, issue 4, Bellcore, July 1989.

[20] K. Sato, H. Ueda, and N. Yoshikai, "The role of virtual path crossconnection," *IEEE Mag. Lightwave Telecommun. Syst.*, vol. 2, pp. 44–54, Aug. 1991.

Control Algorithms of SONET Integrated Self-Healing Networks

Satoshi Hasegawa, Yasuyo Okanoue, Takashi Egawa, and Hideki Sakauchi

Abstract—As the deployment of high-speed fiber transmission systems has been accelerated, they are widely recognized as a firm infrastructure of our information society. Under this circumstance, the importance of network survivability has been increasing rapidly in these days. In SONET, the self-healing networks have been highlighted as one of the most advanced mechanisms to realize SONET survivable networks. Several schemes have been proposed and studied actively due to a rapid progress on the development of highly intelligent NE's. Among them in this paper, a DCS based distributed self-healing network is discussed from a viewpoint of its control algorithms. Specifically, our self-healing algorithm called TRANS is explained in detail, which possesses such desirable features as providing fast and flexible restoration with line and path level restoration applied to an individual STS-1 channel, capability to handle multiple and even node failures, and so on. Both software simulation and hardware experiment verify that TRANS works properly in a real distributed environment, the result of which is shown in the paper. In addition, the combined use of TRANS and the ring restoration control is proposed taking into account the use in a practical SONET.

I. INTRODUCTION

WITH a rapidly growing interest in the enhanced network services, a future wideband network such as B-ISDN has been intensively studied. A synchronous optical network (SONET) [1], [10] is regarded as an infrastructure of a transport network which conveys the broadband services. In order to realize SONET in a cost effective manner, highly intelligent NE's (network element) such as DCS (digital cross-connect system) and ADM (add-drop multiplexer) have been developed due to the advent of digital and networking techniques. A transport network survivability by using the intelligent NE's is a baseline to ensure dependable services to users, since our business and even our daily activities have been heavily involved in networks.

Under this circumstance, several types of self-healing techniques to achieve a high degree of network survivability have been proposed and discussed [2]–[5], where the intelligence is distributed to each NE since a powerful CPU and a large amount of memory can be easily implemented in NE. In this

paper, the self-healing technique is defined as a distributed restoration scheme whose control is initiated and executed locally at each NE. The SONET ring architecture [5] and a DCS based self-healing algorithm [2]–[4] are good examples of the self-healing networks in SONET. The SONET ring architecture is currently being considered as one of the most promising schemes, which has been actively discussed and standardized in T1X1 committee. The ring architecture can perform fast restoration within 50 ms, where ADM should be a key NE. On the other hand, the DCS based self-healing algorithm is applied to a mesh-typed network, and dynamically and flexibly forms restoration routes in a distributed fashion. The DCS based self-healing algorithm would also provide relatively fast restoration within a few seconds. Moreover, the algorithm can handle various kinds of failures such as multiple and even node failures in a cost effective manner, since the spare bandwidth assigned to each link can be shared among a variety of failure scenarios. Each self-healing scheme has its own feature in terms of restoration time, cost, applied networks, a range of failures which can be handled, granularity of restoration unit, and so on. In addition, different user would require different survivability levels on the above mentioned items. Hence, a single network survivability technique cannot always give a satisfactory solution to users in SONET. As a result, the substantial importance on the SONET survivability is the capability to combine several types of the schemes in a coherent way to provide a satisfactory survivability level for users.

This paper is organized as follows. The next section briefly review the conventional self-healing techniques, where the techniques are largely classified into three categories. In Section III, a DCS based distributed self-healing algorithm is highlighted. Our proposed self-healing algorithm called TRANS [4], [9] is explained in detail. Both software simulation and hardware experiment by using T1 Multiplexer were performed to verify and evaluate the algorithm. In Section IV, the integrated self-healing network is proposed, where the ring architecture and the DCS based self-healing are combined to achieve the SONET network survivability covering both an interoffice and an access network. Section V gives conclusions.

II. CONVENTIONAL SELF-HEALING SCHEMES

Several restoration schemes based on distributed control have been proposed and studied, some of which have been already applied to SONET. The conventional self-healing techniques can be largely classified into three schemes, each of which is briefly reviewed in the following.

Manuscript received August 26, 1992; revised February 5, 1993. This paper was presented in part at GLOBECOM'90, San Diego, CA, December 3, 1990, and at NOMS'92, Memphis, TN, April 8, 1992.

S. Hasegawa is with C&C Product Technologies Development Laboratories, NEC Corporation, 1953 Shimonumabe, Nakahara-ku, Kawasaki 211, Japan.

Y. Okanoue is with the Overseas Transmission Division, NEC Corporation, 1953 Shimonumabe, Nakahara-ku, Kawasaki 211, Japan.

T. Egawa and K. Sakauchi are with C&C Systems Research Laboratories, NEC Corporation, 1753 Shimonumabe, Nakahara-kuy, Kawasaki 211, Japan.

IEEE Log Number 9212475.

Reprinted from *IEEE J. Selected Areas Communi.*, vol. 12, no. 1, pp. 110–119, Jan. 1994.

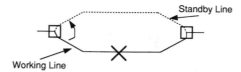

Fig. 1.   Route diversity.

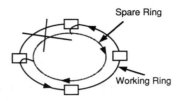

Fig. 2.   Ring architecture.

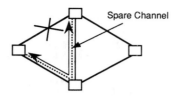

Fig. 3.   DCS based self-healing algorithm.

 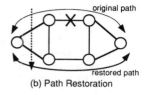

(a) Line Restoration   (b) Path Restoration

Fig. 4.   Line and path restoration.

A. Route Diversity

In the route diversity mechanism [11], a dedicated standby line is prepared for each working line, which is shown in Fig. 1. The transmission signals are bridged to both working and standby lines, and the receiving side switches to the standby line in case that the incoming signals from the working line are lost or degraded. The standby line is generally allocated on a different route from its corresponding working line, and hence, even if a disastrous single line failure such as fiber cut has occurred in a working line, the network can be restored. The route diversity can be regarded as the static restoration and is applied to a point-to-point transmission system, since a standby line is preallocated for each working line in the stage of network installation. This scheme can achieve an instantaneous restoration, although it requires a large amount of spare bandwidth and it is difficult to handle multiple failures.

B. Ring Architecture

The ring architecture is currently being considered as one of the most promising restoration technique in SONET. Several mechanisms such as the unidirectional path restoration, the bidirectional line restoration, two fiber/four fiber configuration, etc., have been actively proposed and discussed in T1X1 committee [6]. Literally, the ring architecture is applied to a ring-shaped network as shown in Fig. 2, and is considered as static restoration since restoration routes are pre-determined. The ring architecture provides fast restoration within 50 ms and can save spare bandwidth because the bandwidth can be shared with various failure scenarios. However, it cannot always handle simultaneous multiple failures. In order to expand an application network of the scheme, an interlocked ring architecture is being studied [7].

C. DCS-Based Distributed Self-Healing Algorithm

A DCS-based distributed self-healing [2]–[4] is applied to a mesh-typed network. The restoration routes are dynamically and flexibly formed around the failed line upon detection of a failure as shown in Fig. 3. A highly intelligent NE with a powerful CPU and memory such as DCS, can perform the

self-healing algorithm. This scheme can provide relatively fast and flexible restoration which would have an ability to handle even multiple failures and a node failure. Moreover, spare bandwidth in a network can be minimized since the bandwidth is shared with all the possible failure scenarios. A detailed mechanism of our self-healing algorithm is explained in the next section.

III. DCS-BASED DISTRIBUTED SELF-HEALING ALGORITHM

Among the restoration schemes, a DCS-based distributed self-healing algorithm provides the most flexible scheme. Several similar algorithms have been proposed, where restoration routes are dynamically formed by controlling each DCS in an autonomous and distributed fashion. The control is actually executed by exchanging messages between adjacent nodes. Specifically in SONET, the section or line overhead can provide the message channels. We have proposed our self-healing algorithm named TRANS (telecommunication restoration algorithms for network survivability). TRANS has such features as fast and flexible restoration, ability to restore multiple and even node failures, capability to handle either line or path restoration on an STS-1/STS-3c basis. In the line-level restoration, restored routes are formed around the failed line upon detection of line alarms such as LOS (loss of signal) and the line AIS (alarm indication signal). The line-level restoration has a possibility to achieve fast restoration and to enable revertive control in a simple way when the failure is repaired. On the other hand, the path-level restoration provides restoration routes between two path terminating equipment of the failed path, and has an advantage to restore the network efficiently since the restoration path length is generally shorter than that in the case of the line-level restoration. The detection of the path AIS should be a trigger for the path-level restoration. In addition, the recovery from a node failure can be achieved only by the path-level restoration. TRANS is applicable to both of them, which can be chosen according to the application or user's requirement on survivability. A schematic outline of the line- and the path-level restoration is depicted in Fig. 4.

A. Distributed Control Algorithms

TRANS is basically composed of three phases such as a broadcast phase, a selection phase and a confirmation phase. Each node has the same state machine algorithm to execute the phases to find a restoration route in a distributed fashion. In TRANS, several assumptions are made, which are listed below.

1) A mesh-typed network is assumed, where each *node* has a plural number of *links* connecting physically adjacent nodes. Each node shall be intelligent enough to perform TRANS. An example of the node is DCS.

2) Each node in a network has the same state machine algorithms to perform TRANS. The state transition of the algorithm is invoked by receiving alarm signals or the TRANS control messages.

3) Each node possesses merely local network configuration data regarding its adjacent nodes.

4) The message exchange channel should be prepared on each link. In SONET, section or line overhead could be used for the channel.

In the following, the detailed TRANS algorithm is explained by taking the line-level STS-1 channel restoration for instance. In the case of the path-level restoration, the basic mechanism is exactly the same as the line-level restoration except that a release phase is added to the above mentioned three phases. The release phase is used to maximize the usage of spare channels, where the available working channels on the failed path are released to reuse as spare channels.

1) Broadcast Phase: The purpose of this phase is to search all the possible candidates for restoration routes. In order to search restoration routes in a fast and an efficient way, the following methods are employed.

• *Help Message Propagation:* In the broadcast phase, help messages are traveling in a network in order to notify all the nodes of the failure occurrence. The help message is initiated at the *sender* and flooded throughout the network, and terminated at the *chooser.* The sender and chooser pair around the failed line should be predetermined in a provisioning stage. A simple rule can be applied to determine the pair, that is, a node with smaller address is a sender and the other node should be a chooser. An example of the sender and chooser pair is depicted in Fig. 5. In TRANS, a help message specifically contains such information as the hop count value (HCV) which indicates the distance from the sender, available spare channels (ASC) on the route, failed line ID (FLI), path trace information (PTI), etc.

• *Selective Flooding:* To avoid the generation of redundant help messages is essential for fast and efficient restoration. To this end, a selective flooding is employed in TRANS, where a loop condition is avoided by using PTI contained in the help message. Furthermore, the help message whose hop count value exceeds a pre-determined limit shall be discarded.

By referring to Fig. 5 as an example, the algorithm of the broadcast phase is explained. In this example, a line between nodes 1 and 5 has failed, which contains three STS-1 channels. A thick solid line which connects nodes stands for a working line and a dotted line is a spare channel in

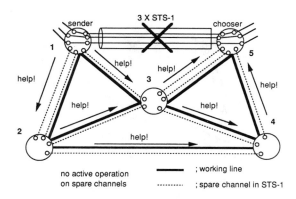

Fig. 5. Self-healing algorithm—broadcast phase.

STS-1. For instance, two STS-1 spare channels are prepared between nodes 1 and 2 in this figure. When a network failure such as fiber cut occurs, a sender and a chooser detect the failure, and the sender initiates broadcasting help messages to its all adjacent nodes, i.e., nodes 2 and 3. At the sender, the number of the lost STS-1's and the number of spare channels in each connecting line are compared. The smaller number of them is set in the ASC field of the help message which is sent to each line. An intermediate node when it receives the help message, stores the received help message and relays the message to its all adjacent nodes. The ASC field is also updated in each intermediate node to specify the available spare channel capacity in STS-1 on the route from the sender to the intermediate node. At this time, HCV is incremented by one. As was mentioned before, the intermediate node discards the message whose HCV exceeds the predetermined limit, so that the number of generated help messages can be extremely suppressed. In the broadcast phase, no active operation such as reservation is done on spare channels.

2) Selection Phase: The purpose of this phase is to select the specific restoration routes for each individual lost STS-1. The routes are selected among the possible candidates searched in the previous broadcast phase. Basically, the routes with the short length can be selected through this phase. In order to select the restoration routes in a fast and a concurrent way, the following method is employed.

• *Multiple Return Messages:* In TRANS, upon receipt of each help message, multiple return messages are simultaneously sent by the chooser to select restoration routes for each lost STS-1. The number of return messages issued should be the ASC value in the received help message, since the ASC value indicates the number of available spare channels in STS-1 on the route. By employing the multiple return message principle, the fast restoration route selection can be achieved, although some confliction to capture resources, i.e., spare channels, could be developed. The confliction can be resolved by employing the first-come first-served principle and the resource release rule as well when the confliction occurs.

Fig. 6 shows a schematic outline of the selection phase. Upon receipt of each help message, a chooser sends return messages back to the sender basically along the route where the corresponding help message was coming through. The total number of return messages issued by the chooser is

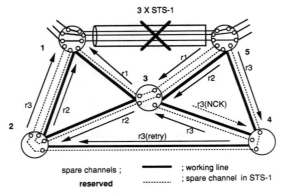

Fig. 6. Self-healing algorithm—selection phase.

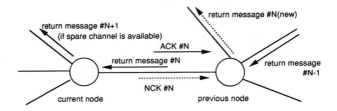

Fig. 7. Self-healing algorithm—ACK/NCK control.

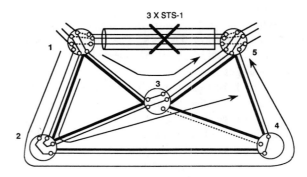

Fig. 8. Self-healing algorithm—confirmation phase.

limited to the number of lost STS-1 channels. In this example, three return messages are supposed to be sent corresponding to three help messages coming through nodes 1-3-5, 1-2-3-5, and 1-3-4-5. Again, since one return message is issued to restore one lost STS-1 channel, more than one return messages could be sent upon receipt of one help message if the ASC value of the message indicates more than one. In each intermediate node when it receives a return message, the minimum hop route from the node to the sender is selected by referring to the stored help messages. Since each stored help message contains HCV which specifies the hop count value from the sender to the current intermediate node, the node can easily find the next node on the route where the help message with the minimum HCV was coming through. If an available route is found, then the current intermediate node returns positive-acknowledgment (ACK) message back to the previous node and sends a return message to the next node on the selected route. At this time, the spare channel between the current intermediate node and the next node is reserved, and the spare channel reservation between the current intermediate node and the previous node is committed. If the node cannot find any available spare channels, that is, all the spare channels have already been reserved or committed for instance, then the negative-acknowledgement (NCK) message is returned back to the previous node and the spare channel reservation is cancelled. A schematic description of the ACK/NCK control is shown in Fig. 7. Another case of the NCK transmission is occurred when two return messages are crossing between two nodes. In order to avoid a duplicated reservation in the case of the message crossing, the node with a smaller address voluntarily returns the NCK message to the previous node. The detection of the message crossing can be easily made if the node receives a return message from the line in question before receiving the ACK message which commits the reservation. In the example of Fig. 6, when node 3 receives $r3$ and if the return messages $r1$ and $r2$ have already been processed, then the node cannot find any available restoration routes to the sender. Hence, NCK message is sent back to node 4. This is based on the first-come first-served principle. When the node 4 receives NCK message, the node reselects another route to the sender again by referring to the stored help messages. In case that NCK is unfortunately returned back to the chooser and the chooser

cannot find any available routes, the request is canceled. Hence, TRANS can achieve a deadlock-free algorithm even though the concurrent resource (spare channel) acquisition is done.

3) Confirmation Phase: Finally, when a return message is received by the sender, a confirmation phase is invoked by initiating a link message at the sender. A link message is also sent corresponding to each lost STS-1. Fig. 8 shows an outline of the confirmation phase. After cross-connecting the first lost STS-1 channel to the selected restoration channel at the sender, the sender sends a link message back to the chooser on the selected route and start transmitting the restored traffic. In this example, at first the route through nodes 1-3-5 can be formed. A prioritized selection can be applied in this phase, that is, the most important STS-1 channel can be restored with the highest priority by cross-connecting the channel and initiating the corresponding link message first. When each intermediate node receives a link message, the node confirms the spare channels, cross-connects the channel, start transmitting the restored traffic through the established cross-connection, and relays the link message to the next node on the selected route. Finally when the chooser receives a link message, it switches over the specified lost STS-1 to the restoration channel, and the restoration process for one lost STS-1 is completed. The above mentioned procedure is repeated for each lost STS-1 in a concurrent fashion whenever a return message is received by the sender. The example shows three restoration routes with nodes 1-3-5, 1-2-3-5, and 1-2-4-5 are formed.

B. TRANS Simulation

Both software simulation and hardware experiment were made in order to verify the TRANS algorithms and to evaluate its performance in terms of the restoration time and ratio.

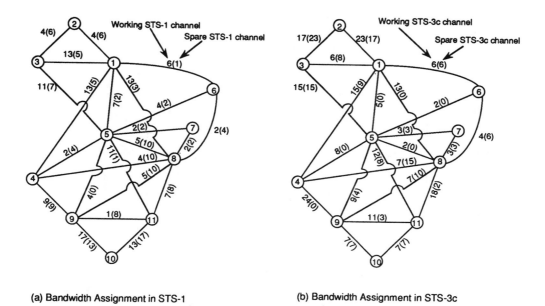

(a) Bandwidth Assignment in STS-1

(b) Bandwidth Assignment in STS-3c

Fig. 9. Sample network and design results.

1) Software Simulation and Evaluation:

• *Outline of Software Simulator:* A developed software simulation is basically composed of three parts, such as a network design, a TRANS algorithm execution, and a simulation output generation. In the network design part, a physical network topology (node and link arrangement) and distance information of any node pairs should be preassigned. Based on the provided physical network and the communication demand, the program automatically calculates both the working channel and the spare channel assignment on each link for STS-1 and STS-3c. A linear programming technique in conjunction with the maximum flow approach was taken to design a network [4]. An event-driven simulation program was developed to simulate TRANS, where the alarm signals or the TRANS control messages should be a trigger for the events. In the simulation, each node has a received message queue and a sending message queue for each connected link. In addition, another internal queue is provided to simulate a cross-connection action. The simulation is actually controlled by a *time-stamp* which is assigned to each message. The program picks up the message with the smallest time-stamp value from all the received message queues, processes the message, and outputs the message to the appropriate sending message queues. Then the message is reinput to the received message queues of the adjacent node. The time-stamp value is updated whenever the message is processed, where the independent incremental values can be set for each of the algorithm execution time, queueing delay, message transmission time, cross-connection time, etc. After the TRANS simulation is completed, the results are generated. Several statistics can be obtained through the program such as restoration time for each lost STS-1/STS-3c, restoration routes, the number of messages processed in each node, etc. The comparison between the optimal restoration and the TRANS restoration in terms of the restoration routes (the number of hops and the distance of the

restoration route) can be also obtained, so that the performance of the TRANS can be totally examined.

• *Results of Software Simulation:* In the software simulation, a sample network with a mixture of STS-1 and STS-3c paths is assumed, although the STS-3c level cross-connection is not currently supported in SONET. The parameters for the software simulation are listed in Table I. The message processing time is largely dependent on the processor and the operating system used in DCS. In this simulation, the processing time is assumed to be a random value whose possible range is between 1 ms and 9 ms as is shown in Table I. The candidate for the message exchange channel is either 192 Kbps SONET section DCC or a dedicated control channel in the SONET overhead field. In the simulation, we assume 64 Kbps message exchange channel with 50% duty, which could be achieved in a practical SONET environment. In the simulation, the map switch typed cross-connection is assumed, where 100 ms is given to the operation. The network design results for each of STS-1 and STS-3c are shown in Fig. 9(a) and (b), respectively, where communication demand for each of STS-1 and STS-3c is given in Table II. In Fig. 9, the working capacity on each link is designed such that the allocated capacity should be as uniformly distributed as possible, while each working route is determined with the minimum number of hops. The spare channels are optimally calculated and assigned to each link in the sense that the total number of the spare channels is the minimum. The designed spare channels theoretically ensure enough capacity

TABLE I
PARAMETERS OF SOFTWARE SIMULATION

Message Processing Time	$1 \sim 9$ ms (random value)
Message Exchange Channel	64 Kbps
Cross-Connection Time	100 ms
Message Length	20 bytes
Hop Limit	7

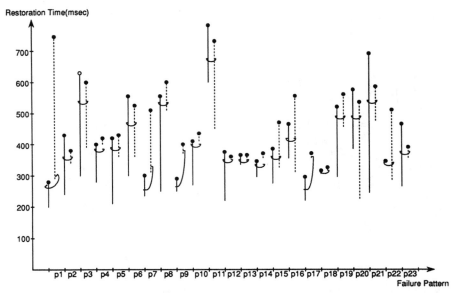

Fig. 10. Range of restoration time—a case of line restoration.

TABLE II
COMMUNICATION DEMAND

	Communication Demand
nodes 1–10	30 × STS-1
nodes 2–11	40 × STS-3c
nodes 3–8 20	× STS-1
nodes 4–9	30 × STS-3c

TABLE III
LOCATION OF FAILURE PATTERNS

Failure Pattern	Failure Position	Failure Pattern	Failure Position
P1	nodes 1–2	P13	nodes 5–7
P2	nodes 1–3	P14	nodes 5–8
P3	nodes 1–4	P15	nodes 5–9
P4	nodes 1–5	P16	nodes 5–11
P5	nodes 1–6	P17	nodes 6–8
P6	nodes 1–8	P18	nodes 7–8
P7	nodes 2–3	P19	nodes 8–9
P8	nodes 3–5	P20	nodes 8–11
P9	nodes 4–5	P21	nodes 9–10
P10	nodes 4–8	P22	nodes 9–11
P11	nodes 4–9	P23	nodes 10–11
P12	nodes 5–6		

for 100% network restoration against any single line failure. The simulation is done for all the possible single failure scenarios which affect services.

Fig. 10 depicts the range of restoration time for the line-level restoration. The horizontal axis shows the pattern of a failure, which is summarized in Table III. In the figure, a solid line and a dotted line show the range of restoration time for the affected STS-1 and STS-3c channels, respectively. For instance, the first STS-1 channel is restored in 240 ms and the last STS-1 channel is restored in 430 ms for the failure pattern 2(p2). The black circle at the top of the restoration time range indicates that 100% restoration can be achieved, while the white circle means that 100% restoration can not be achieved in the simulation. For the failure pattern 3(p3), 11 STS-1 channels out of 13 affected STS-1 channels are restored although there are enough spare channels for 100% restoration. This is because the spare channel is optimally allocated to each link, and hence, if an adequate route is not selected through the TRANS distributed algorithm, 100% restoration cannot always be achieved, which was discussed in [4]. This is regarded as the inevitable limitation of a distributed restoration method, whose execution is based on the local information. However in this example, only one case of the restoration performance degradation is observed out of 22 failure patterns. In a practical sense, TRANS can achieve enough restoration capability. Fig. 11 is the simulation results for the path-level restoration. Compared to the line-level restoration, the path-level restoration requires more restoration time, however,

100% restoration can be achieved for all the failure patterns. For example in this simulation, the worst restoration time for the path-level restoration takes about 2 seconds (for p1), while it takes 770 ms (for p11) in the case of the line-level restoration.

2) Hardware Experiment and Evaluation: In order to verify that TRANS works properly in a real distributed environment, a hardware experiment was performed by using the existing T1 Multiplexers (T1 MUX). Although the line speed of T1 MUX is basically 1.544 Mbps and much smaller than that of SONET DCS, the function of T1 MUX is similar to DCS, where TSI (time slot interchange) is implemented as a cross-connection. Hence, the TRANS algorithm can be verified from a functional point of view through the experiment. In our experiment, three T1 MUX's are connected in a triangle fashion as as shown in Fig. 12. However in a logical sense, six T1 MUX's form a mesh-typed network by a separated use of LIF's (line interface) of T1 MUX. Fig. 13 shows a logical network configuration corresponding to the physical triangle network. The TRANS programs are implemented in personal computers (PC) which reside outside T1 MUX's,

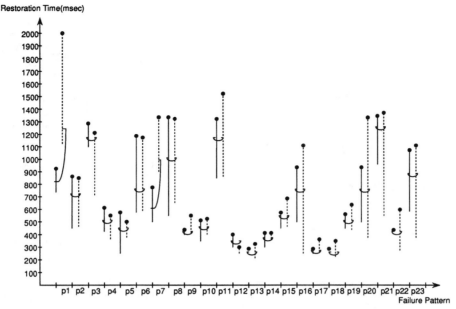

Fig. 11. Range of restoration time—a case of path restoration.

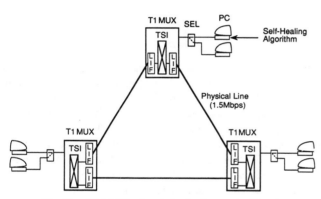

Fig. 12. T1 MUX network for hardware experiment.

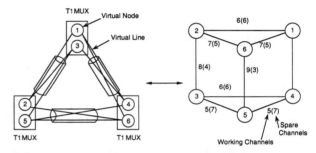

Fig. 13. Logical network configuration of T1 MUX network.

TABLE IV
PARAMETERS OF HARDWARE EXPERIMENTS

Message Length	15 bytes
Message Channel between T1 MUX's	9.6 Kbps (HDLC-ABM)
Processing Clock	8 MHz
Message Channel between PC and T1 MUX	9.6 Kbps (HDLC-ABM)

and the program controls TSI in T1 MUX through a control channel with an HDLC-ABM protocol between PC and T1 MUX. Since one T1 MUX is logically acted as two nodes, two PC's are connected to each T1 MUX as shown in Fig. 12. The experimental parameters are listed in Table IV. The message exchange channel between T1 MUX's also employs an HDLC-ABM data link layer protocol at the rate of 9.6 Kbps. Through the experiment, it was verified that both the TRANS line- and path-level restoration work properly in a real distributed environment. Also, the experiment verified that TRANS was able to work even in a multiple failure situation. In the experiment, we examined the recovery time for the telephone line by connecting two telephone instruments to different T1 MUX's and having the T1 line cut. Clearly in this experiment, since TRANS is applied to low speed transmission networks (1.544 Mbps networks) with 9.6 Kbps message exchange channel, the restoration time should be larger compared to the above mentioned software simulation which is applied to SONET. The result showed that it took about 12 seconds to recover the affected telephone line, where the TRANS execution time was only 1 second and another

11 seconds were consumed for TSI information loading time from PC to T1 MUX and TSI switching time.

IV. INTEGRATED SELF-HEALING NETWORK

In a practical SONET environment, several kinds of NE's must be accommodated in a network. It should be rare that only the intelligent NE's such as DCS, which can perform the TRANS algorithm, are placed in a network. Moreover, a different user would require a different survivability level. Hence, it should be difficult to give a total network survivability by employing a single restoration technique. The integrated use of several restoration schemes could give one of the solution to this problem. This section discusses on the integrated restoration control specifically considering the joint control of the DCS based self-healing and the ring architecture [8], [9].

197

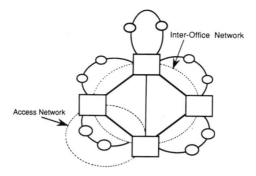

Fig. 14. Example of SONET logical network.

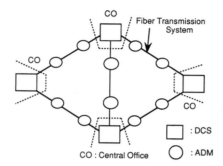

Fig. 15. Example of SONET physical network.

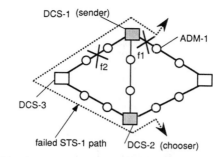

Fig. 16. Sample network for integrated self-healing control.

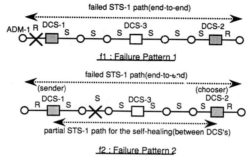

Fig. 17. Failed STS-1 path.

A. Network Configuration

In SONET, a transport network can be largely classified into two categories such as an interoffice network and an access network. A key NE for an interoffice network is DCS which can form a mesh-typed network. On the other hand, a ring-shaped network is one of the strongest candidate for an access network. ADM is considered to be a key NE of the ring-shaped access network. Other configurations are of course possible for SONET, however in this section, the above mentioned ring and mesh combination is assumed in the discussion on integrated restoration. That is, it is proposed that TRANS is applied to an interoffice meshtyped network and the SONET ring architecture is used for an access network. Fig. 14 shows such a network from a logical configuration point of view. Note here that the physical network (fiber network) would be installed as shown in Fig. 15 even if its logical counterpart is the network shown in Fig. 14. In Fig. 15, ADM's are placed between DCS's. In this kind of situation, the line AIS may not reach DCS even in the case of fiber cut, because the AIS signal is terminated at ADM's placed between DCS's. When the line AIS is terminated at ADM, the ADM generates the path AIS and propagates it to the downstream. As a result, DCS's could not detect the line AIS even if a line failure has occurred between the DCS's, but detect the path AIS. Hence, in a practical SONET environment, the line-level restoration is difficult to be initiated in TRANS because the line restoration is triggered by a detection of line alarms such as LOS or the line AIS. The TRANS path-level restoration seems to be preferable compared to the line-level restoration if the local algorithm initiation is performed. In order for the path-level restoration to work, DCS should at least possess the capability to detect the path AIS at the STS-1 level.

B. Integrated Self-Healing Control

Fig. 16 shows a sample network which is composed of both DCS's and ADM's. In the figure, a failed STS-1 path is depicted as a dotted line, and a circle and a rectangle stand for ADM and DCS, respectively. Two failure patterns are included in this figure, such as $f1$ and $f2$. The failed STS-1 path with passing-through nodes is shown for each of the failure patterns in Fig. 17. Clearly, the original TRANS cannot be applied to restore the STS-1 path in this example since the path is not terminated at DCS's but terminated at ADM's. In order to apply TRANS for this kind of STS-1 path, the *partial STS-1 path* is newly introduced, which is defined as the longest DCS to DCS portion of the original STS-1 path.

In Fig. 17, the partial STS-1 path for the corresponding original STS-1 path is defined between DCS-1 and DCS-2 passing through DCS-3. By introduction of the partial STS-1 path, even if the STS-1 path in question is not terminated at DCS's, TRANS can be applied for the partial STS-1 path. However, TRANS cannot restore the failed STS-1 path if a failure has occurred outside of the partial STS-1 path. In this situation, the ring architecture is applied to restore the network. Hence in the proposed integrated control, either TRANS or the ring architecture is selected according to the location of the failure position. For instance, when a failure has occurred at the section with "R" of the STS-1 path in Fig. 17, the failure is restored by using the ring architecture. On the other hand, if a failure has occurred within the partial STS-1 path denoted as the "S" section, the TRANS path-level restoration is used. In the figure, the sections within the partial STS-1 path as denoted as "S", and "R" is put to other sections. In order to enable the integrated control, some information must be provisioned. One information is a sender and chooser pair for each STS-1 path. The sender and chooser pair information should be provided to the DCS's terminating the corresponding partial STS-1 path,

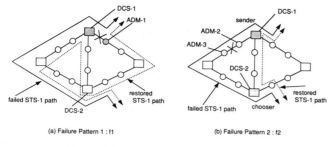

Fig. 18. Restoration examples of integrated self-healing control.

which is used for the TRANS path-level restoration. Another information is a ring initiation flag provisioned to the NE's which should reside outside or the boundary of the partial STS-1 path. The ring architecture is activated at each node with the flag when alarm signals are received. The integrated control mechanisms are summarized as follows.

Control Mechanism:

1) TRANS for the partial STS-1 path is activated in the terminating DCS's, if one of the path AIS, the line AIS or LOS, which is generated within the partial STS-1 path, is received.

2) The ring architecture for each STS-1 path is activated in NE's where a ring initiation flag has been provisioned, if LOS or the line AIS is received.

Fig. 18 depicts examples of the integrated self-healing control for two different failure patterns. In the figure, a failed STS-1 path is shown as a solid line. In the case of the failure pattern 1, the ring architecture is activated in DCS-1 and ADM-1, because the failure position is outside of the partial STS-1 path and a ring initiation flag must be provisioned to both DCS-1 and ADM-1 for the STS-1 path. As a result, the failed segment of the STS-1 path is detoured along another part of the ring, while the route from DCS-1 to DCS-2 for the restored STS-1 path uses a part of the original STS-1 path. In the case of the failure pattern 2, since the failure position resides within the partial STS-1 path, TRANS is activated in DCS-1 as a sender and DCS-2 as a chooser when the DCS's detect the path AIS for the STS-1 path. In this case the ring architecture is not activated in ADM-2 and ADM-3 for instance even if they detect LOS or the line AIS, since a ring initiation flag has not been provisioned to them for the STS-1 path.

V. Conclusion

This paper discusses the self-healing techniques applied to SONET. We have proposed our DCS-based self-healing network called TRANS which is a fully distributed algorithm. Through a software simulation, we could ensure that TRANS has a possibility to achieve network restoration against any single line failure within a few seconds. For further study, it should be important to obtain results with more accurate and realistic assumptions. From a hardware experiment using T1 MUX, it was verified that TRANS works functionally in a real distributed environment. The integrated self-healing control

with TRANS and the ring architecture is also proposed to provide the restoration mechanisms both for interoffice and an access network. The distributed control is an advanced technology in a transmission system mainly due to a rapid progress on highly intelligent NE's. With more advanced intelligence embedded in NE's, the distributed control will be naturally applied not only to the network restoration but also to managing dynamically reconfigurable networks with real-time bandwidth control, route selection, etc., which will create new transmission services to users.

Acknowledgment

The authors wish to thank Dr. Hirosaki for his helpful comments and encouragement. Special thanks are given to Ms. Kawakubo for her effort in drawing figures.

References

[1] *American National Standard for Telecommunications—Digital Hierarchy Optical Interfaces Rates and Formats Specifications (SONET),* ANS1 T1.105-1990, 1990.
[2] W. D. Grover, "The selfhealing network: A fast distributed restoration technique for networks using digital cross-connect machines," in *Proc. Globecom'87,* Nov. 1987.
[3] C. Han Yang and S. Hasegawa, "FITNESS: Failure immunization technology for network service survivability, in *Proc. GLOBECOM'88,* Nov. 1988.
[4] H. Sakauchi, Y. Nishimura, and S. Hasegawa, "A self-healing network with an economical spare-channel assignment," in *Proc. GLOBECOM'90,* Dec. 1990.
[5] T. H. Wu, "High-speed self-healing ring architecture for future interoffice networks," in *Proc. GLOBECOM'89,* Dec. 1989.
[6] J. Baroni *et al.,* "SONET line protection switched ring APS protocol," Rep. T1X1.5/91-026, Feb. 1991.
[7] W. Kremer, "Ring interworking with a bidirectional ring," Rep. T1X1.5/91-043, Apr. 1991.
[8] Y. Okanoue, H. Sakauchi, and S. Hasegawa, "Design and control issues of integrated self-healing networks," in *Proc. GLOBECOM'91,* Dec. 1991.
[9] S. Hasegawa, O. Tabata, Y. Okanoue, and H. Sakauchi, "Integrated self-healing network for STS-1/STS-3c path level restoration," in *Proc. NOMS'92,* Apr. 1992.
[10] R. Ballart and Y-C. Ching, "SONET: Now it's the standard optical network," *IEEE Commun. Mag.,* vol. 27, no. 3, Mar. 1989.
[11] D. J. Kolar and T. H. Wu, "A study on survivability versus cost for several fiber network architecture," in *Proc. ICC'88,* June 1988.

Survivable SONET Networks—Design Methodology

Ondria J. Wasem, *Member, IEEE,* Tsong-Ho Wu, and Richard H. Cardwell

Abstract— A prototype software system that implements a methodology for the strategic planning of survivable interoffice networks is presented. The software system determines strategic locations and ring types for Synchronous Optical Network ring placement. Two types of survivable network architectures are considered—1 : 1 diverse protection and SONET self-healing rings. The software considers three types of SONET self-healing rings—unidirectional, 2-fiber bidirectional, and 4-fiber bidirectional. Hubbing is assumed in all architectures. Inputs include nodes, links, connectivity, facility hierarchy, and multiyear point-to-point demands, together with the costs of fiber material and splicing, route mileage (installation), and multiplexors and regenerators of different rates. The outputs are a set of near-optimal rings based on cost, specifying the ring types and rates, fiber span sizes and counts, regenerator locations and speeds, the topology (set of links to be used), and the network cost. In addition, the software outputs the time in the planning period that each ring and fiber span should be installed.

I. INTRODUCTION

THE major issue in designing survivable Synchronous Optical Network (SONET) physical networks is how to best utilize the unique characteristics of different architectures to meet different demand requirements in a cost-effective manner. In general, when compared to 1 : 1 Diverse Protection (1 : 1/DP) systems, the self-healing ring (SHR) architecture may be more economical because it shares facilities and equipment and has the potential to reduce the number of regenerators required. In particular, the SHR may offer a very cost-effective solution to applications requiring dual-homing protection [1]. A primary task in SONET physical network design is determining how to best utilize the merits of alternative survivable architectures (e.g., ring, point-to-point/diverse protection, hubbing/diverse protection, hubbing/ring) to minimize the cost of survivable network evolution.

The design model in this paper serves as a starting point for developing more powerful SONET network design systems which use 1 : 1/DP with automatic protection switching (APS) and SHR's. Section II describes the overall structure of a prototype software system for survivable SONET network design [2], and then Section III discusses each of the design modules that make up the design system and mentions some alternate approaches. Section IV contains case study results for designing an actual LATA (Local Access Transport Area) network with our methodology. Finally, Section V contains a summary and conclusions.

Manuscript received July 10, 1992; revised February 15, 1993. This paper was presented in part at ICC'92, June 1992.

The authors are with Bellcore, NVC 3X-317, 331 Newman Springs Rd., Red Bank, NJ 07701-7040.

IEEE Log Number 9212484.

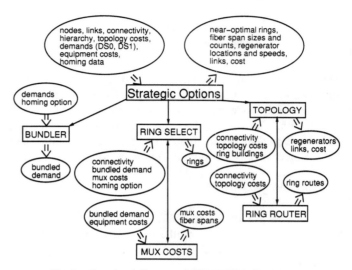

Fig. 1. Functional diagram of STRATEGIC OPTIONS.

II. SOFTWARE OVERVIEW

The purpose of our research prototype software is to test and demonstrate our methods for determining strategic locations and ring types for SONET ring placement in interoffice networks. It is not a commercial package. For ease of reference, we term our research prototype software "STRATEGIC OPTIONS." STRATEGIC OPTIONS chooses a set of SONET SHR's based on a cost comparison with 1 : 1/DP. This section discusses its basic functionality.

Fig. 1 shows the structure of STRATEGIC OPTIONS. STRATEGIC OPTIONS considers two types of survivable network architectures—1 : 1 Diverse Protection (1 : 1/DP) and SONET SHR's. The 1 : 1/DP architecture provides each central office (CO) with both a working fiber span and a diversely routed protection fiber span to its hub. The term "hub" denotes a location where a DCS grooms DS1-level demands into STS-1's and where STS-1 demands are interconnected between rings or 1 : 1/DP systems. STRATEGIC OPTIONS considers three types of SONET SHR's—unidirectional (USHR/P), 2-fiber bidirectional (BSHR/2), and 4-fiber bidirectional (BSHR/4) [3]. It uses a multiyear cost model, and considers the time value of money to decide when to build particular fiber spans and rings.

The top of Fig. 1 shows the input and output of the STRATEGIC OPTIONS software system, while the bottom shows the input and output of individual modules. The user must input the nodes, links, connectivity, facility hierarchy (i.e., hubs and their clusters), topology costs, and multiyear point-to-point demands of the network. The topology costs include fiber (material and splicing), route mileage (installa-

Reprinted from *IEEE J. Selected Areas Communi.*, vol. 12, no. 1, pp. 205–212, Jan. 1994.

200

tion), regenerator cost, and regenerator threshold (maximum regenerator spacing). (Note that we do not include structure cost. We assume that conduit exists.) The demands should be in DSO's (circuits) or DS1's (24 DS0's). The user may optionally input dual-homing data, equipment costs, interest rate of money, and type(s) of ring to be considered. The dual-homing data indicate a foreign hub for each CO to be dual-homed. When no dual-homing data are input, single-homing is assumed. The equipment costs indicate the costs of multiplexors and regenerators of different rates, as well as costs corresponding to the different types of rings. The user may also specify different regenerator thresholds for different signal rates.

The STRATEGIC OPTIONS software assumes that two-connected CO's that are not on SHR's will have 1 : 1/DP to their hubs. Output includes a set of rings, fiber span sizes and counts, regenerator locations and speeds, the topology (set of links to be used), and the network cost. The type (USHR/P, BSHR/4, and BSHR/2) and size of each ring is also output. In addition, the time in the planning period that each ring and fiber span should be installed is output.

STRATEGIC OPTIONS has three main software modules: the bundler, the ring selector, and the topology module. Every module uses the facility hierarchy information; therefore, hierarchy is not listed as input to the modules to reduce the complexity of the figure. The inputs and outputs of individual modules are displayed in Fig. 1 and described in the subsection of Section III corresponding to that module.

III. DESIGN METHODOLOGY

This section describes the methodologies in STRATEGIC OPTIONS module by module. As mentioned above, methodologies described may not be what is actually implemented in the software, although the design philosophy is the same.

A. The Multiplex Costs Module

The survivable SONET network architecture selection problem can be stated as follows.

Given: 1) A set of buildings (or CO's) that potentially can be connected by a ring.

2) A list of link distances on the ring.

3) Candidate survivable network architectures:
- SHR's, hubbing with 1 : 1/DP systems.

4) End-to-end bundled multiyear demands for the considered network (Subsection B).

5) Cost associated with each network component.

6) A list of candidate SONET line rates.

Objective: Determine an appropriate set of survivable network architectures and associated capacities for a predetermined planning period such that the total network present worth cost for the considered area is minimized.

Subject to: End-to-end multiyear demand requirements.

The candidate survivable network architectures considered here include USHR/P, BSHR/2, BSHR/4, and 1 : 1/DP. The network cost in the model is the present worth cost,[1]

[1] The present worth cost takes into account the interest or discount rate in that planning period.

which includes facility and equipment costs, including the working and protection terminal cost, the APS cost, the fiber material cost, and the regenerator cost. The demand requirement considered here is the STS-1 (51 Mb/s) demand requirement. The goal of the multiplex cost model is to evaluate the best combination of available architectures and associated capacities that can be deployed in a cost-effective manner over a planning period.

The mixed use of 1 : 1/DP and SHR architectures carrying the demands is based on the following growth strategy.

Network Growth Assumptions: 1) If 1 : 1/DP is used as a start-up architecture, it is used throughout the entire planning period.

2) If the SHR architecture is used as the start-up architecture, it is used until its capacity is exhausted. The remaining demand is carried by either another SHR or 1 : 1/DP until further capacity is exhausted or the end of the planning period is reached.

Assumption 1 is based on an observation that capacity exhaust does not significantly impact the 1 : 1/DP architecture. As for Assumption 2, if the SHR capacity is enough to carry the growth demand, then it is most economical to let the SHR accommodate these growth demands.

The model is based on the following assumptions.

1) The topology allows the SHR to be built.

2) Incremental demand on any fiber span in any year is not greater than the maximum line rate considered in the 1 : 1/DP option.

3) Similarly, the incremental demand on any SHR in any year is not greater than the maximum line rate considered in the SHR option.

4) Fibers and equipment that are already installed are not rearranged.

The algorithm for solving the capacity expansion problem for a combination of SHR's and 1 : 1/DP architectures, denoted by CapExp R&D, is similar to the capacity expansion algorithm for the 1 : 1/DP architecture alone [4]. At each period, a maximum of $n + 1$ options is considered—1 : 1/DP, and SHR's corresponding to the n line rates. Note that for 1 : 1/DP, the capacity expansion layout can be obtained by the procedure described in [4]. The capacity expansion model for neworks using a combination of 1 : 1 /DP and SHR options is summarized below and corresponds to the growth model depicted in Fig. 2. A dynamic programming approach is used here to find an optimum solution for the architecture selection and capacity expansion problem.

Fig. 3 depicts a decision tree created by using Algorithm CapExp R&D; the detailed algorithm and analysis can be found in [4]. In this example, a 4-year planning period is assumed, and the cumulative demands in each of the planning periods (years 1–4) are 40 STS-1's, 56 STS-1's, 70 STS-1's, and 98 STS-1's. Three candidate OC-N line rates are also assumed: OC-12, OC-24, and OC-48.

First, the algorithm creates a root and then tries the 1 : 1/DP option and the three line rate options for SHR's. The capacity is assumed to be placed at the beginning of the period. If the algorithm starts from the 1 : 1/DP option, it uses that option to carry demands to the last planning year. For each

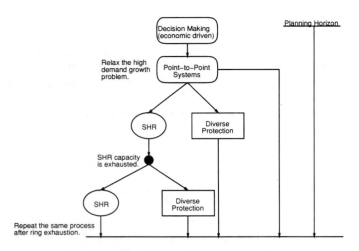

Fig. 2. Multiperiod network growth assumptions.

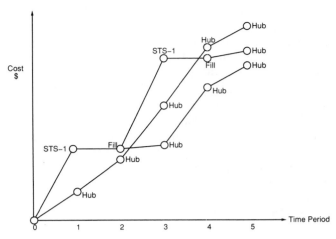

Fig. 4. An example of a multiperiod demand bundling algorithm.

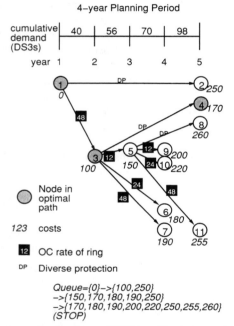

Fig. 3. An example of using a SONET multiperiod capacity expansion algorithm.

SHR option, the system carries the demand to the period in which the capacity is exhausted. For example, the OC-48 ring capacity will be exhausted after the second year. Thus, a child node is created from the root for the OC-48 ring at the second year. The algorithm computes the cost for each line rate option for SHR's or for the 1 : 1/DP option and inserts the costs into a queue. The node cost depends on the total demand on the SHR, the demand routing algorithm (demand is routed around the ring differently for bidirectional and unidirectional rings), and the cost model used. The costs in the queue are then sorted in increasing order, so that the node with the lowest cost in the queue is the next one selected for processing. The program stops when the minimum-cost node is at the end of the planning period. In the example, the program stops when a node with a cost of 170 is found because 170 is the minimum

cost in the queue and the node is in the fifth year. The best solution for the example depicted in Fig. 3 is as follows. An OC-48 SHR is installed to carry demand to the end of the first year, and a set of 1 : 1/DP spans is added in the beginning of the second year to carry demand to the end of the planning period.

The computing efficiency of Algorithm CapExp R&D has been reported in [4]. In most practical applications, Algorithm CapExp R&D can generate the optimum solution in less than 5 s using a VAX 6420 computer.

B. Multiperiod Demand Bundle Algorithm

The multiperiod demand bundle algorithm bundles multiperiod end-to-end DS1 (or VT1.5) demand requirements to multiperiod STS-1 demand requirements in a cost-effective manner. The bundler takes into account any input dual-homing data to route the demand appropriately. This algorithm uses an extension of a single-period demand bundle algorithm discussed in [5]. The criterion used for determining a direct STS-1 (i.e., formed without grooming by a DCS) or indirect STS-1's (i.e., using a SONET wideband DCS) may be either the minimum cost or an STS-1 threshold.

Fig. 4 depicts a conceptual example that shows how this multiperiod demand bundle algorithm works. In Fig. 4, nodes labeled "STS-1" represent direct STS-1 paths, and nodes labeled "Hub" represent indirect STS-1's using DCS's. The cost model is used in this example as the criterion for determining direct or indirect STS-1's. First, the algorithm creates two nodes with two possible options: using and not using a DCS. If the algorithm starts from a node using the DCS, it uses the same process to groom VT1.5 demands over the entire planning period. If the algorithm starts from a node with a direct STS-1, it continues to fill this STS-1 until the STS-1 capacity is full. Then it tries two options (i.e., with and without using the DCS) and repeats the same process until the end of the planning period is reached. Each node has an associated cost, which can be calculated using an algorithm similar to the one described in [5].

The decision tree creating process is similar to the one for architecture selection that has been discussed in Subsection A.

The key idea for building this decision tree is to store the cost of each node being created in a queue that is then sorted in increasing order; the candidate node for the next move is the node at the head of the queue (i.e., minimum-cost node). The algorithm stops when the node considered for the next move is at the end of the planning period. For example, in Fig. 4, the solution is as follows. For the considered demand pair, a direct STS-1 path is formed to carry demands from the beginning of the planning period to the end of the second year, and the DCS is used for grooming demands from the beginning of the third year to the last year.

C. Ring Selection Algorithm

The design problem for ring selection is locating low cost SHR's in the network while considering network topology, network hierarchy, STS-1 demand requirements, the SHR cost model, and the 1 : 1/DP cost model. The network starts as a facility network with a fiber-hubbed architecture using 1 : 1/DP. A group of CO's served by a single hub is called a cluster. Hubs may be fully interconnected or may be grouped to a higher hierarchy. We assume that economies-of-scale considerations dictate that each building will have only one piece of terminal electronics (add-drop multiplexors (ADM's) in the case of rings, or optical line terminating mmultiplexors (OLTM's) in the case of 1 : 1/DP networks). More than one OLTM or ADM is used in situations where the demand level requires higher speed terminal equipment. In that case, the rings or fiber spans used by the CO are overlaid. With this approach, SHR's are utilized to share bandwidth among several CO's and decrease cost from the cost of the 1 : 1/DP network. The ring selection algorithm uses the following simple heuristic to locate potential CO's for ring deployment.

Ring Selection Algorithm: 1) Choose any cluster as a starting point.

2) Find all cycles containing that hub. If dual-homing is used, the cycle must contain the foreign hub of any dual-homing buildings in the cycle considered.

3) For each combination of CO's on each cycle, calculate costs (as in Subsection A) and save if the SHR costs less than 1 : 1/DP.

4) Build the ring that saves the most over 1 : 1/DP.

5) Remove all rings that are now illegal.

6) Repeat steps 4 and 5 until there are no more rings for that cluster.

7) Pick another cluster if there is one, and go to step 2. Otherwise, stop.

Fig. 5 depicts an example of the above ring selection algorithm. This example includes six nodes, with Node 1 serving as the hub (or ring interconnection point). For this example, possible cycles are {12345, 1235}. In Cycle 12345, possible rings are 123, 124, 125, 134, 135, 145, 1234, 1235, 1245, 1345, and 12345. The algorithm does not consider two-node SHR's, under the assumption that it is less expensive to use OLTM's for 1 : 1/DP than ADM's for an SHR when only two nodes are involved. For each possible SHR, the network costs for the SHR and the 1 : 1/DP system are calculated based on a multiperiod multiplex cost model discussed in Subsection

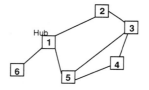

Possible cycles: 12345, 1235

In cycle 12345, possible rings are: 123, 124, 125, 134, 135, 145, 1234, 1235, 1245, 1345, 12345

If ring 145 is chosen, then rings 124, 125, 134, 135, 145, 1234, 1235, 1245, 1345, and 12345 must be removed. Ring 123 remains for further consideration.

Fig. 5.   An example of ring selection.

A. If the SHR cost is less than the 1 : 1/DP cost, that ring is saved for further consideration; otherwise that ring is deleted from the list of candidates. After selecting all SHR's that cost less than 1 : 1/DP for a particular hub, these candidate SHR's are compared, and the SHR that saves the most over 1 : 1/DP within that cycle is chosen. In this example, if SHR 145 is chosen, then all SHR's that include Nodes 4 or 5 must be removed from the candidate list, since a node that is not a hub can be on only one SHR. In this case, SHR's 124, 125, 134, 135, 145, 1234, 1235, 1245, and 12345 must be removed. SHR 123 remains and is retained for further consideration. Selecting final candidate SHR's is based on the comparison between the SHR cost and 1 : 1/DP.

Note that the ring selection algorithm is an exhaustive search within each cluster (or pair of clusters for dual-homing). Thus, this process may become a bottleneck, in terms of computing time, when the network becomes large (e.g., 200 nodes or more). Any computing improvement on this process will certainly speed up the entire process for obtaining the solution. For this purpose, STRATEGIC OPTIONS implements an optional filter on the building's combinations, which restricts consideration to SHR's in which the fill will be high for one of the input OC ring rates. The user chooses whether or not to use this filter.

D. Ring Fiber Routing

The next step in network design is topology design, and that step requires finding the best fiber path for previously determined SHR's. This problem is referred to as the "ring fiber routing problem." Formally speaking, this problem is to route fiber around a ring in a network, when the network nodes, links, connectivity, and offices to be placed on that ring together are known [6].

This subsection shows how the ring routing algorithm, called BUILD_RINGS, works by studying an example, illustrated in Fig. 6. Details of the algorithm can be found in [7]. Fig. 6(a) shows a network with seven offices; offices 1 and 5 are hubs. The example problem is to route a ring through offices 1–4.

First, BUILD_RINGS finds a hub on the ring and the ring office furthest from it, and then uses two iterations of Dijkstra's shortest path algorithm to find the two shortest, link disjoint paths (paths sharing no links) between the hub and the office.

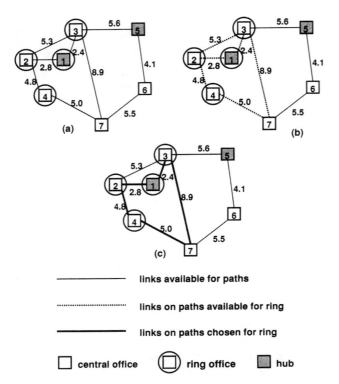

links available for paths

.... links on paths available for ring

links on paths chosen for ring

☐ central office ◻️ ring office ■ hub

Fig. 6.   An example illustrating the ring routing algorithm.

If the two paths are node disjoint (share no nodes), and the cycle formed by them contains all of the ring offices, then the ring is routed that way.

If the paths are not disjoint, or the cycle does not contain all of the ring offices, then BUILD_RINGS performs a depth-first search for paths between all pairs of ring offices, and containing only two ring offices. A threshold upper-bounds the lengths of paths that will be considered in constructing the ring. In the beginning, this threshold is low; however, the threshold increases if necessary. There is also a limit on the number of hops (links) in a path. The default value of the hop limit is three, but the user can override this. In Fig. 6(b), links corresponding to the paths with less than three hops are shown as dotted lines.

After finding paths, BUILD_RINGS tries combinations of node disjoint paths to construct a ring. If, after trying every combination, there is no ring, the distance threshold increases and the algorithm finds more paths. Rings long enough to require two regenerators between a pair of ring offices are considered uneconomical. Therefore, if the threshold grows large enough to allow paths that require two regenerators, BUILD_RINGS stops without routing a ring. The user sets the regenerator threshold. The algorithm stops after the first ring is found. Fig. 6(c) shows the computed ring routing in bold lines.

The algorithm was programmed in C, and run on s SPARC-station. Computation times on 47 examples of feasible and infeasible rings were reasonable. Overall, the average, minimum, and maximum run-times were 0.41, 0.06, and 2.93 s, respectively [6]. The largest example network used in these results (167 offices and 240 links) is the size of a typical large intraLATA network.

E. Topology Module

The last module of the network design model is the topology module, which chooses an economical routing of the fibers for the SHR's and 1 : 1/DP systems. The following descriptions of the methodologies and algorithms to generate and improve the topology of a survivable fiber network are taken from [8]. The topology cost includes route mileage (installation), fiber (material and splicing), and regenerator costs. Multiplexing costs are not considered in topology design because they do not change based on how fibers are routed; they depend on demand between CO's and fiber system rates.

Before beginning to design a topology, it must be determined which offices are to be protected (input to the model) and which of these protected offices ought to be placed on SHR's together (determined by the ring selection algorithm, see Subsection C). After these decisions are made, there are three steps in designing a topology.

The first step is to determine whether the SHR's are topologically feasible, given the conduits available for running the fiber. If the SHR's are feasible, the routings (paths around the rings) must be determined because it may be necessary to route fiber through offices that are not on the SHR's. (Subsection D discussed the ring fiber routing algorithm). The second step is to determine a two-connected topology that includes both the SHR's and all other protected offices. A two-connected topology provides for implementing SHR's and 1 : 1/DP by ensuring that each protected CO has two disjoint paths to its hub. The final step is to improve the topology based on the costs mentioned above.

The topology module includes four algorithms [8]. BUILD_RINGS for routing rings [6]. GREEDY_EARS for determining an initial two-connected network, and ADD_CHORDS and 1_OPT for improving a topology. Because Subsection D discussed BUILD_RINGS, we only discuss GREEDY_EARS, ADD_CHORDS, and 1_OPT, as follows.

1) Initial Topology: GREEDY_EARS, the algorithm used to compute a two-connected topology, is identical to the algorithm used for the same purpose in designing survivable fiber-hubbed networks [5]. Its purpose is to compute an initial topology that has a two-connected subnetwork containing all protected offices, in order to ensure some level of survivability by providing for diverse working and protection spans for protected offices. Unprotected offices will be contained in the resultant topology, but not necessarily in the two-connected

GREEDY_EARS begins by finding a cycle in the network components that contains at least one protected office. It then chooses a protected office not on that cycle, and finds a path ("ear") from an office contained in the initial cycle, through the chosen protected office, to an office contained in the initial cycle. The cycle and the ear together comprise the current solution. The algorithm continues by choosing another protected office not contained in the current solution, and building an ear off the current solution, containing the chosen office. When all protected offices are on the current solution, GREEDY_EARS connects any remaining unprotected offices to that solution via spanning trees.

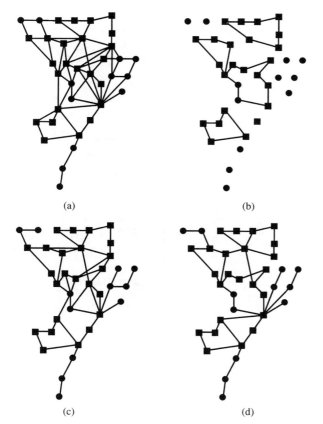

(a)   (b)

(c)   (d)

Fig. 7.   Topology construction. (a) Example network. (b) Initial rings. (c) Greedy Ears solution with rings added. (d) Optimized topology.

To ensure that the initial topology contains the initial ring routings, links contained in the initial ring routings, but not in the GREEDY_EARS solution, are added to the GREEDY_EARS solution. Fig. 7(a)–(c) shows an example network, initial ring routings, and GREEDY_EARS solution with ring routing links added.

2) Topology Improvement: Two algorithms are used to improve the topology: ADD_CHORDS and 1_OPT. These algorithms are essentially the same as the design algorithms for survivable fiber-hubbed networks (see [5]), except that they now account for SHR's, as described below.

ADD_CHORDS simply adds to the topology any links that will result in a lower cost. Candidate links to be added are those input as available for fiber. Although adding links will increase the route mileage cost, it can decrease the fiber and regenerator costs by providing shorter paths between offices and their hubs. Each time the algorithm adds a chord, it checks for lower cost ring routes by calling BUILD_RINGS.

1_OPT seeks to replace links in the solution with links not yet in the solution to lower the topology cost. It replaces a link (u, v) with a link (u, x), where x is one of the W offices closest to u, and W is the search window. Distance between two offices is defined here as the minimum number of links in a path between the two offices. Users input the search window W to limit the range in which 1_OPT searches for a link to replace a removed link. Again, only links input as available for fiber are considered. 1_OPT will not make changes that preclude routing the existing SHR's, although it

will reroute SHR's by calling BUILD_RINGS. If the SHR cannot be rerouted, or if the new route (in combination with the link exchange) increases the topology cost, then the links are not exchanged. Otherwise, both the link exchange and the SHR reroute take place. After each exchange, 1_OPT calls BUILD_RINGS to check for lower cost ring routes, even if no ring routes were disturbed by the exchange. Fig. 7(d) shows an improved topology for the example network.

The topology design programs were written in C and were run on a VAX model 785.[2] Numerical results for these algorithms (run on a variety of example networks) show that the algorithms do reduce the costs mentioned above and run fast enough to be of use to a network planner [8]. Computation times on 23 example rings in 7 example networks were reasonable. For ADD_CHORDS, the average, minimum, and maximum run-times were 2, 0.04, and 14.5 min, respectively. The average, minimum, and maximum run-times for 1_OPT were 16.8, 0.2, and 101.5 min, respectively [8].

F. Alternative Approaches

This subsection describes the limitations of our approach to survivable SONET network design and discusses alternative solution approaches.

Our assumption that economies of scale dictate that each CO put all of its demand on one SHR or one 1 : 1/DP system precludes SHR's that overlap at nonhub locations (that is, locations at which no grooming can occur). This results in more interring traffic, since not all of the demand from a CO is destined for other CO's on the same SHR. The interring demand goes through one or more hubs on the SHR. Although this increases the relative vulnerability of the network, dual-homing (Sections II and III-C) and matched node schemes [9], which require two interconnection points on each SHR, increase the survivability. This was demonstrated in [1] and [9].

Other studies have shown that it is economic to minimize interring traffic [9], [10]. In our methodology, SHR's are interconnected at hubs, but aggregation and grooming (and therefore the costs of the same) are not altered during the ring design phase. This means that neither the ring selector nor the multiplex cost module directly attempts to minimize interring demand. However, this issue is addressed indirectly through the network hierarchy. The user can minimize intercluster demand when defining the network hierarchy and dual-homing arrangement. Since the ring selector obeys the network hierarchy, it thus indirectly minimizes interring demand.

Another limitation of our methodology is that when span exhaust occurs, the rest of the demand for the planning period is carried on 1 : 1/DP spans or on a fully stacked ring (one that visits all of the sites on the exhausted ring). Other feasible options include upgrading the ring, alleviating exhaust with a 1 : 1/DP system that later can be upgraded to a ring, adding a partially stacked ring (one that visits only some of the locations on the exhausted ring), or adding an interlocking ring (one that

---

[2] The computing speed of the VAX 785 is slower than that of the SPARC workstation used to run BUILD_RINGS.

includes some locations on the exhausted ring and some other locations). Reference [9] describes such evolution strategies.

The ring selector is limited in that it is a greedy algorithm that does not consider interactions between SHR placements. Each SHR cost savings over 1 : 1/DP is computed in absence of any other SHR's in the network, and then SHR's are chosen one at a time based on cost savings, without considering the relative costs of combinations of rings. An alternative to a heuristic would be a mixed integer programming approach, as presented in [11], but the model is intractable for networks with more than 20 nodes. That approach designs "logical rings" for single-year demand, without simultaneously considering topology.

Finally, STRATEGIC OPTIONS does not address the engineering tradeoffs between cost and survivability, and between cost and design time. Although STRATEGIC OPTIONS uses survivable architectures and has the capability to compute the survivability metric defined in Section IV, survivability is not explicitly accounted for in the design. Instead, costs of different architectures are compared regardless of survivability. A design tool using suitable survivability metrics, in which a network planner can specify a survivability constraint and obtain the least cost solution within that constraint, or in which the user could specify how much cost is allowed to increase for a given increase in the survivability metric, would facilitate making engineering tradeoffs. This is an area for further work. Reference [11] describes a method by which the network planner can specify how much computation time to spend, and the heuristic returns the best network design meeting the time

IV. CASE STUDY

STRATEGIC OPTIONS was run on a LATA network with 23 nodes and 36 links. A ten-year planning period was used. Over the ten years, DS0 demand grew from 32,993 to 54,922. The network has two clusters and hubs, as well as foreign hubs to dual-home all CO's. The network is shown in Fig. 8. The relative costs of ADM's at various rates for different architecture types are shown in Table I. In the table, the first two columns show the cost of ADM's for the three types of SHR's relative to the cost of a USHR/P OC-3 ADM. Note that although the cost of a USHR/P ADM and the cost of a BSHR/4 ADM are identical, the BSHR/4 architecture requires twice as many ADM's as the USHR/P architecture. The third column shows the ratio of OLTM cost to USHR/P ADM cost for the given rate. The fourth column shows the ratio of regenerator cost to USHR/P ADM cost for the given rate.

First, STRATEGIC OPTIONS was used to design a pure dual-homed 1 : 1/DP architecture. Then, a dual-homed SHR architecture was designed using STRATEGIC OPTIONS. In both cases, network cost and hub survivability were compared. Network cost included ADM's and OLTM's, regenerators, fiber material and splicing, and fiber installation costs. Survivability was the worst-case (minimum) percentage of circuits surviving if a single hub were to go down.

For the network with SHR, STRATEGIC OPTIONS designed five rings; 15 CO's (including the hubs) were on SHR's.

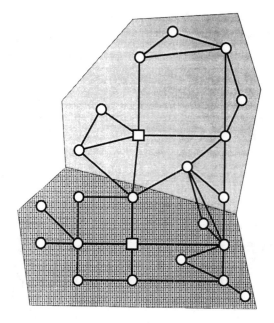

Fig. 8.   Case study network.

TABLE I
ADM RELATIVE COST RATIOS

OC Rate	USHR/P & BSHR/4	BSHR/2	OLTM/ADM Ratio	Regen./ADM Ratio
3	1.00	1.04	0.5	0.51
12	1.92	1.97	0.5	0.73
48	7.13	7.32	0.5	0.52

The cost of the network with SHR's was 10% lower than the cost of the 1 : 1/DP network. The worst-case survivability for 1 : 1/DP with dual-homing was 23%, while the worst-case survivability for dual-homed SHR's was 40%. Thus, for a dual-homed network, survivability increased by 17% while cost went down by 10% when SHR's were used. Each network design took about 15 min.

V. SUMMARY

Future survivable SONET interoffice networks are expected to use both 1 : 1/DP architectures and SHR architectures to provide an integrated and affordable network restoration system. Due to the difficulty of planning such a complex SONET integrated network restoration system, computer-aided design planning tools are essential to ensure a reasonable and cost-effective network design that best utilizes the merits of each restoration architecture and meets required survivability and cost constraints. The philosophy and methods described in this paper primarily serve as a first step toward a more complete and optimized network design tool. Some of the algorithms described here have already been incorporated into a larger commercial tool. The model discussed in this paper can also be used to study survivable architectural tradeoffs and associated network growth strategies. Several enhancement issues for SONET network design have been suggested in [5] for further study.

The case study demonstrates the cost and survivability advantage of using SHR's. The survivability is defined as the worst-case (minimum) percentage of circuits surviving if a single hub goes down. For 10% less expense, survivability increased by 17%. Using more SHR's would increase the survivability more dramatically, but it would also cost more. Since designing each of the two networks took only 15 min, the case study also demonstrates the practicality of automated SONET network design.

ACKNOWLEDGMENT

The authors thank summer students D. Erdmann and P. V. Krishnarao for their work on some of the algorithms described. They also thank D. Ladd for providing the network data.

REFERENCES

[1] T.-H. Wu and M. Burrowes, "Feasibility study of a high-speed SONET self-healing ring architecture in future interoffice fiber networks," *IEEE Commun. Mag.,* vol. 28, pp. 33–42, Nov. 1990.
[2] O. J. Wasem, R. H. Cardwell, and T.-H. Wu, "Software for designing survivable SONET networks using self-healing rings," in *Proc. ICC'92,* Chicago, IL, June 1992, pp. 316.5.1–316.5.7.
[3] T.-H. Wu and R. C. Lau, "A class of self-healing ring architectures for SONET network applications," in *Proc. IEEE GLOBECOM'90,* San Diego, CA, Dec. 1990, pp. 403.2.1–403.2.8.
[4] T.-H. Wu, R. H. Cardwell, and M. Boyden, "A multi-period architectural selection and optimum capacity allocation model for future SONET interoffice networks," *IEEE Trans. Reliability,* vol. 40, pp. 417–427, Oct. 1991.
[5] R. H. Cardwell, C. L. Monma, and T.-H. Wu, "Computer-aided design procedures for survivable fiber optic telephone networks," *IEEE J. Select. Areas Commun.,* vol. 7, pp. 1188–1197, Oct. 1989.
[6] O. J. Wasem, "An algorithm for designing rings in survivable fiber networks," *IEEE Trans. Reliability,* vol. 40, pp. 428–432, Oct. 1991.
[7] O. J. Wasem, "An algorithm for designing rings in communication networks," in *Proc. NFOEC'91,* Nashville, TN, Apr. 1991, pp. 1–21.
[8] O. J. Wasem, "Optimal topologies for survivable fiber optic networks using SONET self-healing rings," in *Proc. IEEE GLOBECOM'91,* Phoenix, AZ, Dec. 1991, pp. 57.5.1–57.5.7.
[9] M. To and J. McEachern, "Planning and deploying a SONET-based metro network," *IEEE LTS,* vol. 2, p. 19–23, Nov. 1991.
[10] T. Flanagan, "Fiber network survivability," *IEEE Commun. Mag.,* June 1990.
[11] M. Laguna, M. Epstein, and E. H. Freeman, "Designing synchronous optical network rings for interoffice telecommunications," presented at the ORSA/TIMS Joint Nat. Meet., San Francisco, CA, Nov. 1992, Presentation WC43.1.

Section 4
Network Performance

SUPERIOR network performance, whether facilitating carrier network management or directly observed by the end-user, is one of the great promises of SDH/SONET. Attaining the goal, however, is a complex matter: it not only involves almost intuitively understood attributes such as reduced error rate and increased service availability, but also an understanding of the most recent applicable standards, the SDH/SONET approach to transmission performance, performance management for new services, and specific technical issues. Those subjects are presented in this section.

The first paper, "Improvements in Availability and Error Performance of SONET Compared to Asynchronous Transport System," by K. Nagaraj, J. Gruber, J. Leeson, and B. Fleury, is an especially lucid introduction to network performance, examining the topic in terms of availability and error rate of SONET networks relative to customary asynchronous networks. Noting that the SONET format affords simplified multiplexing/demultiplexing via direct tributary and payload visibility within the digital hierarchy, add/drop and cross-connection permits a more integrated network, thereby reducing the need for back-to-back circuit packs, and simplifying transmission nodes and associated error-susceptible maintenance procedures. These factors are considered for both local exchange and toll Hypothetical Reference Digital Paths (HRDPs), with analysis of availability and error performance on DS-3 (44.736 Mb/s) paths. The authors show that both predicted and measured network performance reveal that SONET is not only superior to the asynchronous counterpart, but that the latter can fail to meet short-haul and long-haul service objectives.

Where the first paper compares error-rate and availability data for a specific network paradigm, the second paper, "The Impact of G.826" by M. Shafi and P. J. Smith, pursues the subject of error rate and availability more rigorously. Using CCITT Recommendation G.821 as a basis for contrast, the new performance recommendation, G.826 ("Error Performance Parameters and Objectives for International Constant Bit Rate Digital Paths At or Above the Primary Rate"), is discussed in the context of: 1) the relationship between block-based error performance parameters and bit-error rate, 2) how these bit-error rates compare with G.821, 3) apportioning error performance objectives between customer access networks and the domestic interexchange network, and 4) bit-error rates for these apportionments scenarios. A special aspect of this paper is to provide a nonmathematical relationship of block-based error-performance objectives to bit-error rate criteria, and the resulting bit-error rate requirements for terrestrial radio, cable, and satellite systems.

"Transmission Performance in Evolving SONET/SDH Networks" by J. Gruber, J. Leeson, and M. Green, steps back from the details of the second paper and the preliminary performance data of the first paper to provide an especially comprehensive overview of the full scope of SDH/SONET transmission performance. This paper elucidates the challenges engendered by SDH/SONET to assure high network performance, particularly network synchronization and trouble reporting. Performance assessment and performance monitoring are discussed, including a generic monitoring process and its relationship to performance primitives. This paper also describes the Telecommunications Management Network (TMN), the emerging basis for future OAM&P, at three defined architectural levels, and includes two comprehensive appendices on transmission performance in broadband networks and synchronization performance measurements.

As the preceding paper shows, network performance is a substantial topic which coherently integrates transmission monitoring and diagnostic parameters (quality of service) with the related subject of network management. That theme melds well with the next paper, "Traffic Management and Control in SONET Networks" by S. Kheradpir, A. Gersht, A. Shulman, and W. Stinson, which presents a real-time performance management scheme that accommodates multiservice networks. The performance management strategy introduced is made up of a network-level controller that periodically projects multiclass demands and thereby allocates network resources. In so doing, traffic aspects (e.g., call admission and utilization level satisfying specific performance attributes) are emphasized, with special attention to asynchronous transfer mode, ATM.

As several papers in this section make clear, synchronization and jitter are technical dimensions of SDH/SONET which are implicitly related to network performance. The final two papers are thus a fitting conclusion to this section. In "Network Synchronization — A Challenge for SDH/SONET?," M. J. Klein and R. Urbansky provide a tutorial overview of synchronization as it relates to the plesiochronous digital hierarchy and SDH/SONET, with additional observations on an advanced pointer processing implementation. R. O. Nunn, in "SONET Requirements for Jitter Interworking with Existing Networks," discusses broadband jitter due to: 1) signal mapping into SONET, and 2) SONET pointer adjustment to maintain phase-aligned tributaries. Nunn discusses the significance and allocation of jitter, and how it arises due to pulse stuffing and pointer adjustment.

Improvements in Availability and Error Performance of SONET Compared to Asynchronous Transport Systems[1]

K. NAGARAJ, J. GRUBER, J. LEESON,
AND B. FLEURY

BELL-NORTHERN RESEARCH, OTTAWA, CANADA

Abstract—This paper compares SONET and Asynchronous (Async) systems in the context of stringent network error performance and availability requirements for future video, imaging, and other critical services.

The paper asserts that the SONET approach should provide superior network performance owing to integrated functionality, resulting in correspondingly less exposure to operationally induced errors and outages, and fewer component failures. An analysis of preliminary DS3 data supports this assertion. Significant improvements are also expected when the analysis is extended to the DS1 level.

In particular, it is demonstrated by way of representative SONET and Async-based Hypothetical Reference Digital Paths (HRDPs) that the SONET approach significantly reduces the number of circuit packs, manual cross-connect panels and associated jumper cables, patch cords, and connectors. Since these components are prone to failure and procedural errors, reduction in their numbers results in a significant improvement in error performance and availability of the transport network.

1. INTRODUCTION

IT is expected that SONET-based Network Elements (NEs) will be widely deployed in access, metropolitan (metro), and toll networks to transport both today's services (e.g., DS0, DS1, DS3) and tomorrow's services (e.g., high quality video, high resolution imaging, high speed data). Contrary to the existing Async hierarchy, the advantage of the SONET format is the observability of DS0, DS1, and DS3 signals and the consequent ease with which these signals can be manipulated (e.g., multiplexed, add/dropped, cross-connected). The SONET standard also eliminates the back-to-back DS1 and DS3 intermediate interfaces that are required in today's Async transmission networks. The higher level of integration possible with SONET leads to fewer components at transmission nodes, and less susceptability to failures and errors due to both intrinsic causes and maintenance activities. Therefore, this paper asserts that SONET, with inherently fewer equipment components, has the potential to offer improved availability and error performance over Async systems for both current and future services.

An analysis of availability and error performance data indicates the degree to which various equipment components contribute to the degradation in availability and error performance of DS3 paths because of maintenance activities and other causes. This will affect emerging new services (e.g., high quality video,

high resolution imaging, high speed data) which generally require more stringent error performance and possibly availability requirements. This paper examines error performance objectives for these new services, and presents preliminary views of these objectives.

To test the above assertion, this paper estimates the difference in the availability and error performance of SONET and Async-based transport systems, and compares the results to the network objectives as allocated to the relevant components in the network.

2. HYPOTHETICAL REFERENCE DIGITAL PATHS

This section provides the rationale and derivation of the HRDP for both metro (Intra-LATA) and toll (Inter-LATA) networks. The derivation is based on statistical and geographical representations for DS3 service.

2.1 Metro HRDP

Two North American metro networks were analyzed, one a high growth and the other a low growth. The traffic characteristics of the two networks were found to be similar. In both cases, the majority of the DS3 traffic was found to traverse multiple transmission nodes, with 5% of the traffic traversing more than six nodes. Thus, the six-node HRDP represents the near-worst case. Figure 1 illustrates the metro HRDP which is composed of access and metro core portions. The metro core consists of four intermediate and two end nodes. Two types of intermediate nodes are assumed; one type consists of Line Terminating Equipments (LTEs) and Digital Cross-Connect Systems (DCSs), and the other type consists of a back-to-back LTE or an Add–Drop Multiplex (ADM).

2.2 Toll HRDP

The toll core portion of the HRDP is based on a ladder network. The length of the toll core portion is about 4000 mi which covers about 97.5% of the traffic (i.e., 2.5% of the traffic traverses more than this distance), and also covers distances between most major cities in North America. The toll HRDP is shown in Fig. 2, and consists of three portions, viz. access, exchange access, and toll core. The access portions are the same as for the metro HRDP. Each exchange access portion is assumed to be

[1]Revised from an article printed in *PROC. IEEE ICC*, vol. 1, 1990, pp. 248–254.

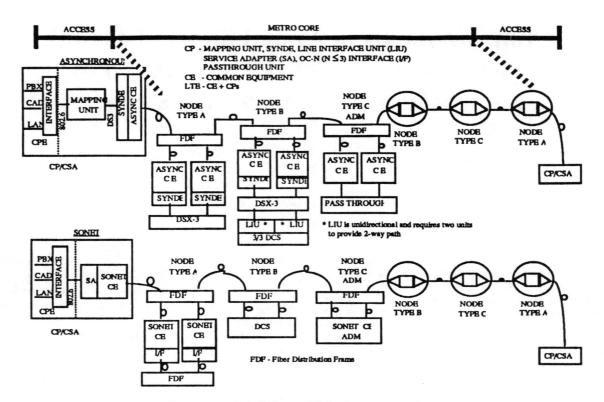

Fig. 1.   Hypothetical reference digital path—metro network.

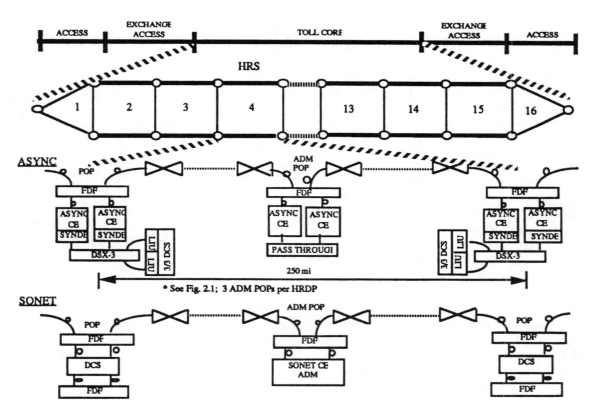

Fig. 2.   Hypothetical reference digital path—toll network.

212

one-half of the metro core. The toll core (based on a ladder network) consists of 15 Hypothetical Reference Sections (HRSs) of length 250 mi. Each HRS consists of three add/drop sites and six repeaters (based on 25 mi repeater spacing) on the average.

Note that while the above provides a general description of the toll HRDP, it is primarily the performance of the nodal equipment that is most relevant to this paper. For this reason, the analysis in this paper is also applicable to connections with ring architectures.

2.3 Comparison of SONET and Async HRDPs

SONET, being a synchronous format, provides signal observability, thereby allowing transit traffic to pass directly through the intermediate nodes without going through intermediate interfaces. For example, an integrated SONET DCS provides both LTE and DCS functions, thus eliminating intermediate interfaces and related equipment such as various Circuit Packs (CPs) and electrical (DSX-3) patch panels. However, in the Async network, the NxDS3 proprietary signal is demultiplexed and multiplexed at every node, and is also patched through a DSX-3 patch panel at many nodes.

Figures 1 and 2 show the CPs used to provide DS3 appearances for Async systems, and equivalently, OC-N (N≤3) appearances for SONET systems. Thus, the CPs are principally SYNchronizer/DEsynchronizer (SYNDE) units for Async systems, and OC-N optical interface (I/F) units for SONET systems (see Fig. 1 for details).

Figures 1 and 2 also show the electrical and optical patch panels (or Fiber Distribution Frames — FDFs) used to interconnect different equipments. The relevant parts of any panel which can lead to failures and/or operational errors are the incoming and outgoing cables, cable connectors at the back of the panel, and the jumpers at the front of the panel. However, for simplicity, these parts are all collectively referred to here as jumper cables.

A comparison of SONET and Async HRDPs (both metro and toll) shows that the number of CPs and jumper cables in an Async HRDP is significantly more than the number of functionally similar components in a SONET HRDP; for SONET, these components are typically only required at end nodes. Thus, the number of CPs and jumper cables increases with the number of

nodes for the Async network, but is independent of the number nodes in the SONET network.

Figures 3 and 4 illustrate the numbers of CPs and jumper cables for the SONET and Async HRDPs. The one-way numbers are used for error performance estimation of SONET and Async HRDPs (both metro and toll) because error performance objectives are always specified on a per-direction basis. On the other hand, the two-way numbers are used for the estimation of availability which is always specified for a two-way path. (The one-way and two-way numbers differ owing to the unidirectionality of certain CPs — see Figs. 1 and 2 for details.)

3. AVAILABILITY COMPARISON

3.1 Approach

The approach to assessing availability is based on the metro and toll HRDPs, as well as the differences in the numbers of relevant SONET and Async components identified above, i.e., components subject to functional integration by SONET. Here, the focus of the SONET/Async comparison is on the numbers of relevant DS3 components (DS3 CPs and DSX-3 jumper cables), and functionally similar SONET components (OC-I or OC-3 CPs and FDF jumper cables).

It is assumed that the performance of functionally similar SONET and DS3 equipment is the same; this applies to Common Equipments (CEs), CPs, and FDF jumper cables. However, it is the different numbers of SONET and DS3 components that affect overall performance. The above assumption is somewhat conservative, as SONET jumper cables provide 1:1 protection, while the DS3 jumper cables do not.

A similar assessment can be done for other hierarchical rates, such as DS1 which is briefly discussed later in this paper.

The steps involved in the availability assessment are as follows:

- Determine end-to-end DS3 short-haul and long-haul network availability objectives applicable to the metro and toll HRDPs, respectively.
- Estimate the availability of the relevant DS3 components based on field data.

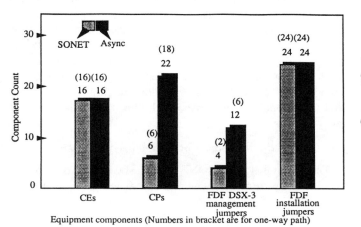

Fig. 3.   Component counts for metro HRDP.

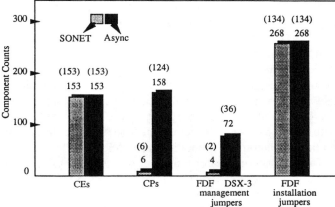

Fig. 4.   Component counts for toll HRDP.

- Estimate the availability of the SONET and Async HRDPs, using the DS3 component availabilities estimated above, and the number of components in the HRDPs.
- Compare the estimated SONET and Async HRDP availabilities to one another, and to the objectives.

3.2 Availability Objectives

Availability is the ability of a network (or network component) to perform its intended function, taking into account reliability (mean time to failure), maintainability (mean time to repair), system architecture (level of protection/redundancy), maintenance/logistic support, and other engineering and environmental factors.

Achieving overall availability objectives for high-capacity fiber transport routes generally requires such capabilities as diverse routing and automatic restoration. However, these capabilities apply largely to the fiber transmission media, and are required to reduce outage time due to cable cuts. Cable cuts are common to both SONET and Async systems, and as such, are not relevant to this paper. In addition, this paper assumes that the restoration capabilities of these systems are the same (although SONET should have advantages owing to enhanced OAM capability and the flexibility of the synchronous hierarchy).

This paper therefore focuses on the differences between SONET and Async equipments in transport nodes, and assumes the usual levels of redundancy associated with transmission equipments.

3.2.1 Short-Haul Objectives.
Short-haul objectives are based on those determined for 250 mi systems [1]–[4]. The design availability objective for DS1 circuits (DSX-I to DSX-I) is 99.98%/yr, or 105 min/yr of total outage time. Of this, 20 min/yr is relevant to this paper, and is allocated to DS3 and higher rate equipment (i.e., LTEs, DSX-3s, FDFs and associated cables, connectors and jumpers, etc., and to repeaters). The 20 min/yr includes hardware failures, software failures, and procedural errors.

It should be noted that the equivalent allocation for a SONET-based implementation is relaxed at 26 min/yr [1]. This is due to the advantage of functional integration which permits a larger allocation to higher rate equipment. However, for the purpose of this paper, the more stringent 20 min/yr is used as a reference objective to evaluate both DS3 and SONET-based systems.

3.2.2 Long-Haul Objectives.
Availability objectives for long-haul fiber transport systems are not as well established as for short-haul systems. The traditional design objective for other long-haul technologies such as analog and digital radio is 99.98%/yr, or 105 min/yr total outage time [5]. However, owing to cable cuts, 99.98%/yr is generally not considered feasible with high-capacity fiber systems, unless accompanied by diverse routing and automatic restoration capabilities.

Thus, an alternative for long-haul objectives is to consider end-user to end-user requirements which range from 99.5 to 99.9%/yr [6]–[9]. In this paper, an availability objective for long-haul DS3 circuit (DSX-3 to DSX-3) of 99.92%/yr, or 420 min/yr of total outage time, is used as the basis. This objective satisfies the above-noted service needs and is achievable with a wide

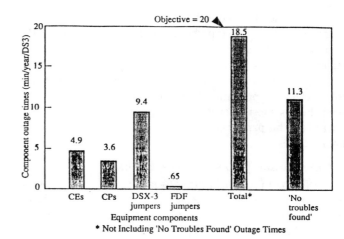

Fig. 5. Outage data for Async DS3 circuits.

range of restoration capabilities [9].

Of the 420 min/yr, an equipment plus operational error allocation of 210 min/yr is relevant to this paper.

3.3 Field Data for DS3 Circuit Availability

The field data were gathered from hundreds of DS3 circuits over many weeks, and involved over 100 outage events. As such, the data are assumed to be representative of the operating company's network. Before using these data to estimate the performance of the HRDPs, some useful observations are made as follows. The data are summarized in Fig. 5, and for comparison with the short-haul objective of 20 min/yr/DS3, has been prorated to a 250 mi reference model [1]. The prorating is done by multiplying the observed per-component outages by the average number of such components (6.5 each) in the reference connection.

As indicated in Fig. 5, observed outages from the CPs and DSX-3 jumpers (i.e., the equipment which will be subject to functional integration with SONET) account for about half of the total equipment outage budget of 20 min/yr/DS3. Note also that, although the total observed outage time due to known causes satisfies the objective, a large amount of outage time was also observed due to "No Troubles Found" (NTF), which when included in the total, exceeds the objective.

Extensive subsequent analysis has indicated that over half of the NTF outage time can be attributed to troubles with DSX-3 equipment. Thus, in addition to significantly reducing outage time of known causes, SONET integration should also reduce outage times due to NTF, and in addition, eliminate the effort and operational expense associated with tracing the cause of such troubles.

3.4 Estimated Availability of SONET and Async HRDPs

The availability of the SONET and Async HRDPs is estimated by scaling the observed outages for each type of equipment component in Fig. 5 by the ratio of the number of such components in the HRDPs to the number of corresponding components in the reference connection for Fig. 5 [1]. The results are summarized in Figs. 6 and 7 for the metro and toll HRDPs, respectively. Note

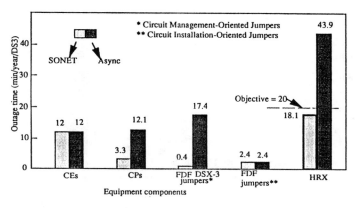

Fig. 6.   Intra-LATA HRX: predicted components outage times.

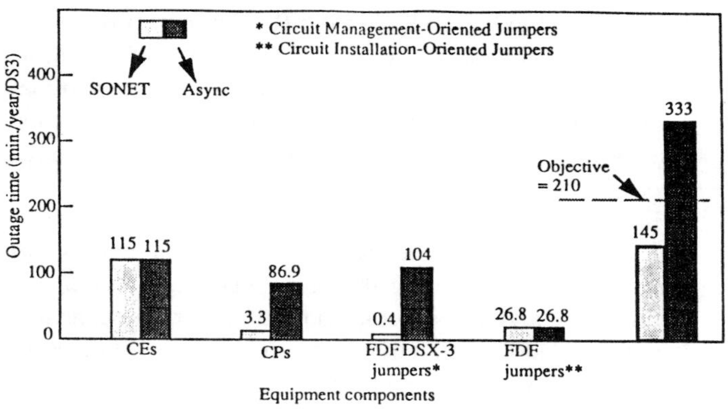

Fig. 7.   Inter-LATA HRX: predicted components outage times.

again that equivalent performance is assumed for functionally equivalent SONET and Async components.

Thus, in Fig. 6, the SONET-based metro HRDP, having about three times fewer CP and circuit management-oriented jumper components, also has considerably less outage time than the Async HRDP. In addition, the SONET HRDP meets the short-haul objective of 20 min/yr, whereas the Async HRDP does not.

Similarly, in Fig. 7, the SONET-based toll HRDP, having many times fewer CP and jumper components, also has much less outage time than the Async HRDP. Again, the SONET HRDP meets the long-haul objective of 210 min/yr, whereas the Async HRDP does not.

A further observation from Figs. 6 and 7 is that the outages due to SONET CPs and management-oriented jumpers is the same for the metro and toll HRDPs. The reason is that these components only appear at the periphery of SONET-based networks where service streams originate and terminate. Because of the synchronous hierarchy, these components (physical appearances) are not necessary for transiting traffic through the core part of SONET networks, and these components can be eliminated from the core network. Thus, with SONET-based networks, availability performance, as affected by the above CP and jumper components, will not degrade with distance or the number of nodes in the circuit. However, the opposite is true for Async-based networks and is apparent from Figs. 6 and 7.

The above analysis is conservative in that it does not include the beneficial effects of SONET in reducing outages due to NTF, nor its advantages in terms of enhanced operations and maintenance capabilities, and its improved redundancy.

4. ERROR PERFORMANCE COMPARISON

4.1 Approach

The approach to assessing error performance is based on the metro and toll HRDPs developed in Section 2, as well as the difference between the relevant SONET and Async components (CPs and jumper cables). It is assumed that the performances of functionally comparable SONET and Async CPs and jumper cables are identical. However, it is the different numbers of SONET and Async components that affect overall performance.

The steps involved in the assessment are as follows:

- Derive the end-to-end performance objectives for transport networks, based on new broadband services (e.g., HDTV, High Resolution Imaging, and ATM). The objectives provide benchmarks to which estimated SONET and Async network performance can be compared.
- Separate the DS3 error performance data obtained from the operating company's network into intrinsic and activity-related causes.
- Estimate the performance of the SONET and Async-based HRDPs by scaling the activity-related performance data by the component ratio (components in the HRDP divided by components in the measured DS3 path) and adding the intrinsic performance data.
- Compare the estimated SONET and Async HRDP error performance capabilities to one another, and to the objectives.

4.2 Error Performance Objectives

It is a reasonable expectation that error performance objectives for future networks will be dominated by the needs of high-rate, high-quality communications services such as video and High Resolution Imaging. Another factor is the methodology for implementing Broadband ISDN Networks: the Asynchronous Transfer Mode, or ATM. A view of future network performance objectives may be developed based on these considerations.

The services mentioned above require relatively stringent performance levels compared to today's narrowband services. Video, being a real time service, is sensitive to Short Interruptions (SIs) in information, which may lead to loss of picture synchronization. The incidence of such SIs (brief error burst events with very high error density) must be kept to a very low level. While performance criteria for video services are still under international consideration, and may be influenced by codec implementation, the following are proposed for planning purposes.

- For short interruptions, a threshold for degraded performance of more than one SI in an hour should be exceeded for only a small proportion of the time (e.g., of any month).

215

- With regard to bit error ratio (BER), the impact of even a single bit error can cause visual impairment where information compression is used. Therefore, it is expected that future requirement for BER performance will be on the order of 10^{-10}. This performance level also avoids costly error correction methods.
- High bit-rate image transfer must minimize retransmissions as these services may be carried using very large information blocks. Considering that the round trip delay of the connection may be a substantial portion of image transmission time, Go-back-N retransmission protocols would result in practically the entire image being discarded. A simple error-handling protocol implementation may be to initiate the retransmission of the entire image on encountering an error. Under the worst case assumptions, a Percentage Error Free Second (PEFS) requirement of 99% yields an image transfer efficiency (image throughput) of 99%.

ATM may introduce performance constraints, but at the same time may implement compensating mitigation. This may take place at processing layers within a network or in adaptation at network edges. An example is Cell Header Forward Error Correction (FEC) to reduce discarded cells due to cell header error. Another example is adaptation layer processing on lost cells (e.g., sequence numbers and filling in of lost cells), which minimizes the timing misalignments that affect video services.

A current view of performance objectives, supporting the more stringent of future services, is shown in Table 1 for the Metro and Toll HRDPs. Objectives for the three performance parameters, BER, PEFS, and SI, are given in terms of a threshold value to which performance is compared over a measurement period (1 hr). In addition, performance is permitted to be lower than the threshold for a certain small percentage of all measurement periods in a longer evaluation interval (e.g., one month).

4.3 Error Performance Data

Table 2 gives the measured performance data of a DS3 path for any month. The path includes DSX-3 patch panels and LTEs (DS3 CPs are part of the LTE). Three different paths (of lengths 80, 81, and 390 km) were monitored for a period of several months and the data correlated with maintenance activities to separate the intrinsic performance from craft-related activities. The results are assumed to be representative of the operating company's network.

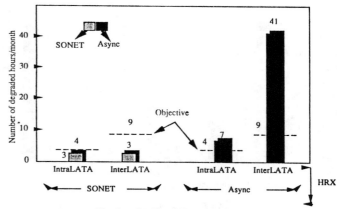

Fig. 8. Predicted SI performance.

4.4 Estimated Error Performance of SONET and Async HRDPs

The performance of the SONET and Async-based HRDPs is determined by simply multiplying the above performance data by the ratio of the number of components in the HRDP to the number of components in the operating company's network.

Both SONET and Async HRDPs meet the BER and PEFS performance requirements, but only the SONET HRDP meets the SI performance requirements. Figure 8 shows the SI performance of SONET and Async HRDPs. Thus, the SONET-based metro HRDP, having about three times fewer components, also has about half the SI hours of the Async HRDP.

The SONET-based metro and toll HRDPs have the same number of CPs and jumper cables, and thus have the same performance, and meet objectives. On the other hand, the Async HRDP SI performance degrades with length, and does not meet the toll HRDP objectives by a factor of four.

5. AVAILABILITY AND PERFORMANCE OF DS1 PATHS

The above analysis of DS3 services indicates a substantial improvement in both availability and performance with SONET as compared to Async networks. The main reason for the improvement is the significant reduction in CPs and jumper cables in SONET-based networks. If the above analysis is extended to DS1 paths, even more CPs and jumper cables are introduced in Async networks, but no extra components are added to a SONET network providing DS1 services. Because of the further functional integration provided by SONET, it offers even greater opportunity to provide increased availability and performance for DS1 services. Table 3 illustrates the increase in these components over the DS3 case for the Async network. Note that the values given in the table are based on the assumption that the DS1 service is carried in a DS3 from the originating node to the

TABLE 1. METRO AND TOLL HRDP ERROR PERFORMANCE
OBJECTIVES FOR ANY MONTH

HRDP	BER > 10^{-10}/hour	PEFS < 99%/hour	SI > 1SI/hour
Metro	< 6 h	< 14 h	< 4 h
Toll	< 9 h	< 18 h	< 9 h

TABLE 2. PERFORMANCE DATA OF A 40 KM DS3 PATH FOR ANY
MONTH

Cause	BER > 10^{-10}/hour	PEFS < 99%/hour	SI > 1SI/hour
Intrinsic	0.072 h	1 h	1 h
Activity	0.072 h	0.16 h	1 h

TABLE 3. PERCENT INCREASE IN THE DS1 CPS
AND JUMPER CABLES OVER THE DS3 PATH FOR
THE ASYNC NETWORK

Type of Component	Metro	Toll
CPs	82%	12%
Jumper Cables	34%	6%

216

terminating node without demultiplexing to the DS1 level at any of the intermediate nodes.

6. Conclusions

In conclusion, the number of SONET CPs and FDF connectors that are functionally similar to those at DS3 do not increase with distance. Thus, the SONET-based metro and toll HRDPs have the same number of CPs and jumper cables, and hence the same availability and error performance. Thus, based on component count alone, SONET has an inherent advantage over Async.

Based on the measured data, the Async HRDPs do not meet the objectives by significant amounts, whereas the equivalent SONET metro and toll HRDPs will meet the availability and error performance objectives:

SONET meets the error performance objectives for existing and future services, and has an advantage over Async systems, particularly for short interruptions.

SONET is expected to provide further improvement in availability and error performance at the DS1 level because of the reduction in DS1 CPs and DSX-1 jumper cables.

Thus, SONET provides superior performance for both today's and tomorrow's services.

7. Acknowledgment

Appreciation is due to many people at BNR who made this work possible. In particular, special thanks are due to I. Ebert for suggesting this work, to G. Goddard and G. Williams for their support and comments, to W. Johnson (then of BNR) who assisted in the data analysis, and to P. Reddigari (then of BNR) who helped in formulating the error performance objectives.

References

[1] Bellcore TR-TSY-000418, "Generic reliability assurance requirements for fiber optic transport systems," issue 2, Dec. 1991.

[2] Bellcore TR-TSY-000233, "Wideband and broadband cross-connect system (B-DCS) generic requirements and objectives," issue 2, Sept. 1989.

[3] Bellcore TA-TSY-000917, "SONET regenerator generic criteria," issue 2, Mar. 1990.

[4] Bellcore TA-TSY-000253, "SONET transport systems: Common generic criteria," issue 8, Oct. 1993.

[5] H. Kostal *et al.*, "Advanced engineering methods for digital radio route design," in *ICC Rec.*, vol. 2, Seattle, WA, June 1987, pp. 19B.6.1–19B.6.6.

[6] AT&T PUB 54014, "ACCUNET T45 services description and interface specifications," issue 1, June 1987, adden. 3, Mar. 1990.

[7] AT&T PUB 62411, "ACCUNET T1.5 service description and interface specifications," issue 2, Dec. 1988, adden. 1, Mar. 1990, issue 3, Dec. 1990.

[8] AT&T PUB 62310, "Digital data system channel interface specifications," issue 2, Nov. 1987, adden. 3, Dec. 1989.

[9] D. F. Saul, "Constructing Telecom Canada's coast to coast fibre optic network," in *ICC Rec.*, vol. 3, Toronto, Canada, June 1986, pp. 53.1.1–53.1.5.

The Impact of G.826

MANSOOR SHAFI

TELECOM NEW ZEALAND LIMITED
WELLINGTON, NEW ZEALAND

PETER J. SMITH

INSTITUTE OF STATISTICS AND OPERATIONS RESEARCH
VICTORIA UNIVERSITY OF WELLINGTON
WELLINGTON, NEW ZEALAND

AT their June 1993 meeting, International Consultative Committee for Telephone and Telegraph (CCITT) Study Group (SG) 13, formerly SG 18, approved a draft of a new performance recommendation on the subject of error performance for digital systems operating at or above the primary rate [1]. This recommendation has now been submitted for formal approval by the accelerated approval procedure. The new performance recommendation is a significant departure from its precursor, G.821 [2], in the sense that objectives are media-independent, block-based, and suitable for doing in-service measurements (ISMs). In addition, the apportionment of the performance objectives to the various parts of the network is left up to the operators.

The motivation for the preparation of G.826 is the move towards providing broadband services to customers; for example, primary rate access (PRA), integrated services digital network (ISDN), and broadband ISDN (B-ISDN). Because synchronous digital hierarchy (SDH) systems will transport the broadband services, it is necessary that the performance of SDH systems comply with G.826.

In this article, we specifically address the following questions:

- What is the relationship between block-based error performance parameters and the bit error rate (BER)?
- How do the BERs compare with G.821?
- How should the error performance objectives be apportioned between the customer access network (CAN) and the domestic interexchange network (IEN)?
- What are the resulting BER requirements for the various apportionment scenarios?

The format is as follows. The transition from G.821 to G.826 is discussed, including a brief statement of the main points found in G.826. The conversion of block-based error performance objectives (EPOs) to BER criteria is discussed in a nonmathematical way. We discuss the implications of G.826, and some conclusions are then presented.

THE TRANSITION FROM G.821 TO G.826

G.821 — Error Performance of an International Digital Connection Forming Part of an ISDN

Recommendation G.821 has been in place for nearly ten years.

It specifies EPOs for a 27 500 km hypothetical reference path (HRP) operating at 64 kb/s. The heart of the recommendation can be summarized by discussing its guidelines for error performance parameters, EPOs, apportionment, and availability.

Error Performance Parameters

An errored second (ES) is a 1 s interval containing one or more errors. A severely errored second (SES) is a 1 s interval with BER $\geq 10^{-3}$. Note that the degraded minute criterion has recently been abolished.

Error Performance Objectives

During a fixed measurement interval (one month recommended), the following objectives must be met concurrently: % of ES < 8% and % of SES < 0.2%.

Availability

Unavailable time commences at the start of a block of ten consecutive seconds, each experiencing a BER greater than 10^{-3}. The unavailable time finishes (and available time begins) at the start of a block of ten consecutive seconds, each experiencing a BER less than 10^{-3}.

Apportionment

The apportionment rules in G.821 are shown in Table 1.

TABLE 1. G.821 APPORTIONMENT RULES

	% of EPO		
	Local Grade	Medium Grade (IEN)	High Grade (International Section)
ES	15% to each originating/ terminating country	15% to each originating/ terminating country	40%
SES* (50% of the EPO only)			

*The remaining SES of SES EPO allowance is a block allowance to the medium- and high-grade classification to accommodate the occurrence of adverse network conditions as follows:
- 0.05% to a 2500 km hypothetical reference digital path (HRDP) for radio relay systems.
- 0.01% to a satellite HRDP.

**Irrespective of location, 20% of the ES EPO is allocated to a satellite HRDP. In addition, a block allowance of 0.02% SES is also allocated to a satellite HRDP.

The Development of G.826

Recommendation G.826 makes use of block-based measurements to make ISMs more convenient. In particular, SDH transport systems may be more easily adapted to block-based ISMs. There is also a desire to dispense with media-dependent block allowances (i.e., the EPOs should be media-independent); there is no provision for a block allowance to be made for radio relay systems or satellite systems in order to account for disturbances during periods of anomalous propagation.

Before presenting the concrete recommendations given in G.826, it is useful to introduce some terminology concerning the relationship between bits and blocks. This is shown schematically in Fig. 1 and in the following definitions:

- M = Number of blocks per second
- R = Number of bits per second
- $\Delta t = 1/M$.

The error performance events and parameters in G.826 are:

- Errored block (EB) — A block in which one or more bits are in error.
- Errored second (ES) — A 1 s period with one or more errored blocks. The SES, defined below, is a subset of the ES.
- Severely errored second (SES) — A 1 s period that contains $\geq$ 30% EBs or at least one severely disturbed period (SDP). For out-of-service measurements, an SDP occurs when, over a minimum period of time equivalent to four contiguous blocks, either all the contiguous blocks are affected by a high binary error density of 10^{-2}, or a loss of signal information is observed. For in-service monitoring purposes, an SDP is estimated by the occurrence of a network defect. The term "defect" is defined in the relevant annexes (2, 3, or 4/G.826) for the different network fabrics — PDH, SDH, or cell-based, respectively.
- Background block error (BBE) — An EB not occurring as part of an SES.
- ES ratio (ESR) — The ratio of ESs to total seconds in available time during a fixed measurement interval.
- SES ratio (SESR) — The ratio of SESs to total seconds in available time during a fixed measurement interval.
- BBE ratio (BBER) — The ratio of EBs to total blocks during a fixed measurement interval, excluding all blocks during SESs and unavailable time.

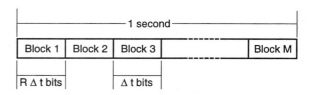

Fig. 1. Block notation.

EPOs

The EPOs proposed in G.826 are summarized in Table 2. All objectives are measured over "available" time in a fixed measurement interval (one month recommended). All three objectives (ESR, SESR, and BBER) must hold concurrently to satisfy G.826, and apply end-to-end for a 27 500 HRP.

Availability

The concept of available time is as defined in G.821, except that the BER > 10^{-3} criterion is replaced by an SES criterion. Hence, unavailable time commences at the start of a block of ten consecutive SESs. The unavailable time finishes (and available time begins) at the start of a block of ten consecutive seconds, each of which is not severely errored.

Apportionment

The apportionment rules contained in G.826 are more complex than those in G.821, even though they do not break allocations down to smaller units than the national sections. Table 3 gives details of the apportionment rules. Figure 2 shows a schematic HRP with the different sections of the HRP discussed in the apportionment rules.

It is not immediately obvious that the EPO percentage figures in Table 3 yield 100%. However, under the assumptions of four

TABLE 2. ERROR PERFORMANCE OBJECTIVES FOR G.826

Rate (Mb/s)	Bits/Block	ESR	SESR	BBER
1.5–5	2000–8000	0.04	0.002	3×10^{-4}
>5–15	2000–8000	0.05	0.002	2×10^{-4}
>15–55	4000–20 000	0.075	0.002	2×10^{-4}
>55–160	6000–20 000	0.16	0.002	2×10^{-4}
>160–3500	15 000–30 000	**	0.002	10^{-4}
>3500	FFS*	FFS*	FFS*	FFS*

*FFS: for further study.

**No objective given due to the lack of available information.

TABLE 3. APPORTIONMENT RULES FOR G.826

% of EPO				
National Portion*			International Portion*	
Block Allowance	Distance Allowance	Transit Allowance**	Distance Allowance	
17.5% to both terminating countries	1% per 500 km	2% per intermediate country 1% per terminating country	1% per 500 km	

*Satellite hops each receive 35%, but the distance of the hop is removed from the distance allowance.

**Four intermediate countries are assumed.

intermediate countries and no satellite hops, we can obtain the following breakdown:

Terminating countries:
$$2 \times 17.5 + 2 \times 1\% \quad \Rightarrow \quad 37\%$$
Intermediate countries:
$$4 \times 2\% \quad \Rightarrow \quad 8\%$$
Distance allowance:
$$27\,500 \text{ km} = 55 \times 500 \text{ km} \quad \Rightarrow \quad 55 \times 1\%$$
Total: 100%

If one satellite hop is used, it uses 35%, corresponding to a nominal hop distance of 17 500 km.

The identification of an SES event during ISM is not straightforward. This is because the definition of an SES involves SDPs. The SDP events can only be measured in an out-of-service condition. Hence, "equivalent" ISM events need to be specified if SDPs are to be detected in service. It is recognized that there is not an exact 1:1 correspondence between SDP events measured in service and those measured out of service. The aim of annexes 2, 3, and 4 is to provide ISM events reasonably close to the G.826 error events.

As an example, annex 2 of G.826 provides details of converting ISM events for PDH paths into equivalent G.826 events. In particular, events such as errored frame alignment signals, EBs, loss of signal, alarm indication signals, and loss of frame alignment are related to the basic ESR, SESR, and BBER parameters of G.826. Annexes 3 and 4 contain similar guidelines for SDH and cell-based networks.

THE RELATIONSHIP BETWEEN G.826 EPOS AND EQUIVALENT BER

It is useful to express the G.826 EPOs in terms of BER values. This is necessary for comparing the stringency of G.821 and G.826 EPOs, checking to see if equipment specifications (which often quote receiver sensitivities in terms of BER) will comply with G.826, and assessing network performance as per G.826 from BER data and distributions.

There are two ways to relate the G.826 EPOs to BER values. The most simple approach is to assume a constant BER and to evaluate the threshold BER which just meets the most stringent of the three EPOs. This is the approach used here. Hence, most of the conclusions discussed below are valid for the constant BER case. A more accurate technique is to derive in some sense a worst case BER against time distribution (or mask) which yields the maximum EOP allowed [3]. This is useful in that it models the real life situation where a low BER is achieved for the majority of the time, but there may be short periods of time where high BER values occur. Since masks allow a spread of BER values (including some high values), the result is that the lower end will be below the simple BER threshold discussed above. Hence, the use of BER masks will suggest the need for lower BER floors in devices aiming to satisfy G.826. Where different conclusions are reached by the use of BER thresholds and BER masks, then both results are given in the rest of the article.

The relationship between G.826 EPOs and BER is dependent on the choice of models to describe the occurrence of errors. The

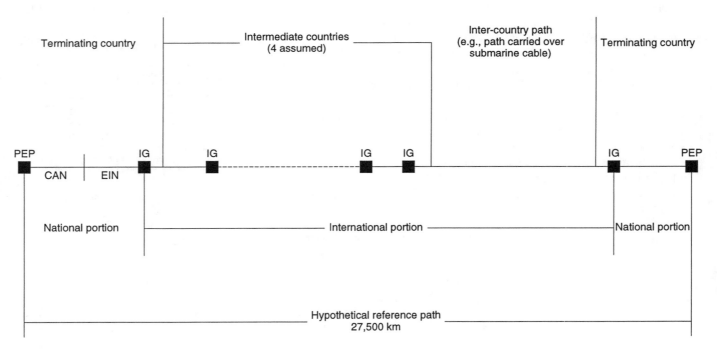

PEP Path end point
IG   International gateway

Fig. 2.   Hypothetical reference path.

errors can occur randomly or in bursts. Systems using forward error correction (e.g., satellite systems and digital radio systems) and/or adaptive equalization (e.g., digital radio systems) are particularly susceptible to burst errors.

The derivation of relationships between the EPOs and BER is quite complex for both error occurrence mechanisms. For the sake of brevity, we only show the results of this derivation. The interested reader can find the mathematical work in [3].

Figures 3–8 show some results based on these models. Figures 3 and 6 show the ES probability versus BER for random and burst errors, respectively. Figures 4 and 7 show the SES probability versus BER for random and burst errors, respectively. Figures 5 and 8 show the corresponding BBER versus BER relationships for the two types of error mechanisms.

Some conclusions can be drawn directly from these figures. From Fig. 4, we can observe that whenever the BER exceeds 1.7×10^{-5} for more than 1 s, an SES event is created. Also, the $\geq 30\%$ EBs event (denoted as event E_1) is the most dominant

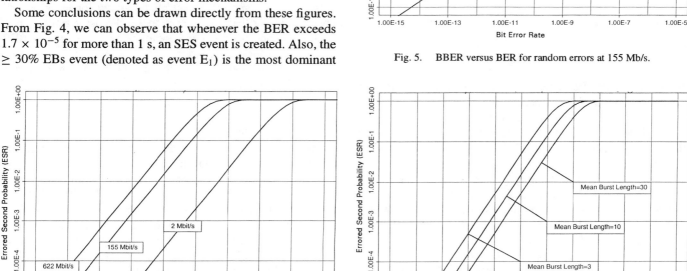

Fig. 5.   BBER versus BER for random errors at 155 Mb/s.

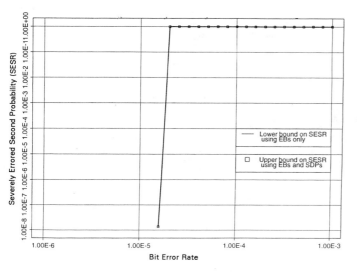

Fig. 3.   ESR versus BER for random errors.

Fig. 4.   SESR versus BER for random errors at 155 Mb/s.

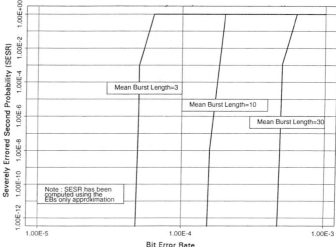

Fig. 6.   ESR versus BER for burst errors at 155 Mb/s.

Fig. 7.   SESR versus BER for burst errors at 155 Mb/s.

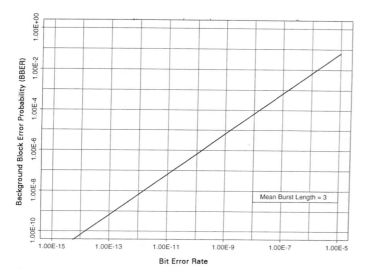

Fig. 8.   BBER versus BER for burst errors at 155 Mb/s.

event causing an SES. This is because the E_1 plus SDP curve lies almost on top of the E_1-only curve.

As radio relay systems experience the occurrence of SESs that corresponded to a BER of 10^{-3} in G.821, the G.826 requirement of a BER threshold of 1.7×10^{-5} appears considerably more stringent.

The results of ES probability versus BER for burst errors is shown in Fig. 6. It may be seen that the BER requirements ease as μ (the mean errors per burst) increases. For example, corresponding to an ESR value of 10^{-4}, the equivalent BER is 2×10^{-12} ($\mu = 3$), 6.5×10^{-12} ($\mu = 10$), and 2×10^{-11} ($\mu = 30$).

In the case of the SESs, we assume that, like the random case, the SESs are dominated by the E_1 event ($\geq 30\%$ EBs). This

assumption is discussed further in [3]. Hence, the E_1 probability shown in Fig. 7 can be used as an approximate SES probability. It can be seen that the threshold BER that will certainly cause an SES is 6.2×10^{-5} for $\mu = 3$; this compares to 1.7×10^{-5} for the random case. As μ increases to 30, the threshold BER increases to a value quite close to the 10^{-3} threshold of G.821.

APPORTIONMENT SCENARIOS

Terrestrial Radio and Cable Systems

Draft Recommendation G.826 does not define apportionment rules for the originating and terminating country block allowances. The stringency (or otherwise) of G.826 can only be assessed after an apportionment methodology for the block allowances is clarified. In this section, we discuss various scenarios for the apportionment and comment on the equivalent BER requirements for the scenarios.

We consider three types of domestic IENs according to the length of the longest call distance:

$$\begin{aligned}
\text{short} &= 1000 \text{ km}, d = 2 \\
\text{medium} &= 4000 \text{ km}, d = 8 \\
\text{long} &= 6000 \text{ km}, d = 12.
\end{aligned}$$

The d parameter value gives the number of 500 km increments contained in the IEN distance.

Now, we choose a part, Z, of the originating/terminating country block allowance to be apportioned to the IEN. Therefore, the total IEN allocation is $(Z + d)\%$ of the end-to-end EPO. The rest, $(17.5 - Z)\%$, is the apportioned CAN budget. We arbitrarily choose Z from a set of $\{0, 2, 10, 15\}$. Therefore, when $Z = 0$, all the 17.5% block allowance part is allocated to the CAN; the IEN only receives d% of the end-to-end EPOs.

TABLE 4. APPORTIONMENT SCENARIOS

	End to End			% of Allocation		
	ESR	SESR	BBER	ESR	SESR	BBER
G.826 CAN	0.16	0.002	2×10^{-4}	$(17.5 - Z)\%$	$(17.5 - Z)\%$	$(17.5 - Z)\%$
G.826 IEN	0.16	0.002	2×10^{-4}	$(Z + d)\%$	$(Z + d)\%$	$(Z + d)\%$
G.826 Satellite	0.16	0.002	2×10^{-4}	35%	35%	35%
G.826 Per transit country	0.16	0.002	2×10^{-4}	$(d + 2)\%$	$(d + 2)\%$	$(d + 2)\%$

TABLE 4a. APPORTIONMENT SCENARIOS

	EPO Values for 500d km		
	ESR	SESR	BBER
G.826 CAN	$(17.5 - Z) \times 16 \times 10^{-4}$	$(17.5 - Z) \times 2 \times 10^{-5}$	$(17.5 - Z) \times 2 \times 10^{-6}$
G.826 IEN	$(Z + d) \times 16 \times 10^{-4}$	$(Z + d) \times 2 \times 10^{-5}$	$(Z + d) \times 2 \times 10^{-6}$
G.826 Satellite	5.6×10^{-2}	7×10^{-4}	7×10^{-5}
G.826 Per transit country	$(d + 2) \times 16 \times 10^{-4}$	$(d + 2) \times 2 \times 10^{-5}$	$(d + 2) \times 2 \times 10^{-6}$

TABLE 5. BER REQUIREMENTS FOR ESR

| d → | 2 | | 8 | | 12 | |
Z ↓	Prob	BER	Prob	BER	Prob	BER
0	1.60E-04	1.04E-12	1.6E-04	1.04E-12	1.6E-04	1.04E-12
2	3.2E-04	2.09E-12	2.0E-04	1.3E-12	1.87E-04	1.21E-12
10	9.6E-04	6.34E-12	3.6E-04	2.35E-12	2.93E-04	1.91E-12
15	1.36E-03	9.01E-12	4.6E-04	3.02E-12	3.6E-04	2.35E-12

TABLE 6. BER REQUIREMENTS FOR SESR (RANDOM CASE)

| d → | 2 | | 8 | | 12 | |
Z ↓	Prob	BER	Prob	BER	Prob	BER
0	2.00E-06	1.69E-05	2.00E-06	1.69E-05	2.00E-06	1.69E-05
2	4.00E-06	1.70E-05	2.50E-06	1.69E-05	2.33E-06	1.69E-05
10	1.20E-05	1.73E-05	4.50E-06	1.70E-05	3.67E-06	1.70E-05
15	1.70E-05	1.73E-05	5.75E-06	1.71E-05	4.50E-06	1.70E-05

TABLE 7. BER REQUIREMENTS FOR SESR (BURSTY ERRORS CASE WITH A MEAN OF THREE ERRORS PER BURST)

| d → | 2 | | 8 | | 12 | |
Z ↓	Prob	BER	Prob	BER	Prob	BER
0	2.00E-06	4.93E-05	2.00E-06	4.93E-05	2.00E-06	4.93E-05
2	4.00E-06	4.94E-05	2.50E-06	4.94E-05	2.33E-06	4.94E-05
10	1.20E-05	4.96E-05	4.50E-06	4.94E-05	3.67E-06	4.94E-05
15	1.70E-05	4.96E-05	5.75E-06	4.95E-05	4.50E-06	4.94E-05

TABLE 8. BER REQUIREMENTS FOR BBER (RANDOM ERRORS)

| d → | 2 | | 8 | | 12 | |
Z ↓	Prob	BER	Prob	BER	Prob	BER
0	2.00E-07	1.03E-11	2.00E-07	1.03E-11	2.00E-07	1.03E-11
2	4.00E-07	2.07E-11	2.50E-07	1.29E-11	2.33E-07	1.20E-11
10	1.20E-06	6.24E-11	4.50E-07	2.33E-11	3.67E-07	1.90E-11
15	1.70E-06	8.86E-11	5.75E-07	2.98E-11	4.50E-07	2.33E-11

TABLE 9. BER REQUIREMENTS FOR BBER (BURSTY CASE WITH A MEAN OF THREE ERRORS PER BURST)

| d → | 2 | | 8 | | 12 | |
Z ↓	Prob	BER	Prob	BER	Prob	BER
0	2.00E-07	2.95E-11	2.00E-07	2.95E-11	2.00E-07	2.95E-11
2	4.00E-07	5.93E-11	2.50E-07	3.70E-11	2.33E-07	3.45E-11
10	1.20E-06	1.79E-10	4.50E-07	6.67E-11	3.67E-07	5.43E-11
15	1.70E-06	2.53E-10	5.75E-07	8.53E-11	4.50E-07	6.67E-11

Using these scenarios, Tables 4 and 4a may be derived for STM-1 (155.52 Mb/s) rate systems.

For each of the apportionment scenarios, the per-hop EPO for a 50 km hop may be found. Taking the case of the IEN ESR value as an example, the 50 km objective would be

$$\frac{(Z + d) \times 16 \times 10^{-4}}{10d}.$$

This ESR objective may then be converted into an equivalent BER, assuming a constant BER, using Figs. 3 and 6. Likewise, a similar procedure may be adopted for the SESR and the BBER parameters. The BER requirements, for a 50 km 155 Mb/s repeated span, for all three G.826 error parameters for the range of Z and d values are shown in Tables 5–9 for the IEN.

Now, the following comments may be made about Tables 5–9.

The BER requirements are only mildly sensitive to the values of Z and d. For example, the biggest variation in BER requirements is only one decade. Recall that this variation is due to changes of 1000–6000 km in call distance and 0–15% in the block allowance given to the IEN.

The BER requirements for the SESR objective shown in Tables 6 and 7 are not sensitive at all to the values of Z and d. This is obvious from the steepness of the curves in Figs. 4 and 7. In other words, whenever the BER is on the order of 10^{-5}, there is a high degree of certainty that an SES will occur.

The BER requirements for the BBER objective shown in Tables 8 and 9 are higher for the corresponding BER requirements for the ESR parameter. Therefore, if the ESR objective is respected, the BBER objective will also be met. This conclusion is only valid for constant BER. Further research in [3] using BER masks shows that the BBER is the critical parameter, and satisfying the BBER objective usually satisfies the ESR objective.

Error rates on the order of 10^{-12} are required to meet the ESR criteria. In order to meet this error rate, the BER floor of the modem should be on the order of 10^{-13}. Both radio and fiber systems in use today can meet this BER floor requirement; however, terrestrial radio systems will need to use the forward error correction technique.

Error rates on the order of 10^{-5} are required to meet the SESR criteria. While fiber optic systems do not experience SESs, radio systems often do. Relative to G.821, the threshold BER is lower (approximately 10^{-5} as compared to 10^{-3}). The lower BER will result in increased outage. Comparing 10^{-3} and 10^{-5} signature curves of typical 16-quadrature-amplitude-modulated (QAM)/64-QAM modems, it can be said that outage will increase by about 30%. However, the deployment of sophisticated adaptive equalizers and space diversity may compensate for the increase of outage.

The BER requirements in Tables 5–9 are not particularly sensitive to the value of Z. Thus, it may be appropriate to accord most of the block allowance to the CAN, while the IEN receives the balance of the block allowance and distance-based apportionment. This way, at least the IENs of originating and terminating countries and transit countries will work to similar performance standards.

Satellite Systems

Satellite systems receive a 35% block allowance, and are not given an additional distance-based apportionment. Under this scenario, the EPOs for satellite systems are shown in Table 10. The example is for a 2 Mb/s satellite link. The BER column gives the equivalent BER threshold value for the given EPO. This is computed from Figs. 3–8.

From an examination of Table 10, the following observations may be made:

- Error rates on the order of 10^{-9} are required for the ESR

TABLE 10. BER REQUIREMENTS FOR SATELLITE SYSTEMS (2 Mb/s)

Parameter	Probability	BER
ESR	1.4×10^{-2}	6×10^{-9}
SESR (random errors)	7×10^{-4}	3×10^{-5}
SESR (burst errors ($\mu = 3$))	7×10^{-4}	9×10^{-5}
BBER (random errors)	1.05×10^{-4}	1.3×10^{-8}
BBER (burst errors ($\mu = 3$))	1.05×10^{-4}	3.9×10^{-8}

criterion. In order to meet this requirement, the modem error floor should at least be on the order of 10^{-10}. In fact, the more detailed approach of using masks suggests that the error floor may need to be lower at around 10^{-11}. Compliance with such a low error rate floor would require the use of very sophisticated forward error correction techniques, possibly in conjunction with trellis coding. The forward error correction requirements will also increase the gross bit rate to be transported.

- Error rates on the order of 10^{-5} are required for the SESR criterion. Relative to G.821, where the equivalent error rate was 10^{-3}, this could represent a potential loss of about 2 dB of fade margin. This is assuming that, for every decibel change in E_b/N_o, the corresponding BER changes by a decade.

- If the ESR requirement is met, the BBER requirement is automatically respected. Again, this observation does not continue to hold when variable BER values are considered. The BBER becomes the critical parameter when masks are used.

Conclusions

The block-based performance objectives of G.826 can be converted to equivalent BER values for both random and bursty error processes. This is the basis of the construction of BER equivalents, and the more detailed BER masks, which can be used to check adherence to G.826. An in-depth study of G.826 should be based on masks and not simple BER equivalents since the use of masks can lead to different conclusions. For example, lower error floors and a different understanding of the relative importance of the ESR and BBER objectives are gained by the use of masks. However, even the simple approach of using BER equivalents contains most of the detail necessary to evaluate the broad implications of G.826.

Compliance with the new recommendation G.826 will impact on transmission systems in the following ways. Radio relay systems will need to use forward error correction techniques to meet the BBER and ESR requirements, and sophisticated adaptive equalization techniques to meet the SESR requirements. Despite receiving a 35% block allowance, satellite systems will need to employ very sophisticated forward error correction techniques to comply with the BBER and ESR objective. This could also result in a significant increase of the gross bit rate to be transported. Compliance with the SESR objective could be at the expense of a potential 2 dB loss of fade margin. Fiber optic systems will comply with all three objectives. However, receiver sensitivities are often quoted at a BER of 10^{-10}. Compliance with a BER requirement on the order of 10^{-12} for the ESR objective is possible, but perhaps at the expense of a system margin which is reserved for system aging.

The error performance parameters in G.826 are based on doing ISMs. However, the definition of the SES parameter is quite complex (particularly the severely disturbed period), and there is no simple correspondence between some in-service and out-of-service conditions leading to an SES.

The BER requirements are not particularly sensitive to the apportionment of the block allowance between the IEN and the CAN. Therefore, it may be appropriate to accord most of the block allowance to the CAN, while the IEN receives the balance of the block allowance and the distance-based apportionment. This way, at least the IENs of originating and terminating countries and transit countries will work to similar performance standards.

References

[1] CCITT Study Group XVIII, "Recommendation G.826: Error Performance Parameters and Objectives for International Constant Bit Rate Digital Paths At or Above the Primary Rate," Geneva, Switzerland, June 1993.

[2] CCITT Study Group XVIII, "Recommendation G.821: Error Performance of an International Digital Connection Forming Part of an ISDN," Geneva, Switzerland, June 1980.

[3] P. J. Smith and M. Shafi, "The impact of performance recommendation G.826 on transport systems," Technical Report No. 38, Institute of Statistics and Operations Research, Victoria University, Wellington, NZ, 1994.

Transmission Performance in Evolving SONET/SDH Networks[1]

JOHN GRUBER, JIM LEESON,
AND MIKE GREEN

BELL-NORTHERN RESEARCH, OTTAWA, CANADA

WORLDWIDE, telecommunications network providers are beginning to deploy fiber transport systems based on Synchronous Optical Network/Synchronous Digital Hierarchy (SONET/SDH) standards — the first North American and international standards for fiber-optic telecommunications. For example, high-speed network elements (NEs) operating at speeds up to 2.4 Gb/s, and exemplified by a number of equipment vendors' products, are now being deployed in major networks. Telecommunications network providers are deploying SONET/SDH systems to gain the substantial benefits they offer, including:

- standardized optical interfaces, which considerably reduce the cost and complexity of engineering multivendor networks;
- enhanced operations, administration, maintenance, and provisioning (OAM&P) capabilities which, among other things, enable networks to be monitored and maintained on an end-to-end, as well as a per-hop, basis;
- cost-effective delivery of existing narrowband telephony and special services; and
- an evolution path toward future broadband services, emerging operations architectures such as the Telecommunications Management Network (TMN), and new transport technologies such as asynchronous transport mode (ATM). (For more information on ATM, see Appendix 1.)

PERFORMANCE CHALLENGES

In evolving to new SONET/SDH transport networks — while maintaining high network performance during and after the transition — network providers may face a number of challenges, ranging from network synchronization to trouble reporting.

The first challenge — network synchronization — minimizes the timing differences among switches in various parts of the SONET/SDH network. Timing differences could cause impairments, such as reduced throughput in data services and lost scanning lines in facsimile printouts. Unlike traditional technologies, which typically locate intelligence only at the end points of a connection, SONET/SDH also distributes intelligence to network midpoints, introducing additional points where timing must be reestablished as signals are routed through the network. (For more information on synchronization, see Appendix 2.)

[1] Adapted from an article printed in *TELESIS*, issue no. 95. ©Bell-Northern Research, 1993.

The second challenge — trouble reporting — involves the detection of transmission impairments, or errors. Trouble reporting is more important in broadband networks than in voice-optimized networks because some emerging services are highly vulnerable to transmission errors. Because video services, for example, are delivered in real time, errors can have an immediate and significant impact on service quality — typically appearing as streaks on the images or as blank areas on a screen.

The success of network providers in addressing these technical challenges will determine their ability to support existing and future broadband services. The goal of network providers' monitoring and maintenance strategies is to enable the detection and repair of network troubles before customers become aware of them.

To help network providers gain a competitive edge through rapid trouble reporting, equipment vendors can adopt a proactive planning strategy that addresses transport performance troubles in two stages:

- performance assessment, where engineers evaluate — through the use of mathematical models and laboratory tests on network equipment and on simulated transport networks — performance criteria, such as the allowable number of errors in a specific time interval of transmission; and
- monitoring, where network providers check the performance of their networks to ensure performance criteria are being met.

These performance criteria and monitoring capabilities are incorporated into the development of SONET/SDH NEs, to support provisioning, monitoring, maintenance, and testing within transport networks.

PERFORMANCE ASSESSMENT

During performance assessment, engineers — working with network providers — examine service needs through planning studies. Then, using analytical methods and laboratory tests, they evaluate the sensitivity of these services — which range from voice to very-high-speed data and video services — to transmission impairments.

In addition to observing the effects of impairments on services (for example, by noting a corrupted portion of a video image), it is necessary to compare those effects to detailed transmission data observed both in simulated network equipment and in actual equipment. The advantage of this approach is that it directly links end-users' service-quality requirements to the actual performance indicators that network providers must monitor in their networks to ensure those levels of quality.

During this assessment process, the sensitivity of each voice, data, and video service is classified according to four general types of impairment:

- subtle transient effects, caused by static discharges or electrical surges, which may result in minor audible "clicks" on voice channels;
- intermittent timing impairments, such as frequency variations in received signal clocking rates, that are related to network synchronization. In data services, for example, these impairments may cause data errors, requiring retransmission of the garbled information. If many errors occur, these retransmissions could reduce data throughput noticeably;
- gradual degradation, which results from "drifting" components. For example, temperature changes may cause timing circuits used in network regenerators to "drift," or become mistuned, which in turn may cause a buildup of errors; and
- hard impairments, such as regenerator failures.

Performance assessment shows that voice services are the least vulnerable to these impairments because the end-user's ear and brain compensate for occasional impairment-related noise by reconstructing the sense of the telephone conversation. Data services, on the other hand, are more vulnerable because garbled data bits appear as discrete errors that require correction, and that may result in slower transmission rates.

Impairments, however, have the greatest impact on high-bandwidth video services because their effects are immediate and, if severe, can be disruptive. Performance assessment identifies three levels of impact and relates them to standard performance parameters. These parameters are defined by the International Telecommunications Union — Telecommunications Standardization Sector (ITU-T) in Recommendation G.826, which covers transmission rates of 1.5 Mb/s and higher. In order of increasing severity, these impairments are:

- errors that affect a single video scan line (typically, dispersed or scattered errors), and are measured most efficiently by the errored second (ES) parameter;

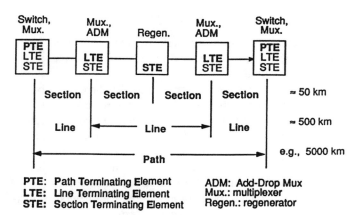

PTE: **Path Terminating Element**   ADM: **Add-Drop Mux**
LTE: **Line Terminating Element**   Mux.: **multiplexer**
STE: **Section Terminating Element**   Regen.: **regenerator**

Fig. 1.   SONET/SDH section, line, and path monitoring.

- bursts of errors that affect one or several video fields, and are measured by the background block error (BBE) parameter; and
- brief interruptions of the bit stream integrity, which may affect many video fields and momentarily cause loss of the image or a loss of horizontal or vertical synchronization.

These impairments are measured by the severely errored second (SES) parameter.

Because video-related impairments represent a "worst-case" impairment scenario, engineers used these criteria to define a set of performance objectives for a path in a SONET/SDH transmission system that will ensure that the quality of *all* services (including voice, data, and video) will not be compromised. These stringent transmission performance objectives are:

- less than 0.5% of 1 s intervals may contain errors;
- the background block error ratio must be less than 10^{-5} — that is, less than one in 100 000 transmission blocks may be errored; and
- less than 0.02% of seconds may be severely errored.

Having established these transmission performance objectives for current and future services, they can be factored into the development of SONET/SDH NEs.

PERFORMANCE MONITORING

Using such NEs, network providers can begin the second stage of the performance strategy: monitoring their transport networks to ensure that these performance objectives are met.

These NEs enable network providers to meet performance objectives by monitoring error checks, framing, and maintenance signals transmitted through the SONET/SDH transport network.

If this monitoring function indicates that impairments are serious enough to cause imminent failure, NEs can autonomously invoke survivability functions, such as automatic protection switching, without involving the network provider's operations systems. At the same time, however, NEs report this information to an operations system so that the problem can be located and repaired quickly. The NEs and the operating systems are designed to work together to perform this trouble location task — even if the network extends for thousands of kilometers and contains hundreds of NEs.

This trouble locating function uses a capability, called sectionalization, which examines the transmission network in progressively finer detail — from the highest sublayer of the SONET/SDH physical network to the lowest. In order of increasing detail, the three physical sublayers defined by SONET/SDH standards are (Fig. 1):

- a digital path layer, which is the logical end-to-end connection between path terminating equipment, such as SONET/SDH-based central office switches;
- a transmission line layer, which is the fiber link between consecutive line terminating equipments, such as SONET/SDH-based multiplexers; and
- a regenerator section layer, which is the fiber link between consecutive section terminating elements, such as SONET/SDH-based regenerators.

THE TMN ARCHITECTURE

The interfaces over which NEs will report trouble location information to an operations system are in the process of being defined in such international and North American standards bodies as the ITU-T and the American National Standards Institute (ANSI). In addition to supporting transmission network maintenance, the reporting function will enable network operations staff to access network performance records and verify the quality-of-service delivered to a particular end-user.

Thus, network providers will have a "window" on such transmission monitoring and maintenance information by designing NEs to be compatible with the Telecommunications Management Network (TMN); see Fig. 2. The emerging foundation of future OAM&P networks, TMN provides standardized interfaces to interconnect network providers' operations systems and intelligent NEs in the transport network.

TMN defines three architectural levels — the physical transport network, the operations system, and an interconnecting operations network — that interwork to manage the flow of performance information. In the transport network, NEs perform monitoring and generate performance data, which is consolidated by a mediation device. This device can then transfer these data across the operations network to the operations system.

Engineers have described this logical flow of information in a generic monitoring process (Fig. 3), which has been factored into equipment design. This process includes: the detection of performance primitives; the generation of standard parameters and failures; the storing and thresholding of performance information; and the reporting of performance data, alerts, and failures to the operations system.

DETECTION OF PRIMITIVES

The first stage of this generic monitoring process is the detection of performance primitives for all sections, lines, and paths supported by NEs in SONET/SDH networks. Primitives relate impairments to standardized error-checking codes, such as bit

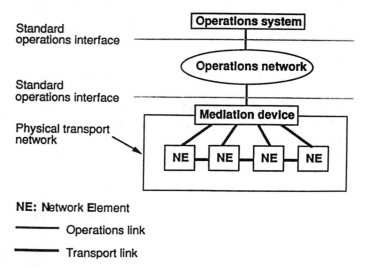

NE: Network Element

—— Operations link

—— Transport link

Fig. 2.   Telecommunications network architecture.

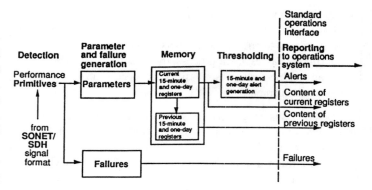

Fig. 3.   Generic NE monitoring process.

interleaved parity (BIP).

(BIP is a byte of data that is computed at the sending end of the path, transmitted with the information (i.e., the payload which includes the end-user's voice or data information) to the receiving end, and computed again to check for errors in the received payload. If the transmitted and received parity bits do not match, a block error has occurred during transmission.)

In addition to BIP, examples of other performance primitives include:

- a severely errored frame (SEF) primitive, which indicates when valid framing cannot be found and represents a significant degradation in performance;
- a loss-of-pointer (LOP) primitive, which indicates that a valid pointer cannot be found, and also represents a significant degradation in performance (pointer information identifies where the end-user's data begin in the SONET/SDH frame); and
- an alarm indication signal (AIS), which is sent downstream (toward the recipient) in response to an LOP primitive.

An AIS is typically sent from an NE where a failure is detected, to the downstream connection end-point to warn of an impending failure. The downstream end-point then sends a remote defect indication (RDI) to the other end-point, so that both ends are aware of the imminent failure and can take corrective action — by reconfiguring the path, for example.

GENERATING PARAMETERS AND FAILURES

In the second stage of the generic monitoring process, these performance primitives are used to generate performance parameters and failures.

Parameters, which are industry-standardized measures of impairments, provide end-users and network providers with a common reference for relating impairments to quality-of-service — avoiding confusion, for example, when end-users claim that network providers have not met performance levels defined in service agreements. On the other hand, network providers can use the same standard units to verify or dispute such claims.

There are two basic types of monitored parameters: quality-of-service parameters and diagnostic parameters.

Quality-of-service parameters — such as errored second (ES), severely errored second (SES), and background block error (BBE) parameters — are operational counterparts of the parameters defined earlier in the performance assessment process.

The second type of parameter, the diagnostic parameter, is typically used by network providers who — in order to correct network problems — need more detailed information than do end-users. For example, network providers may need to examine an errored second parameter more closely to determine whether it has been caused by errors related to BIP, framing, or pointers. Examples of diagnostic parameters include:

- a severely errored frame second (SEFS) parameter, which captures events related to framing trouble, and contributes to errored second and severely errored second parameters; and
- a protection switch count (PSC) parameter, which indicates that protection switching activity is generating errored second and severely errored second parameters.

In addition to parameters, failures (such as those derived from persistent AIS) are generated in this second stage. Failures indicate a component failure or serious performance degradation — for example, the persistence of an LOP primitive — that is severe enough to require the network operator to take rapid corrective action.

MEMORY, THRESHOLDING, AND REPORTING

In the third stage of the generic monitoring process, parameters are stored in registers as counts. For example, NEs provide enough memory to record performance data on each monitored parameter, and for each section, line, and path the NE supports. For each parameter that is monitored, there are two types of registers:

- one-day registers, which store information for the current 24-h period and for at least one previous day; and
- 15-min registers, which store information at quarter-hour intervals. NEs provide these registers for the current 15-min period and for an additional 32 periods — representing 8 h, or one working shift.

The daily register contains performance information that is transferred regularly from the 15-min registers. This 15-min and one-day information enables network operators to track performance, by examining the short- and long-term histories of network performance, and to detect impairments that occur over time.

As a result, components that are gradually degrading can be replaced before they fail. These long-term records also enable network providers to verify customer trouble reports and to ensure that the network performance specifications required by service agreements have been met.

Another important function of the current 15-min and one-day registers is to support thresholding — the fourth stage of the monitoring process. This process involves comparing parameter counts for each section, line, and path against programmable threshold limits. When a current parameter count reaches or exceeds a threshold limit, NEs generate an alert, so that proactive maintenance activities can be invoked. Alerts warn the operator that network performance has degraded, which could occur over time in the case of a "drifting" component.

In the fifth stage of the monitoring process, the alerts (as well as failures and stored performance data) are sent to the operations system, which may then initiate corrective action. Because both failures and alerts are applied to sections, lines, and paths, they support the sectionalization capabilities of the SONET/SDH network when they are reported to the operations system (see Fig. 1).

MONITORING AT NETWORK INTERFACES

In addition to detecting, locating, and reporting network troubles, NEs also help assign responsibility for repair to individual network providers by pinpointing problems in relation to network interfaces. These interfaces represent boundaries between local exchange carriers and interexchange carriers, between public and private network providers, between national and international network providers, and between network providers and end-users.

This interface monitoring capability is important because typically at least three carriers are involved in signal transmission for long-distance calls — the caller's local exchange carrier, an interexchange carrier, and the called party's local exchange carrier. Just as each carrier contributes to the quality of end-to-end services, it is also responsible for maintaining the network within its jurisdiction.

To determine responsibility for impairments between jurisdictions, network providers can use two capabilities — far-end and intermediate monitoring — to enhance more traditional interface monitoring practice.

In the traditional practice, called near-end monitoring, network providers have access only to the end of the system within their jurisdiction. For example, near-end monitoring assesses the quality of a signal sent from the end-user's equipment to the network by examining the received BIP code.

Using this technique, network providers can determine the quality of signals coming from the end-user (Fig. 4), but cannot assess the quality of signals received by the end-user. The end-user has similar limitations in assessing signal quality.

SONET/SDH enables network providers and end-users to overcome these limitations through the use of far-end monitoring, which transmits — along with the BIP code — a far-end block error (FEBE) code located in the path overhead of the SONET/SDH frame.

When a BIP code is received at the end-user's location, for example, it registers one of two states: error-free data or impaired data. The end-user then transmits a FEBE code to the network provider, indicating whether the end-user received error-free or impaired data. The network provider also sends FEBE codes to the end-user to indicate whether the network provider received error-free or impaired data.

Although far-end monitoring determines that a problem exists between the network provider and the end-user, it may not reveal

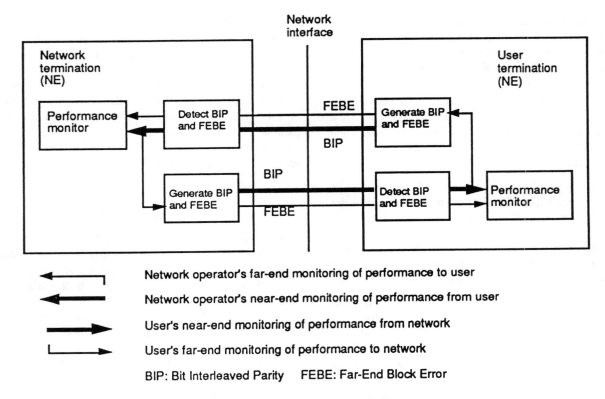

Fig. 4. Near-end and far-end monitoring.

on which side of the interface the problem occurred. To pinpoint these impairments, network providers can deploy intermediate monitoring units near network interfaces.

Configured as standalone units, or integrated within NEs, intermediate monitoring units evaluate the performance of incoming and outgoing signals. For example, consider an intermediate monitoring unit at the network interface in Fig. 4. Assume that — on the left-to-right link from the network provider — the monitor receives error-free BIP codes from the network provider, but — on the right-to-left link from the end-user — the monitor receives errored FEBE codes (representing errors from the network); this means there is trouble in the left-to-right direction on the end-user's side of the interface.

Such monitoring capabilities represent important advances in assessing and eliminating impairments in emerging SONET/SDH fiber-based networks. In the future, similar performance monitoring capabilities will also be applied to an evolving ATM cell-based broadband network that supports multimedia services.

Worldwide, research and development engineers are playing an active role in the standards bodies that are defining the advanced monitoring capabilities that will enable network providers to achieve higher SONET/SDH network performance; to deliver more accurate voice, data, and video services; and to respond proactively to end-users' service expectations.

APPENDIX 1. TRANSMISSION PERFORMANCE IN BROADBAND NETWORKS

Researchers are developing a performance strategy that will help establish critical cell-based transmission performance criteria and will provide the required monitoring capabilities for future broadband networks based on such technologies as asynchronous transfer mode (ATM).

ATM — which provides more flexible bandwidth allocation than is available in SONET/SDH networks — is one of the key technologies driving broadband services. ATM makes it possible to transmit simultaneously both continuous services (voice calls and video conferencing) and noncontinuous services (bursty high-speed data and image file transfer), in cells of fixed length, at whatever bandwidth the service requires. ATM will also support multimedia services, which combine voice, text, data, and visual imaging.

ATM, for example, can transfer data between two supercomputers — a very low-distribution, low-interaction application that requires very high bandwidth from the network. It can also be used for applications that require rapid interaction, wide distribution, and high bandwidth — for example, for an application that links a publishing firm with graphic designers, printers, and other suppliers.

Another emerging transmission technology that will be used for broadband services is frame relay, which enables computers or local area networks (LANs) to communicate cost-effectively with each other using short, high-speed bursts of data. In contrast to ATM, frame relay is an evolution of existing packet-switching technology, which transfers and routes information in discrete frames.

Both ATM and frame relay technologies can build on the SONET/SDH architecture using a configuration known as the Broadband Integrated Services Digital Network (B-ISDN) protocol stack. ATM is a set of higher layer functions located between the Open Systems Interconnection (OSI) end-use application layers and the SONET/SDH physical layer. Frame relay is one higher-layer application that maps onto the ATM functionality.

From the highest to the lowest level of functionality, the B-ISDN stack includes (Fig. A-1):

- Open Systems Interconnection (OSI) higher layers, which support end-use applications;
- an ATM adaptation layer, which provides convergence, segmentation into cells, and reassembly functions that support various services, such as frame relay;
- an ATM virtual path and virtual connection layer, which is largely service-independent, allowing networks to support a variety of voice, data, and video services;
- a transmission convergence sublayer, which maps cells into and out of the SONET/SDH payload; and
- a SONET/SDH physical layer, which is responsible for basic transport functions.

The monitoring capabilities described here apply to the three lowest layers of the stack — the physical, transmission convergence, and ATM virtual path and virtual connection layers.

To support the transport of ATM-based services on future SONET/SDH networks, a two-phased performance strategy can be used. First, laboratory simulation techniques can be used to determine cell-based performance parameters — for parameters such as errored cell, severely errored cell block, and lost cell ratio parameters — for the transmission of such sensitive broadband services as video.

These performance parameters, which are among those established for B-ISDN technology by the ITU-T in Recommendation I.356, define the accuracy of information transfer at the ATM virtual connection layer.

The second phase in this performance strategy is to monitor the various layers to ensure that these performance criteria are met for all services. In addition to monitoring the physical layer, it is also necessary to monitor the ATM layer, and the transmission convergence sublayer, which is situated between the ATM and the physical layers of the B-ISDN architecture (Fig. A-1).

Transmission Convergence Monitoring

The transmission convergence sublayer requires monitoring in order to detect impairments that occur when cells are mapped into and out of the SONET/SDH payload. This mapping occurs on each SONET/SDH path (link) in ATM connections, and is aggregated across all ATM connections carried by each link. For implementation and operational consistency, this monitoring uses the same process — of the detecting, storing, thresholding, and reporting functions — provided in the SONET/SDH generic monitoring process (Fig. 3).

Impairments that occur during this mapping process are associated with three transmission convergence functions: cell delineation, cell header integrity, and cell priority.

The first function — cell delineation — locates the start of the ATM cells within the SONET/SDH payload by detecting no

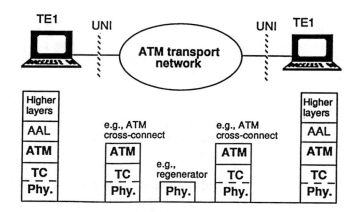

UNI : User-Network Interface
TE1 : Terminal Element Type 1 (B-ISDN)

AAL : ATM Adaptation Layer
ATM : Asynchronous Transfer Mode layer
TC : Transmission convergence sublayer
Phy. : SONET/SDH physical layer

Fig. A-1. B-ISDN/ATM protocol stack.

errors in cell headers. These headers provide cell identification and routing information. If this function detects errors in consecutive headers — a condition called out-of-cell delineation — all virtual connections are affected.

The second function — cell header integrity — verifies that header formats are valid and that identifiers are assigned for virtual channels and virtual paths. Also, if a single transmission error is detected in a cell header, it often can be corrected automatically by the receiving NE, which examines special bits included in each transmitted cell header — a process called forward error correction. The cells with corrected headers are then delivered to their destination.

Multiple errors, however, are too severe to correct using forward error correction, so cells with more than one error in the header are discarded. For certain high-priority services, such as bank transaction services, the information can be replaced through the use of recovery capabilities located in the ATM adaptation layer.

The ability to assign cell priority — that is, to transmit cells of high-priority applications and to discard, if necessary, those of low-priority applications (such as text files) — is the third transmission convergence function. Overall, this function mitigates impairments caused by system congestion that can occur when memory (buffering) is overutilized.

ATM Virtual Connection Monitoring

Although transmission convergence monitoring checks the integrity of the mapping between the SONET/SDH and ATM layers on SONET/SDH links, monitoring and maintenance functions may also be needed to check the performance of ATM virtual connections.

This monitoring is useful because ATM services are transported across virtual connections that can consist of more than one SONET/SDH link.

Because many virtual connections — generally, too many to be monitored simultaneously — are available for transporting

ATM services, this monitoring process can be applied to a selectable subset of connections. Again, for implementation and operational consistency, the same generic monitoring process that was applied for the SONET/SDH physical layer (Fig. 3) is also being considered for ATM connections.

In addition, the ITU-T and T1 standards forums propose the use of operations, administration, and maintenance (OAM) cells for ATM maintenance. These cells contain such information as error checks, and may also include node identifiers, fault descriptions, and time stamps.

Using OAM cells, network providers can detect and respond proactively to a wide range of impairments, including virtual connection failures related to either physical layer impairments or node hardware and software failures. For example, three maintenance functions that use OAM cells are: alarm surveillance, continuity check, and loopback.

The first function — alarm surveillance — detects and reports virtual connection failures caused by network impairments or node failures. This capability rapidly warns the downstream connection end-point (toward the recipient) of an impending failure by sending an alarm indication signal (AIS), contained in an OAM cell, from the location where the failure is detected. The downstream end-point then sends an OAM cell containing a remote defect indication (RDI) to the other end-point, so that both are aware of the imminent failure and may take corrective action.

The second maintenance function — continuity check — ensures that virtual connections are established by detecting such troubles as corrupted address tables. These tables, which are located in each ATM node, are needed to establish virtual connections.

When no cell traffic has been sent for a predetermined period of time, this maintenance function is activated by launching a continuity check OAM cell from the originating end-point. If the other end-point receives no cells of any kind for a certain period, it detects loss of continuity. It then sends an RDI in an OAM cell back to the first end-point — again, so that corrective action can be taken.

The third maintenance function — loopback — is an OAM cell-based maintenance capability to provide proactive in-service fault analysis. This loopback capability accomplishes fault analysis by looping an OAM cell back at the end of successively smaller segments of a connection until the fault is located.

Such maintenance and monitoring capabilities for ATM-based transport networks will support the future delivery — over the public switched network — of a dynamically changing mix of services, such as multimedia services, using a much broader range of flexible bandwidths than is now available in SONET/SDH networks.

APPENDIX 2. SYNCHRONIZATION PERFORMANCE MEASUREMENTS

Digital Network Synchronization

As an introduction to SONET/SDH synchronization, it is useful to examine the basic approaches that have been developed in global standards forums for maintaining accurate synchronization throughout digital networks.

To ensure virtually error-free performance when interconnecting autonomous synchronous networks, standards bodies have defined an average "clocking" tolerance of less than 1 s in 3000 yr. This level of accuracy is typically provided by a cesium-beam atomic clock (also termed a primary reference source), which is located at the pinnacle of a synchronization hierarchy, called a master–slave synchronization distribution network (Fig. A-2). In this hierarchy, all slave, or downstream, clocks are ordered into performance or stratum levels. In addition to the primary reference source location, which is designated a stratum 1 office, this hierarchy — in descending order of accuracy — consists of:

- stratum 2 nodes, which are typically major network nodes, such as toll switches;
- stratum 3 nodes, such as end offices and large private branch exchanges; and
- stratum 4 equipment, which includes channel banks and small private branch exchanges.

All nodes track the signal received from higher-level nodes unless failures make tracking impossible. In fault-free conditions, all timing signals (and the traffic signals they retime) are ultimately traceable to the primary reference source.

This hierarchy also implies that the network can operate at optimum efficiency only when more accurate, higher-level nodes are driving peer or lower-level nodes. It is acceptable, for example, for a stratum 2 node to control the timing of a stratum 3 node. However, if these accuracies are reversed — as a result of redirected traffic after a timing path failure, for example — network synchronization will be compromised.

This disallowed state is not detected by traditional DS1-based schemes, even though the local node evaluates the acceptability of the signal from the upstream node that is providing it with timing. This evaluation is limited to a "good-or-bad" scenario, which assumes that if a received signal is acceptable for traffic, it is also acceptable as a timing reference source.

In contrast, SONET/SDH signals provide the local switch with a range of information about an incoming signal. Specif-

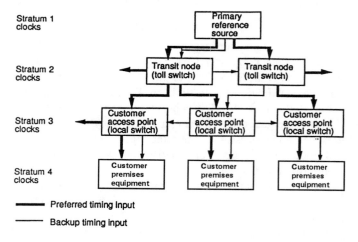

Fig. A-2.   Priority master–slave synchronization network.

ically, this information will typically include an indication of the stratum level traceability (usually, stratum 1) of the clock that is timing the signal. Under fault conditions, when stratum 1 traceability is lost, the local clock can compare its own timing with the stratum level of the remote clock that replaces the original source, and with other potential timing signals. It can then choose the best timing source.

This improved communication is possible because SONET/SDH signals incorporate timing status messages in their maintenance, or overhead, channels. Specifically, the system assigns a synchronization message to the STS-1 (synchronous transport signal, level 1) bit stream overhead.

This messaging capability was developed in the Committee T1 forum and in ITU-T SG 13, and has been incorporated into SONET/SDH equipment. This capability provides a much more robust network than is possible with traditional timing systems. (The benefits of SONET/SDH synchronization messaging have prompted standards forums to adopt a similar scheme for the distribution of timing in traditional DS1-based networks. This development has enabled network providers to improve the timing of their networks, which in turn has allowed them to gracefully evolve their networks to SONET/SDH technology.)

This robustness is possible because intelligent messaging enables SONET/SDH equipment to accommodate a wider range of network failure conditions — for example, when a SONET/SDH survivable ring rapidly redistributes service and timing as a result of a central office failure. In such conditions, SONET/SDH standards provide for messages to allow downstream synchronization nodes to select better input timing signals (which can still be traceable to a stratum 1 source) and to preserve the integrity of the synchronization hierarchy.

To ensure an extra margin of robustness in case the main signal input (also called the preferred, or priority, input) is unsuitable for timing, traditional approaches provide at each slave node an additional input, called the secondary, or backup, input. To reduce the likelihood that the local node will be deprived of both inputs by equipment failure along a single path, the backup input is typically transported along a different physical route from that of the preferred input.

In the unlikely event that neither signal arriving at the local node is acceptable, the node can fall back on a redundant pair of clocks of a defined stratum level. This condition, called "holdover" mode, provides stable, although less accurate, timing until traceability to the stratum 1 source can be restored.

Performance Assessment

Even when a suitable stratum 1 source is controlling timing, every receiver that interfaces DS1 or SONET/SDH signals to an NE must deal with timing impairments that are passed on from upstream nodes. These timing impairments are classified into two types, according to the range of the frequencies of their deviations in phase:

- higher-frequency (10 Hz and above) phase deviations, called jitter, which do not appreciably affect network timing because they are usually compensated for in the receiver circuitry; and

- lower-frequency (less than 10 Hz) deviations, called wander, which are important causes of network timing problems because, unlike jitter, they are not readily accommodated by the receiver circuitry. Thus, they can propagate and accumulate in the network.

Both jitter and wander can be caused by deficiencies in NEs located along the DS1 transmission path, or by delay variations in that path. These problems can stem from environmental causes, such as the sensitivity of fiber-optic transmission systems to temperature changes; or from such systematic causes as the mapping functions of multiplexers, and protection switching activities.

Compensation for these deviations is required along two paths: the timing signal path, which carries the reference signal; and the traffic signal path, which carries service information.

Timing reference signals are commonly routed through a standalone clock system, called a Building Integrated Timing Supply (BITS). Deployed in central offices since the late 1980s, BITS extracts timing from the upstream node's reference frequency, filtering out jitter and some higher-frequency wander. BITS then provides "clean" timing reference signals to all NEs within the central office, such as switches and channel banks. These NEs then automatically pass this timing on to downstream NEs by using it to "reclock" all transmitted DS1 and SONET/SDH signals. In effect, every output signal acts as a potential timing reference source.

On DS1 traffic signal paths, jitter and higher-frequency wander are filtered out at NE receivers in dynamic storage devices, called slip buffers. If wander accumulates to the point where it exceeds the capacity of the buffer, it causes an error condition in which a single frame of data transported in the DS1 bit stream could be lost or processed twice. However, the sequence of the framing bit contained in that frame — and therefore the integrity of subsequent data — is preserved by the conventional receiver circuitry.

This action, called a controlled slip, is the only degradation of DS1 traffic that is caused by inadequate synchronization. The effect of a controlled slip on the customer depends on the service carried by the DS1 carrier. For voice services, the effects are minimal because the end-user's ear compensates for the occasional noise generated. In services that rely on sophisticated data-compression or encoding techniques, such as video or encrypted services, the corruption is more severe because it could involve large blocks of the end-user's traffic.

Such effects of timing impairments will not be limited to the local node because marginal timing conditions spread from node to node as signals are transported through the network. Thus, a problem at any location in the reference signal distribution path can be detected in a signal received at an NE not directly located on that path.

For this reason, it is necessary to assess the effects of synchronization performance from a networkwide perspective. Thus, to acquire a full picture of the synchronization performance of their networks, network providers need to collect data about timing accuracy at many points in the network and over extended periods of time.

In monitoring performance impairments in actual operating networks, typical synchronization test instrumentation concentrates mainly on wander. Synchronization test sets simultaneously collect continuous records of wander for different signals for a predetermined time period. This multiple-signal assessment allows timing analysis of more than one NE. The collected data are then processed by data reduction tools to compare timing performance with industry-standardized parameters.

Traffic Management and Control in SONET Networks

SHAYGAN KHERADPIR, ALEXANDER GERSHT,
ALEXANDER SHULMAN, AND WILLIS STINSON

GTE LABORATORIES INCORPORATED
40 SYLVAN ROAD, WALTHAM, MA 02254

Abstract—Emerging multiservice networks will exploit SONET to integrate technologies such as real-time configurable SONET digital cross-connect systems (DCSs), ATM and narrowband circuit switches, and intelligent remote units (RU) in providing a variety of user services. In this paper, we introduce an efficient performance management scheme to satisfy end-to-end performance requirements of various services, particularly focusing on the traffic management and control aspects of the scheme. The suggested traffic manager is a multilevel system comprising network, call, and cell level controllers. The network level controller (NLC) periodically provisions an "optimal" and "fair" (across source destination pairs) bandwidth allocation for meeting the demand and quality of service (QOS) requirements. The NLC directs SONET DCSs to configure end-to-end SONET "pipes" accordingly. The NLC also downloads to the call and cell level controllers parameters of traffic management policies for traffic admission and regulation within the allocated bandwidth. The call and cell level controllers ensure provisioning of the NLC. The call level controller efficiently fills the SONET pipes with the accepted calls and rejects the demand excess at source nodes. The cell level controller regulates cell transmission and provides congestion-free high-speed data transmission within the SONET pipes.

1. INTRODUCTION

In this paper, we present a real-time performance management scheme for SONET-based multiservice (SBMS) networks [1]. These networks will exploit such emerging technologies as SONET transmission, broadband switching, and intelligent remote units to transport a variety of user services. As such, SBMS networks will consist of new high-capacity intelligent network elements embedded within the existing network infrastructure.

Upcoming SONET-based networks can be characterized by the following features: integrated transmission based on the SONET standard; heterogeneous switching technologies, circuit switching, and ATM; local access provided by "intelligent" RUs acting as service gateways; and services constructed from basic multimedia and intelligent network capabilities. Figure 1 displays a suggested SBMS network architecture [3], [4]. The RUs map user-services into network services and control access to network resources. In an SBMS network, traffic classes are switched using a combination of circuit and ATM switches. Real-time configurable SONET DCSs enable integrated transmission of multiclass traffic on fiber optic links (FOLs). In the proposed scheme, the tandem office merely acts as a SONET virtual tributaries (VTs) cross-connection point; it does not perform call level switching for traffic not terminating there.

In Fig. 1 two source destination (SD) SONET route pipes are allocated between central offices (COs) A and B. One pipe for SD AB is accommodated within FOL AB; the second one is accommodated within FOLs AC and CB following path ACB. The traffic is not switched at CO C. Thus, the SONET pipes represent a set of (logical) one-hop routes from the source to the destination COs. Although there has been a recent flurry of

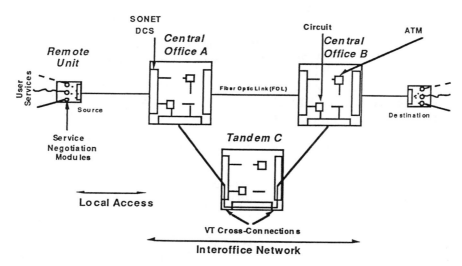

Fig. 1. Simplified SBMS network architecture.

235

activity in standardizing the network management (NM) protocols for SBMS networks [2], notably absent are methodologies for management of services in such environments. In this paper we introduce an end-to-end performance management strategy for multiclass services in an SBMS network environment.

Due to its flexibility and high bandwidth, an SBMS network can accommodate a wide variety of user services. However, to simplify network transport and control operations, it is desirable to aggregate "similar" user services into generic network services [4]. We map user services according to the following attributes: switching technology (ATM, circuit), bandwidth demand, revenue generation, and QOS requirements for cell traffic (maximal end-to-end delay, information loss, delay variations, etc.). Table 1 shows the mapping of some user services.

Given the above definition of network services, the function of performance management then becomes to optimize network performance (i.e., maximize throughput, revenue, etc.) and to provide the QOS expected by the customer at all times, regardless of the operating conditions. The performance management strategy introduced in this paper is composed of two parts: a top-level manager responsible for end-to-end performance across all services and network elements, and task-oriented managers responsible for preventing and reacting to specific network anomalies, such as faults and congestion. This paper is focused on the traffic aspects of performance management.

The proposed traffic manager is a multilevel system comprising network, call, and cell level controllers. The NLC is exercised across the entire network on a periodic basis (say every 5 min). For each control period, it projects multiclass demands per SD pair and FOL loads and optimally allocates network resources for meeting the projected demand.

The multiclass demand projection scheme developed in [3] is used for this purpose. In each control period, the NLC sizes the SD SONET pipes by allocating, for each pipe, the optimal number and/or type of VT containers. It maximizes the projected residual capacity of the network under the condition that very long duration calls should only be admitted when the residual capacity of SD routes is greater than a specified threshold. The NLC optimizes the projected network throughput while avoiding freezing-out routes with long duration calls. For SONET pipes carrying cell traffic, the NLC then computes the maximum utilization level that satisfies cell loss and maximal end-to-end delay constraints [5]. SONET pipes are realized through VT cross-connect commands to SONET DCSs and Add/Drop Multiplexers (ADM) [1]. The amount of projected SD

bandwidth demand that cannot be accommodated is then split across SD service classes in a reverse proportion to a measure that we define as service "profitability." This measure is proportional to the ratio of the revenue generated by the service to its resource consumption level.

At the end of the optimization procedure, the NLC downloads control parameters to the call and cell level controllers. The call level controller handles call processing and ensures that the bandwidth allocation policy of the NLC is not violated. Call processing takes place during call admission and setup in the form of SONET pipe assignment and bandwidth reservation. If there is a sufficient bandwidth available on one of the SD SONET pipes, the call is admitted and assigned to this pipe. The pipe's available bandwidth is adjusted accordingly. Otherwise, the call is blocked. In this process, the call controllers ensure that the bandwidth allocated during an update period to SD SONET pipes is not exceeded.

The call level controller implements routing decisions by assigning to each admitted call a VT mapped into the selected pipe. This is done at the call's source node. Consequently, circuit/ATM switches located at the call's source node perform call/cell switching according to the routing tables that map traffic into the VTs.

Based on the parameters downloaded by the NLC, the cell-level controller regulates the cell flows to ensure that the delay and loss requirements are met [5]. This is also done at the "entrance" of each SD SONET pipe. Table 2 displays the functions, performance measures, and operational time frames of three levels of traffic control.

It is important to emphasize that three control levels are highly coupled, and that the parameters of all three levels are readjusted at the beginning of each control period to meet global performance requirements. The control hierarchy and interactions between different levels are shown in Fig. 2.

The rest of this paper is organized as follows: Section 2 describes the network, call, and cell level control schemes; Section 3 overviews how the control strategies can be realized in an SBMS network; and Section 4 summarizes the paper.

2. Three-Level Control Strategy

In this section, we present the network, call, and cell level control for the SBMS network .

2.1 Network-Level Control

The NLC is designed to simplify call processing, reduce switch loading, and be compatible with existing switching technologies.

TABLE 1. SBMS USER SERVICE ATTRIBUTES

| User Services | Attributes | | | |
Examples	Switching	Bandwidth Demand	Relative Revenues	Real-Time Requirements
64 kb/s voice	Narrowband circuit (ckt)	64 kb/s	Low	N/A (since ckt switched)
Full motion video	ATM	155 Mb/s	Medium to high	Delay <D0; loss <L0
Interactive graphics	ATM	Varies	Medium to high	Delay <D1; loss <L1
Multimedia	ckt, ATM	Varies	High	Delay <D2; loss <L2

TABLE 2. FUNCTIONS, PERFORMANCE MEASURES, AND OPERATION TIME FRAMES
OF THREE LEVELS OF TRAFFIC CONTROL

Control Level	Functions	Performance Measures	Frequency	Class
Network	All SDs • BW demand forecast • BW allocation and provisioning • Optimization of call admission • Congestion control parameter assignment	• Accuracy in BW demand forecast • Call/cell throughput • Fairness in access across SD pairs • Profitability in access • Fairness in access within a class • Quality of service support Maximal delay for express traffic, average delay for first-class traffic, and cell loss	Slow: per update period (e.g., 5 min)	All
VC/Call	Implementation of call/VC control according to NLC provisioning	Accuracy in implementation of NLC policies	Faster: within update period on a call-by-call basis	All
Cell	Choking/relieving, bandwidth enforcement	Accuracy in implementation of NLC policies	Fastest: within update period, triggered by cell queues	ATM

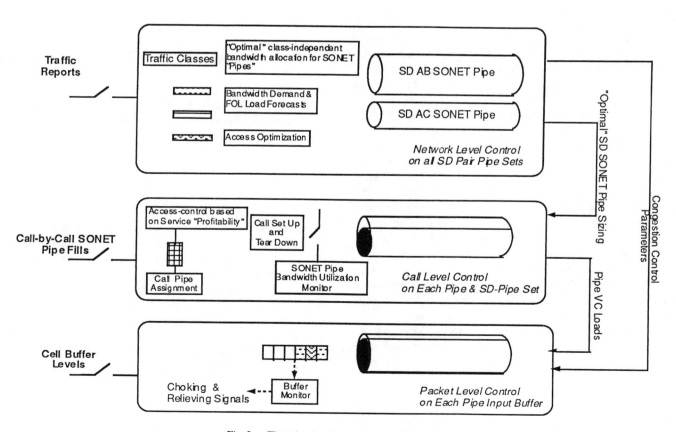

Fig. 2.   Three levels of network control hierarchy.

It is invoked periodically, say every 5 min, to optimally allocate bandwidth adapting to multiclass demand and FOL load dynamics.

The NLC has two objectives. The bandwidth allocation objective is to maximize the minimal FOL residual capacity. A special consideration is given to the calls of very long duration. To prevent the situation when these calls monopolize the network capacity (at the expense of all other calls), the SD bandwidth demand generated by long duration calls should be rejected when an SD SONET pipe set is saturated up to an engineered threshold. For all other calls (we call them regular calls), bandwidth allocation is limited by available network capacity. The access optimization objective is a fair rejection of bandwidth demand excess across all SD pairs at the minimum possible level. As the consequence of our formulation, the access optimization strategy equalizes the fraction of rejected bandwidth demand among all SD pairs contending for the same resources.

There are four distinct phases involved in the determination and implementation of the optimal bandwidth allocation strategy. In phase 1, the NLC obtains aggregate bandwidth reservation attempts and reservation/release counts for each SD class during the preceding control interval. In addition, the NLC obtains the FOL loads at the end of the preceding interval. In phase 2, an on-line multiclass load projector filters and translates these measurements into projected SD bandwidth demands for the upcoming control interval. These bandwidth demands and FOL loads are then translated into projected FOL loads using the bandwidth allocation variables. The bandwidth allocation variables represent the proportion of the projected SD bandwidth demand to be assigned to each SONET pipe or to be rejected at source during the upcoming interval. The optimal bandwidth allocation policy is class-independent if all traffic classes of each SD pair have the same route-set (see Section 2.1.1). In phase 3, the NLC expresses the multiclass access optimization and bandwidth allocation objectives as a function of projected FOL loads. The NLC then jointly optimizes these two objectives to yield the optimal SD bandwidth allocation and access optimization policy. The bandwidth demand to be rejected is split across SD service classes in reverse proportion to their profitability. In phase 4, the NLC implements the new network bandwidth allocation strategy by downloading access-control parameters, switch routing tables, and DCS commands (see Section 3).

2.1.1 NLC Optimization. In this section, following [3], [4], we formulate the joint bandwidth allocation and access optimization problem for the SBMS network and present the property of the optimal solution. We construct two class-independent objective functions. The first function represents allocation of projected SD bandwidth demands in the SBMS network. The second function represents rejection of excess SD bandwidth demands at the source. The excess SD bandwidth demand to be rejected is split across SD service classes according to service profitability.

Mathematical formulation of the tradeoff between the bandwidth allocation and access optimization objectives leads to an Equilibrium Programming Problem (EPP) [3], [4]. These objectives are tied by the following constraints: i) the projected offered bandwidth demand for regular calls is rejected when there is insufficient network capacity, ii) the bandwidth demand for long duration calls is rejected when the SD route set is saturated up to an engineered threshold, and iii) bandwidth demand conservation. The optimization algorithm represents a simple extension of the solution algorithms developed in [6], [7]. It yields a set of bandwidth allocation variables which form the optimal bandwidth allocation strategy for the upcoming control period. Our formulation and forecasting scheme [3] are applicable to a broad range of call holding and interarrival times.

We use the following notation.

$K = \{k | k$ is a network service class$\}$;

$K = [K_1 \cup K_2]$, where K_1 and K_2 consist of regular and long holding time calls correspondingly;

$P_i = \{p \,|\, p$ is a defined route for SD pair $i\}$;

$Q = \{q | q$ is a network FOL$\}$;

$C_q =$ Capacity of the FOL;

$(i, k) =$ A set of class k calls of SD pair i;

$BW(i, k) =$ Projected bandwidth demand of the set (i,k) computed using the algorithm in [3];

$h^0 = \{h_i^0(k) =$ Fraction of $BW(i,k)$ to be rejected at source$\}$;

$h = \{h_i^p(K) =$ Fraction of $(1 - h_i^0(K))\, BW(i,k)$ to be offered to the pth route, $p > 0\}$;

$H = (h^0, h)$, where h satisfies flow conservation constraints:

$$\sum_{p>0} h_i^p(k) = I, \quad h_i^p(k) \geq 0;$$

$F_q(H) =$ Residual capacity of the FOL q at the end of the updating period;

$F^i(H) =$ maximum (over route set P_i) of route residual capacities.

The total fraction of SD bandwidth demand to be rejected for regular and long-duration calls is computed by the following formula:

$$h_i^{01} = \sum_{k \in K_1} h_i^0(k);$$
$$h_i^{02} = \sum_{k \in K_2} h_i^0(k)$$

The bandwidth allocation objective function $J_R(H)$ is the minimum of FOL residual capacity. The access optimization objective functions $J_A^r(H)$ are defined separately for bandwidth demands generated by regular and long-duration calls. For each class of calls, this function represents the minimum (over all SD pairs) fraction of accommodated demand.

Given (i) the physical network topology, (ii) FOL capacities C_q, (iii) service classes k, (iv) projected SD's bandwidth demand $BW(i, k)$ for an upcoming interval, and (v) the instantaneous FOL bandwidth occupancy levels, we formulate the NLC problem as follows:

Bandwidth Allocation Objective: $\max_{h \in H} J_R(H)$,

$$J_R(H) = \min_{q \in W} F_q(H)$$

Access Optimization Objectives: $\max\limits_{h^0 \in H} J_A^r(H)$,

$$J_A^t(H) = \min\limits_i \{(1 - h_i^{0r})\}, \qquad r = 1, 2.$$

The above objectives are coupled through conservation and capacity constraints, and long-duration calls accessibility conditions:

$$\min\limits_{q \in Q} F_q(H) \geq 0,$$

$$h_i^{02} > 0 \quad \text{if and only if } F^i(H) \geq \gamma,$$

$$h_i^{01} > 0 \quad \text{if and only if } F^i(H) = 0.$$

The engineered parameter γ determines the balance between long- and short-term bandwidth allocation goals. The access-control variables $h_i^0(k)$ are defined as follows:

$$h_i^0(k) = \left(z(k)^{-1} / \sum_{k \in k_1} z(k)^{-1} \right) h_i^{01} \quad \text{if } k \in K_1$$

$$h_i^0(k) = \left(z(k)^{-1} / \sum_{k \in k_2} z(k)^{-1} \right) h_i^{02} \quad \text{if } k \in K_2.$$

The variable $z(k)$ denotes the profitability of class k calls. One way to measure profitability is as follows:

$$z(k) = a(k)u(k)/m(k)$$

= (revenue generated) f(servicing rate) / (bandwidth required)

where $a(k)$ is the expected revenue generated by servicing a class k call, $m(k)$ is the bandwidth required to serve a class k call, and $u(k)$ is a function proportional to the servicing rate of class k calls. For data traffic, $m(k)$ is defined as bandwidth sufficient under this scheme to meet delay and cell loss requirements (Section 2.3). Thus, computed SD demand excess levels are split across service classes in reverse proportion to service profitability. The proof of the following proposition is an extension of the proofs given in [6] and we omit it here.

2.1.2 Proposition. If all traffic classes of an SD pair have the same route-set, the optimal control variables $h_i^p(k)$, h_i^{02}, and h_i^{01} in the EPP formulation have the following properties:

$$h_i^p(k) = h_i^p \quad \text{for all } k \text{ and } p > 0$$

Thus, the allocation of bandwidth for demands to be admitted calls is class independent.

$$h_i^{02} = h^{02} \text{ and } h_i^{01} = h^{01}$$

for all SD pairs i contending for the same resources; that is, fairness in bandwidth demand allocation exists among all SD pairs.

From this point on, we assume that all traffic classes of each SD pair have the same route set and, thus, based on the proposition, will only consider class-independent bandwidth allocation. If the route set of each SD pair is selected in such a way that the difference in propagation delay is practically negligible for all routes of each SD pair, the service call setup procedures are simplified through class-independent bandwidth allocation.

Under the suggested architecture (see Fig. 1), the end-to-end bandwidth allocation, determined by the variables h_i^p, has the following advantages. First, only source and destination switches are involved in processing the arrived calls. Consequently, call admission and setup procedures are greatly simplified. Moreover, class independent bandwidth allocation makes possible class-independent call level routing. Second, since each route consists of only one hop, cells do not incur queueing delays at intermediate COs. This results in reduced delay variation. One-hop routes of sufficiently the same end-to-end propagation delay will simplify the cell level controller job in meeting QOS requirements essential for call routing. By properly sizing buffers and controlling line utilization levels at the source and destination nodes, high-speed cell level performance requirements such as cell loss and maximal delay can potentially be met uniformly for all routes.

2.2 Call Level Control

Call level control implements the NLC policy at the call level. It limits SONET pipe fills and provides access control across SD pairs according to call class "profitability" on a call-by-call basis. Call level control is done in the form of bandwidth reservation. It is performed at the source node during the call setup. To prevent congestion inside the pipe, the control limits, by call blocking, the maximum bandwidth reservation levels on the SONET SD pipes. For pipes filled with Virtual Circuit (VC) traffic, this also prevents cell congestion. The routing for each admitted call is done at its source node. The long-duration calls are rejected by the access control when the SD route set is saturated up to an engineered threshold. For regular calls, the admission is limited by available network capacity.

An ATM user call request specifies the desired transmission rate and transport mode selection. The amount of logical bandwidth reservation for a call is based on the peak rate for express mode or the reduced rate for first-class mode (see the cell level control section for more detail). For ATM-switched calls, lower level controls are necessary to prevent cell congestion in SONET pipes. For circuit-switched calls, call level control is sufficient.

2.3 Cell-Level Control

The function of the cell level controller is to provide congestion-free high-speed data transmission within the SONET pipes. Following [5], we suggest that the network provides two different modes for data transport: express and first-class. The express mode is appropriate for real-time applications, whereas *first-class* service is used for nonreal-time applications. In the express mode, the network provides guaranteed throughput with tightly bounded CPE-to-CPE cell delay; for first-class calls, the network provides flow-controlled service with guaranteed throughput and bounded average delay. The amount of bandwidth reservation for express and first-class calls is based on the peak (denoted by P_k) and reduced (denoted by R_k) rates, respectively.

Once the VCs are established, the source switch directs the traffic to the SONET pipe. In this process, express cells have priority over first-class cells. In case of pipe congestion, which would be detected at the SD Pipe Buffer (SDPB), cell-level control is exercised at the source backbone node in the form of choking/relieving of first-class VC traffic. In contrast, express VC traffic is not subject to any flow control. Thus, traffic control

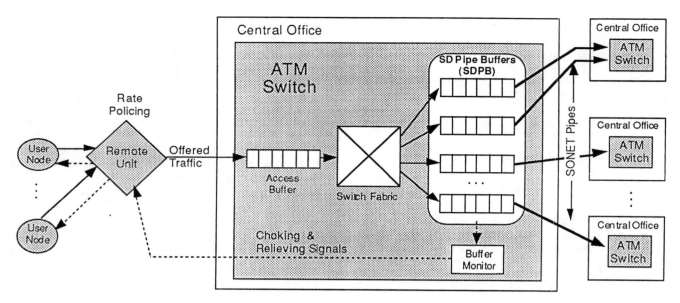

Fig. 3.   ATM node architecture for SONET pipe congestion management.

is exercised only at the source backbone node to regulate the input flow of first-class cells to each pipe. The block diagram describing this regulation is shown in Fig. 3. The scheme also requires rate regulation by the user. In addition, a bandwidth enforcement mechanism is needed to ensure that users do not exceed the requested rate; the mechanism terminates all calls violating the requested rate.

The parameters of the congestion control scheme are adjusted at the beginning of each control period to meet the cell level QOS requirements: maximal delay for express calls, average delay, and choking/relieving frequency limitation for first class, and cell loss for both transmission modes. The adjustment is caused by resizing of SONET pipes according to new demand and FOL load forecasts. Call and cell level congestion control for ATM traffic is exercised as follows.

Each CO ATM switch maintains SDPBs for high- and low-priority cells (express and first class, respectively). For each incoming call, logical bandwidth m_k is reserved along a SD pipe. It is calculated by the following equations:

$$m_k = P_k/\rho_e \text{ for express service; and}$$
$$m_k = R_k + \sigma_k P_k \text{ for first-class service.}$$

The reservation rates P_k and R_k are determined according to the call request. The parameter $0 \leq \sigma \leq 1$ is used to control the average delay and the choking and relieving frequency in an SDPB for first-class service and ρ_e is the maximum utilization level for express traffic computed by the NLC.

- Cells flow from SDPBs to each SONET SD pipe at the total reserved rate Σm_k of all VCs using each particular route. In this process, express cells have priority over first-class cells.
- On the outgoing links of MUXs, DEMUXs, and destination backbone nodes, cells are transmitted in a first-come-first-serve manner (i.e., no priority processing).
- Express VCs can transmit cells at their P_k rates at all times. First-class VCs can transmit cells at their peak rates only

when the SDPBs at the source node are not congested. When an SDPB at the source is congested, first-class VCs using the pipe receive choking cells from the SDPB monitor, and reduce their cell transmission rates to 0. Choking of first-class VCs is relieved when the congestion in the particular SDPB has abated.

- The congestion status of each SDPB is determined by the occupancy level of its first-class buffers as follows. First-class cell buffers at each SDPB have two threshold levels: A (congestion abatement) and T (congestion onset) where $A < T$. When the buffer occupancy level exceeds T, the switch sends choking cells to all first class VCs using this particular SDPB. This choking process stops the first-class cell arrival, thereby decreasing the buffer occupancy level. When the buffer occupancy level drops below A, the congestion is considered abated and the switch sends relieving cells to all first-class VCs using the particular SD SONET pipe. For the SDPB to become congested again, the buffer occupancy level must exceed T again.

Dynamic parameter sizing is performed as follows. At the beginning of each control period, NLC sizes maximal permitted utilization level ρ_e to satisfy maximal delay and cell loss requirements. Furthermore, it sizes (using the technique introduced in [5]) threshold parameters A and T of SDPB and parameter σ for first-class traffic in such a way as to minimize $\sigma > 0$ under the average cell delay and choking/relieving frequency constraints.

3. APPLYING THE STRATEGY

The NLC should be embedded in Operations Systems (OSs) that are responsible for implementation of its functions. Fig. 4 summarizes the basic performance management procedure implemented by the OS.

The scheme requires SBMS network elements (NEs) to provide various measurements for performance management, com-

TABLE 3. SBMS NE MEASUREMENTS, COMMANDS, AND LOCAL CONTROL FUNCTIONS

Network Elements	Measurements	Management Commands	Local Control Functions
Local access nodes	• SD demand request counts (per class)	• Access control parameter updates	• Forward choke/relieve msgs.* • Police contracted cell rate* • Call admission
Switches	• FOL leads • SDPB queue lengths, etc.	• Routing table updates • SDPB updates*	• Generate choke/relieve msgs.*
DCSs		• Reconfiguration commands	• Reconfiguration of SONET pipes

* ATM traffic only.

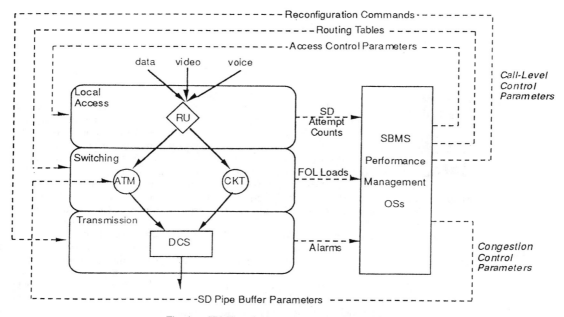

Fig. 4. SBMS performance management procedure.

mands, and local control functions; Table 3 summarizes the NE requirements.

In addition to measurements and controls provided for the OS, some NEs must provide local control functions. In particular, to support cell level traffic management, ATM switches must be capable of issuing choking and relieving messages; and local access nodes must be capable of forwarding them to user nodes. Fig. 3 illustrates a switch architecture suitable for VC congestion management. Applying the scheme also places requirements on user CPE. User equipment must be capable of throttling offered data traffic on demand. If the user does not comply with choking requests, the strategy will enforce the allowed rate by rejecting offered traffic at the local access node.

4. SUMMARY

This paper described a comprehensive performance management scheme for SBMS networks. The developed scheme efficiently integrates network, call, and cell level controls to satisfy end-to-end performance requirements of multiclass services. For an upcoming period, the NLC projects multiclass demands per SD pair and optimally allocates network resources according to

the solution of the EPP. Call level control handles call-setup and ensures that the bandwidth allocation policy of the NLC is implemented. Cell level control regulates the cell flows (according to the parameters downloaded by the NLC) to ensure that the maximal delay and loss requirements are met.

References

[1] Bellcore, "Synchronous Optical Network (SONET) transport systems: Common generic criteria," TA-TSY-000253.

[2] L. N. Cassel et al., "Network management architectures and protocols: Problems and approaches," *IEEE J. Select. Areas Communi.*, vol. 7, no. 7, Sept. 1989.

[3] A. Gersht and S. Kheradpir, "Integrated traffic management in SONET-based multiservice networks," in *Proc. 13th ITC*, Copenhagen, Denmark, June 1991.

[4] S. Kheradpir, A. Gersht, and W. Stinson, "Performance management in SONET-based multiservice networks," in *Proc. IEEE GLOBECOM '91*, Phoenix, AZ, Dec. 1991.

[5] A. Gersht and K. J. Lee, "A Congestion control framework for ATM networks," *IEEE J. Select. Areas Communi.* vol. 9, no. 7, Sept. 1991.

[6] A. Gersht, S. Kheradpir, and A. Friedman, "Real-time traffic management by a parallel algorithm," *IEEE Trans. Communi.*, vol. 41, no. 2, Feb. 1993.

[7] A. Gersht and S. Kheradpir, "Real-time decentralized traffic management using a parallel algorithm," in *Proc. IEEE GLOBECOM '90*, San Diego, CA, Dec. 1990.

Network Synchronization — A Challenge for SDH/SONET?

It is a historical misassumption that SDH/SONET networks require syncronization for compatibility with PDH equipment and networks.

Michael J. Klein and Ralph Urbansky

MICHAEL J. KLEIN is engaged in the field of systems planning at Philips Kommunikations Industrie AG.

RALPH URBANSKY is responsible for the system architecture and ASIC design of SDH products at Philips Kommunikations Industrie AG.

SDH/SONET has been rapidly acknowledged as a world-wide transmission standard replacing the existing PDH infrastructure. As the actual introduction of SDH equipment commences, deficiencies in the context of network synchronization and timing are becoming apparent. In view of the increasing activities and discussions on this issue in the international standardization bodies, the question arises: is the synchronization issue a threat for SDH/SONET which may hold back the implementation of this technique or is it just a challenge to improve some implementation-specific deficiencies?

This article first summarizes the basic characteristics and requirements of existing networks, focusing on network synchronization aspects and related parameters. It then describes the SDH/SONET-specific pointer-based multiplexing technique. It is shown that the need to synchronize the SDH/SONET network elements is not an inherent characteristic of these "synchronous" networks but rather a result of the generally applied ("classical") pointer processor implementation. Finally the principles of an advanced pointer processor implementation that is fully compatible with the existing telecommunication environment is outlined. This pointer processor does not require synchronization (i.e., allows plesiochronous operation!) and additionally provides timing transparency. Finally, measurement results demonstrating the feasibility of this approach are presented.

Network Synchronization

Network synchronization has been required since the introduction of digital exchanges in the Public Switched Telephone Network (PSTN) and their interconnection by digital primary rate signals (2 Mb/s/1.5 Mb/s). This is due to the synchronous, byte-oriented data structure of the primary rate signals where each byte in a frame represents a telephone circuit. This structure is optimized with respect to the cost-efficient implementation of time-multiplex-based switch fabrics, as well as the cost-efficient implementation of primary-rate multiplexing equipment, because no bit-rate adaptation (justification) needs to be performed.

While the digital telephone switches had to replace their analogue counterparts and a cost-efficient implementation of the switching function was the major concern, the digital transmission equipment was expected to replace analogue transmission lines characterized by low delay, almost no time-dependent delay variations (wander), and low cost per telephone channel. The low delay was achieved by employing bit rate justification instead of frame buffers. The low cost could only be achieved by the use of cheap free-running clocks (accuracy 15 . . . . 50 x 10^{-6}) which leads to the label plesiochronous digital hierarchy (PDH). The use of bit rate adaptation results in a decoupling of the bit rate of the aggregate signal (e.g., 140 Mb/s) and the tributary signals (e.g., 2 Mb/s), respectively, which allows the transport of both data and timing information via the payload of high-bit rate transmission signals.

Clock Characteristics

The objective of network synchronization is to keep to a minimum the occurrence of byte slips due to frame buffer overflows or underflows in digital exchanges. The slip performance is covered by the CCITT Recommendation G.822. In order to avoid frequent byte slips, the network nodes of a telecom service provider (e.g., British Telecom, France Telecom, DBP-Telekom, AT&T, . . .) is generally operated synchronously: the clocks of the synchronous switching equipment (exchanges) are locked to a common clock, the Primary Reference Clock (PRC, Stratum 1).

The quality of PRCs is determined by the slip performance of connections between synchronous islands and corresponds to one slip in 70 days. This results in a stability requirement of 1 x 10^{-11} as per CCITT-Rec. G.811 with respect to Universal Time Co-ordinated (UTC).

Reprinted from *IEEE Communi. Maga.*, vol. 31, no. 9, pp. 42–50, Sept. 1993.

The performance of slave clocks (Stratum 2, 3, and 4) is specified in CCITT Rec. G.812. Compared to PDH equipment clocks, they are characterised by a higher stability as well as by their reliability. The higher stability is required in case of loss of synchronization reference (holdover mode).

Network Wander

In the current digital networks based on the Plesiochronous Digital Hierarchy (PDH) both data and timing is transported via 2-/1.5-Mb/s links. The timing content is degraded by jitter and wander. While the high frequency jitter can be reduced by filtering, wander, once generated, cannot be eliminated due to its low frequency spectrum. The maximum wander amplitude is an important design parameter for exchanges determining the size of the wander buffers necessary to prevent byte slips and related data corruption.

CCITT Recommendations G.823/824 [1] specify network limits for wander: Network nodes in a synchronous network, i.e. exchanges, must be able to cope with the permissible wander tolerance of 18 µs between synchronization and data inputs. The underlying assumption, based on a wander reference model, is that the synchronization link wander is limited to 7 µs (including cascaded slave clocks) and the data link between two synchronized nodes may contribute up to 4 µs.

Frame Alignment for Time Multiplex Switching

T he need to increase the flexibility of transmission networks creates a demand for cost-effectively implemented, bit rate-independent, switching fabrics capable of switching all kinds of signals.

Switch fabrics can be efficiently implemented using time multiplex switching techniques presently used to implement digital PSTN switches. This type of switch fabric requires a "synchronous" frame structure providing direct access to each tributary channel (just like the access to 64 kb/s channels in a primary rate signal of 1.5/2 Mb/s) in contrast to the well-known hierarchical multiplexing structure of the PDH where a separate demultiplexing function is required in addition to the bit rate specific switch matrix.

The Buffer Approach

The technique used in the digital PSTN exchanges requires a synchronization of the switching nodes, and buffers at the input of the nodes to compensate the different transmission delay of the links connecting the switching nodes. Two types of buffers can be distinguished: a frame buffer to compensate the static delay differences and a wander buffer for the time dependent delay variations.

Applying this concept to transmission nodes (e.g., digital cross-connects: DXC) introduces an additional delay to the service channels (e.g., telephone circuits) due to the buffers which could result in the need for echo cancellers. This is not acceptable.

SDH/SONET Pointer Alignment

The solution to this problem is the introduction of the well known SDH/SONET multiplexing technique utilising pointers that allow the flexible allocation of the payload channels within a transport frame so that the payload frames may have almost any position with respect to the transport frame. Frame alignment is performed by generating transport frames in a node synchronous to the node clock and mapping the tributary (payload) channels from the received unaligned transport frames into these new synchronous frames. Payload channels occupy well defined time slots in these transport frames. The payload frame (the phase of the payload signal) is indicated by the pointer residing at fixed locations in the transport frame. Switching is performed by moving time slots representing a tributary channel from one synchronous transport frame to another. This technique introduces only a marginal delay compared to the frame buffer delay.

While a constant delay would only require a static pointer corresponding to the frame buffer, substituting the wander buffer requires the implementation of a dynamic pointer.

The advantage of such a (dynamic) pointer solution is the lower delay and the fact that phase deviations larger than 18 µs can be accommodated without any corruption of data. Even in the case of plesiochronous operation (e.g. a connection between networks of different operators) — where in the classical synchronous networks, implemented with frame alignment and wander buffers, slips are unavoidable — the introduction of dynamic pointers eliminates data corruption due to slips.

Pointer Justification Events

S DH and SONET standards specify that pointer values may change frequently (every 4th frame i.e. after 500 µs/2 ms depending on the VC) which corresponds to a maximum frequency offset of approx. 3×10^{-4} that can be handled by this scheme. From this it is evident that the dynamic pointer performs bit rate adaptation: the pointer is actually used to perform byte justification!

Effects of Pointer Justification Events

The change of a pointer value (increasing or decreasing the value as required by the phase drift) is called a Pointer Justification Event (PJE). Justification events produce a well known effect — phase hits that translate into jitter when the payload is demultiplexed [4]. In the case of PDH, each time a tributary signal is demapped and subsequently multiplexed into another aggregate signal the intermediate signal appears as a physical signal that has to fulfil stringent jitter requirements at well defined interfaces. This avoids excessive jitter accumulation along a PDH trail.

PJEs correspond to phase hits of the payload of about 150 ns (3 VC-4 bytes) in the case of 140 Mb/s transported in a VC-4, approximately 50 ns (1 VC-3 byte) in case of 45 Mb/s and 3.5 ms (1 VC-12 byte) in case of 2 Mb/s transported via VC-12. Within the SDH/SONET network there are no physical interfaces where jitter due to PJEs becomes apparent. From an equipment perspective it is therefore not necessary to eliminate the phase hits or PJEs representing jitter as in the case of PDH. This may be why the problem was not recognised by the SDH community and the proposal to specify a "pointer adjustment jitter transfer function" [2] was not supported at that time.

Is the synchronization issue a threat for SDH/SONET that may hold back its implementation, or just a challenge to improve some implementation-specific deficiencies?

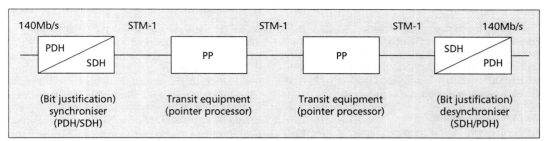

■ **Figure 1.** *Timing relevant functions in an SDH/SONET path: PDH/SDH synchronizer, SDH/PDH desyn-chronizer and transit equipment comprising pointer processors (PP).*

At gateway network elements of the SDH net-work the PDH payload is demultiplexed by the SDH/PDH desynchronizer (Fig. 1). The STM sig-nal and the Virtual Container (VC) comprise overhead and stuff bytes in addition to the 140-Mb/s data. By means of a PLL the desynchronizer removes the jitter (high frequency components) asso-ciated with all regular gaps in the PDH payload result-ing from the SDH overhead and stuff bytes. However, PJEs representing irregular gaps are superimposed on the PDH signal. The corre-sponding jitter amplitude requires appropriate filtering to achieve compatibility with existing PDH transmission equipment (input jitter toler-ance). The elastic store in the desynchronizer has to be optimised with respect to the resulting delay (small buffer required) and jitter amplitude (large buffer required). CCITT recommendation G.783 specifies that SDH/PDH desynchronizers have to provide a PDH compatible output jitter (and no data corruption) in the case of "single" and "dou-ble" pointers.[1] To achieve this jitter requirement the time constant of the desynchronizer has to be of the order of at least 0.5 second.

Effects Generating PJEs

A variety of effects contribute to the creation of PJEs. Two basic scenarios can be distinguished con-cerning the quality of the synchronization net-work: the synchronous mode of operation with 18 µs wander and the plesiochronous mode of opera-tion. The plesiochronous mode of operation is the typical case for a connection between net-works synchronized to their individual PRCs.

Plesiochronous mode of operation — In the plesiochronous mode of operation the frequency offset introduces PJEs independently of the buffer size. Assuming a maximum frequency off-set of 1×10^{-11} as specified for primary reference clocks (PRC, G.811) about 1 PJE will occur every second day in the case of 2 Mb/s and 1 PJE within two hours in the case of 140 Mb/s.

Synchronous mode of operation — In this case both buffer size and clock noise have to be considered.

If the buffer size is > 18 µs a static pointer value is sufficient to adjust the constant transmis-sion delays due to different distances.

When the buffer size is reduced below 18 µs down to about 1 µs the wander of the synchronization network and the wander of the transmission links create PJE. Due to the factors responsible for the wander (i.e. changing temperature of the transmission media, . . .) these PJEs occur fairly infrequently: a couple of PJEs per day (18 µs max. wander vs. 3.5 µs and 150 ns, respectively).

The clock noise becomes predominant when the buffer size is reduced below approximately 1 µs, because the high quality clocks (G.812/Stratum 2) in current use are characterised by a very narrow and low frequency noise spectrum combined with a well limited noise amplitude (MTIE <1 µs). In con-trast, cheap clocks (and SDH equipment clocks are expected to be cheap) exhibit a wider noise spec-trum with random (unlimited) amplitudes (see reference 3 for more information on the clock noise problem). The probability or mean frequen-cy of PJEs depends mainly on the short term stability of the related slave clocks. As the clock noise is a ran-dom process the PJEs are also randomly dis-tributed.

Pointer Accumulation

In order to minimize the transmission delay of the payload, buffer sizes in SDH/SONET equipment should be as small as possible; the pointer, imple-mented as a dynamic pointer, performs the required bit rate adaptation within the SDH/SONET network. As outlined above, in the case of a syn-chronized network PJEs occur randomly due to the non-deterministic nature of the clock noise. Each transit equipment in a chain (Fig. 1) randomly creates PJEs. This may lead to an accumulation of PJEs, i.e., consecutive pointers in the same direction within a certain (short) time interval, if PJEs are simply passed through by the equip-ment. Whether a PJE is fed through or absorbed depends on the actual buffer fill when the PJE arrives. Pointer accumulation is not critical inside of the SDH/SONET network, but becomes a problem when the PDH payload is recovered. Depending on the implementation of the desynchronizer, the big phase step of several consecutive pointers (in the same direction within the time constant of the desyn-chronizer) results either in a buffer overflow/under-flow or a jitter amplitude unacceptable for PDH equipment. Therefore, special precautions have to be taken to prevent pointer accumulation.

The Classical Pointer Processor

The classical SDH/SONET approach is to keep the number of PJEs to a minimum in order to reduce the probability of accumulation.

This pointer processor is specified in terms of a buffer (elastic store), a certain hysteresis threshold spacing, and a clock with a certain short term sta-bility (Fig. 2). The data is written into the buffer in accordance with the clock of the preceding equip-ment and is read from the buffer using the clock of the current equipment. The buffer fill varies according to the noise of the two related clocks, the wander of the connecting link and the wander of the synchronization signals synchronizing the

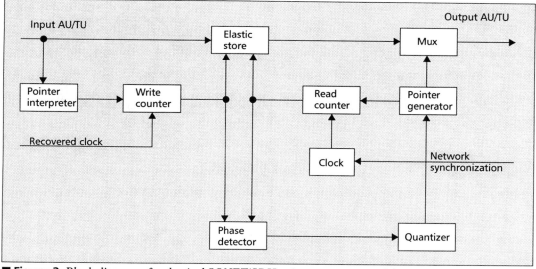

■ Figure 2. *Block diagram of a classical SONET/SDH pointer processor model.*

clocks. (The impact of the frame structure — gaps due to overhead bytes — on the buffer fill is not considered here). The phase detector recovers the phase from the buffer fill. The quantizer monitors the phase with respect to the thresholds and generates PJE requests.

The number of PJEs per time interval depends on the threshold spacing, the probability density function of the clock noise and the noise spectrum. Simulation results, based on clock noise characteristics assuming a certain short term stability, indicated that pointer accumulation is reduced to acceptable values, if the buffer threshold hysteresis spacing is at least 12 bytes in the case of VC-4, 4 bytes in the case of VC-3 and 2 bytes in the cases of VC-11 and VC-12 (G.783).

This implementation approach solves the problem of pointer accumulation with respect to PDH transmission equipment requirements. However, it does not consider that the buffer in conjunction with the threshold hysteresis spacing generates not only delay but also wander due to the time-dependent buffer fill.

The wander contributions of the individual pointer processors in a chain accumulates. The resulting wander of the SDH transmission link is limited by the synchronization network wander.

Pointer Processor Designs

Optimizing the classical pointer processor approach with respect to pointer accumulation and wander generation means to reduce both the buffer threshold hysteresis spacing and the clock noise in addition to the reduction of the synchronization network wander. Obviously, this has a severe cost implication.

As an alternative, a pointer processor design is required which allows the independent optimisation with respect to pointer accumulation and wander generation. This has been already realised in the PDH and requires just to adopt the related PDH principles.

To prevent pointer accumulation and the resulting corruption of data in the desynchronizer, filtering of incoming PJEs is required in pointer processors. This principle is visualised by Figure 3a which shows the boxes labelled transit equipment in

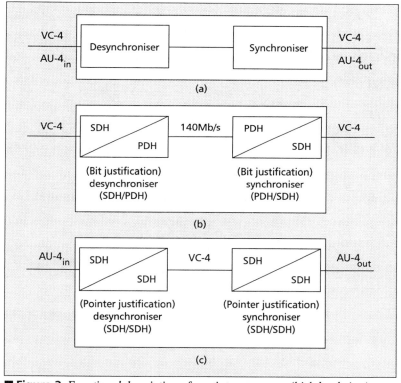

■ Figure 3. *Functional description of a pointer processor (high level view).*

Figure 1 as a back to back combination of a desynchronizer and a synchronizer. The desynchronizer eliminates/reduces PJE-related phase hits prior to the generation of outgoing PJEs in the synchronizer thus preventing pointer accumulation.

This can obviously be achieved by an implementation as described by Figure 3b where the input of the synchronizer is a 140 Mb/s signal (or any other PDH signal) and this signal is filtered as required to get compatibility with PDH equipment. The bit rate adaptation is performed by means of bit justification as specified in the context of the mapping of PDH signals into the Virtual Containers (G.709). This requires a bandwidth of the desynchronizer of somewhere between 100 and 500 Hz. This solution is applicable only for trans-

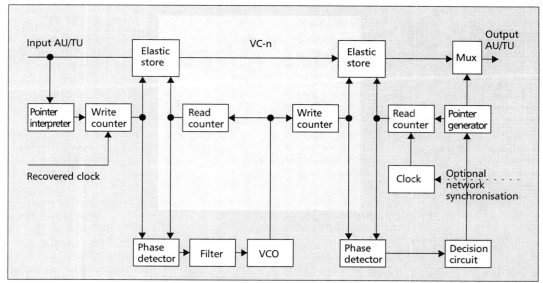

Figure 4. *Block diagram of a desynchronizer/synchronizer pointer processor model (2 control loops).*

__B__it rate adaptation by use of PJEs as specified for SDH/SONET is a +0-justification scheme, that is the pointer may move in either direction.

parent 140 Mb/s (or PDH) connections where VC path integrity is not required.

If VC path integrity is required the intermediate signal has to be a VC (Figure 3c). In this case the bit rate adaptation requires pointer processing.

Alternative Pointer Processor Designs

Figure 4 represents a straightforward implementation of a pointer processor as described by Figure 3c showing the separated synchronizer/desynchronizer functions as a hypothetical model. The VC-n clock is recovered by the desynchronizer PLL comprising the phase detector, filter and a VCO. The PLL exhibits a large time constant (more than 0.5 s) and corresponding buffers similar to SDH/PDH desynchronizers. A disadvantage is the high effort for this hypothetical implementation: two elastic stores with their control circuitry and a VCO.

To reduce this effort the functions which perform only internal intermediate signal processing (indicated by the shaded area in Figure 4), can be combined. A simple combination, however, has the effect that the filter, which filters incoming PJEs, also acts on the pointer generation process. This can be avoided by introducing a feed-back loop from the

decision circuit to the filter compensating the impact of the filter on the pointer generation process (Figure 5). The same effect can be achieved by inserting an equivalent filter function between write counter and phase detector instead of the filter between phase detector and decision circuit.

Therefore, Figure 5 describes a simplified model with the same characteristics as the desynchronizer/synchronizer combination of Figure 4. If the decision circuit represents a pure quantizer, this design is improved with respect to the classical pointer processor (Figure 2) by the use of a filter (recovering the VC-n phase) combined with a quantizer having a finer granularity (e.g. 3 bytes corresponding to a PJE step at the AU-4 level) instead of a coarse quantizer (12 bytes).

The Advanced Pointer Processor

Bit rate adaptation by use of PJEs as specified for SDH/SONET is a +0-justification scheme, that is the pointer may move in either direction. Even in the case of a reduced threshold hysteresis, this justification scheme may create very low frequency jitter components with amplitudes of about a pointer step size which contribute to wander [4].

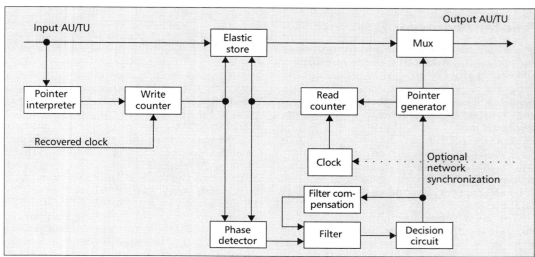

Figure 5. *Block diagram of a desynchronizer/synchronizer pointer processor model (1 control loop).*

Methods on the basis of threshold modulation have been proposed to shift these low frequency jitter components to higher frequencies in order to allow efficient filtering [5, 6].

Threshold modulation can be realized by adding a fixed modulation wave form, e.g. a saw tooth signal, to the input signal before quantisation (Fig. 6a)

Instead of a fixed wave form, a modulation wave form derived from the quantisation error, i.e. the difference of the quantizer input and output, can be added (Fig. 6b). This is termed adaptive threshold modulation and corresponds to sigma-delta modulation.

Figure 6b represents a first order sigma-delta modulator which may be considered as a numerical controlled oscillator.

In conjunction with the pointer processor feed-back loop (phase detector, decision circuit, pointer generator, read counter) a second order sigma-delta modulator characteristic is achieved. This provides improved spectral characteristics of the quantisation error compared to the threshold modulation methods which employ fixed wave forms [7, 8].

The advanced pointer processor generates alternating PJEs. The corresponding mean value represents the phase of the VC which provides timing transparency. These high frequency PJEs generated by the advanced pointer processor allow the implementation of desynchronizers with a shorter time constant/higher cut-off frequency, which considerably reduces the effort in the desynchronizer function of the pointer processor (Figure 3c).

This pointer processor implementation improves the dynamic behaviour of a chain of equipment due to the reduced time constant. A big advantage is the option (not the need!) to operate the SDH network elements unsynchronized, i.e., plesiochronously: this allows the use of cheap equipment clocks (just like in PDH equipment) and obviates the need of an SDH synchronization network. Besides this, the timing transparency provided by this pointer processor allows the provision of the synchronization links required by clients of the SDH network, e.g. the PSTN, in the conventional manner, i.e. via traffic carrying primary rate signals.

A first implementation of the adaptive threshold modulation scheme in the 140 Mbit/s-VC-4 mapper of the Philips SDH line transmission equipment (SLE-4 and SLE-16) was reported in 1991 [9].

Recent measurement results from a first pointer processor employing the adaptive threshold modulation scheme have been published [10]. The next section discusses these results and compares them with results from traditional implementations.

Consequences for the Network

Although at first glance the pointer processor seems to be a fairly unimportant function in the SDH equipment which can be easily implemented by means of well-known simple components, the impact of the pointer processors on the network as a whole is enormous.

Classical pointer processor — Network synchronization via primary rate signals is not possible in a transmission network comprising the classical SDH/SONET pointer processor. While the relevant

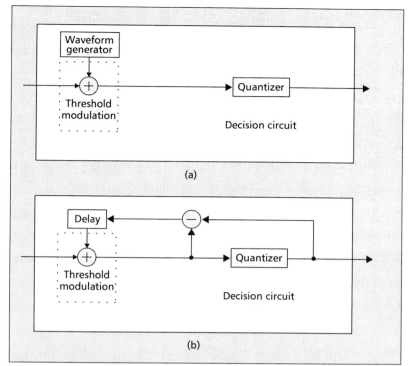

■ **Figure 6.** *(a) threshold modulation; (b) adaptive threshold modulation.*

CCITT recommendations G.823/824 specify that a synchronization link must not introduce more than 6 μs wander, a 1.5/2 Mb/s link via SDH may generate wander of up to 18 μs corresponding to the maximum wander amplitude of the synchronization network.

On the other hand, synchronization of all SDH/SONET network elements is required to prevent excessive pointer accumulation if the classical SDH/SONET pointer processor is employed.

Therefore, the CCITT (G.803) recommends to use STM-N timing; i.e., timing is transported via the line bit rate where no pointer processors degrade the timing. This work around introduces a series of problems.

At present, synchronization is in the domain of the switching (PSTN) operators. They are responsible for the expensive G.812 clocks (Stratum 2) residing in switches or separate synchronization supply units (SSU). The performance of these clocks is optimized to the requirements of the PSTN and the synchronization network is mapped on the hierarchy of the PSTN to optimize the cost. With the approach recommended by the CCITT, the clocks are shifted into SDH equipment, i.e. the operation and maintenance of the synchronization network becomes the responsibility of the transmission network operators which may result in problems.

Operating problems can be expected from the proposed SDH/SONET synchronization network which is no longer based on the Stratum concept that a lower quality clock never clocks a higher quality clock. To prevent timing loops, a set of additional measures (timing marker, . . .) has to be developed and implemented to cope with these problems.

In addition, the considerable number of clocks connected in tandem (up to 60 in a synchronization path including 8 G.812 clocks with 1 μs wander each) is not likely to meet the 6 μs requirement for a synchronization path [1].

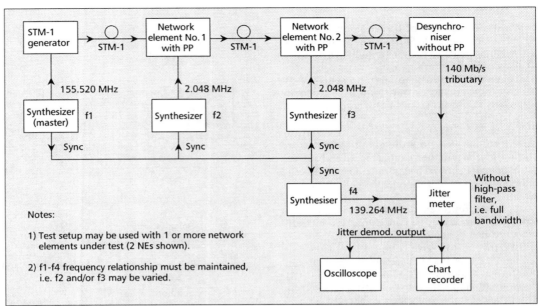

■ Figure 7. *Basic measurement set-up to determine jitter and wander generated by pointer processors (PP) in SDH/SONET network elements.*

To reduce the overall cost of the synchronization of the SDH network elements, the existing synchronization networks based on G.812 clocks (Stratum 2) have to be improved. This is not accepted by the users of these synchronization networks. To overcome this problem, ETSI started work on a specific SDH synchronization network specification that will result in an additional expensive synchronization network. Furthermore, expenditures related to the need to have such a synchronization network (i.e., planning, operating, and maintaining the network) have to be considered.

Finally, the problem of "third party timing," i.e., the task of transporting synchronization information across an SDH/SONET network as presently extensively performed using PDH transmission networks, cannot be solved using this approach. This is of growing importance with the increasing number of private and public service networks having a wider geographical coverage than the supporting synchronous transmission networks.

Advanced pointer processor — SDH/SONET networks that employ advanced pointer processors performing PJE-related jitter reduction and providing timing transparency do not have these problems. Even considering the cost of SDH equipment clocks alone, cost savings of one to two orders of magnitude are achieved by implementing an advanced pointer processor instead of a classical pointer processor employing a synchronized clock with a good short-term accuracy. Furthermore, the SDH transmission network becomes more reliable because there is no synchronization network whose faults may cause SDH network failures.

The data and timing transparency behave like PDH networks, while providing the expected added value of reduced OAM cost.

Considering that the probability of pointer accumulations in the classical SDH/SONET approach is kept low by keeping the probability

for PJEs low, it may be concluded that the high-frequency PJEs of the advanced pointer processor employing adaptive threshold modulation may give rise to pointer accumulation when a combination of both pointer processor types is used. This conclusion, however, does not take into account that the high frequency pointers are absorbed by the buffer threshold hysteresis in the classical pointer processor such that no harm for the network results from this implementation.

Broadly speaking, the advanced pointer processor makes SDH/SONET fully compatible with the PDH environment, as it provides timing transparency and requires no synchronization of the SDH network elements.

Measurement Results on Pointer Processor Implementations

*I*n the following section the effects of PJEs due to clock noise are studied in a best case scenario. Different pointer processor implementation approaches are compared with respect to the PJE accumulation mechanism and the timing transparency characteristics. Although pointer processors for both AU and TU levels are available, these examples only refer to 140-Mbit/s signals and AU-4 pointer processing for simplicity reasons.

Figure 7 shows the basic measurement setup that consists of an SDH network represented by a synchronizer (STM-1 generator), a number of transit NEs (pointer processors), and a desynchronizer in addition to the synchronization network and measurement equipment.

The test signal generator acts as a synchronizer, which maps the 140-Mb/s test signal via VC-4 into an STM-1 signal, each with nominal frequency. For measurement purposes all NEs derive their timing from external reference signals which are generated by separate synthesisers locked to the master clock.

The NEs represent the devices under test, which may introduce (superpose) phase errors corrupting the phase transfer of the transmission chain. As these phase errors associated with PJEs pass through further NEs accumulation may occur.

The STM signal and the Virtual Container (VC) comprise overhead and stuff bytes in addition to the 140-Mb/s data. The desynchronizer, in case of Philips equipment implemented as a linear low-pass filter, removes the jitter (high frequency components) associated with all regular gaps in the PDH payload resulting from the SDH overhead and stuff bytes. Irregularities due to PJEs or phase detector errors (resolution, nonlinearity) are fed through as phase deviations of the 140-Mb/s signal with respect to the 140-Mb/s input test signal.

In a chain of NEs employing the classical pointer processor with a PJE reduction filter, most of the PJEs will be absorbed in the next node, but some will feed through. If the phase deviation or wander occurring at the NE pointer processor buffer exceeds the threshold limit, PJEs will adjust the output phase by a phase step that is always directed towards the centre value. Only a small fraction of phase hits due to PJEs are generated by the last NE.

Therefore, the observable wander at the desynchronizer output is not caused by PJEs, but instead represents the sum of buffer fills in a chain of pointer processors changing in time according to the local SDH Equipment Clock (SEC, G.81s) noise and synchronization network wander. As the desynchronizer removes only the jitter of regular stuff bytes and bits, the output phase response is directly derived from the phase shape of the last NE clock preceding the desynchronizer.

The desynchronizer output phase shown in Fig. 8 is obtained with a network configuration consisting of a chain of 7 NEs ideally synchronized (i.e., no synchronization network wander) to a common clock; the phase deviations are only due to the clock noise of the local SDH equipment clock. Phase deviations are scaled down to the 140-Mb/s rate.

As PJEs in a synchronized network occur due to random noise effects of the SDH equipment clock and the synchronization signal, measurement results are difficult to reproduce. Deterministic clock signals (e.g., frequency offsets) in a simple network configuration (test setup) will generate predictable test results with respect to PJEs.

In Fig. 9 the desynchronizer output phase of a network comprising a single transit NE is shown. This NE is operated with a constant frequency offset of 0.01 ppm with respect to the synchronizer reference frequency, to generate regular PJEs. The NE under test is a Philips Add/Drop multiplexer (ADM of the PHASE[2] series), which employs a classical pointer processor upgraded by an improved phase detector that reduces frame structure effects due to SOH gaps. In this measurement, regular PJEs are expected from an ideal phase detector in conjunction with a standard pointer processor. Figure 9 shows small deviations of about 2 UI from the expected ideal saw tooth shape. This discrepancy is due to the limited resolution of the improved phase detector.

The maximum value in Fig. 9 is associated with the upper stuff threshold. Applying a frequency offset of opposite sign would generate PJEs of the opposite polarity due to crossing the lower stuff threshold. Therefore, wander generation is

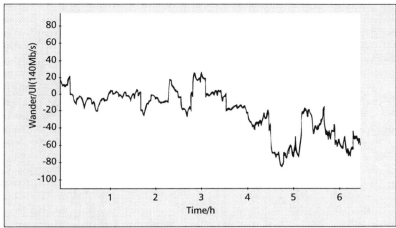

■ **Figure 8.** *Measurement result on the wander of a chain of seven ideally synchronized SDH network elements equipped with classical SDH/SONET pointer processors.*

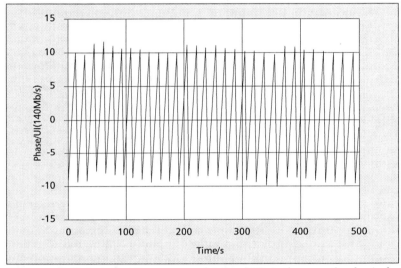

■ **Figure 9.** *Output phase inaccuracy related to PJEs in the case of a classical pointer processor upgraded by an improved phase detector which reduces frame structure effects due to SOH gaps (frequency offset 0.01 ppm).*

related to the stuff threshold hysteresis (12 and 4 bytes in the cases of AU-4 and AU-3, respectively) in conjunction with the synchronization quality. It is not directly related to the occurrence of PJEs.

It should be noted that by employing a simple phase detector that does not eliminate the frame structure effects, an additional phase deviation of approximately 9 bytes (approximately 70 UI) would be observed.

Figure 10 shows a measurement result obtained with the same network configuration and clock parameters as in Fig. 9. In this configuration the PHASE-ADM is operated in the advanced mode employing adaptive threshold modulation in conjunction with a PJE jitter reduction filter. As the PJE jitter only consists of high frequency components as a result of adaptive threshold modulation, the jitter can be easily removed by the desynchronizer. As no stuff threshold spacing effects are involved, the result is independent of the sign of the frequency offset or clock phase noise in the case of the synchronized mode. The small saw tooth shape is associated with the limited resolution of the phase detector and can be further reduced if required.

[2] PHASE is an acronym for Philips advanced SDH equipment.

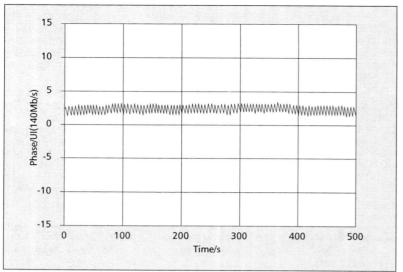

■ **Figure 10.** *Output phase of an advanced pointer processor employing adaptive threshold modulation and a PJE jitter reduction filter (frequency offset 0.01 ppm).*

In an extended network configuration, the PJE jitter reduction filters act as desynchronizers with a higher cutoff frequency and thus provide a good estimate of the original phase. This filter prevents PJE accumulations in the network. The phase estimate is corrupted by the phase detector error of preceding nodes. Only this residual error of the phase detectors will propagate through the network.

Standards on SDH Synchronization and Timing

*T*he international standards concerning synchronization and timing are far from being finalized. The quality of PDH primary rate signals currently used for synchronization is no longer specified by the CCITT since the Blue Book version of G.823/824 on jitter and wander of PDH signals has been modified due to problems with PDH outputs from SDH/SONET equipment.

As no international standards on SDH synchronization and timing currently exist that could be used to specify SDH equipment, the European Telecommunications Standards Institute (ETSI) has started work on this issue on order to create a standard specifying SDH synchronization network parameters (Draft title: DE/TM-3017: Generic Requirements for Synchronization Networks). This standard is restricted to the SDH synchronization network, as it was felt that the resulting performance and cost will probably not be accepted by people responsible for existing synchronization networks.

Lately, G.825 was adopted in which jitter and wander for STM-N interfaces are very stringently specified, because the underlying assumption is that these signals generally transport timing information. This implies the general use of fairly expensive SDH equipment clocks.

Conclusions

*T*he simple implementation of the classical pointer processor, which works well within the SDH/SONET network, results in considerable problems when implementing a network based on this type of pointer processor that is obliged to fulfil the requirements of client networks (jitter, wander, timing transparency) presently using the PDH network infrastructure.

Equipment using a new type of pointer processor with improved timing transparency is available and has been tested in the lab. The use of this pointer processor, providing the combined advantages of both the plesiochronous and synchronous techniques, will result in a more reliable network and lower cost.

References

[1] CCITT Blue Book Volume III — Fascicle III.5 (Nov.1988) Recommendation G.823: "The control of jitter and wander within digital networks which are based on the 2048 kbit/s hierarchy" and G.824: "The control of jitter and wander within digital networks which are based on the 1544 kbit/s hierarchy" and the revisions CCITT: COM XVIII-R 106-E, July 1992.
[2] M. Robledo and R. Urbansky, "Fixed and Adaptive Pointer Processing Schemes," ETSI-STC-TM3 Meeting, Brussels, April 24-28 1989, Temporary Document 92.
[3] J. E. Abate, *et.al*, "AT&T's New Approach to the Synchronization of Telecommunication Networks," *IEEE Commun. Mag.*, vol. 27, no. 4, April 1989, p. 35-45.
[4] D. L. Duttweiler, "Waiting Time Jitter," *Bell Systems Technical J.*, vol. 51, no 1, January 1972.
[5] W. D. Grover, T. E. Moore, and J. A. McEachern, "Waiting Time Jitter Reduction by Synchronizer Stuff Threshold Modulation," *Proc. IEEE GLOBECOM*, 1987.
[6] G. L. Pierobon and R. P. Valussi, "Jitter Analysis of a Double Modulated Threshold Pulse Stuffing Synchronizer," *IEEE Trans. on Commun.*, vol. 39, no. 4, April 1991.
[7] R. Urbansky, "Synchronization of Synchronous Networks," Philips Innovation 2/1991, p. 23-32.
[8] R. Urbansky, "Pointer Processing for Synchronization Signals in SDH Equipment," ETSI-STC-TM3, The Hague, April 1991, Temporary Document 2/39.
[9] R. Urbansky, "Simulation Results and Field Trial Experience of Justification Jitter," Proc. 6th World Telecommunication Forum, Technical Symposium, Geneva, October 10-15, 1991, part 2, vil. III, pp. 45-49.
[10] M. Klein, "Results from Calculations and Measurements on Different Pointer Processor Implementations," ETSI-STC-TM3 Meeting, Madrid/Spain, March 29-April 2, 1993, Temporary Document 68.

SONET REQUIREMENTS FOR JITTER INTERWORKING WITH EXISTING NETWORKS

Robert O. Nunn

AT&T Bell Laboratories
Holmdel, New Jersey

Abstract

Existing asynchronous transport networks have been engineered to maintain jitter at levels low enough that it causes essentially no errors. SONET will interwork with existing networks by transporting signals originating in those networks (eg DS1- and DS3-rate signals) as part of its payload. As SONET is introduced, it is important to ensure that jitter on these signals remains within acceptable levels.

Both high-frequency jitter that affects clock recovery circuits, and broad-band jitter that affects circuits in multiplexers and demultiplexers present interworking issues. This paper addresses broad-band jitter. There are two major mechanisms by which SONET generates broad-band jitter on a payload: pulse-stuffing used to map a signal into SONET, and the pointer mechanism employed by SONET to maintain phase alignment of tributaries. The effect of pointer adjustments depends on equipment characteristics, network architecture, and characteristics of the signals that provide timing to SONET devices. This paper discusses the requirements for controlling jitter from both mapping and pointer adjustments.

1. Introduction

SONET is a fiber optic transmission system with bit rates defined from the basic rate of 51.84 Mbs up to 2.488 Gbs. SONET has standardized interfaces that ensure that equipment from multiple vendors will interwork. An important aspect of interworking is control of jitter. Jitter specifications have been developed for SONET optical and electrical signals (OC-N and STS-N). In addition, specifications are being developed for the jitter performance of SONET payloads[1], so that SONET will interwork correctly with existing networks.

The jitter requirements for SONET OC-N signals are very similar to those for line systems of existing networks. The requirements for SONET payloads, however, require careful analysis because of the phase-alignment mechanism of SONET, which has no analogue in today's networks. The phase-alignment mechanism is the pointer adjustment. SONET employs a pointer to indicate the position of the payload in its frame, and frequency differences of tributaries are accommodated by allowing the tributary to float within the frame. An adjustment of the tributary position (by one byte) is referred to as a pointer adjustment. The jitter that results from pointer adjustments will combine with jitter from other sources, and the total must be controlled to ensure that existing networks will continue to function correctly. The analysis of

this accumulation of jitter from several sources is the subject of this paper. The analysis of SONET jitter has taken place over several years through contributions by many companies in the working group T1X1 of the Exchange Carriers Standards Association. This paper cites a few of these contributions.

2. Significance of Jitter

Digital jitter is defined to be the result of high-pass filtering the phase of a signal. The steps in determining jitter are shown in Figure 1. A phase extractor compares the signal with a

Figure 1. Definition of Jitter

reference signal to produce a difference signal. This difference signal is then high-pass filtered. The output of the filter is jitter.

The purpose of the filter is to represent the effect of phase noise on devices that will receive the signal[2]. Such devices include clock-recovery circuits and phase-smoothing circuits that are part of demultiplexers. Both of these devices have a low-pass characteristic.

The effect of phase error on a device with a low-pass characteristic is shown in Figure 2. If the transform of

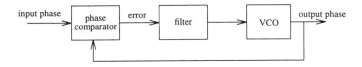

Figure 2. Phase-locked loop

the input phase is ϕ_i, the transform of the output phase is ϕ_o, and the transfer function is H_l, then

$$\phi_o = H_l \times \phi_i$$

The transform of the circuit's error signal is

$$E = \phi_i - \phi_o$$
$$= (1 - H_l) \times \phi_i$$

Since H_l is the transfer function of a low-pass filter, $(1-H_l)$ is the transfer function of a high-pass filter. Therefore if the corner frequency of the jitter filter is the same as the corner

Reprinted from *IEEE GLOBECOM '93 Conf. Rec.*, pp. 1501–1505, Dec. 1993.

frequency of the device receiving the signal, then jitter is approximately the same as the error signal. For a phase-smoothing circuit in a demultiplexer, the error signal represents the offset from center of the buffer. Thus if the amplitude of jitter exceeds the distance from the center to the edge of the buffer, the buffer will spill. For this reason, while jitter is generally expressed as a peak-to-peak measurement, the peak amplitude is the correct measure for determining performance.

Procedures for measuring jitter include a measurement time, which is often taken to be 30 seconds. The significance of this time has primarily to do with testing convenience. Since (current) network jitter has energy primarily in tens to hundreds of Hertz, a measurement of 30 seconds yields many hundreds of effectively independent samples. For Gaussian distributed samples, the difference between 30 seconds and one hour of measurement is likely to be small. With SONET, however, it is possible for lower frequency energy components to be present. As explained above, a jitter peak corresponds to a buffer spill, so ensuring that errors do not occur may require a longer measurement.

3. Network Model

Predicting performance of a network requires a network model. Jitter requirements for SONET have been based on a network model that was developed by information provided by network operators[3] concerning their existing networks and plans for deploying SONET. The model is shown in Figures 3 and 4. Figure 3

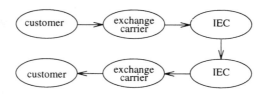

Figure 3. Network model

shows that it consists of two customer networks, two exchange carrier networks, and two inter-exchange carrier networks. Each of these parts is considered to be a combination of existing devices, plus SONET devices, so that a signal goes into and out of a collection of SONET devices (an island) a number of times, as shown in Figure 4. The number of islands was taken to be 2 for each customer, 6 for each exchange carrier, 8 for one inter-exchange carrier and 4 for the other, plus an extra 4 for restoration. The total for the network is 32 islands. Each island was taken to have ten pointer processors.

This model is quite large. Most DS3 circuits will be much simpler than the model. The model is intended to represent the most complex connection that is likely to occur, so that SONET will work correctly for even the longest routes.

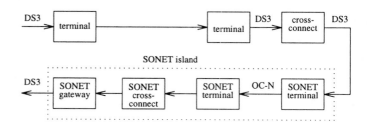

Figure 4. Illustration of SONET Island

4. Allocation of Jitter

Network jitter requirements consist of a requirement for high-frequency jitter corresponding to clock recovery circuits, plus a broad-band requirement corresponding to multiplexing circuitry. The method of deciding requirements for high-frequency jitter is similar to that for existing networks. This paper considers broad-band jitter, for which new analysis methods were needed. Because of the nature of the pointer adjustment process, jitter on DS3-rate signals (44.736 Mbs) is considered more critical than that for DS1-rate signals (1.544 Mbs). This paper therefore discusses jitter for DS3-rate signals. Analysis for DS1-rate signals will be similar.

The purpose of jitter requirements for SONET is to ensure that accumulated jitter does not cause buffers in the existing network to overflow or underflow. The buffers are required to tolerate 5.0 UI of jitter. The method used by T1X1 is to allocate the 5.0 UI network limit to each of three sources of jitter[4].

The first of the three sources is the asynchronous mapping process, for which the allocation is 2.0 UI. This allocation includes mapping jitter from both existing equipment, and SONET. In today's network, jitter levels are allowed to be as large as 5.0 UI. In order to ensure compliance with this lower level of mapping jitter, the requirements for SONET devices will have to be more strict than those for existing devices.

The remainder of the 5.0 UI allocation is for jitter due to the pointer adjustment process, and is divided into two parts. One part is an allocation for synchronized islands (2.4 UI), and the other is an allocation for one island that has lost synchronization (0.6 UI). SONET is required to be operated as a synchronized system. That is, terminals and cross-connect devices are to be timed from a source that has been derived from a network clock of highest quality (Stratum 1). However, on rare occasions, a device may lose synchronization. T1X1 expects the network to continue operating correctly when this occurs. Therefore an allocation is made for a portion of the network to be operating in a condition of synchronization loss.

Jitter from the above three sources will combine to yield total network jitter. T1X1 did not investigate the manner in which jitter will combine, but rather simply required the sum of peak-to-peak levels to be no more than the allowed 5.0 UI of peak-to-peak jitter. The combined jitter will likely be lower, but modeling the combined jitter is difficult because of the difference in rates between the pulse-stuffing process of mapping jitter and the pointer adjustment process.

5. Analysis of SONET Payload Jitter

SONET devices used to transport a DS3-rate signal are shown in Figure 5.

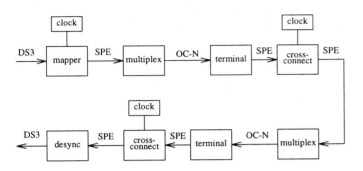

Figure 5. Elements of a SONET system

The payload is first mapped into the SONET frame. This process employs pulse-stuffing for synchronization. Once this is done, synchronization of SONET tributaries is accomplished by the process of pointer adjustments. Pointer adjustments occur at an element that performs cross-connecting, or perhaps at an element that drops a tributary. At the point where the payload is extracted from SONET, the phase of the original signal will have been altered by overhead, the phase steps due to pointer adjustments, and the phase steps due to pulse-stuffing. Since overhead is deterministic, resulting phase steps can be smoothed quite thoroughly. The phase steps due to pulse-stuffing and pointer adjustments are random and more difficult to smooth.

5.1 Pulse Stuffing

This process is employed in existing asynchronous systems, and the resulting jitter is well understood. Jitter accumulation is affected by the stuffing ratio, and by the characteristics of the phase-smoothing filter in the desynchronizer. For mapping a DS3-rate signal into SONET, the nominal stuffing ratio is two thirds.

The amplitude of mapping jitter varies considerably with stuffing ratio[5]. The variation for a particular phase-smoothing filter is shown in Figure 6, which represents results of a computer simulation of the pulse stuffing and filtering processes. Jitter amplitude has peaks at stuffing ratios equal to a ratio of small integers. The peak at two thirds is about one third UI. This is much larger than the jitter amplitude of most existing proprietary systems for transporting DS3-rate signals.

In order to determine the effect on accumulation of jitter, computer simulations were made of a series of SONET connections[6]. The simulations were performed by modeling the phase errors introduced by the stuffing process, and numerically performing the low-pass filtering operation corresponding to the gateway phase-smoothing filter. Stuffing

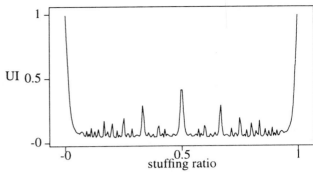

Figure 6. Jitter dependence on stuffing ratio

ratios randomly distributed near the nominal value were used. Simulations were performed for several types of phase-smoothing filters. Results are shown in Figure 7.

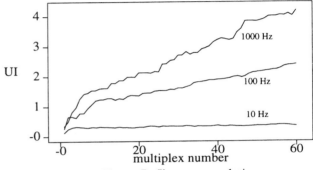

Figure 7. Jitter accumulation

It is desired to keep jitter at low levels through 32 multiplex-demultiplex pairs. Based on the results shown in Figure 7, the phase-smoothing filter for DS3-rate signals transported by SONET is required to have a low-pass filter with a corner frequency no more than 40 Hz. This compares with a requirement of 1000 Hz for asynchronous systems.

5.2 Pointer Adjustments

The other primary contributor to jitter is the pointer adjustment process. Pointer adjustments are due to differences in reference timing signals. The difference can be due to noise on a signal, or to a frequency error of a device that has lost its reference. For developing jitter requirements, it was assumed that one island could have a synchronization loss, while the others would be synchronized.

5.2.1 Synchronized Case Pointer adjustments for the synchronized case occur due to noise on the reference timing signals. In order to predict pointer adjustments, a noise model is needed. The model used by T1X1 was developed from data provided by network carriers from their existing networks. From these data, a noise mask for DS1 reference signals was developed. The mask exceeds almost all the data. The intention of this procedure is to ensure that SONET will work correctly when timed from the existing synchronization network.

The parameter used by T1X1 to describe noise on a reference timing signal is TDEV. Its use is a result of a lengthy effort to find a proper measure for noise on a synchronization signal. Before work began on SONET, the parameter used by T1X1 to characterize reference timing signals was MTIE, which is useful for slip performance of network switches. For SONET pointer adjustment performance, however, MTIE does not provide a very good characterization of noise. In the first version of SONET requirements, a new parameter called rms TIE was employed. However, it also proved deficient. T1X1 then turned to the National Institute of Standards and Technology (NIST) for help in choosing a parameter. Experts in characterization of precision oscillators at NIST suggested the parameter TDEV, which is derived from the modified Allan variance[7]:

$$TDEV = \sqrt{\frac{1}{6} < (\Delta_2 \overline{\phi})^2 >}$$

where $<>$ is the expected value, and $\Delta_2 \overline{\phi}$ is the second difference of averaged phase. TDEV distinguishes the noise processes (such as flicker phase and white frequency) that are commonly seen in oscillators.

T1X1 specifies a mask for DS1 reference signals[1] together with a second mask for SONET OC-N signals[1] that is lower at short observation times than the DS1 reference mask. SONET clocks will have to produce an output complying with the OC-N mask, and hence may have to provide filtering. The purpose of the lower mask is to ensure that the number of pointer adjustments is properly limited.

Modeling reference timing noise was accomplished by generating noise that just meets the OC-N mask. (Modeling noise based on TDEV requirements can lead to a noise signal that violates the existing MTIE requirement. Calculation of jitter based on noise signals compliant with both clock specifications has not yet been performed.) Computer simulators then calculated pointer adjustments occurring through an island by modeling the pointer adjustment process.

The method used to ensure that network jitter from pointer adjustments for the synchronized case will comply with the 2.4 UI allocation was to place a limit on the jitter due to a single pointer adjustment. This limit, together with the statistics of pointer adjustments and the network model, imply a network jitter level. Since jitter that has accumulated across a network is a random process, jitter amplitude depends on observation time. T1X1 interprets the 2.4 UI limit as a level that is to be exceeded no more than once per day. (That is, the peak magnitude should exceed 1.2 UI no more than once per day.)

The limit on jitter from a single pointer adjustment was decided by analyzing jitter accumulation through the reference connection described in Section 3. Jitter levels were predicted by both computer simulation[8] and analytical methods[9]. The analytical methods employed statistical arguments that used the results of the pointer adjustment modeling process.

The computer simulations modeled the phase smoothing process of the SONET desynchronizer. The smoothed phase from an island was added to that produced by pointer adjustments in previous islands. (Phase from each island is encoded by the pulse-stuffing process into the succeeding island.) This sum of smoothed phase steps represents the result of the pointer adjustment process. Jitter was calculated by simulating the process of passing the output phase from the final island through a 10 Hz high-pass filter.

Simulations were performed using a variety of phase-smoothing devices. Results of simulation of a device that releases the 8-UI phase step of a pointer adjustment one UI at a time are shown in Figure 8. The time between the 1-UI

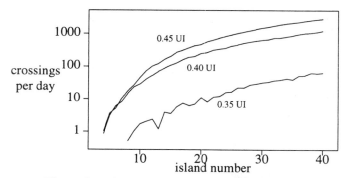

Figure 8. 1.2 UI crossings due to pointer adjustments

phase steps was taken to be 0.05 seconds. Following the device that produces 1-UI phase steps is a linear phase-smoothing filter. Figure 8 shows threshold crossings for three filters, whose single pointer adjustment jitter values are 0.45 UI, 0.40 UI, and 0.35 UI. The results are from a simulation of 6 days of network operation.

Figure 8 shows that the number of threshold crossings decreases rapidly with a fairly small reduction in the jitter due to a single pointer adjustment. Simulation results for a filter generating 0.30 UI showed almost no threshold crossings. These results were matched very closely by the results of statistical modeling of jitter accumulation. Based on such results the requirement for jitter performance of a desynchronizer was taken to be no more than 0.30 UI peak-to-peak for the response to a single pointer adjustment.

5.2.2 Synchronization Loss Case Jitter due to pointer adjustments for an island that has lost synchronization is accounted for by laboratory tests. That is, since only one island with synchronization loss is required to be accommodated, the performance of a device under synchronization loss conditions is measured. Several tests have been defined that are considered to represent worst-case combinations of pointer adjustments. These include regularly occurring pointer adjustments that correspond to a clock running at offset rates from zero up to the maximum allowed rate of 4.6 ppm. In addition, the tests include added and deleted pointer adjustments that correspond to randomly arriving pointer adjustments altering the pattern resulting from a frequency offset.

5.2.3 Testing Methods The limits described above for jitter due to a single pointer adjustment and due to synchronization loss cannot be tested directly, since an operating system includes other sources of jitter. (Mapping jitter is always present.) The limit for the result of a test must account for the other sources.

Measurements have shown considerable variation between results from differing measuring devices, and measurement procedures. This indicates a need for standardization in test methodology.

6. Conclusion

Network operators have recently begun installing SONET devices in their networks. Good jitter performance of these networks must be ensured as SONET devices become more widely deployed. The basis for jitter performance requirements has been a combination of analysis, computer simulation, and laboratory measurements. Developing requirements has required investigation of several areas about which little was previously known. In all steps of the process, conservative estimates have been made to ensure that jitter levels will not be excessive.

REFERENCES

1. Draft American National Standard for Telecommunications, "Synchronous Optical Network (SONET): Jitter at Network Interfaces," T1X1.3/93-006R2, 1993.

2. R. O. Nunn, "Effect of Pointer Adjustments on a Downstream Asynchronous Network," AT&T-Communications, T1X1.6/90-015, May, 1990.

3. R. O. Nunn, "Discussion of Number of SONET Islands in an IEC Network," AT&T-Communications, T1X1.2/92-036, November, 1992.

4. R. W. Cubbage, "Hybrid Network Jitter Allocation Proposal," Alcatel Network Systems, T1X1.3-92-019, February, 1992.

5. D. L. Duttweiler, "Waiting Time Jitter," Bell System Technical Journal, Vol. 51, pp 165-207, January, 1972.

6. R. O. Nunn, "DS3 Jitter Control in a Mixed SONET-Asynchronous Network," AT&T-Communications, T1X1.3/91-129, October, 1991.

7. D. Allan, M. Weiss, J. L. Jespersen, "A Frequency-Domain View of Time Domain Characterization of Clocks and Time and Frequency Distribution Systems," 45th Annual Symposium of Frequency Control, May, 1991.

8. B. Powell, D. O'Connell, T. Onofrio, "Multiple SONET Islands Jitter Simulation with Jitter Thresholds and Frequency Offset," Alcatel Network Systems, T1X1.3/91-128, October, 1991.

9. Z. Luan, F. McAllum, "DS3 Jitter Allocation, Budget, and Specification Proposals," Northern Telecom Incorporated, T1X1.3-92-062, June, 1992.

Section 5

Operations, Administration, and Management

PERHAPS the three most oft-cited carrier benefits identified with SDH/SONET are: 1) optical midspan meet, providing a basis for consistent multivendor product assessment, and a necessary (although insufficient) condition for management interoperability; 2) bandwidth formats and mapping options that accommodate existing services while facilitating introduction of new ones, and an important ingredient for a simplified network architecture that improves performance and facilitates agile robust networks; and 3) comprehensive network management. With regard to this last point, unlike the plethora of disparate networks extant today, the expectation for coherent network management was articulated early and has been a prominent theme throughout years of SDH/SONET standardization.

The seven papers that make up this section provide an excellent basis for understanding the management of SDH/SONET-compliant equipment, including the layered overhead signal structure and relevant standards and consortia initiatives (Telecommunications Management Network, TMN; Telecommunications Information Networking Architecture, TINA; and SDH Management Network, SMN).

With seven papers in this section, some topical overlap is to be expected. This should be viewed as a strength in that it introduces different perspectives while reinforcing the most fundamental points. The first paper, "SDH/SONET — A Network Management Viewpoint" by R. F. Holter, affords an excellent introduction to the layered overhead structure, performance monitoring data, maintenance signals, and orderwire, user, and data communications channels (DCCs). Section and line DCC protocols are reviewed, and TMN and SDH/SONET Management Network (SMN) configurations are discussed.

The second paper, "SDH Management" by J. F. Portejoie and J. Y. Tremel, builds upon the first paper by offering additional detail in the area of management processes as reflected in relevant ITU standards. Specific protocol standards are considered, the TMN object-oriented approach to modeling managed resources is discussed, and very specific information regarding fault and performance management is provided.

A layered approach to network management in the TMN context is elaborated by L. H. Campbell and H. J. Everitt in "A Layered Approach to SDH Network Management in the Telecommunications Management Network." Four network management control layers are discussed, followed by a discussion of TMN functional layering and network management, with ample reference to applicable standards. The authors include a discussion of the relation between physical and logical network views, using service restoration as an example.

The following paper differs substantively from the preceding three in that it takes a much more product-oriented view toward SDH management of network elements. J. Blume *et al.*, in "Control and Operation of Network Elements," provide a thorough review of basic SDH network management, and then maps those requirements to specific network element functions and hardware/software partitions, with special emphasis on a graphical interface, processing, and control internal to an SDH-compliant multiplexer.

The fifth paper provides still another perspective on operation, administration, and management. As the beginning of this section preface should make clear, SDH/SONET standards create a more even playing field for both vendors and service providers. Consequently, telecommunication carriers must obtain a competitive edge with efficient, imaginative services that address customer needs. In "A Synchronous Digital Hierarchy Network Management System," T. Kunieda *et al.* propose the application of TMN and offer insights into Nippon Telegraph and Telephone (NTT) network operations and maintenance. Protocol and message data based on standard management interfaces are seen as a means to facilitate the introduction of value-added services in the marketplace.

The TINA Consortium (TINA-C) was formed to promote dialogue among network operators, the R&D community, and suppliers for the purpose of converging telecommunications and computing. The sixth paper, "SONET Operations in the TINA Context" by W. J. Barr *et al.*, draws upon perspectives gleaned from TINA workshop contributions to describe an information networking architecture. The power of the architecture is illustrated by considering an application example for SONET maintenance and operation.

The seventh and final paper in this section examines Recommendation G.784 in the context of implementing routing and addressing schemes for Synchronous Digital Hierarchy Management Network (SMN). The authors note that one implementation strategy is to use both X.25 networks and Data Communications Channels (DCCs). They argue that implementing the protocol specified in G.784 can then result in serious deficiencies. Instead, they endorse a second option, to use the DCC exclusively, with the protocol specified in G.784.

SDH/SONET — A Network Management Viewpoint

RONY HOLTER

NETWORK PLANNING, MARKETING

ALCATEL NETWORK SYSTEMS, INC.

RICHARDSON, TX

1. INTRODUCTION

SDH, Synchronous Digital Hierarchy [1]–[7] and SONET, Synchronous Optical NETwork [8], [9], define feature sets and functionality for the next generation lightwave transmission systems to be used in the public telephone networks. This new technology provides three major benefits to the network for future applications, i.e., optical midspan meet, flexible bandwidth for future services, and extensive network management features. The importance and priority of each of the three major benefits have changed with the evolution of the specifications and the understanding of the standard itself. The midspan meet benefit allows consistent multivendor product evaluations, while flexible bandwidth features can be demonstrated by applying format and mapping options to new service offerings. Unlike the other two benefits, the extensive network management features reside in the architecture, and must be understood as applied to the existing transmission network and its operation support systems. Understanding of these features and how to apply them to managing a network are the mysteries that network operators must solve to reap the wealth of the new technologies.

Another plot in this mystery is the standardization of network management applications addressing the telecommunication management network, TMN[10]. Understanding how SDH/SONET management features build upon this overall TMN architecture is another challenge for the network operators to solve. Network management standardization covers all aspects of the telephony network, from information architectures to the determination of an errored second of transmission. An overview of the TMN standards which support SDH/SONET will be addressed to understand the application of the infrastructure of the transport's management features.

2. SDH, SONET, AND TMN SPECIFICATIONS

The first challenge of solving the mystery of optical transport and network management of the future is to locate the precious sets of information that describes these standards or recommendations. ANSI T1.105-1991 [9] contains the necessary information to understand SONET, while ITU-TS (formally CCITT) Recommendation G.781 [4] outlines the set of specifications required to understand SDH. Management standards for SDH begin with ITU-TS Recommendation G.784 [7] which applies the TMN standards to SDH and defines the control and monitoring functions relevant to SDH network elements. It also defines the SDH subnetwork architecture, embedded control channel functions, and control channel protocols. ANSI's T1.105-1991 contains the management information which is similar to G.784. Each standards body has addressed important management message sets, and these standards and recommendations are just becoming available in draft forms[11]. Consistency of management message sets and future interface specifications between the two standards bodies is a major concern for world network interoperability.

In addition to the new transport technology and its specific management requirements, the overall network management architecture requires new approaches. This new challenge is requiring new features in operations support systems (OSSs) and new methods of collecting information from network elements (NEs). The new OSS features will have to support fault locating with continuous in-service performance monitoring to be able to reduce reactive dispatches for network repairs. The management features available in SDH/SONET provide the means to assist the OSSs, but mediating this information and managing the NEs will require more management intelligence to be distributed in the network.

This new management architecture to address managing the new technologies is specified in a set of draft ITU-TS recommendations in the M.3000 series[10], i.e., M.3010 Principles for a Telecommunications Management Network, M.3020 TMN Interface Specification Methodology, M.3100 Generic Network Information Model, M.3200 TMN Management Services: Overview, and M.3400 TMN Management Functions. ANSI's set of documents [12]–[15] in the TMN area have very little information that would add to the M.3000 series; therefore, ITU-TS documents will be used to address SDH/SONET management.

3. SDH/SONET — MANAGEMENT IN THE NEW TRANSPORT SYSTEMS

All the features of the next generation lightwave transmission systems have been summed up with just one of two words, SDH/SONET. This new transport technology has been evolving since the early 1980s, with Bellcore, ANSI, and ITU-TS providing the standards forum to develop consensus on features. A base rate, format, and overhead structure coupled with a multiplexing scheme have been designed to incorporate a modular family of optical interfaces to support flexible bandwidth allocations, vendor compatibilities, and extensive network management features.

Why are there two different ways to spell the same set of standards, or are there really two different standards for the new transport network? A new draft technical report, ANSI T1X1, "A Comparison of SONET and SDH" [16], describes the differences in the two standards, and from a network management viewpoint, only the terminology differs. Some minor management features have not been defined in one or the other, but all in all, we can consider the standards as one. Continued standardization efforts will hopefully provide a migration from ANSI's SONET to ITU-TS SDH, but only time and cooperation in the future will determine this success.

The initial deployment of products in the network has provided the question of how to manage bandwidth allocations and use network management features to support existing and future network services. SDH/SONET contains management features that are not obvious to the layman in providing visibility of the level of performance of the telecommunication system. For example, the layered architecture of the synchronous format defined in the specification provides the foundation required to determine fault location. The other management features standardize some existing management operations, while others present new ways of managing future networks. For discussion of these features and their applications, we will categorize them into the following areas:

- Layered overhead structure
- Performance monitoring data
- Maintenance signals
- Orderwire channels
- User channels
- Data communication channels

With these management features built into the format, the next challenge is understanding their applications and building networks to take advantage of these features.

3.1 Layered Overhead Structure

One of the basic concepts of the synchronous signal format is the layered structure of the signal and the overhead supporting the signal processing. The four[1] basic layers of the signal format and how they relate to equipment and specifications are as follows:

- Photonic layer — defines the optical pulse shape, power levels, and wavelength
- Section layer — provides signal framing and basic level of performance monitoring of the payload (referred to as

[1]ANSI's SONET specifications have divided the section overhead layer into two separate layers, "section" which contains the RSOH (regenerator section overhead), and "line" which contains MSOH (multiplex section overhead). The SONET terms to differentiate the overhead will be used for clarity.

regenerator section in SDH)
- Line layer — provides protection switching and multiplexing functions for the information payload (referred to as multiplex section with an administrative unit in SDH)
- Path layer — provides signal labeling and tracing for end-to-end payload management.

In the specifications, the layered structure was extended to the overhead functionality which supports the synchronous signaling format. All but the photonic layer have supporting overheads, as shown in Fig. 1, with the path layer having two overhead functions, the synchronous transport path overhead and the virtual tributary/container (VT/VC) path overhead. From the network management viewpoint, the importance of the layered structure resides in the types of information provided at each layer and the ability to use that information in determining network performance.

3.1.1 Overhead Layers. The three overhead layers contain information that can be used to manage the network in a hierarchical manner. The lightwave network elements shown in Fig. 2 denote the generic naming convention used throughout the SONET specification. Each network element that performs the path (PTE), line (LTE, section multiplexer in SDH), and section (STE, section regenerator in SDH) termination function on the payload must be capable of processing the overhead information for signal processing and for network management support.

The section overhead must be processed by each network element to accomplish the basic transport function of framing on the payload signals. In addition to signal framing information, the overhead also contains the basic level of performance monitoring data, local orderwire channel, user channel, and a 192 kb/s data communication channel. Since every network element must process this overhead layer, these management features will be the basic offering on initial compliant products.

Functions supported by the line overhead layer for payload management consist of payload pointer storage and automatic protection switching commands to be processed by line termination or multiplex elements. The network management functions supported at the line layer consist of additional performance monitoring data, express orderwire channel, and a 576 kb/s data communications channel.

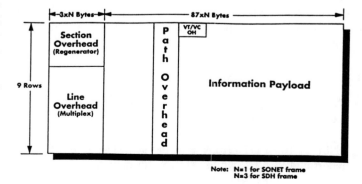

Fig. 1.   Layered overheads.

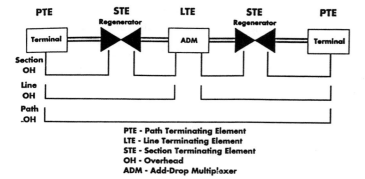

PTE - Path Terminating Element
LTE - Line Terminating Element
STE - Section Terminating Element
OH - Overhead
ADM - Add-Drop Multiplexer

Fig. 2. Termination of overheads.

The path overheads are used to provide end-to-end management of the payloads at the service terminating location of the lightwave networks. Path overhead provides performance monitoring, signal labeling, status feedback, user channel, and a tracing function. The virtual tributary (VT) or virtual container (VC) path overhead provides performance monitoring, status feedback, and signal labeling. Fig. 2 can be used as a reference in determining network elements required to process the path, as well as the section and line overhead bytes of information.

3.2 Performance Monitoring Data

Performance monitoring is a term used in the telecommunication industry to measure the performance of the network. Until recently, performance data could consist of any information available from a network element used to determine how well it performed its function, i.e., alarms and statuses. Telecommunication standards organizations have defined performance monitoring primitives, parameters, and failure criteria for various signal rates provided by network elements. ANSI has documented these standards in ANSI T1.231[17], while the ITU-TS version is contained in several recommendations that can be summarized in tables contained in G.784. The next subsections will briefly discuss the primitives and parameters that are available from the basic information provided in the SDH/SONET overhead architecture.

3.2.1 Performance Primitives and Defects.

The set of primitives and defects defined in the standards document are the basics used at each overhead layer in building the parameters to provide meaningful management information. The following list briefly describes the primitives and defects which are used to determine the performance of the network.[2]

- Bit Interleaved Parity (BIP) — Parity error code generated for comparison at the receiver to determine transmission integrity.
- Path Far-End Performance Report (FEPR) — Path status message sent from receiver to transmitter.
- Loss of Signal (LOS) — Occurrence of no transitions on the incoming signal for a defined period of time.
- Section Severely Errored Frame (SEF) — Four consecutive

[2]The international agreement on these terms and the definitions is still in progress. The latest ANSI terms are used which are contained in ANSI T1.231-1993.

errored frame alignment signals followed by two successive error-free frames.

- Section Loss of Frame (LOF) — Occurrence of a severely errored frame defect persists for a period of 3 ms.
- Loss of Pointer (LOP) — Occurrence of a valid pointer not being detected in eight contiguous frames or when eight contiguous frames are detected with a new data flag set.
- Alarm Indication Signal (AIS) — Defect occurs with the reception of an AIS signal for a set number of frames, five at the line layer and three at the path layer.
- Remote Defect Indication (RDI) — Defect occurs with the receiving of the RDI signal for five frames defined at each layer. At the line layer, a remote failure indication (RFI) is derived with the persistence of the RDI signal at the path layer, previously referred to as a path yellow signal.

Other indicators and signals for performance consist of laser bias current, optical power transmitted and received, protection switched events, and others that may be available, but are now seen as optional and manufacturer specific in measurement or for future standardization. Equipment performance and network users will determine if the standardized set of defects and primitives will suffice to support the services on the network or if more measurements are needed to support new broadband services.

3.2.2 Performance Parameters.

Performance parameters are derived by the processing of performance primitives and defects. These parameters are essentially counts of the various impairment events accumulated by the network element in 15 min intervals. These parameters, rather than the primitives and defects, are usually used in measuring quality of services on transport systems. The following list briefly describes the parameters.

- Coding Violations for Section (CV-S), Line (CV-L), Payload Path (CV-P), and VT/VC Path (CV-V) — Bit interleaved parity errors detected on incoming signals at each layer.
- Errored Seconds for Section (ES-S), Line (ES-L), Payload Path (ES-P), and VT/VC Path (ES-V) — A second during which at least one coding violation has occurred at that layer.
- Severely Errored Seconds for Section (SES-S), Line (SES-L), Payload Path (SES-P), and VT/VC Path (SES-V) — A second with a variable number of coding violations with the variable value set in relation to the layer having the errors.
- Severely Errored Framing Seconds for Section (SEFS-S) — A count for 1 s intervals containing one or more SEF events.
- Unavailable Seconds for Line (UAS-L), Far-End Line (UAS-LFE), Payload Path (UAS-P), Far-End Payload Path (UAS-PFE), VT/VC Path (UAS-V), and Far-End VT/VC Path (UAS-VFE) — A count of 1 s intervals where the associated layer is not available.
- Alarm Indication Signal Second for Line (AISS-L) — A count for 1 s intervals containing one or more AIS defects for the line layer.

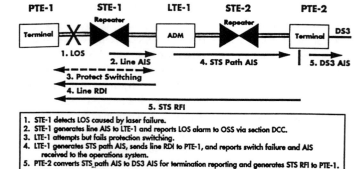

1. STE-1 detects LOS caused by laser failure.
2. STE-1 generates line AIS to LTE-1 and reports LOS alarm to OSS via section DCC.
3. LTE-1 attempts but fails protection switching.
4. LTE-1 generates STS path AIS, sends line RDI to PTE-1, and reports switch failure and AIS received to the operations system.
5. PTE-2 converts STS path AIS to DS3 AIS for termination reporting and generates STS RFI to PTE-1.

Fig. 3.   SDH/SONET maintenance signals.

- Alarm Indication Signal/Loss of Pointer Second for Payload Path (ALS-P) and VT/VC Path (ALS-V) — A count of 1 s intervals containing one or more AIS or LOP defects at the particular layer.

Errored second, severely errored second, and unavailable second parameters seem to be the most agreed upon set of data in determining required network performance. The other parameters are described to show other ways in determining network performance which may be used between carriers and may not be applicable to all network applications.

3.3 Maintenance Signals

The standardizing of maintenance signals requires signal terminating elements to make decisions on conditions of the received payloads, and the specific actions taken by the network elements should be reported as status information to the management system. The information available consists of:

- Unequipped Indications — Provides status reporting on partially equipped network elements.
- Alarm Indication Signal (AIS) — Indicates loss of signal condition on upstream network elements.
- Remote Defect Indication (RDI) — Message returned to transmitting network element of the AIS being received or of a loss of pointer defect.

- Remote Failure Indication (RFI) — Indicates receipt of AIS to upstream network elements of the same layer peer level.

An example of the maintenance signal sequence of events caused by a loss of signal at a section terminating element is illustrated in Fig. 3.

3.4 Orderwire Channels

Orderwire channels are reserved in the section and line overheads for voice communications. The section orderwire, just like the rest of the section overhead information, is to be terminated at all network elements, thus identifying this orderwire as local. The line orderwire is not terminated at section terminating equipment (STEs or regenerators) and is denoted as express orderwire. The signal format for these 64 kb/s channels has not been specified in the standard and is for future study.

3.5 User Channels

Two user channels are reserved in the standard for use by the network providers. These 64 kb/s channels are contained in the section and path overheads, which provide for terminations at each network element and at the path terminating elements, respectively. The specific access method for each of these channels is not specified and is for future study.

Network providers could use these channels, if access is defined, to backhaul additional alarm information or to provide additional orderwire channels.

3.6 Data Communication Channels

The two data communication channels (DCCs) have been defined to allow passing of network element information to another network element or with additional protocol conversion to operations support systems. These channels, one required in the section layer and the other optional at the line (multiplex) layer, provide connectionless datagram service across the network. The protocols selected for the DCCs, Fig. 4, use existing international standards which provide end-to-end message reliability. These channels are used by each network element to transmit alarm, status, control, and performance information to other network elements. Figure 5 illustrates connectivity of the

International Standards Organization Open Standards Interconnect Seven-layer Protocol Stack	SONET Section and Line Data Communication Channel Protocol Standard
Application layer	CMISE - ISO 9595/9596 ACSE - X.217/X.227 ROSE - X.219/X.229
Presentation layer	CCITT X.216/X.226 ASN.1 Basic encoding rules: X.209
Session layer	CCITT X.215 and X.225
Transport layer	ISO 8073/8073-DAD2
Network layer	ISO 8473
Link layer	CCITT Q.921
Physical layer	Section and line DDC

Fig. 4.   SDH/SONET section and line DCC protocols.

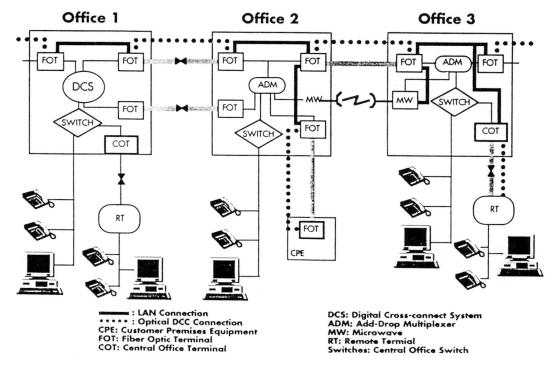

Fig. 5.　DCC connectivity.

: LAN Connection
····· : Optical DCC Connection
CPE: Customer Premises Equipment
FOT: Fiber Optic Terminal
COT: Central Office Terminal

DCS: Digital Cross-connect System
ADM: Add-Drop Multiplexer
MW: Microwave
RT: Remote Terminal
Switches: Central Office Switch

DCCs in a network application with other network element types and in conjunction with the standard IEEE 802.3 LAN [12] for intraoffice connectivity.

The section DCC provides a 192 kb/s channel which is required to be terminated at every network element for processing of messages. The addressing and routing of the messages, along with the list of required messages, has not been completed in the standard and must be addressed before the DCC can be used in midspan meet applications.

The line DCC provides a 576 kb/s channel that uses the same standard protocol stack, but is only terminated at line terminating elements. This channel is not required, but can be provided as an option on the equipment. Applications which may require access and usage for this channel could include cross-connect communication to provide network restoral and software downloads for network element upgrades.

4. NEW MANAGEMENT CHALLENGE

Once transport specifications begin to develop into products, the understanding of features defined in the specification becomes an interpretation of fiction (a paper product) into reality (a live traffic system). The questions from operations staffs usually begin with, "How can network providers take advantage of SDH/SONET's extensive management features in providing reliable and cost effective network services?" To begin answering this question, an understanding of the existing management structure is required. The TMN (Telecommunication Management Network) as shown in Fig. 6 provides a model to pictorially define the six entities and five interfaces used to describe a telecommunication management network. The

following section will be an attempt to describe today's management systems in terms used in describing the forward-looking management systems for the next generation transport systems.

4.1 TMN Entities

Management of transport systems has evolved from the responsibility of each central office to observe and manage each rack of equipment to the collection of summary and detail information from each office transmitted serially to computer systems. Operation support systems (OSSs) consist of these computer information gathering systems which are used to collect network equipment performance and provide visual and historic information to network operations staffs. In today's network, most management information consists of a few major and minor alarms on each transport system which is transmitted serially using AT&T's predivestiture standard TBOS (telemetry

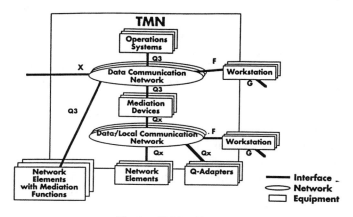

Fig. 6.　TMN architecture.

263

byte oriented serial) and E-telemetry protocols. Some network providers have collected detailed alarm and status information on equipment, but a proprietary data communication network and operations system had to be provided to collect and display the information. As described in Section 3, SDH/SONET provides the detailed information on an open standard interface for operations systems collection.

Workstations provide access to all the other TMN entities in viewing and controlling the management system. In today's management network, two types of workstations have prevailed as standards, VT-100 terminal emulation and the IBM PC workstation. These two de facto standards are replacing vendor-specific workstations in accessing network elements.

SDH/SONET uses data communication networks (DCNs) to provide management information to operations systems. Transport management systems use two types of DCNs to support operation support systems today, direct or dial-up data circuits and packet networks (X.25). Data rates of 1.2–56 kb/s are deployed in providing information to the operations systems. SDH/SONET products are required to support the standard interfaces described in Section 3 which use packet networks (X.25) with defined common management information service elements (CMISEs) for standard functions and common management information protocol (CMIP) for implementation of messages.

The main function of mediation devices (MDs) is to provide collection and protocol conversion of simple messages from central office equipment to compact and reliable data messages transmitted to operation support systems. This conversion process allows the transport equipment to support a simple protocol that does not have to provide modem control for data reliability. As transport equipment begins to supply more detailed equipment information, i.e., performance monitoring, alarm thresholding, etc., mediation devices will be required to support information conversion, data handling, decision making, and data storage.

Another data communication network, sometimes referred to as a local communication network (LCN), is specific communication networks used in central offices or on specific equipments to provide collection of messages. These communication networks support wiring of contact closures, connecting serial lines, or the use of proprietary overhead channels in collecting equipment alarms. SDH/SONET's data communication channels (DCCs) provide an LCN for the operations systems which will be discussed in the following sections.

Finally, to the entity that provides the service the network providers are deploying and the reason for the management system, network elements (NEs). Most transport network elements have several options in producing management information to operation support systems, i.e., office alarms and proprietary solutions. The office alarms are used for local central office alarm notifications, but can be transmitted to the OSSs for high level summary equipment alarms. SDH/SONET products will provide standard messages via the packet (X.25) and LAN (IEEE 802.3) networks to support detailed performance information.

4.2 SDH/SONET's Management Network (SMN)

Standardization of entities and interfaces required in management networks has provided a standard local communication network (LCN) that connects many central offices with a single gateway to the operation support systems. The data communication channels (DCCs) provide a routing mechanism to network operations, administrative, maintenance, and provisioning (OAM&P) information from one element to another.

4.3 TMN and SMN Configurations

The TMN as described in the model shown in Fig. 7 contains the operations system using a data communication network, and then having two possible paths to network elements. The two paths provide for different levels of communication interfaces to be supported on the network element; one level has the network element containing all protocol conversion and message routing functions, and the other level allows a separate mediation device to provide communication functions for the subtended set of network elements. The criteria for selection of using mediation devices or providing the communication functions in a gateway network element are determined by the amount and types of information desired to manage a network. Another important item in the selection is the methods used in integrating management of SDH/SONET network elements with existing management systems of asynchronous transport networks. These two selection criteria are discussed in exploring the functions of mediation required for the SMN and support of the TMN in managing the complete network.

4.3.1 Mediation for Element Management. Back to the question of how to take advantage of the extensive management features. If SDH/SONET equipment is deployed in the existing network and contact closures or serial interfaces are supplied, it will provide the same limited set of information the operation support systems have today. This solution does not take advantage of SDH/SONET management features. But if we begin to use the overhead information and the data communication channels, then the SMN provides the path to OSSs and the amount of performance information can be enhanced to support proactive management.

Since the operation support systems are just beginning to come of age in supporting proactive data and SDH/SONET technology is being deployed, new management features must be supported by distributing the operation functions. The TMN entity for this purpose is the mediation device which can have the functionality to support element management. These element management functions would support SDH/SONET subnetworks providing vehicles to support root cause analysis, NE software management, alarm filtering, message migration, etc.

5. CONCLUSIONS

Understanding the management features in SDH/SONET is just one piece of a large puzzle planning engineers must be concerned with when deploying new products. Developing an evolutionary plan to take advantage of the flexible bandwidth, midspan meet, and the extensive network management features requires

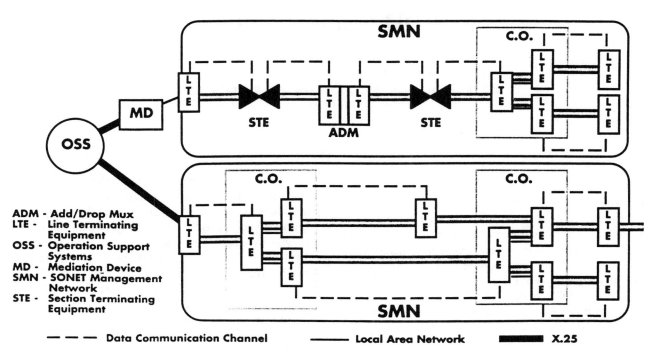

Fig. 7. Data communications channel usage.

planning for the evolution of SDH/SONET product features and the growth of management systems as new transport systems are deployed.

The evolution of the telecommunication network will determine the requirements in managing that network. If new and powerful management features of SDH/SONET can be used in providing reliable and cost effective services, then ways to implement them will most certainly follow. SDH/SONET, being that new technology to provide a graceful migration in being compatible with the network of today and providing a controllable path towards the future, will require new solutions in managing tomorrow's networks. Operation support systems must also develop a graceful migration plan in managing the existing and new networks. Mediation devices that were simple protocol converters must now develop into distributed processors for operation support systems. To take advantage of SDH/SONET's robustness, we must begin to plan for this new OSS migration.

References

[1] G.707 — "Synchronizer Digital Hierarchy and Bit Rates."

[2] G.708 — "Network Node Interface for the Synchronous Digital Hierarchy."

[3] G.709 — "Synchronous Multiplexing Structure."

[4] G.781 — "Structure of Recommendations on Multiplexing Equipment for SDH."

[5] G.7F2 — "Types and General Characteristics of SDH Multiplexing Equipment."

[6] G.783 — "Characteristics of SDH Multiplexer Equipment Functional Blocks."

[7] G.784 — "SDH Management."

[8] Bellcore's TA-NWT-000253, Issue 6, "Synchronous Optical Network (SONET) transport systems: Common generic criteria."

[9] ANSI T1.105 — 1991, "Digital Hierarchy — Optical Rates and Formats Specifications (SONET)."

[10] ITU-TS Recommendations M.3010 "TMN Principles," M.3020 "TMN Methodology," M.3100 "TMN GNM," M.3200 "TMN Management Services," M.3400 "TMN Management Functions."

[11] ANSI T1 Letter Ballot 341, "Synchronous Optical Network (SONET): Operations, Administration, Maintenance, and Provisioning (OAM&P) Communications."

[12] ANSI T1.204-1988, "OAM&P Lower Layer Protocols for Interfaces Between OSs and NEs."

[13] ANSI T1.208-1989, "OAM&P Upper Layer Protocols for Interfaces Between OSs and NEs."

[14] ANSI T1.210-1989, "OAM&P Principles of Functions, Architectures and Protocols."

[15] ANSI T1.215-1990, "OAM&P Fault Management Messages for Interfaces Between OSs and NEs."

[16] ANSI T1X1 Letter Ballot T1X1/LB93-06 "Technical Report — A Comparison of SONET and SDH."

[17] ANSI T1.231-1993, "Layer 1 In-Service Digital Transmission Performance Monitoring."

SDH Management

J. F. PORTEJOIE AND J. Y. TREMEL

France Telecom/CNET 2, Route de Trégastel 22301 - Lannion
Cedex France

1. INTRODUCTION

THE development of Synchronous Digital Hierarchy (SDH) is one of the most significant events in the Telecommunication era in recent years. An opportunity presents itself to include sophisticated processes in these new equipments, thereby offering efficient facilities for their management. The SDH can be seen as a first application of the Telecommunications Management Network (TMN) concept. The following sections describe the management processes for SDH which have been recorded in the relevant ITU standards.

2. SDH NETWORK ARCHITECTURE

A characteristic feature of the SDH transmission systems is their ability to automate the cross-connect function at each multiplexing level. A logical network structure is therefore crucial if the network is to be managed using one or more operation systems. The formalization of this structure has resulted in the standardization of a functional architecture of the SDH transport network (Fig. 1).

The functional architecture is based on three main concepts, namely, layering, partitioning, and client–server relationship. Each of these is individually described below.

2.1 Layering

The complexity of the architecture of a telecommunication network and the relationship between its components could not be understood globally without a structured decomposition. The layer concept is used to structure the transport network. A layer is defined as a set of points of the same kind that can be interconnected due to the very same nature of the signal carried. The point where the signal enters in a layer is called an access point. A layer network is divided into topological elements called links and subnetworks. A link is a fixed route between two subnetworks, and a subnetwork provides the flexibility to route the signal between two links (Fig. 2).

The transportation of a telecommunication signal within a layer network is provided by a trail (Fig. 3). A trail is defined as the means to provide a qualified transport service across a layer network. In the SDH, the analysis of the overhead information is used to qualify the transport function.

Another transport entity is the connection which conveys information transparently. Unlike a trail, a connection provides no supervision of information between input and output.

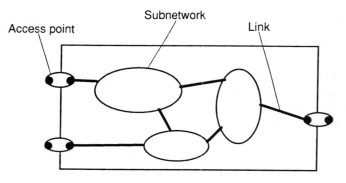

Fig. 2. Layer network topology.

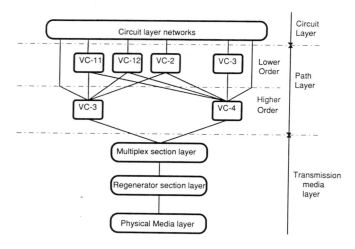

Fig. 1. SDH layer structure.

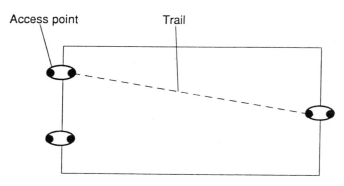

Fig. 3. Trail in a layer network: a trail connects two access points.

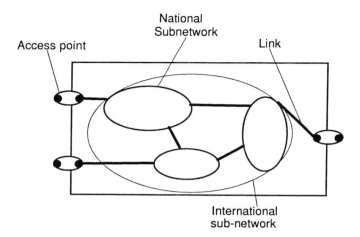

Fig. 4. Administrative partition of a network.

2.2 Partitioning

Partitioning is the process of recursively dividing a subnetwork into other subnetworks and links. The last division is the matrix to be found in a switch or a distribution frame. The partitioning concept (Fig. 4) could represent different levels of administrative boundaries between international and national networks.

2.3 Client–Server Relationship

Client–server relationship (Fig. 5) describes how a client network uses an underlying server layer. This relationship is expressed by the fact that a link connection of a client layer network is provided by a trail of a server layer network.

3. SDH MANAGEMENT NETWORK

The management of an SDH network uses a distributed management process as shown in Fig. 6. The transport facilities provided by an SDH Network Element (NE) are represented by the functional blocks. The Synchronous Equipment Management Function (SEMF) contains management application functions which communicate with peer NEs and mediation devices and/or Operation Systems (OSs). The communication process is provided via the Message Communication Function (MCF) within each entity. The SDH NE is connected to the TMN components through standard interfaces :

- the F interface to a workstation, or
- the Q interface to a mediation device or an OS.

The Embedded Control Channels (ECCs) may be used to carry management information between two SDH NEs.

4. MESSAGE COMMUNICATION FUNCTION

The Telecommunication Management Network (TMN) concept provides an architecture for interconnecting management services and telecommunication services. This results in the exchange of management information using standardized protocols and interfaces

TMN concepts introduce a functional architecture based on function blocks. Standardized interfaces are defined between machines implementing function blocks.

OSI protocol standards were defined for communicating information between management processes contained in TMN functions. Those standards, including SMASE, CMISE, ROSE, and ACSE, are used in the SDH environment.

Association control service element (ACSE) is used to establish associations between system management applications entities (SMAEs). Once this is done, a system management application service element (SMASE) is used to exchange informations between the associated SMAEs.

The SMASE relies on other standard ASEs to effect communications. Notably, it relies on the common management information service element (CMISE) which, in turn, implies the presence of the remote operation service element (ROSE).

Three kinds of protocols have been selected for the management of SDH equipment: QAx, QBx, and Q.ecc.

4.1 QAx and QBx Protocol

QAx and QBx management information communications means is provided as a selection of protocol suites. QA1 and QA2, as shown in Fig. 7, are defined as a short stack of protocol, and QB1, QB2, and QB3 as a complete stack.

4.2 Qecc Protocol

Q.ecc uses dedicated communication means defined in SDH frames, called Data Communication Channel (DCC).

Q.ecc protocol stack depiction is given in Fig. 8.

4.3 Management Services

Alarm surveillance functions are used to monitor network elements. Events are generated by the equipment upon detection of an abnormal condition seen through the "S" reference point. Event data can be reported, logged, or both.

Performance monitoring functions are used to monitor the quality of network entities. Performance data could be reported, logged, or both.

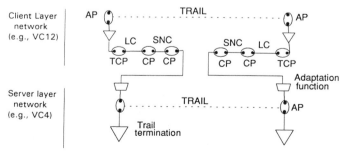

TCP: Termination Connection Point
AP: Access Point
LC: Link Connection
SNC: Sub-Network Connection

Fig. 5. Client–server relationship.

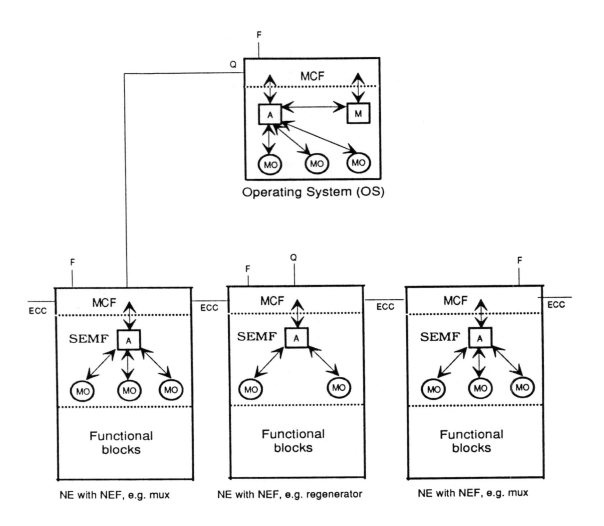

A : Agent
M : Manager
MCF : Message Communication Function
MO : Managed Objects
NE : Network Element
NEF : Network Element Function
SEMF : Synchronous Equipment Management Function

Fig. 6.   Example of an SDH management network.

5. MODELING

5.1 Object-Oriented Approach

According to the TMN (Telecommunication Management Network) philosophy, managed resources are modeled in order to exchange management information in a standardized manner. The effective definition of managed resources makes use of the OSI systems management principles and is based on an object-oriented paradigm.

Thus, management processes exchange information modeled in terms of managed objects. Managed objects are conceptual views of the resources that are being managed or may exist to support certain management functions. For a specific associa-tion, the management processes can take either a manager role, which is the part of the application process that issues management operation directives and receives notifications, or the agent role, which is the part of the application process that manages the associated objects by responding to directives issued by the manager and issues notifications. Fig. 9 shows the interaction among manager, agent, and objects.

The SDH management function usually contains an agent, while the manager role is performed by the Operating System (OS).

5.2 Information Model

Information model specifications contain objects defined for

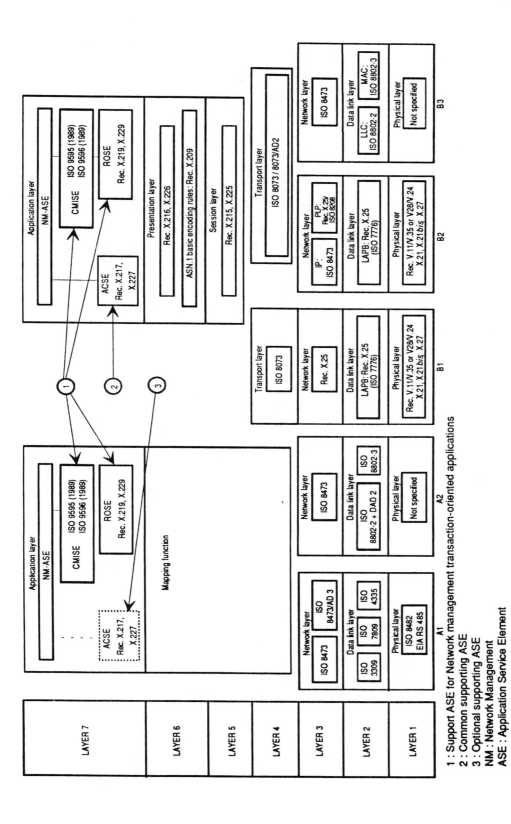

Fig. 7. Protocol suites for the Q interface.

1 : Support ASE for Network management transaction-oriented applications
2 : Common supporting ASE
3 : Optional supporting ASE
NM : Network Management
ASE : Application Service Element

the management of network elements. To provide for the reuse of specifications, managed objects are organized into object classes. Managed objects are defined in terms of attributes, behaviors, and operations.

5.2.1 Managed Object Properties.

5.2.1.1 Attributes. Attributes represent an observable value that could be directly modified by the resource or by a management action. The relationship between the value and the property of a resource is defined in the behavior of the attribute.

5.2.1.2 Behaviors. Behavior defines relationships between an object and the resource it represents.

5.2.1.3 Operations.

Read–write: Read or write operation could be addressed to attributes contained in managed objects

Action: Action is an operation applied to an object. An example of action is "connect" directed to the "fabric" object; this action enables the connection of two termination points in a cross-connect.

Notification: Notification is a spontaneous message sent by an object due to a change in the resource status. An example of notification is the "Communications Alarm" sent by a physical termination point when a loss of signal is detected.

Creation/Deletion: Creation/deletion enables a manager to modify the configuration of the managed system.

5.2.2 Object Addressing. In order to be able to send operation to an object instance, a naming structure has been defined. As shown in Fig. 10, the object instances within a managed system are organized in a tree where the root is the managed system itself.

5.2.3 SDH Object Model. The information model to be used for the management of SDH network elements is provided in

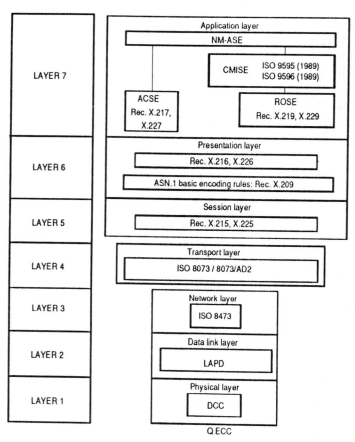

Fig. 8.   Protocol suites for the Q.ecc interface.

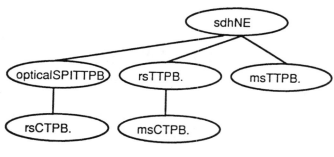

sdhNE : SDH Network Element
optical SPITTPB. : Optical Synchronous Physical Interface Trail Termination Point Bidirectional
rsCTPB. : regenerator Section Connection Termination Point Bidirectional
rsTTPB. : regenerator Section Trail Termination Point Bidirectional
msCTPB. : Multiplex Section Connection Termination Point Bidirectional
msTTPB. : Multiplex Section Trail Termination Point Bidirectional

Fig. 10.   Example of an SDH naming tree.

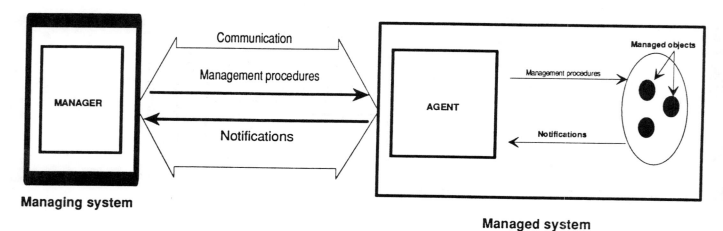

Fig. 9.   Interaction among manager, agent, and objects.

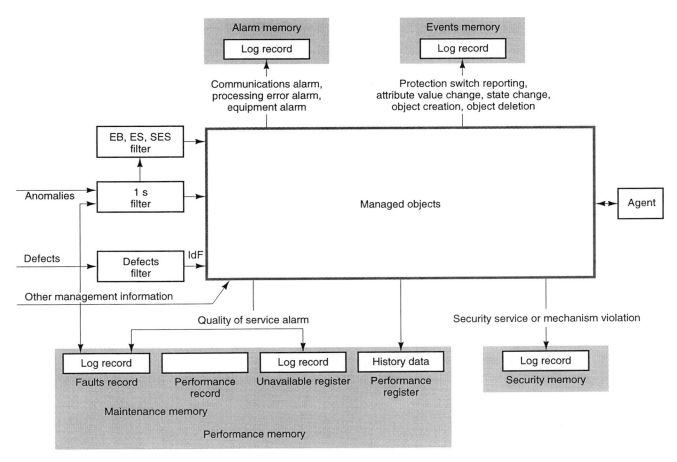

Fig. 11. Synchronous equipment management function.

ITU Recommendation G.774. Managed objects for the SDH environment are derived from the generic information model defined in the Generic Information Model.

During the modeling process, it was first decided to separate the implementation aspect from the functional aspect. The physical aspect covers the physical entities such as equipment, cards, and software. The functional aspect covers telecommunication functions provided by equipment or groups of equipment. The functional aspect of the generic information model was derived from the layering concept.

6. SYNCHRONOUS EQUIPMENT MANAGEMENT FUNCTION

The Synchronous Equipment Management Function (SEMF) provides the means through which the Network Element Function is managed by an internal or external manager (Fig. 11). It interacts with the other functional blocks by exchanging information across the S reference points. Managed objects provide event processing and storage, and represent the information in a uniform manner. The agent converts this information to CMISE messages, and responds to CMISE messages from the manager performing the appropriate operations on the managed objects. This information to and from the agent is passed across the V reference point to the Message Communication Function (MCF). Two management domains provided by the SEMF are clearly

identified: the alarm management and the performance management.

6.1 Fault Management

6.1.1 Alarm Surveillance. Alarm surveillance is concerned with the detection and reporting of relevant events and conditions, called defects, which occur in the network. The defects are detected within the equipment and the incoming signals. An associated defect indication is transmitted to the SEMF through the virtual S interface. Table 1 gives the required defect indications as defined in ITU Recommendation G.784.

Alarm indications are generated by the SEMF and reported to the OS as a result of the failure filter.

Three parameters are associated to the alarm indication and can be set or retrieved by the OS :

- detection period: period of time in which a defect shall be present to generate an alarm indication,
- reset period: period of time in which a defect shall not be present to generate an end of alarm,
- severity grade: critical, major, minor, warning, and undetermined.

6.1.2 Alarm Reporting. When an alarm indication is generated, an alarm report is generally forwarded to the OS. However, the OS has the ability to define which defects generate autonomous reports, and which are reported on request.

271

TABLE 1. REQUIRED DEFECT INDICATIONS

Defect	Physical Interface	Regenerator Section	Multiplex Section	Path HOVC	Path LOVC	PPI/LPA	Timing Source	Required for Performance
Transmit Fail	R					R		
Loss of Signal	R					R		R
Loss of Frame		R				R+		R
Loss of Pointer				R	R			R
Far-End Receive Failure			R	R	R			R
Trace Identifier Mismatch				R	R#			
Signal Label Mismatch				R	R			R
Loss of Multiframe				R*				R
Alarm Indication Signal			R	R	R			
Excessive Errors			O					
Loss of Timing Input							R	
Signal Degrade			O					

HOVC: High Order Virtual Container.

LOVC: Low Order Virtual Container.

PPI: Plesiochronous Physical Interface.

LPA: Low Order Path Adaptation.

O: Optional.

R: Required.

SETS: Synchronous Equipment Timing Source.

+: For byte synchronous mappings only.

#: Provided that use of the J2 byte in the VC-11 12 and 2 is confirmed.

*: Only for payloads that require the multiframe indication.

6.1.3 Alarm History Management. Alarm reports are stored in registers in the NE with all its parameters. For practical reasons, the number of registers is calibrated to store information over 24 h.

6.2 Performance Management

Transmission performance is evaluated by estimation of performance parameters based upon the detection of defects and the measurement of errored blocks. A block is a set of consecutive bits monitored by means of an error detection code, e.g., Bit Interleaved Parity (BIP).

Table 2 gives block size and the associated error detection code for each Virtual Container (VC), multiplex section, and regenerator section level. The last column of Table 1 indicates the defects involved.

Three performance parameters are estimated during a fixed measurement interval :

- Errored Second Ratio (ESR): This is the ratio of Errored Second events (ES) to the total seconds. The ES is defined

TABLE 2. ERROR DETECTION CODE

Level		Nb bits/Block	Nb Blocks/s	Error Detection Code
VC	VC-11	832	2000	BIP-2
	VC-12	1120	2000	BIP-2
	VC-2	3424	2000	BIP-2
	VC-3	6120	8000	BIP-8
	VC-4	18 792	8000	BIP-8
Multiplex Section	STM-1	19 224	8000	BIP-24
	STM-4	76 896	8000	BIP-24x4
	STM-16	307 584	8000	BIP-24x16
Regenerator Section	STM-1	19 440	8000	BIP-8
	STM-4	77 760	8000	BIP-8
	STM-16	311 040	8000	BIP-8

Nb : Number.

Note : This table is applicable for the receive side. The FEBE (Far-End Block Error) is used for the performance monitoring of the reverse side of a path.

as one second period with one or more errored blocks, or a defect.

- Severely Errored Second Ratio (SESR): This is the ratio of Severely Errored Second events (SES) to the total seconds. The SES is defined as one second period which contains a percentage equal to or greater than 30% of errored blocks, or at least one defect.
- Background Block Error Ratio: This is the ratio of errored block events (BBE) to the total blocks excluding all blocks during SES.

Performance is usually evaluated by the OS over a long period of time (e.g., one month). The end-to-end performance objectives for a path are deduced from those defined in ITU recommendation G.826 for a path of 27 500 km at or above the primary rate.

6.2.1 Maintenance Thresholding. A thresholding mechanism provides an operator with information on the transmission performance level, and particularly to send an alarm notification to the OS when the transmission performance is degrading.

Performance events are counted separately over two fixed 15-min and 24-h windows of time. The count is compared at any second to a threshold. As soon as a threshold is crossed, a notification is sent to the OS. Moreover, events continue to be counted to the end of the current window, at which time the count is reset to zero after being transferred to the historical registers. A notification of reset is forwarded to the OS as soon as there has been a fixed window with a count below the reset threshold. The process is illustrated by Fig. 12.

The implementation of the second thresholding mechanism is optional, but it offers the advantage of limiting the number of notifications in case of intermittent events. Table 3 gives an example of threshold values associated with performance events counted across a 15-min windows.

6.2.2 Performance Monitoring History. Performance history data are necessary to assess the recent performance. Such information can be used to sectionalize faults and to locate sources of intermittent errors.

Performance events are counted and stored in 15-min and 24-h registers, per direction and type of event.

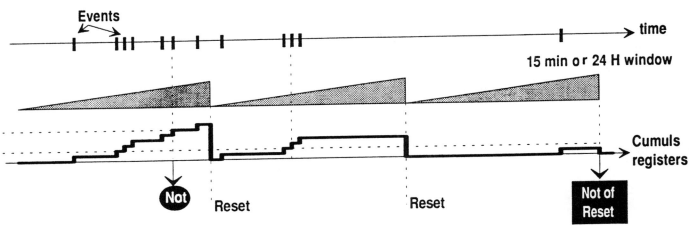

Not : Notification forwarded to the OS

Fig. 12.   Thresholding mechanism.

TABLE 3. EXAMPLE OF 15 MIN THRESHOLDS ASSOCIATED WITH PERFORMANCE EVENTS

Type of Path	Nb ES		Nb SES		Nb BBE	
	Set	Reset	Set	Reset	Set	Reset
VC-11 or 12	150	5	15	0	9000	50
VC-2	150	10	15	0	9000	50
VC-3	150	10	15	0	36 000	200
VC-4	180	20	15	0	36 000	200

Nb : Number.

There are two 24-h registers, one current and one recent. The current register is cleared after a 24-h period after the data have been transferred to the recent register.

Seventeen 15-min registers are at least required, one current and 16 recent. If not empty, the content of the current register is transferred to the first of the recent registers after being time-stamped at the end of a 15-min period.

6.2.3 Unavailability. Performance monitoring and the storage of performance events are inhibited during unavailability periods. An unavailability period begins at the onset of ten consecutive SES events. These ten seconds are part of the unavailability period. An unavailability period ends at the onset of ten consecutive non-SES events. These ten seconds are not considered to be part of the unavailability period. This definition is applicable to a single direction. In particular, a path is bidirectional, so it is in the unavailable state as soon as one direction is in the unavailable state.

6.3 Configuration Management

Configuration management could be divided into two aspects, one concerned with equipment management and the second with transmission resources management.

6.3.1 Equipment Management. During the life of an equipment, its configuration may change. It is therefore of major interest to be able to remotely modify the hardware configuration of an SDH system. This could be achieved using the creation and deletion facilities offered by CMISE. Because card failure and replacement could occur, the operator must be warned immediately; likewise, when a card is removed or inserted from the SDH network element.

6.3.2 Transmission Resources Management.

Protection Function: SDH network element incorporates automatic protection switching at the multiplex section layer, but also at the path layer. Those functions are to be controlled by the management system.

Cross-Connect Function: Management functions enable the setup or release of connections in add/drop multiplexers or cross-connect systems through the remote modification of their fabric.

7. CONCLUSION

Unlike PDH, SDH offers many capabilities for configuration, fault, and performance management that may be considered complete and stable regarding the standards. However, work should be carried out to complete the information model at the Q interface, in particular for the specification of network level management services. The paper sets one's heart to give an overview on the management of SDH networks. The interested reader will consult the referenced ITU Recommendations for more details.

Bibliography

[1] CCITT Recommendation G.826: Error Performance Parameters and Objectives for International, Constant Bit Rate Digital Paths at or above the Primary Rate, Jan. 1993.

[2] CCITT Recommendation G.773: Protocol Suites for Q Interfaces for Management of Transmission Systems, 1990.

[3] CCITT Recommendation G.774: SDH Management Information Model, Nov. 1992.

[4] CCITT Recommendation G.783: Characteristics of Synchronous Digital Hierarchy (SDH) Equipment Functional Block, Nov. 1992.

[5] CCITT Recommendation G.784: Synchronous Digital Hierarchy (SDH) Management, Nov. 1992.

[6] CCITT Recommendation G.803: Architectures of Transport Networks Based on the SDH.

[7] CCITT Recommendation M.2120: Digital Path, Section and Transmission System Fault Detection and Localisation Procedures, 1992.

[8] CCITT Recommendation M.3010: Principles for a Telecommunications Management Network, Nov. 1991.

[9] CCITT Recommendation Q.821: Management Functions at the Q3 Interface — Alarm Surveillance Management, 1993.

[10] CCITT Recommendation Q.822: Management Functions at the Q3 Interface — Performance Management, 1993.

A Layered Approach to SDH Network Management in the Telecommunications Management Network

L. H. CAMPBELL AND H. J. EVERITT

TELECOM AUSTRALIA RESEARCH LABORATORIES

1. INTRODUCTION

THE objective of defining control layers for network management is to provide a consistent control scheme which is (largely) independent of underlying telecommunications technology and which is easily upgradable as new decision-making methods are implemented. A full achievement of this objective may not be possible because control functions may often depend on several processes working in concert. We report the outcomes of a functional analysis of network management control (supported also by an information modeling view [13]), and we aim to indicate a hierarchy of control layers, each of which could provide a uniform view to the layer above, and each of which has a consistent set of control actions that it can exercise on the layer below. We take an object-oriented view of each layer, in which data necessary at each layer cannot be separated from the methods used to maintain it. Our emphasis here is on control and decision-making, not data.

By this proposal for control layering, we aim to provide a smooth transition into the full management capabilities of SDH, especially self-healing functions and bandwidth management, while retaining current capabilities. In this paper, we first describe our layering scheme in Section 2. We then give, in Section 3, a brief overview of the Telecommunications Management Network (TMN) recommendations and their relationship to SDH control. In Section 4, we describe service restoration as an application of layered control, and then indicate its relationship to SDH standard layers in Section 5. Section 6 outlines how SDH bandwidth management is related to traditional network traffic management, while Section 7 discusses some connections between SDH management and existing PDH management. Finally, Section 8 summarizes our conclusions.

2. LAYERING OVERVIEW

Our functional analysis of network management control produces a model of four main control layers, with further elaboration into sublayers at the important Network Layer.

At the bottom, elementary level (see Fig. 1) is the *Network Element Layer*. This consists of the standardized managed objects including network elements and operations systems. We shall have cause to say little more about this layer: it is an area of TMN standardization and is well understood. Control is exercised through communicating changes to relevant network elements or operations systems.

Above the network elements comes the *Network Layer*. At this layer, the connections between network elements and their use for providing telecommunications services are recognized. This is clearly of key importance to network management, and hence we wish to elaborate the Network Layer further into sublayers. The sublayers provide bundles of common functionality used in one or more applications.

The most basic is the *Physical Configuration Sublayer*. This consists of a representation of the physical network, made up of nodes — switches, cross-connects, service control nodes, etc. — and links — transmission bearers, signaling links, etc. This corresponds to the "transmission media layer network" of [20] and the "physical facilities" layer of [7]. This sublayer describes physical constraints on the controls applied in the sublayer above it. An example of the interaction between the Physical Configuration sublayer and the Logical Configuration sublayer is described in Section 4.

Above this sublayer is the *Logical Configuration Sublayer*. This sublayer contains a view of the network consisting of paths

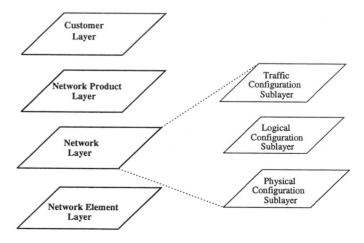

Fig. 1. Network management control layers.

between network nodes. This indicates, for example, the usage of physical facilities to provide telecommunications services. This sublayer corresponds to the "path layer network" of [20] and the "logical network" of [7]. This level of abstraction of the physical network is useful for defining usage patterns for traffic management and the reconfiguration of digital cross-connects.

At the top of the Network layer is the *Traffic Configuration Sublayer*. Here, the routing of actual calls or services is determined, perhaps through distributed controllers, given the path data from the sublayer below. The functions that monitor and collect useful traffic statistics, such as call-holding times and answer-seizure ratios, also reside at this sublayer. An example of the interaction between the Traffic Configuration sublayer and the Logical Configuration sublayer is described in Section 6.

Above the Network Layer is what we choose to call the *Network Product Layer*. This layer includes the definition of all the service products — long-distance dialing, freephone service, etc. — that are provided by the network. The details of the network implementations are hidden by this layer from the service users. A freephone service consists of features like number translation, direct updating of translations by service subscriber, charging to called party, and guaranteed peak calling rate: the user need not know the methods by which these features are provided. Each network product may be managed individually by the network provider in accordance with a business plan. Thus, for example, priority may be given to emergency service calling over ordinary long-distance calling. As network management becomes more automated at the Network and Network Element layers, more attention will be paid to control at the Network Product Layer.

Above the Network Product Layer is the *Customer Layer*. This is where the customer obtains a customer-level view of services for management and control purposes. Network products are assembled to provide a consistent view to the customer who may then, for example, be permitted to manage a total inventory of services, such as in a virtual private network.

3. THE TELECOMMUNICATIONS MANAGEMENT NETWORK (TMN)

The Telecommunications Management Network (TMN) is described in the M.3000 series of recommendations from the Telecommunications Standardization Sector of the ITU (ITU—TS, formerly CCITT). Recommendation M.3010, "Principles for a Telecommunications Management Network," is said to be "the general architectural requirements . . . to support the management requirements . . . to plan, provision, install, maintain, operate and administer telecommunications networks and services" [2]. At the present state of development, however, this and related recommendations provide only a framework within which coordinated network and service management could be achieved. Many details still need to be worked out, and not all of them will be amenable to standardization. It is generally the case, however, that all network providers are seeking to simplify their present network management environments through the use of consistent, TMN-like frameworks.

Functional layers for the TMN are described in Appendix B

of [2]. It indicates five functional layers:

- Business Management Layer;
- Service Management Layer;
- Network Management Layer;
- Network Element Management Layer;
- Network Element Layer.

This is indicative text only, and will be subject to further modification. As we have argued elsewhere [1], this text largely reflects the views of a single network provider (see, for example, Walles [18]) and is not sufficiently generalizable. In particular, the network provider's own business priorities should be reflected in our Network Product Layer, reducing the need for a business management layer. Issues of enterprise networking and interconnection between TMNs should probably be handled at the Customer Layer.

Recommendation M.3010 also identifies five major areas for network management:

- Performance Management;
- Fault Management;
- Configuration Management;
- Accounting Management;
- Security Management.

These areas are not meant to be mutually exclusive or exhaustive: for example, information collected from network elements may be used in more than one area. For the SDH, the emphasis will be particularly on the first three — performance, fault, and configuration management — but, in all areas, coordination between SDH management functions and other network management functions will be required to implement uniform management.

The general relationship between the wider TMN and the SDH Management Network (SMN), for managing the SDH specifically, is indicated in Recommendation G.784 [3]. Here, the concern is mainly to define a "gateway network element" for interaction between the SDH and the TMN and the use of SDH embedded control channels for the transfer of information within SDH networks. The recommendation defines an SDH manage-

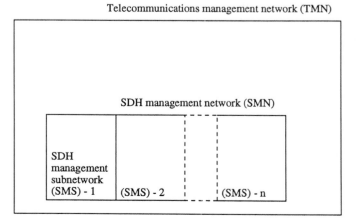

Fig. 2. Relationship among SMN, SMS, and TMN (Fig. 3-3/G.784).

ment subnetwork (SMN) that is only accessible through a gateway network element. The relationships among SMSs, SMNs, and the TMN are described in Fig. 3-3/G.784, reproduced here as Fig. 2. The necessity for gateway network elements in the SMN may provide a useful construct for differentiating the network element layer and the network element management layer in the TMN, although *functionally* this distinction seems to be obscure.

Recommendation M.3010 defines a number of reference points at which standardization will occur. The most important are:

q3—between operations systems functions and other functions, including network elements;

f—between workstations and operations systems;

x—between operations systems in different TMNs.

Interfaces for the q3 reference point now exist in several domains, but work on the F and X interfaces is still preliminary. Development of standards for the X interface may be important for SDH applications involving "midspan meets" between different carriers.

Recommendation G.784 indicates the use of the F interface (as in Fig. 3-8/G.784 on reference configurations) between a workstation and a network element, which would be additional to the F interface between an operations system and a workstation. The implied function here is to permit a common interface for both craft access and management access. This would be a desirable outcome, since it would facilitate remote working across an SDH network and simplify remote testing. At the current state of development, however, craft and test access to SDH network elements is proprietary to the equipment vendor, and progress on specification of an F interface is slow. This presents some current problems for network providers with multiple vendors, in that network-wide or interworking tests require substantial special development for their achievement.

An important aspect of integration between SDH management and the wider TMN is naming and addressing of network elements. An appropriate address structure is available in OSI standards [12].

4. SERVICE RESTORATION

Service restoration is the process by which transmission bearers are assigned (or reassigned) for service to restore connections after a transmission failure. Service restoration may be achieved through the operation of a separate service protection network [5] or the SDH [15]. The example of service restoration shows the interaction between the physical and logical views of the network.

Service restoration is triggered by a failure alarm from one or more network elements. The alarm indicates that a bearer (the classical "back-hoe through the cable") or a transmission channel (perhaps through channel-bank problems) has failed. The alarm is generated at the Network Element layer and is communicated to the Physical Configuration sublayer.

There are two possible types of control action taken in response to the alarm. Action is either in the Physical Configu-

ration sublayer or in the Logical Configuration sublayer. One possibility is that the bearer can be restored by local protection switching if, say, 1:N protection is provided and the spare channel is available. In this case, service is restored after a brief interruption. Further alarms are generated to trigger maintenance actions to repair the original fault. The bearers can be changed back to their original configuration, if required, after the problem has been repaired. In this case, action can be taken solely at the Physical Configuration sublayer, since the connection topology of the network is not altered. (It may still be worthwhile to verify network configuration, as in [21].)

A more complicated possibility is that local protection switching is not available for service restoration. In this case, it is necessary to search for unused paths with adequate capacity. In the SDH, an appropriate number of tributary streams would need to be rearranged. In particularly serious failures, it may not be possible to restore all the failed capacity: in this case, priority should be given to high-value traffic or services.

Because of the need to examine usage data and possibly to make value judgments, this case of restoration must be performed at the Logical Configuration sublayer. Here, the Physical Configuration sublayer generated a notification that the bearer or channel is no longer available. This then triggers a reconfiguration action in the Logical Configuration sublayer because the assignment of logical channels to physical channels is no longer feasible.

Another possibility is that there has been much local Physical Configuration sublayer activity, for reasons of speedy restoral, but this has led to a suboptimal global network configuration. In this case, reconfiguration at the Logical Configuration sublayer could restore network balance and facilitate any further necessary actions at the Physical Configuration sublayer. For example, in a four-fiber bidirectional ring, if a link is cut, fast restoral at the Physical Configuration sublayer can simply loop the ring back upon itself on either side of the failed link. A more considered reaction at the Logical Configuration sublayer would produce paths which made more efficient use of the remaining transmission facilities.

Some methods for restoration at the Logical Configuration sublayer have been proposed (see [20], [9], [17] for distributed schemes and [10], [11] for an optimal, centralized scheme). Most such methods (as in [20]) assume that only spare capacity will be used for service restoration. When not all capacity can be restored ([20] assumes only 40% of needed capacity is available), some priority judgments are needed to determine which tributaries should be given preference. Because of the need for priorities, service restoration in this instance is very closely related to network traffic management optimization (see Section 6) and takes place at the Logical Configuration sublayer.

5. SDH LAYERS

Recommendation G.803 [4] has an SDH transport network model which is based on the concepts of layering and partioning. It is illustrated in Fig. 3.10/G.803 and is reproduced here as Fig. 3. The layering of the transport network collects similar functions

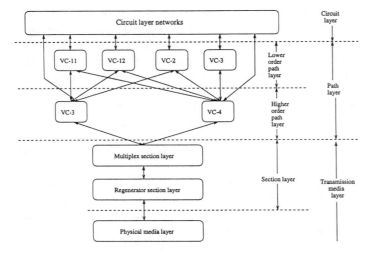

Fig. 3. SDH-based transport network layered model (Fig. 3.10/G.803).

into each layer, and these can be designed, activated, and altered independently from other layers. The layers also facilitate the definition of TMN managed objects.

The SDH-based transport layers are classified into three classes of layer network: the *circuit*, *path*, and *transmission media* layer networks. The *circuit layer networks* provide telecommunication services such as circuit-switched, packet-switched, and leased-line service. *Path layer networks* support different types of circuit layer networks through either the *lower-order path layer* or the *higher-order path layer*, depending on the service network requirements.

Transmission media layer networks are dependent on the actual transmission medium used. They are divided into the *section layer networks*, which provide for the transfer of information between two nodes in path layer networks; and *physical media layer networks*, which deal with the details of the transmission media. There are two section layer networks. The *multiplex section layer network* handles the point-to-point transfer of information between locations which route or terminate paths. It can perform some local automatic protection switching. The *regenerator section layer network* deals with the transfer of information between network elements, which can be regenerators or

locations which route or terminate paths.

From a management control point of view, the SDH transport layer networks reside in the Network or Network Element layer. We will now associate each of the SDH transport layers with the appropriate management control layer.

Within the Transmission media sublayers, the physical media layer lies within the Network Element layer because it deals with basic transmission functions. Both section network layers lie within the Physical Configuration sublayer. The regenerator section layer has a limited network view and clearly resides here. The multiplex section layer provides an automatic protection switching function where spare capacity is hunted for and reserved in the local working links. Reconfiguration is performed in a local, distributed fashion since speed is of the essence [9], [8]. Management control lies within the Physical Configuration sublayer, and path layer networks are not affected.

Path layer network control either belongs to the Physical Configuration sublayer or the Logical Configuration sublayer, depending upon the topology of the SDH network. Figure 4 illustrates our current view on which SDH network topology belongs to which sublayer.

In a linear SDH network topology with $1 + 1$ path protection switching, shown in Fig. 4(a), the spare capacity is reserved for use under failure, and is appropriated on a first-come first-served basis on failure. As no priorities are taken into consideration, this is Physical Configuration sublayer control. The two-fiber unidirectional ring in Fig. 4(b) is similar in the Physical Configuration sublayer. For the mesh and the four-fiber bidirectional ring topologies shown in Fig. 4(c) and (d), under link failure, considered negotiation takes place before sensible paths are restored. This control takes place in the Logical Configuration sublayer. (Remember that simple looping of the ring to avoid the failure is line restoral and not path reconfiguration.) This assignment of sublayers is not a one-to-one relationship between the management control hierarchy and the SDH transport layers because the SDH protocols operate differently depending on the network topology.

Note that the internal relationship between transport network layers allows a recursive relationship. For example [4], a layer network is made up of subnetworks and links between them,

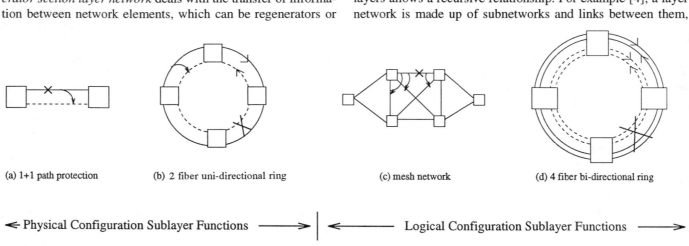

(a) 1+1 path protection   (b) 2 fiber uni-directional ring   (c) mesh network   (d) 4 fiber bi-directional ring

⟸ Physical Configuration Sublayer Functions ⟶ | ⟸ Logical Configuration Sublayer Functions ⟶

Fig. 4. Path layer protection.

while a subnetwork in turn can comprise lower level subnetworks and links between them. This recursion can also be found in the management control hierarchy, although not all layers necessarily will be present. For example, to a telephone network user, the network looks like one large network element (a switch) with lines attached; a network provider, however, requires a more detailed view.

6. NETWORK TRAFFIC MANAGEMENT OPTIMIZATION

Layering assists with the identification of common functions, and should permit new control functions to be inserted without wholesale redevelopment of the control hierarchy. Here, we look at the effects of the introduction of SDH on network traffic management. Network traffic management occurs at the Logical Configuration sublayer, which is affected by SDH technology (see Section 4), and the Traffic Configuration sublayer.

The network traffic management functions determine two sets of related decisions:

- the paths that are available to carry traffic for each origin–destination (O–D) pair;
- the traffic management controls to be applied at each switching node to control the traffic on the chosen paths.

The first of these, the choice of available paths, takes place at the Logical Configuration sublayer. It is generically an optimization problem, as described by Warfield and McMillan [19], to make best use of network resources, although heuristic approximations are often used. The second set of decisions, of traffic management controls given the appropriate path configuration, is taken at the Traffic Configuration sublayer. The division into sublayers represents a formal decomposition of the optimal traffic management problem, but all proposals appear to make this separation. Figure 5 represents the functions in layers and the relevent data/decision flows.

We have described elsewhere [1] the data required for the path optimization problem at the Logical Configuration sublayer and how this data is obtained. We have also identified this problem with the equivalent optimization problem in service restoration, as described in Section 4. That is, the optimal selection of paths

may be done by a single function, and used by both those processes traditionally classified as network traffic management and those processes classified as transmission management. In effect, SDH bandwidth management adds new control functions to an existing decision process.

The path assignments from the optimization problem in the Logical Configuration sublayer are used for detailed control of network traffic in the Traffic Configuration sublayer. The path assignments could be used, for example, as capacities in a state-dependent routing scheme, such as in [16], or could be thought of as desirable set-points to be achieved by an expert system (e.g., [6]) for traffic management. All of these traffic management methods work in the presence of SDH transmission, although it is likely that one or another will be preferred, given SDH bandwidth management.

7. COORDINATED SDH/PDH MANAGEMENT

SDH systems will be introduced into the network gradually. In Australia, they will first appear in high-value intercapital networks and major business centers. The coexistence of SDH and PDH networks will require joint management of SDH and PDH features if seamless service is to be provided to customers. We present here some of the consequences and opportunities of this situation.

Existing management systems generally place management intelligence into central operations systems, with network elements communicating data on their state, but not controlling themselves. These data are collected from the remote network elements using leased lines or packet-switched networks. In contrast, the SDH protocols provide the means to collect and transport detailed management information within the SDH paradigm. This environment facilitates distributed management at a local level as part of an equipment vendor's gateway network element. The network arrangements to permit both centralized and distributed management will require careful study: one proposal is contained in [14].

SDH and PDH equipment should be managed in a way which allows the service networks using them to be oblivious to the differences between them. The TMN provides the means to achieve this through its common modeling of network elements as managed objects, where this is implemented. However, for many applications of fault management and some configuration management, it will be appropriate to implement separate SDH and PDH domains of management at the Network Element layer and the Physical Configuration sublayer.

Providing an agreed level of service across PDH and SDH networks requires common management functions for performance management and some configuration management. Hence, separate SDH and PDH management domains are generally not applicable at the Logical Configuration sublayer or above. At the Network Product layer of management, the underlying transmission networks should not be visible except for services that can *only* be provided by SDH networks (such as on-demand, leased line services). A clear and consistent hierarchy of management control is required to make this possible.

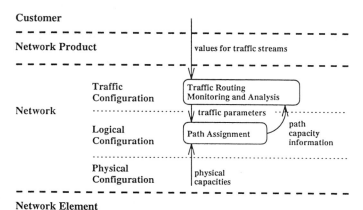

Fig. 5. Network traffic management layers.

As the SDH network expands, there will be a transition to more distributed management and automated reconfiguration. Some decision-making will remain centralized (e.g., [10]), while many restoration actions will be distributed [8]. A consistent control hierarchy is important for this transition, so that functions can be progressively distributed without disruption to end-to-end management. Sections 4 and 5 have indicated how this can be achieved in our hierarchy.

8. SUMMARY

We have described a control hierarchy for network management into which new SDH-enabled management functions may be inserted without wholesale reconstruction of existing high-level functions. Such a hierarchy is necessary if SDH capabilities are to be fully exploited. Our hierarchy is somewhat different from the current text of the TMN recommendations.

SDH transmission will particularly affect fault and configuration management. We have indicated how our layered approach is useful for these cases and, in particular, how SDH restoration capabilities fit into the overall scheme. In addition, we have argued that SDH bandwidth management should be considered as adding capability to management functions that already exist in rudimentary form for network traffic management.

The common management of SDH and PDH networks will be important for performance management and end-to-end configuration management. We have suggested that this commonality can be achieved in our control hierarchy, especially through integrated management at the Network Product layer.

9. ACKNOWLEDGMENT

We wish to acknowledge the generous assistance of our colleague Jeremy Ginger whose knowledge of SDH restoration techniques was invaluable to our understanding. The permission of the Managing Director, Research and Information Technology, Telecom Australia, to publish this paper is hereby acknowledged.

References

[1] L. H. Campbell and H. J. Everitt, "A layered approach to network management control," *J. Network and Syst. Management*, vol. 1, no. 1. pp. 41–55, 1993.

[2] CCITT, "Principles for a Telecommunications Management Network," Recommendation M.3010, *CCITT Com IV-R 28*, Sect. 11.1, Nov. 1991.

[3] CCITT, "Synchronous Digital Hierarchy (SDH) Management," Recommendation G.784, Geneva, 1990.

[4] CCITT, "Architectures of Transport Networks Based on the Synchronous Digital Hierarchy (SDH)," Draft Recommendation G.803, June 1992.

[5] D. Dias, "Digital Service Protection network system," Telecom'91, Geneva, Oct. 1991.

[6] M. Georgeff and A. Rao, "Intelligent real-time network management," in *Specialized Conf. on Artificial Intell., Telecommun., and Comput. Syst.*, held as part of the *10th Int. Conf. Expert Syst. and their Appl.*, Avignon, 1990, pp. 87–101.

[7] G. Gopal, C.-K. Kim, and A. Weinrib, "Algorithms for reconfigurable networks," in A. Jensen and V. B. Iversen, Eds., *Teletraffic and Datatraffic in a Period of Change (ITC-13)*, Elsevier Science, Copenhagen, 1991, pp. 341–347.

[8] W. D. Grover, B. B. Venables, M. H. MacGregor, and J. H. Sandham, "Development and performance assessment of a distributed asynchronous protocol for real-time network restoration," *IEEE J. Select. Areas Commun.*, vol. 9, pp. 112–125, Jan. 1991.

[9] S. Hasegawa *et al.*, "Integrated self-healing network for STS-1/STS-3c path level restoration," NOMS'92, paper 9.1, Memphis, TN, Apr. 1992.

[10] M. Herzberg, "Bandwidth management in reconfigurable networks," presented at the Cracow Int. Workshop on Network Management, paper 5.3, Kraków, Poland, May 1993.

[11] M. Herzberg, "A decomposition approach to assign spare channels in self-healing networks," presented at IEEE Globecom'93, Houston, TX, Nov./Dec. 1993.

[12] H. Katz, "TMN addressing — The choices," *Telecommun. J. Australia*, vol. 43, no. 1, pp. 22–25, 1993.

[13] H. Katz and A. Bridge, "The information layered architecture — Model, layers, methodology and their application for service activation," presented at the Cracow Int. Workshop on Network Management, paper 1.4, Kraków, Poland, May 1993.

[14] H. Katz, G. F. Sawyers, and J. L. Ginger, "SDH management network: Architecture, routing and addressing," presented at IEEE Globecom'93, Houston, TX, Nov./Dec. 1993.

[15] J. Kovess, H. Sabine, and V. Sura, "Towards a synchronous transmission network," *Telecommun. J. Australia*, vol. 40, no. 1, pp. 3–10, 1990.

[16] K. R. Krishnan, "Adaptive state-dependent traffic routing using on-line trunk-group measurements," in A. Jensen and V. B. Iversen, Eds., *Teletraffic and Datatraffic in a Period of Change (ITC-13)*, Elsevier Science, Copenhagen, 1991, pp. 407–411.

[17] H. Sakauchi, Y. Nishimura, and S. Hasegawa, "A self-healing network with an economical spare-channel assignment," presented at IEEE Globecom'90, San Diego, CA, paper 403.1, Dec. 1990.

[18] A. Walles, "Functional descriptions of network management," *BT Technol. J.*, vol. 9, pp. 9–17, July 1991.

[19] R. Warfield and D. McMillan, "A linear program model for the automation of network management," *IEEE J. Select. Areas Commun*, vol. 6, pp. 742–749, May 1988.

[20] J. Yamada and A. Inoue, "Intelligent path assignment control for network survivability and fairness," ICC'91, paper 22.3, Denver, CO, June 1991.

[21] Y. Yasuda and N. Yoshikai, "Automated network connection tracing and data gathering methods using overhead bytes in the SDH frame structure," presented at IEEE ICC'91, paper 2.1, Denver, CO, June 1991.

Control and Operation of SDH Network Elements

JOHAN BLUME, LEIF HANSSON, PEDER HÄGG,
AND LEIF SUNDIN

ERICSSON TELECOM AB

Abstract—Increasingly, flexible and powerful network management solutions allowing Network Elements and Operations Systems to work in a multivendor environment are becoming a key issue to network operators. The Synchronous Digital Hierarchy currently being standardized provides the required capabilities.

   The authors describe the main characteristics of SDH management and discuss some implementation aspects of control systems developed for SDH Network Elements.

BACKGROUND AND DRIVING FORCES

TODAY'S transmission network typically consists of inflexible equipment without provision for remote reconfiguration, and fixed hard-wired point-to-point connections. This means that each change of configuration — when supplying a 2 Mb/s leased line, for example — requires hard-wiring, which is time-consuming and therefore costly.

The Synchronous Digital Hierarchy (SDH) eliminates these disadvantages by providing flexible Network Elements (NE) capable of being configured remotely. This makes it possible to provide new broadband services — such as 2 Mb/s leased lines — to customers in a short time and at low cost.

Another characteristic of today's transmission networks is that each vendor has his own management system with proprietary interfaces and functionality. This situation often necessitates adaptations when new equipment is introduced on a market, which is costly in terms of time, resources, and money, both for the supplier and the operator.

One of the major driving forces behind SDH is improved and standardized management interfaces and functions, allowing the SDH Network Elements and Operations Systems (OS) to interwork in a multivendor environment [2], [3].

BENEFITS OF SDH AND ETNA

Functional Overview

Introducing SDH in the transport network will improve operation and maintenance, and so reduce the operational cost for the Telecom operator. SDH also enables the operator to control the network more efficiently, in comparison with the conditions afforded by existing transmission systems.

SDH makes it possible to set up new connections from a remote site within a few seconds. This enables a Telecom operator to respond quickly to customer demands for new or higher capacity. It also reduces the operational cost because less manpower is required.

Self-healing networks can be configured in such a way that faults in the network—e.g., cable breaks—will not affect the traffic for more than a few milliseconds, or seconds, depending on the size of the network and the restoration principle applied. At present, it can take hours, or even days, to locate the fault and take appropriate actions.

Two different principles are applied to protect the network against the effects of a fault: protection switching and protection routing.

Protection switching is performed by the SDH NE without assistance from a central network management system. This means very fast restoration, but utilizes network resources quite inefficiently.

It is anticipated that rings will be a commonly used network topology in SDH networks, especially in local networks. In the event of a cable break, the traffic can be restored by sending it the opposite way round the ring.

Protection routing requires assistance from an OS and takes a somewhat longer time (5–10 s), but can be used for any type and size of network. In this case, alarm information is sent to the OS from all affected NEs. The OS analyzes the fault situation and calculates a new, optimized way through the network. Cross-connect commands are then sent to the NEs to set up the new connection. If desired, the operator can set conditions for rerouting: that the route should not pass through a particular node, for example.

The Performance monitoring parameters enable the operator to identify potential problems before they cause degradation of end-user service. They also offer a tool for verifying the quality of the connection. This is important since many customers require a high guaranteed quality level, which makes it necessary to be able to measure that level. The parameters used conform to CCITT Recs. G.821, G.826, and G.784, Appendix 1.

An important issue is protection of the SDH NEs from unauthorized access. This becomes important in cases where available functions may cause serious problems if used incorrectly. Each operator must have a unique Userid and password, issued by the system administrator. He is also assigned one of several "user categories." The user category determines what functions the user is allowed to access. User categories — of which some examples follow — can be configured as desired:

- System manager: handling of user categories and database management
- Read access: only read access to management information

- Configuration manager: access to installation and configuration functions
- Data communication manager: handling data communication facilities.

These functions and a number of other functions are described in greater detail and from the point of view of the SDH NEs in the section "SDH Management."

Mode of Operation

The SDH NE can be managed from:

- OS, the management system for the Transport Network
- A Local Operator Terminal (LOT).

The OS is necessary for the management of large transport networks. The LOT is required during installation, and can also be used to manage a single NE or a small transport network.

All SDH NEs use a standardized interface to the OS. The protocols used are TMN (Telecommunications Management Network) Q3 interfaces as defined in CCITT Recs. G.811 and G.812, Appendix 2.

All SDH NEs also use a common Information Model (IM). The IM defines the syntax of the messages that are sent between the SDH NEs and the OS.

SDH MANAGEMENT

General

SDH Management is based on TMN and OSI principles to allow for the building of an open network architecture. The basic concept behind TMN is to provide an organized architecture to achieve interconnection between various types of Operations System (OS) and/or telecommunications equipment for the exchange of management information, using standardized protocols and interfaces.

From a management point of view, the SDH NE can be considered from three different perspectives:

- A functional perspective
- An information model perspective
- A data communication perspective, each of these perspectives defining some of the aspects necessary for standardized multivendor operations.

Functional Perspective

The Functional perspective defines the management services that a single SDH NE can provide to a local operator, or to a network management system. In the TMN context, these functions are referred to as TMN management functions.

The TMN management functions belong to different management functional areas. Those relevant for SDH NEs are: Configuration Management, Fault Management, Performance Management, and Security Management. In addition, a set of Data Communication Network management functions dealing with configuration of data communication resources have been defined.

Configuration Management (CM). Compared with traditional transmission equipment, the SDH NEs also include a switch which provides for the setup of broadband semi-permanent connections. The main CM task is to control this switch, but CM also controls other aspects of the configuration of the NE:

- Termination Point Provisioning: The different types of termination point, i.e., physical interfaces, trail termination points, and connection termination points, can be configured in different ways, e.g., assigned identities, alarm thresholds, enabling and disabling of the laser, etc. The termination points are automatically created when the related printed board assemblies are inserted.
- Equipment Configuration: The SDH NEs are in many ways self-configurable after installation or extension of the hardware, e.g., when new access ports are installed. The equipment configuration functions keep track of the equipment currently installed, e.g., printed board assemblies and software, and report to the OS if changes have been made.
- Cross-Connect: The cross-connect functions set up connections through the switch and keep an up-to-date list of the cross-connections currently being ordered.
- Protection Switching: The SDH NEs can be configured to perform different types of autonomous protection switching (fast restoration) of paths or sections to dedicated standby network resources following a network failure.
- Synchronization Configuration: Each NE must be synchronized from a valid synchronization source, e.g., a 2 Mb/s signal, a 2 MHz reference, or an STM-N signal. The synchronization source configuration functions define the synchronization source to be used, and what actions should be taken when the primary source fails.
- NE-Recovery: The SDH NEs contain a lot of data which must be administered, e.g., regularly backed up. During certain trouble conditions, it may also be necessary to perform restarts at different levels.

Fault Management (FM). Fault management provides functions for detection, isolation, and correction of abnormal states in the network. This includes both network-related faults, resulting from cable breaks or deteriorating line systems, and abnormal conditions within the NEs themselves. The main FM task is to report to the OS upon detection of a serious fault in the network, but it also controls diagnostic and test routines:

- Alarm Surveillance: The SDH NEs have the capability to send alarm reports to the OS upon detection of a failure, and to store the alarms in an event log.
- Fault Localization and Testing: The SDH NEs can be ordered to perform loopbacks, error injection, self-diagnostics, etc.

Performance Management (PM). Performance management deals with the functions necessary for an NE to collect, store, threshold, and report performance data associated with its monitored PDH and SDH trail terminations. All application-specific and optional parameters, as specified in CCITT Rec. G.784, are supported.

- PM Data Attribute Setting: Basic PM data attributes, such as threshold values, can be defined.
- PM Data Reporting: The SDH NEs can send PM data reports to an OS, either when a defined threshold is exceeded (degraded or unacceptable performance level) or in accordance with predefined schedules.
- PM Data Logging: PM data can be stored in logs within the NEs and fetched by the OS when demanded.

Security Management (SM). Security management functions deal with user access control to protect the network against unauthorized access to resources and services.

- Logon/Logoff: A local operator trying to access the NE is checked against a user identity and a password.
- User Categories: A user category defines different levels of function access privileges that can be assigned to a user. The lowest privilege level is read access, and the highest level is the super-user category.
- Users: New users can be defined and assigned a user identity, password, and user category. Users can also be deleted.

DCN Management. DCN management provides the functions necessary to control and configure the data communication resources which allow communication to take place within the SDH Management Network.

- Network Node Configuration: Each SDH NE can be configured as a data communication node in the OSI environment, which means that addresses, application entity titles, and names used locally must be defined.
- Network Route Configuration: Network routes in the OSI context can be defined by means of routing tables and route priorities.

The Information Model Perspective

Another, more formal way of describing the operation and control functionality is in terms of an Information Model (IM). An IM is an object-oriented description, independent of the actual physical realization of the Network Element resources and how these are managed.

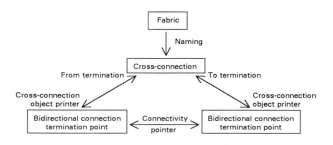

Fig. 1. Model of Duplex Cross-connection. Ptr: Pointer, CTP: Connection Termination Point, Bi: Bidirectional.

The Information Model consists of a collection of object classes, e.g., equipment, software, trail termination point, and SDH switch fabric. The characteristics of an object class are specified in terms of

- read/write attribute values in objects, e.g., values of configuration parameters and relations to other objects, represented as lines between objects in Fig. 1 [2], [3].
- create/delete operations of objects
- actions that can be performed on the object
- notifications (i.e., spontaneous messages) sent from the object.

In an executing system, manageable resources are represented as instances of these object classes. The collection of instantiated objects is referred to as the Management Information Base (MIB) [2]. There are two types of object in the MIB: Managed Objects and Support Objects. A Managed Object (MO) represents a physical or logical resource in the Network Element. A Support Object (SO) represents a log or an alarm filter, for example.

The link between the TMN management functions and the IM is the implementation of each management function as one or more operations, actions, or notifications in the objects that build up the MIB.

A Network Element has what is called a TMN Agent, which can be seen as a process (function) acting on behalf of the managing system(s), relaying messages in both directions (Fig. 2).

A generic Information Model is essential for the generation of

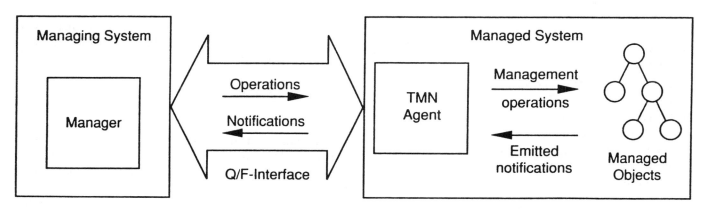

Fig. 2. Interworking among manager, agent, and managed objects.

management standards concerning configuration, fault, performance, and security management functions. A common network model, identifying the generic resources in the network— in this case the Transport Network — and their associated attribute types, events, actions, and behavior, provides a basis on which to explain the interrelationships between these resources and the network management system. Without this common view, a multivendor telecommunication network will not be achieved.

The Information Model provided by Ericsson and describing the SDH NE complies with the information model developed by CCITT's SG XV SDH Model (G.774) [6], and with the SG IV Generic Network Information Model (M.3100) [5]. In general, the configuration management part is derived from G.774, while the fault management and performance management parts are derived from M.3100 and Q.821. By and large, SG IV follows the CCITT X.700 series of recommendations.

The Data Communication Perspective

As well as providing telecommunications services, the SDH NEs also provide powerful data communications and network layer routing functions. The TMN function block that performs these functions is called the Message Communication Function (MCF).

The MCF is based on the OSI reference model, which makes it possible for an SDH NE to work as a data communication node in an open network architecture [7]. Each of Ericsson's SDH NEs can be equipped with a Q-interface and provide for Embedded Control Channel (ECC) access, which means that each NE can be connected to any Operations System that conforms to OSI and TMN standards, without the need for additional Mediation Devices or Q-adapters.

The MCF performs network layer routing functions between the Q-interface and any ECC subnetwork, or between any of the ECC subnetworks, Appendices 2 and 3.

GRAPHICAL USER INTERFACE

General

One of the most important aspects of SDH NE management is the ease with which it can be operated.

It is true that a prime driving force behind the deployment of the Transport Network is to facilitate the management of network resources by way of centralized management—that is, with Q-interfaces and OS—but NEs will nevertheless be operated and maintained from the local site. Some of the reasons for local operation are:

- Backup when the OS, or the communication link to OS, is down;
- The network operator may wish to adopt a more decentralized management philosophy;
- Certain management functions are more easily performed on site because they require physical manipulation of the equipment.

For this purpose, a local operator's terminal with a Graphical User Interface (GUI) can be connected to the NE. The GUI is a

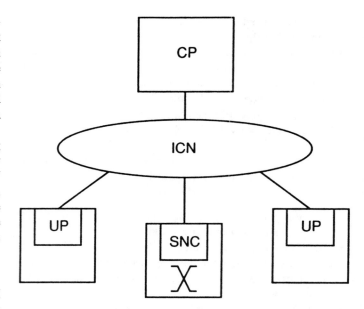

Fig. 3. Control system architecture. CP: Central Processor, UP: Unit Processor, ICN: Internal Communication Network, SNC: Switching Network Controller (only SDXC).

window-based and mouse-operated interface through which the operator has access to all the management functions.

In order for an operator to gain access to the functionality, he has to prove his legitimacy by supplying an identification code and a password. This falls under Security Management, which also includes the possibility of a Super-User assigning operators to particular user categories with extensive or more limited privileges.

Configuration Management covers both physical configuration in an SDH NE — typically Printed Board Assemblies such as the Termination Access Unit — and a logical configuration of the capacities for switching and multiplexing. These different types of Configuration Management are supported by different graphical views, e.g., a physical view and a logical view.

When there is a fault of some kind, e.g., a Transport Network related alarm such as Loss of Frame Alignment or excessive Bit Error Ratio, or a fault related to HW or SW in the actual NE, the details and possible consequences are reported to the GUI. Fault Management, which deals with these functions, also covers testing and diagnostics.

Performance is continuously monitored, and the GUI presents the statistics graphically and/or in tabular form. These functions fall under Performance Management.

CONTROL SYSTEM ARCHITECTURE

Introduction

The control system has to control Network Elements which vary in size from only a few connected cables to over 8000. The functionality associated with each connected cable makes heavy demands on the control system processing capacity. Another important factor to consider is cost, which has to be kept very low for small systems.

A distributed computer architecture has been chosen to meet these requirements. Each unit in the system has a powerful microprocessor, and a central master computer coordinates all the unit processors and provides input/output.

Unit processors enable the unit itself to perform a large amount of processing, and so reduce the load on the central processor.

All processors are connected to an Internal Communication Network (ICN). Depending on the size of the Network Elements, different internal communication network structures have been implemented (Fig. 3).

The central processor houses the MIB, on which the OS and the local operator perform management operations. It will also implement parts of the MCF functionality, such as a network layer routing function between the Q- interface and the ECC subnetworks.

The central processor can vary in size from small and inexpensive one-board microprocessor-based systems, to high-capacity redundant computers.

Each of the different printed board assemblies under the control of the Central Processor contains its own Unit Processor. The Unit Processor is a building block consisting of a microprocessor, memories, communication controllers, and A/D and D/A converters when required.

The Unit Processors perform routine tasks on each printed board assembly, such as alarm surveillance, collection of performance parameters, and self-diagnostics. The Unit Processors also control the lower layers of the ECC protocol suite.

SDXC Control System Architecture

General. The SDXC control system may be composed of a central processor and up to a couple of 100 Unit Processors. It uses a packet-switched Internal Communication Network (ICN) which is integrated with the switch.

Purchased hardware and software are used, the Central Processor being a UNIX computer, for example.

The control system software modularity is ensured through the use of a layered structure combined with object-oriented techniques.

Programs can be downloaded from an Operations System all the way down to a Unit Processor.

Control of Switching Network. Besides Unit Processors and the Central Processor, the control circuitry of the SDXC switch, SNC, is also connected to the Internal Communication Network. This means that all processors can set up and release cross-connections in the SDXC. This facility comes into use, for example, when a rapid switch has to be performed due to a transport network fault discovered by an access unit. In this case, the Unit Processor on the access unit immediately reconfigures the switch according to a predefined configuration previously communicated to the Unit Processor by the Central Processor. The Central Processor is always informed of the resulting configuration.

Unit Processors. Unit Processors typically occupy part of a board in the SDXC. A unit in the SDXC normally consists of only one board. A Unit Processor includes a microprocessor chip and memory.

A Unit Processor continuously performs tasks such as monitoring of hardware and calculation of bit errors, but at the same time it must react quickly to events such as incoming alarms from the transport network. The unit processor is therefore equipped with a real-time operating system kernel, OS 68, which gives response times on a microsecond level.

Central Processor. The Central Processor is the coordinator and master of the complete control system. It continuously monitors the Internal Communication Network to find new Unit Processors. Local operators have access to the SDXC system via graphical workstations. These are connected to the Central Processor via an Ethernet Local Area Network (Fig. 4). Both the Central Processor and the operator workstations are UNIX machines based on SPARC architecture, the Central Processor being a SUN SPARC2 and the operator terminals SUN IPX.

The Central Processor is connected to the Internal Communication Network by a separate Ethernet. It is separate to ensure that there is adequate capacity and security for communication with the Unit Processors. Ethernet, being a standardized interface, enables a change of Central Processor supplier without any interface boards having to be redesigned.

The next version of the CP will consist of basic and optional plug-in units, forming a modular system connected to a high-speed VME backplane and mounted in a standard subrack suitable for telecommunication purposes.

Internal Communication Network. All processors are connected to the Internal Communication Network. It enables all processors to communicate with each other and with the switch control circuitry.

The Internal Communication Network employs packet-switching techniques which enable processors to communicate

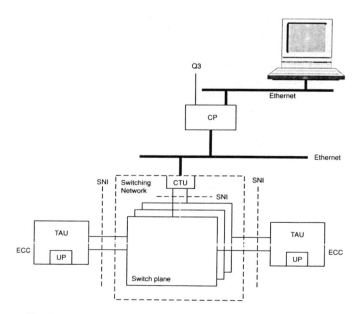

Fig. 4. The CP is connected to the ICN by a separate Ethernet. All control signals are routed through a centralized, triplicated switch and embedded in the traffic signal. CTU: Control Termination Unit, Traffic signal, Control signal.

using different data rates. The Central Processor uses 2 Mb/s and Unit Processors 0.5 Mb/s.

The internal cabling that carries the SNI signals used for transport of traffic information within the SDXC is also used as a part of the Internal Communication Network [4]. Thus, expansion of the SDXC with new equipment at the same time increases the capacity of the communication network to cater for new Unit Processors. Ongoing communication is not affected by this expansion.

By utilizing the SNI signals, the Internal Communication Network benefits from the reliable triplicated structure of the switch. A single failure cannot cause a complete failure of the Internal Communication Network.

All control signals are routed through a centralized packet switch. This is triplicated and each part resides on a switch plane (Fig. 4).

The Central Processor gives or refuses permission to communicate within the Internal Communication Network and keeps a record of all ongoing communication. In fault situations, all the processors involved stop communicating and the fault is cleared by the Central Processor.

Communication is normally between the Central Processor and the Unit Processors, but for time-critical operations, direct communication between Unit Processors can be used, e.g., for fast protection switching.

Broadcast messages can be distributed over the Internal Communication Network. In this case, a message is sent from a processor, and then duplicated by the communication network and sent to all access points where a processor may be connected. This facility is utilized by the Central Processor to find out whether any new Unit Processors have been connected to the Internal Communication Network. This happens, for example, when a magazine is equipped with a new interface unit. The Central Processor sends a broadcast message which is answered by all new Unit Processors. Another possibility with the broadcast facility is for the Central Processor to distribute calendar date and time to all Unit Processors simultaneously.

The use of broadcast greatly reduces the load on the Central Processor, since the single message sent by the Central Processor is multiplied and distributed by the packet-switched nodes within the Internal Communication Network.

A set of rules defines the way communication between the application programs in the Central Processor and the Unit Processors is accomplished. The rules specify the size of packages, priority handling between different messages, actions to be taken when messages or parts of messages are lost, etc.

All these rules are designed into four protocol layers. Together, these four layers hide all Internal Communication Network implementation details from the application. The protocols are implemented as separate software modules, and one layer can therefore be modified without affecting the others. The same protocol software is used both in the Unit Processors and in the Central Processor. It is written in the C programming language.

Software Architecture

General. By definition, all software in the SDXC belongs

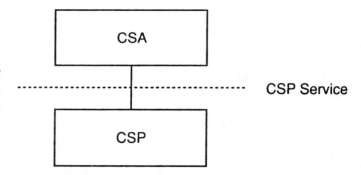

Fig. 5. Software architecture. CSA: Control System Application, CSP: Control System Platform.

to the Control System. Its purpose is to provide Operation, Administration, Maintenance, and Provisioning functions via the external control interfaces, namely, the Graphical User Interface and the Q-interface. The software is physically distributed between the Central Processor and the Unit Processors.

There is a basic architectural distinction between platform-oriented and application-oriented software (referred to in Fig. 5 as CSP and CSA, respectively).

Control System Platform. The Control System Platform (CSP), together with the computer hardware and its Operating System, provides a platform that offers the following services:

- Internal Communication: IPC, i.e., Inter-Process Communication (inter- as well as intraprocessor) by way of "sockets."
- External Communication: Provision of Application Programmer's Interface (API) for communication with external management applications. These services use open, standardized, seven-layer OSI stacks.
- Inventory: Services for checking consistency between software and hardware configuration.
- Program Loading: Services for loading of software packages from the local sites as well as from the centralized Network Management System (FMAS).
- Process Handling: Functions to enable supervision of processes.
- Run-Time Fault Reporting: Functions to enable the reporting of internal DXC Control System faults to a Management System.
- Restart: Upon the detection of a Control System fault, the nature and seriousness of the fault are evaluated, and the system is subsequently restarted from a well defined point of execution.

Control System Application. The DXC CS application software constitutes all Transport Network-oriented functionality. For a static description of the CSA, three aspects of the functionality are taken into account, thereby creating a layered structure, in order to isolate different dependencies and stimulate a modular design (Fig. 6).

- The User Layer: Contains the functionality of the Graphical User Interface. This layer of software will develop and

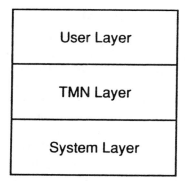

Fig. 6. The CSA architecture, with its layered structure.

change considerably due to different market needs and new tools for graphical presentation. It is therefore separated from the underlying software by an interface called the Controller Interface (CI) which can be said to represent the functionality provided by the TMN Layer and offered to management systems, such as centralized OS or local GUI.

- The TMN Layer: Consists of functions for managing Network Elements specified in an object-oriented Information Model. An Information Model is becoming the standard way of specifying the manageable resources of a Network Element in the Transport Network. CCITT Recommendation G.774 describes the basis for the SDH NE Information Model, and thus forms the foundation for the DXC CS Telecommunication Management Layer (TMN). It will necessarily develop over the years, particularly since different customers will require their SDH NE deployment in stages not coherent with CCITT Recommendation G.774 releases. For the purpose of isolating these dependencies, the Information Model aspects are singled out in the TMN Layer.

- The System Layer: For SDH NEs, CCITT Recommendations G.781–G.783 specify — through the use of Functional

Block descriptions — the functionality that must be provided. The descriptions are a reasonably stable set of requirements, and are singled out in the DXC CS in the System Layer so as to be distinguished from the management aspects. The object-oriented approach is used here too, i.e., the System Layer consists of a number of objects.

The software described above is mapped onto the computer platform as shown in Fig. 7.

From an application programmer's point of view, the Q-interface and the F-interface are handled in a similar way. The protocols are specified in the Controller Interface. OSI standards for service specifications (CMISE) are utilized. The Q-interface is a full seven-layer OSI stack, while the F-interface uses the IPC mechanism provided by the Control System Platform.

Functional requirements. Certain DXC Control System functional requirements must be taken into account when designing the software:

- Several Users: A DXC may be equipped with more than one GUI plus a Q-interface, and all of these interfaces must be able to operate simultaneously. This requires concurrency, which is implemented through several processes working on the objects (the MIB).

- Real-Time Characteristics: The DXC Control System must be event-driven, in the sense that alarm detection mechanisms and subsequent evaluation and filtering are in some cases required to result in autonomous reconfigurations within a certain time.

- Data Storage and Consistency: The information in the objects contained in the MIB — representing everything of relevance to the management system — adds up to a considerable amount of data, the volume of which requires the use of disk storage. The information on the disk is also used for backup purposes. Typically, 1 Gbyte is used in a normal configuration. Since the MIB is the image used by the management system to represent the Network Element's manageable resources, it is of course of extreme importance that the data are consistent with the current configuration.

- Availability and Robustness: It is through the DXC Control System that a management system exerts its network control. This control might, for example, involve the setting up of a digital path through the Transport Network to provide end-users with data communication capacity. Unless the DXC CS has very high availability — a system which is robust, i.e., resistant to faults — the end-users will not receive their services, which ultimately leads to reduced revenues for the network operator.

Since different operators are allowed to operate the system simultaneously, it is essential to ensure that a consistent MIB is maintained. The operations are regarded as one or several transactions. Should a system fault occur, preventing a transaction from being carried out, a rollback to a well-defined state is performed. This means, in some cases, that part of the MIB must be locked during a Transaction. What constitutes a Transaction and what has to be locked is the responsiblity of the TMN Agent functionality.

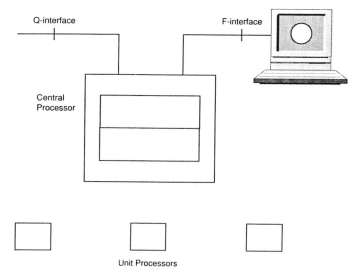

Fig. 7. Mapping of the CSA software onto the computer platform. User Layer, TMN Layer, System Layer.

Also, to allow for the shortest possible reaction times, i.e., to provide the event-driven real-time functionality, lengthy executions must be divided into Transactions.

Data related to the Information Model can be stored by using one of several possible techniques:

- Object DataBase Management System, ODBMS: This solution is very attractive in the sense that it provides persistent objects, which is exactly what the Information Model specification suggests. The difference between the implementation of the system and the way it is described in an Information Model is less than in the other solutions. Also, the amount of design and implementation work is considerably less.
- Relational DataBase Management System, RDBMS: This solution also limits the amount of design and implementation work because it provides mechanisms — such as an ODBMS — for data storage, transaction handling, rollback, etc. However, a translation (design) must be made from the object-oriented specification to the world of tables in a relational database.
- Ordinary Files: This solution is "the hacker's choice." It is straightforward in that it uses only what the Operating System and the programming language provide.
- Class Library:This represents something of a compromise. It provides basic data persistence functionality for objects.

Implementation Structure. The network resource represented by a Managed Object is partly implemented in hardware, and the software is divided between the Central Processor and Unit Processors. To utilize the distributed processor structure, which in a DXC of normal size means one CP and some 100 UPs or more, as much as possible of the functionality is delegated to the UPs. The Internal Communication Network, ICN, allows UPs to communicate directly without involving the CP, which means that real-time functions such as protection switching and network synchronization can be handled in the UPs.

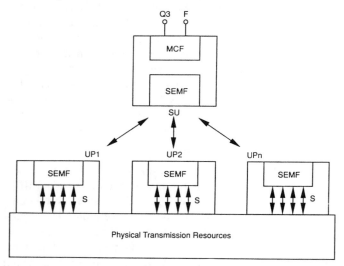

Fig. 8. The SMUX Control System architecture. MCF: Message Communications Function, SEMF: Synchronous Equipment Management Function, S: S-interfaces as specified in CCITT Recs. G.782–G.783.

SMUX Control System Architecture

General. The SMUX Control System is implemented mainly by programs executed on a Central Processor (the Support Unit), as well as Unit Processors (UPs) distributed on each transmission printed board assembly within the NE. The SU is a one-board processor optimized for small NEs. The SU may be common to a number — normally two — of SDH multiplexers.

The SU has an overall responsibility for management of the NE, and receives and evaluates management operations issued from an OS or a local operator. As a response to events detected in the network, the SU issues notifications, e.g., alarm messages, to the OS. The Q3 and ECC protocol suites, and the F-interface, are also controlled by the SU.

Functions of a simple nature but with a high repetition rate, e.g., scanning of binary indications, alarms, calculation of PM parameters, and operations in close connection with transmission hardware, are performed by UPs.

The implementation of the SMUX Control System mapped onto CCITT Recs. G.782–G.783 is shown in Fig. 8.

The SEMF is a function block which sends and receives data on low-level management functions to and from transmission-oriented function blocks.

The MCF is implemented as a protocol machine on the SU, and the SEMF is implemented as software both on the SU and on UPs. The Management Information Base (MIB) is located on the SU, while the S-interfaces, as specified in CCITT Recs. G.782–G.783, are implemented as an internal processor bus between the UP and hardware registers on the transmission printed board assemblies.

Commercially available products supplied by Retix are used to implement the Q3 and ECC protocol suites.

Hardware Platform. The management subsystem hardware platform consists of processors at two different levels:

- Central Processor (SU)
- Unit Processors (UP).

In addition, the following equipment may be required for control and operation:

- IBM-compatible 386/486 PC (Local Operator's Terminal)
- MAU (Ethernet transceiver) for connection to a DCN of LAN type.

The SMUX Control System Hardware Architecture is illustrated in Fig. 9.

The SU communicates with the UPs via an internal ISO 8482 bus, which is similar to RS-485.

The SU implementation is mainly based on the following circuits:

- CPU
- Ethernet controller (Q3-interface)
- RS-232 Communication controller (F-interface)
- LAPD Communication controllers (internal communication)
- Relay contacts (station alarm interfaces)

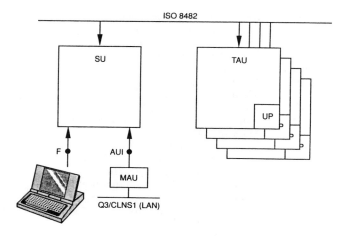

Fig. 9. SMUX Control System hardware implementation. TAU: Termination Access Unit, MAU: Medium Attachment Unit, AUI: Attachment Unit Interface, LAN: Local Area Network.

- Detection logic (external alarm interfaces)
- Program memory
- Data memory
- Backup memory for nonvolatile data.

The UP is a general hardware building block common to all transmission printed board assemblies. The UP implementation is mainly based on the following circuits:

- CPU
- LAPD communication controllers (ECC and internal communication)
- A/D and D/A converters, for measurement of laser characteristics, such as input power and laser bias current
- Test interface (gains access to UP software for an authorized user)
- Program memory
- Data memory
- Backup memory for nonvolatile data.

Not all UP circuits have to be present on every transmission printed board assembly.

The CPU used both for SU and UPs is the Motorola 68 302, which is a microprocessor optimized for data communication (ISDN) purposes.

An IBM-compatible 386/486 PC is used as Local Operator Terminal. The LOT provides the operator with a Network Element view. However, it is possible to manage small networks from the LOT, but of course without the network view.

To manage a network from an LOT without assistance from the OS, communication over the ECCs is used, Appendix 2. By using ECC, it is possible for an LOT to exchange messages with any other SDH NEs within the SDH network. This possibility (of accessing remotely located SDH NEs) is referred to as Remote Login. It will be a valuable feature, especially for the earlier field trials and installations without a complete Network Management System.

Software Architecture. The SU and UP software in an SMUX forms a loosely coupled, distributed software system organized into a layered structure, where each layer has a well-defined task (Fig. 10).

Additionally, PC software which is not indicated in Fig. 10 is required for the Graphical User Interface.

The SU software layers and their tasks are as follows:

- User Access Layer: Provides external access for an OS or local operator to the management view of the SDH multiplexer. Contains the data communication services for the F, Q3, and ECC interfaces.
- TMN Layer: Provides the generalized TMN management view of the SDH multiplexer in the form of a Management Information Base with the managed objects, their attributes, actions, and emitted notifications.
- SMUX Layer: Contains a logical SDH multiplexer as specified in CCITT Recs. G.782–G.783. This logical multiplexer can be controlled from the TMN layer, and has the standardized automatic behavior for protection switching and change of synchronization source.
- Magazine Layer: Manages all hardware units in the magazine so that they provide the logical SDH multiplexer transmission services requested by the SMUX layer using the available units, and signal interconnections in the magazine.
- Unit Layer: Manages individually each hardware unit in the magazine, ordering changes in the unit, receiving events from the unit, and ensuring that the UP software is consistent with the SU software.
- Base Layer: Provides process management and communication, drivers for SU I/O ports, and communication services between the SU and UPs.
- Virtual Machine Layer: Provides a real-time, multitask virtual machine on bare machine hardware. Contains Operating System kernel and low-level hardware interfacing to the SU hardware.

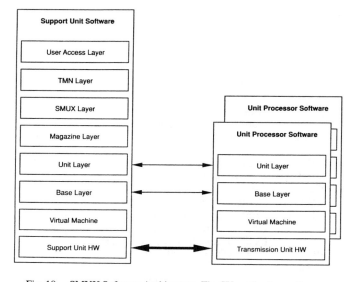

Fig. 10. SMUX Software Architecture. The SU application software communicates peer-to-peer with the UPs' application software. The SU communicates with the UPs by using an ISO 8482 backplane bus.

The UP software layers and their tasks are as follows:

- Unit Layer: Manages each hardware unit individually, making changes on the unit according to orders from the SU, and reporting events from the unit.
- Base Layer: Provides process management and communication, drivers for I/O ports on the unit, and communication services between the UP and SU.
- Virtual Machine Layer: Provides a real-time, multitask virtual machine on bare machine hardware. Contains Operating System kernel and low-level software interfacing to the transmission circuits.

The PC application software and its purpose are as follows:

- Graphical User Interface (GUI): Provides the local operator with a graphical user interface.

SUMMARY

The control and operation of SDH Network Elements is and will continue to be adapted to the evolving TMN standards. This facilitates their connection to centralized Operations Systems.

The implementation of the control system takes into consideration the various demands of Network Elements, ranging in size and complexity from small SMUXs up to large SDXCs.

References

[1] H. Tarle, "FMAS — An operations support system for transport networks," *Ericsson Rev.* vol. 67, no. 2, pp. 163–182, 1990.

[2] W. Widl, "CCITT standardisation of telecommunications management networks," *Ericsson Rev.* vol. 68, no. 2, pp. 34–51, 1991.

[3] W. Widl and K. Woldegiorgis, "In search of managed objects," *Ericsson Rev.*, vol. 69, no. 1/2, pp. 34–56, 1992.

[4] J. A. Bergkvist, G. Evangelisti, and J. Hopfinger, "AXD 4/1, A digital cross-connect system," *Ericsson Rev.*, vol. 69, no. 3, pp. xx–vv, 1992.

[5] CCITT Draft Rec. M.3010, "Principles for a Telecommunications Management Network."

[6] CCITT Rec. G.774, "SDH Network Information Model for TMN."

[7] CCITT Rec X.200, "Reference Model of Open Systems Interconnection for CCITT Applications."

ABBREVIATIONS

ACSE	Association Control Service Element
API	Application Programmer's Interface
AUI	Attachment Unit Interface
CLNP	Conectionless Network Protocol
CMISE	Common Management Information Service Element
CP	Central Processor
CPU	Central Processor Unit
CS	Control System
CSA	Control System Application
CSP	Control System Platform
DCC	Data Communications Channel
DCN	Data Communications Network
ECC	Embedded Control Channel
GNE	Gateway Network Element
GUI	Graphical User Interface
ICN	Internal Communications Network
IM	Information Model
IPC	Inter-Process Communication
ISDN	Intergrated Sevices Digital Network
LAPD	Link Access Protocol on D-Channel
LAN	Local Area Network
LLC	Logical Link Control
MAC	Medium Access Control
MAU	Medium Attachment Unit
MCF	Message Communications Function
MIB	Management Information Base
MO	Managed Object
NE	Network Element
OS	Operations System
OSI	Open Systems Interconnection
PDH	Plesiochronous Digital Hierarchy
PI	Physical Interface
PM	Performance Monitoring
ROSE	Remote Operations Service Element
SDH	Synchronous Digital Hierarchy
SDXC	SDH Digital Cross-Connect
SEMF	Synchronous Equipment Management Function
SMS	SDH Management Subnetwork
SMUX	SDH Multiplexer
SNI	Switching Network Interface
SNPA	Sub-Network Point of Attachment
SU	Support Unit
TAU	Termination Access Unit
TMN	Telecommunications Management Network
UP	Unit Processor

APPENDIX 1

Performance Parameters

In the future, the demand for high-quality connections will be even greater than today. It is therefore important to use relevant and accepted parameters when measuring and verifying the quality of connections. The quality parameter currently in use is Bit Error Ratio (BER), with alarm thresholds at (normally) 10^3 or 10^6. This is not good enough for data traffic. Another drawback of BER is that it does not give any information as to how faults are distributed in the time domain. Normally, faults are not distributed uniformly, but "burstily."

To meet these new requirements, CCITT has defined quality parameters in Rec. G.821:

ES	Errored Seconds	No. of errors during 1 s > 0
SES	Severely Errored Second	BER, measured during 1 s > 10^{-3}
DM	Degraded Minutes	BER, measured during 1 min > 10^{-6}
UAS	Unavailable Second	10 consecutive SES gives 10 UAS.

These parameters are initially intended for 64 kb/s connections. Annex D to Rec. G.821 therefore defines how to deal with higher bit rates. The G.821 parameters — after rather animated discussions — have not been found to be the solution to new requirements imposed on quality parameters. For example, they are still based on BER.

A draft, Rec. G.826, defines new quality parameters for bit rates higher than 64 kb/s. The G.826 parameters will be used within SDH when the recommendation has been approved. The G.826 parameters are based on Errored Blocks (EB) instead of BER. One EB is a block that contains one or more errored bits.

The following parameters were defined in the G.826 draft of June 1992:

ES	Errored Second	≥ 1 EB during 1 s.
ESR	ES Ratio	The ratio of ES to the total number of seconds in available time during a specified measurement interval.
SES	Severely Errored Seconds	$\geq Y\%$ EB during 1 s ($Y = 30$ provisionally).
SESR	SES Ratio	The ratio of SES to the total number of seconds in available time during a specified measurement interval.
BBER	Background Block Error Ratio	The ratio of errored blocks to the total number of blocks excluding all blocks during SES and unavailable time.

APPENDIX 2

Q3-Interface

The Q3 interface provides for standardized communication and exchange of management information between an NE and an Operations System. The protocol suite and the information model must be defined when specifying a Q3 interface.

Gateway Network Element (GNE)

A GNE is connected to an OS via a Q3 interface (Fig A2.1). The GNE has an attached subnetwork of SDH NEs, and provides remote access to these NEs by means of Embedded Control Channels (ECC). The GNE performs Intermediate System (IS) network layer routing functions for ECC messages destined to any NE within the subnetwork.

When considering implementation, there is no difference between a GNE and any other SDH NE. They simply perform different roles in the OSI environment.

Embedded Control Channel (ECC)

The ECCs provide a high-capacity data communication network between SDH NEs, utilizing dedicated bytes (DCC) in the STM-N Section Overhead as the physical layer. Two types of ECC have been defined in the SDH standards:

- ECCr: A 192 kb/s data communications channel accessible by all NEs, including the intermediate regenerators.

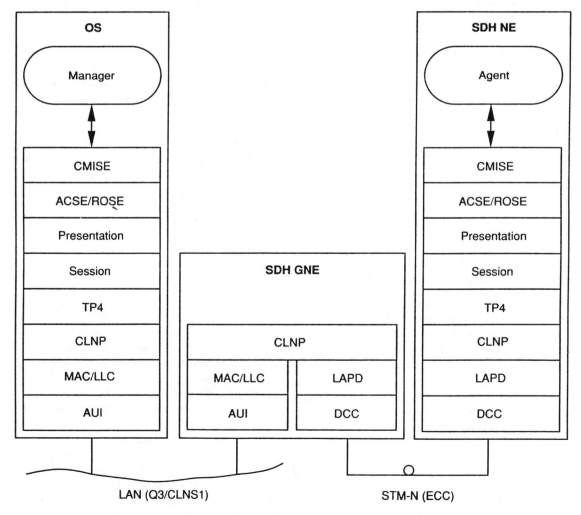

Fig. A2.1. Gateway Network Elements (GNE) may be connected to an OS. The GNE has an attached subnetwork of SDH NEs.

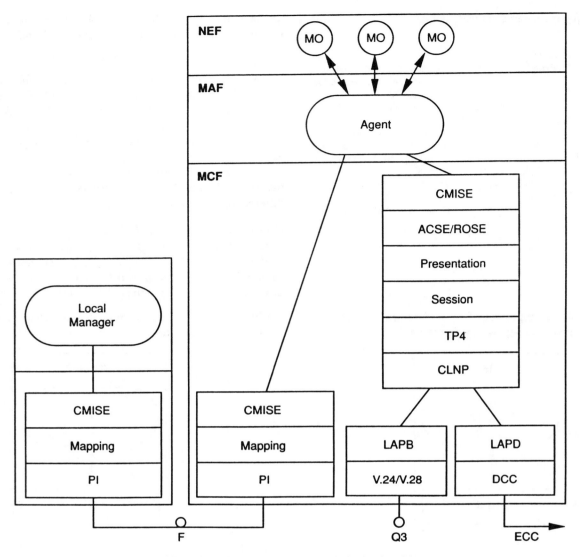

Fig. A3.1.   SDH NE management organizational model.

- ECCm: A 576 kb/s data communications channel accessible by all NEs, excluding the intermediate regenerators.

The ECC network is logically created by defining ECC network routes in the SDH transport network. Network Protocol Data Units (NPDU) are then routed according to address and routing information held locally in the NEs as routing tables, or terminated within the NE.

In the absence of standards, a set of Ericsson proprietary DCN management functions has been defined for the purpose of managing the routing tables and DCN resources.

TABLE 1. SDH GNE LOCAL
ROUTING TABLE

NPDU Destination Address	Next Hop (SNPA)
"SDH NE"	"STM-N ECC"
"OS"	"Q3"
"SDH GNE"	"Own Agent"

The SDH NE Management Organizational Model

Management of SDH NEs is based on the management organizational model as outlined in [6]. The model consists of the following TMN function blocks and components (Fig. A3.1): The Network Element Function (NEF) including the Management Information Base (MIB), the Message Communication Function (MCF), and the Management Application Function (MAF) including the agent.

In addition, an MAF functional component containing a manager for local control of the NE has been defined. The local manager is housed in the local operator's terminal.

Agent: Part of the MAF which is capable of responding to network management operations issued by a manager, and of issuing notifications, e.g., event reports, on behalf of the managed objects.

292

Manager: Part of the MAF which is capable of issuing requests for network management operations, e.g., request performance data, set thresholds, receive event reports, etc.

Local Manager: A manager which is housed in a local operator's terminal and is capable of managing a single network element.

Management Application Function (MAF): An application process providing TMN services. The MAF includes an agent, or a manager, or both. The MAF is the origin and termination of all TMN messages.

Managed Object (MO): The manager's view of a resource within the telecommunications environment that may be managed via the agent. Examples of MOs residing in an SDH NE are equipment, software, trail termination point, SDH switch fabric, alarm log, etc.

Message Communication Function (MCF): Provides facilities for the transport of TMN messages to and from the MAF, as well as network layer routing functions.

Network Element Function (NEF): The entity within an NE that supports transport network-based services, e.g., multiplexing, cross-connection, regeneration, etc. The NEF is represented to the manager as a set of managed objects.

A Synchronous Digital Hierarchy Network Management System

An equipment-independent TMN architecture can avoid backlogs and the problems of multivendor environments.

Toshinari Kunieda, Satoru Sugimoto, and Noriyuki Sasaki

TOSHINARI KUNIEDA heads the new operation systems project in the Transmission Systems Project Group, NTT Customer Systems Development Department.

SATORU SUGIMOTO is senior engineer, and NORIYUKI SASAKI is staff engineer, in the Transmission Systems Project Group, NTT Customer Systems Development Department.

Telecommunication carriers in Japan are competing in the relatively new and competitive environment created by the liberalization of the communications market. At the same time, the introduction of the first phase of Synchronous Digital Hierarchy (SDH) equipment based on fiber optics and SDH interfaces in transmission-line networks has improved transmission quality [1-3]. Therefore, an important strategy for telecommunication carriers in obtaining a competitive advantage is to establish enhanced value-added services and to operate those services efficiently. It is essential to speed up the handling of service orders and service restoration after failure by enhancing the performance of the Operation System (OpS).

In this article, we discuss critical issues in multivendor environments and OpS performance as the software backlog grows. To solve these problems, we propose the application of a Telecommunications Management Network (TMN) architecture [4-6] and the introduction of an Object-oriented Network Resource (ONR) Model [7]. We also examine the second phase of the SDH network management system, where these measures have been practically applied.

The Current Network and Future Systems

Nippon Telegraph & Telephone (NTT) networks are, in general, long distance networks that carry primarily inter-prefecture telecommunication traffic, with associated local networks that handle intra-prefecture traffic (Fig. 1). SDH first-phase equipment, which uses SDH interfaces as

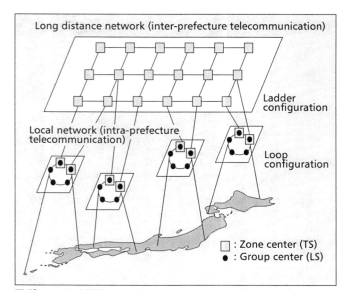

■ **Figure 1.** *NTT networks comprise inter-prefecture telecommunication networks and multiple intra-prefecture networks.*

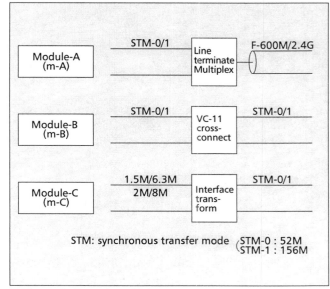

■ **Figure 2.** *First-phase SDH equipment.*

Reprinted from *IEEE Communi. Maga.*, vol. 31, no. 11, pp. 84–90, Nov. 1993.

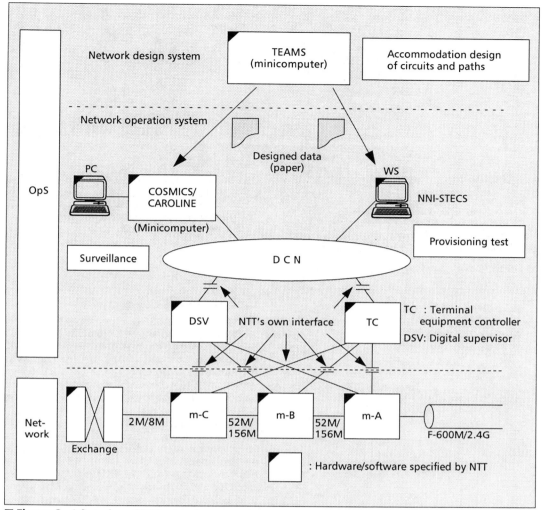

■ Figure 3. *A first-phase SDH network management system.*

shown in Fig. 2, have been introduced in both types of networks since 1989, comprising 30 to 35 percent of all circuits as of the end of the 1992 fiscal year.

As part of the OpS, the network design system, the Telecommunication Engineering and Management System (TEAMS) carries out the design of the circuits and paths. The network operating system NNI-STECS (Network Node Interface-Synchronous Terminal Equipment Control System) handles path provisioning and testing. The Centralized Operating Support and Management Information Control System/Centralized maintenance Administration and Operation system for Local Integrated transmission Network (COSMICS/CAROLINE) takes charge of surveillance [8]. Designed data is printed and passed to these systems, as shown in Fig. 3. These operations are carried out at approximately 50 network centers, each covering a limited local area for path provisioning, testing, and surveillance of equipment. In the future, it is likely that a few tens of thousands of VC-3 paths (52 Mb/s, the equivalent of 672 telephone channels) and several hundreds of thousands of VC-ll paths (1.5 Mb/s, the equivalent of 24 telephone channels) in the long-distance network will be controlled for nationwide operation from a single center, and that in the multiple local networks, local

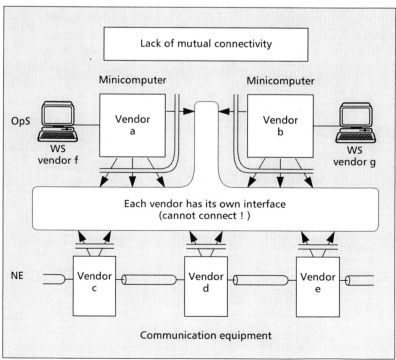

■ Figure 4. *Subjects of multivendor environment.*

295

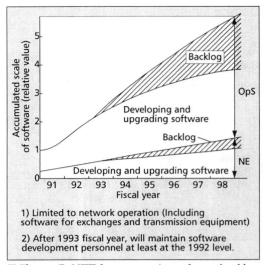

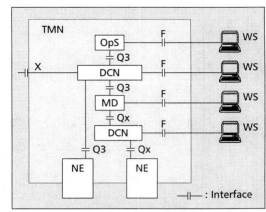

■ Figure 6. *A Telecommunications Management Network (TMN) architecture.*

■ Figure 5. *NTT faces a growing software backlog.*

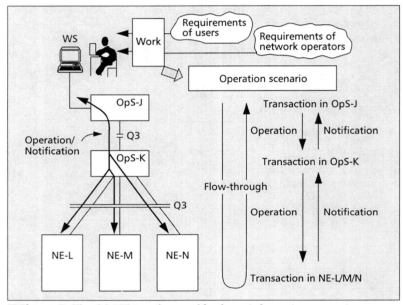

■ Figure 7. *The OS-NE interface enables faster information processing.*

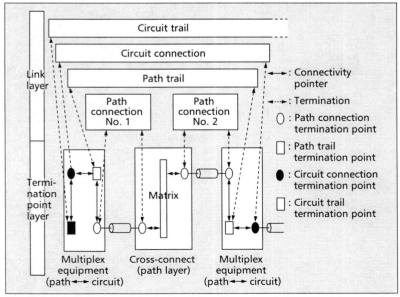

■ Figure 8. *An example of a transmission-line network ONR model.*

operations will be controlled from a single site in each area.

Lack of Mutual Connectivity

"Network operation" is a type of work in which multiple OpSs and network elements (NEs) such as transmission equipment, are closely connected. For instance, the work may be to construct networks with linkages between both the network design system and the network operating system, based on the circuit-path provisioning orders from users or networks, or to restore circuits and paths in real time by triggering a switchover command in the event of failure. Because of the strategies of telecommunication carriers and the limited production capacity of vendors, the number of OpS and NE vendors has increased. However, because each OpS and NE has its own independent interface protocol and data formats, fewer mutual connections are available (Fig. 4). Historically, OpSs were often provided by computer vendors, while the NEs were provided by communication equipment vendors, each using vastly different interface specifications.

The Growing Software Backlog

To gain the advantage in a changing marketplace, NTT is introducing enhanced value-added services such as frame relay and ATM services, as well as modifying its organizations and work procedures. The need for new OpSs and the upgrading of existing OpSs is increasing, and NTT faces a steadily growing backlog, as shown in Fig. 5. As a result, the introduction of new services and the enhancement of network operation has sometimes been delayed. Timely and appropriate responses to the requirements of users and network operators is no longer possible.

One promising solution in a multivendor environment is the concept of TMN architecture, currently being discussed in CCITT. In TMN, protocols and message data are specified as functions of each element (such as OpS and NE) and the interfaces between them based on a unified architecture (Fig. 6). Interfaces in particular use the Common Management Information Protocol (CMIP), which was standardized in (OSI). The OpS-NE interface is a manager-agent relationship; information exchange between the two is carried out using macro commands and responses. Thus, each OpS and NE is assured of individual development, and

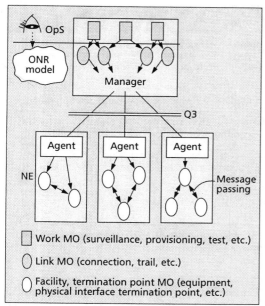

■ Figure 9. *Allocation of managed objects (MO) in NE and OS.*

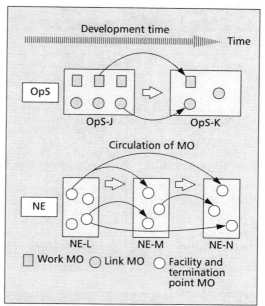

■ Figure 10. *Software circulation by managed objects.*

even if they are provided by multiple vendors, mutual connections are possible. Because work involving multiple OpSs and NEs can be performed in a flow-through fashion, the requirements of users and network operators can be processed much faster (Fig. 7).

Introduction of ONR Model

NTT is studying the introduction of the ONR model (Fig. 8) to resolve the software backlog. In this model, a network is recognized as a set of managed objects (MO) such as facilities, termination points, and links; that is, an object-oriented virtual network model allocates these MOs to the NE and OpS (Fig. 9). With this arrangement, any work can be performed by the OpS, regardless of the type of telecommunication equipment in the network. When OpS performs path provisioning and testing according to a given work scenario, the work is completed by applying an operation, as a manager, to the MO of the agent NE. In addition, as shown in Fig. 9, the work scenario implemented in the OpS can be expressed and installed as an MO by analyzing its work, and thus the entire OpS can be recognized as a set of MOs in the same way as the NEs. Moreover, software can be made for each MO and exchanged between NEs or OpSs with identical functions (Fig. 10). Applying this approach enhances software productivity, and thus averts development backlogs.

The SDH Second-Phase Network Management System

*T*he SDH second-phase transmission line network OpS surveillance concentrated control and evaluation system for SDH network (SUCCESS) is based on the TMN architecture and ONR model. SUCCESS is an OpS that manages nationwide surveillance and control of the SDH network in the long-distance network from a remote center. The configuration of SUCCESS is shown in Fig. 11. SUCCESS consists of control equipment

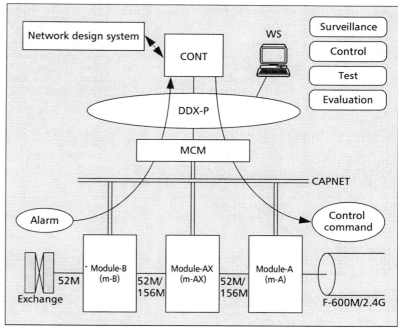

■ Figure 11. *Configuration of SUCCESS.*

(CONT) to perform OpS functions, a message communication module (MCM) having a mediation function, and work stations (WSs) to provide network operators with a UI (user interface). This equipment is connected to inter-office data-communication networks, and multiple NEs are connected under MCM through the intra-office data communication network in the network center.

Control Equipment—CONT consists of five minicomputers in all, one of which is responsible for system management of CONT itself and nationwide SDH path management (Fig. 12). Three other systems manage transmission equipment, each covering one of the three area "blocks" that cover the entire country. Each of these five sys-

Type of work	Functions
Surveillance	Processing of transmission equipment alarms and notifications to operators
Path test	Drop, insertion, and monitor tests on arbitrary sections of transmission line, path, or circuit
Path provisioning and	Control of cross-connect equipment and provision deletion and deletion paths
Protective action against network failure	Control of cross-connect equipment for automatic network protection on 52 M and 156 M path levels, to cope with major failures such as a route failure.
Schedule path configuration change	Control of cross-connect equipment for scheduled path configuration change (switchover or switch back) on 52 M and 156 M path levels due to failure transfer
Network management	Establishment and management of database for link information such as paths and transmission lines
Facility management and operation	Establishment and management of facility-information database and operational control of redundancy switchover

■ **Table 1.** *SUCCESS CONT functions.*

Parameters	Contents
Redundant configuration	Duplicated configuration
Performance	65 MIPS (85 TPS)/ 1 CPU (2-CPU configuration for four systems; 1-CPU configuration for protection system
Memories	512 MB or 256 MB
Disk	2.7 GB (mirrored)
Backup unit	2.0 GB (DAT)

■ **Table 2.** *Major performance parameters of CONT minicomputers.*

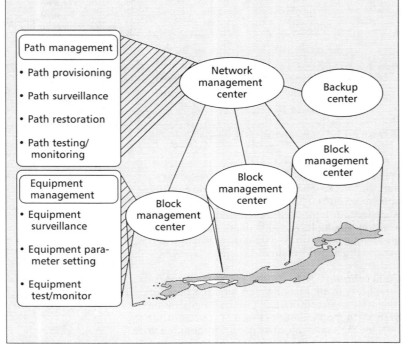

■ **Figure 12.** *Function assignments in SUCCESS CONT (control equipment).*

tems is connected to the others through a LAN to exchange information associated with paths and equipment. Working together, these systems manage the SDH transmission line network. Each system guarantees 24-hour continuous operation through a hot standby redundant configuration using two minicomputers. The fifth system is a remote one that serves as protection against failure of other CONT systems.

The major functions of CONT include surveillance, testing, path provisioning, and path-network switchover (Table 1). In the event of a total route failure such as a disconnection of fiber optics, the path network switchover function detours associated paths through other routes by combining standby transmission lines previously installed in a dispersed fashion. CONT is able to design a route in real time after detecting a transmission line failure taking into account a predetermined order of priority of accommodated circuits and path survivability for the designated destination. In addition, through control of cross-connect equipment, the path can be quickly restored. In a total failure of a 2.4 Gb/s transmission line, the processing time required for the switchover of 30 high-priority paths is approximately 10 seconds.

A minicomputer with better than 100 MIPS is used to manage the network of the size described above and to carry out path network switchover. The major performance parameters of the minicomputers used in CONT are listed in Table 2.

Message Communication Module — MCM is installed in network centers and is used to convert protocol between the inter-office data communication network and the intra-office data communication network. This is done to collect operational information (such as alarm and performance-monitoring data) from NEs and to deliver their signals to CONT after processing alarm priority, suppressing any minor alarms, and generating trigger information to activate network switchover.

Workstations — SUCCESS provides four kinds of WSs, each corresponding to one of the four major operations: path provisioning, surveillance, maintenance, and switchover. Each WS has these specialized functions to perform the allocated work together with common functions to search for CONT

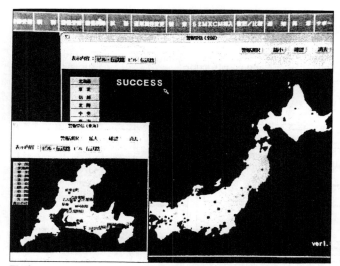

■ Figure 13. *An example of the user interface on a SUCCESS workstation.*

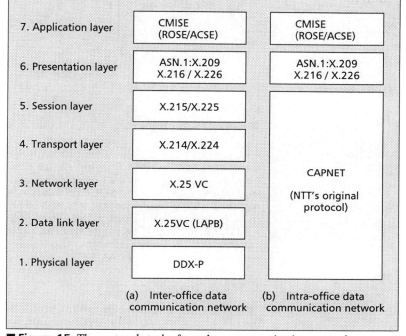

■ Figure 14. *A typical network element configuration.*

data or network status, thus allowing nationwide access from a remote center. To be able to respond to any changes in the given work, the program structure is made up of numerous independent program modules. A highly intelligent work station with a superior cost-performance ratio provides operators with a graphical user interface (Fig. 13).

Configuration of NE

To realize the ONR model, MOs are installed in an NE to make it intelligent(Fig. 14). The network element function (NEF), which consists of both firmware and hardware, contains the basic NE functions for transmitting its main signal. Synchronous equipment management function (SEMF), which consists of MOs, contains the operation, administration, and maintenance (OAM) functions for the main signal and NE. By adopting CTRON [9] as the extended OS (operating system) interface common to vendors and specifications of the S interface, vendor-specific characteristics of the NEF portion are contained. In other words, discrepancies between various vendors' hardware can be eliminated through the installation of identical MOs into different types of vendor NEs. Furthermore, by specifying MOs in accordance with the ONR model and by eliminating specific NE characteristics as far as possible, software commonality on an MO basis can be performed. NEs to which MOs can be applied include 52 M cross-connect equipment (module-AX), line terminate and multiplex equipment (modified module-A), and small capacity 1.5M cross-connect equipment (small capacity module-b), all of which are treated as SDH second-phase transmission equipment. Approximately 70, 60, and 70 MOs are defined for each of these respectively, and about half of these MOs conform to international standards such as M.3100. MO commonality is possible between different types of equipment, and in terms of software size, 70 percent of the small capacity module-b and 90 percent of the modified module-A can be circulated from module-AX software.

CONT-NE Interface

The Q-interface is used as the CONT-NE interface via the inter-office data communication net-

	(a) Inter-office data communication network	(b) Intra-office data communication network
7. Application layer	CMISE (ROSE/ACSE)	CMISE (ROSE/ACSE)
6. Presentation layer	ASN.1:X.209 X.216 / X.226	ASN.1:X.209 X.216 / X.226
5. Session layer	X.215/X.225	CAPNET (NTT's original protocol)
4. Transport layer	X.214/X.224	
3. Network layer	X.25 VC	
2. Data link layer	X.25VC (LAPB)	
1. Physical layer	DDX-P	

■ Figure 15. *The protocol stack of two data-communication networks.*

work and the intra-office data communication network. The protocol stack of each data communication network is shown in Fig. 15. Both interfaces apply CMIP in their application layers. The inter-office data communication network has a Q3 interface with an X.25 protocol in its lower layers; an original protocol called Control and Access for Plant Network (CAPNET) is used in the lower layers of the intra-office data communication network to ensure consistency with existing facilities. In addition, CMISE commands are matched with established standards whenever possible. To achieve functions not specified in international standards, however, approximately 50 original M-ACTIONs are specified, such as hitless-path switchover and switchover after confirmation of path ID.

To achieve flow-through path operation from design to provisioning and surveillance, SUCCESS is responsible for on-line data exchange

The network-management system described here may also be effective in the future for the operation of networks that apply ATM technology.

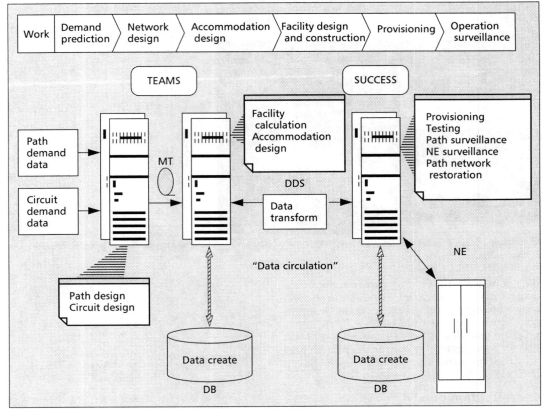

■ **Figure 16.** *The relationship of TEAMS, the network design system, to SUCCESS.*

with regards to path provisioning and deletion through linkage with TEAMS. Database mismatches between the two systems are prevented by sending information from the SUCCESS database back to TEAMS. In addition, because path names in both systems are different, a data delivery system (DDS) is installed between both systems for Path ID conversion. Figure 16 shows the linkage between SUCCESS and TEAMS.

Conclusion

To avoid growing backlogs and the problems inherent to multivendor environments, we have proposed the introduction of TMN architecture together with ONR model, independent of telecommunication equipment type. The SDH second-phase network-management system makes this approach practical. In the future, it is expected that this approach will also be effective for the operation of networks that apply ATM technology.

Acknowledgments

We would like to thank Mr. Miura, project manager of the Customer Systems Development Department, for his guidance and helpful discussions from the study of this concept to the introduction of the SDH second-phase network management system.

References

[1] Y. Yuji, I. Tokizawa, and N. Terada, "Basic Considerations to Deflne Broadband Network Interfaces," GLOBECOM '87, pp.13.2.1-13.2.5, Nov. 1987.
[2] K. Maki *et al.*, "Implementation and Application of Equipment with Network Node Interface," GLOBECOM '87, pp. .13.2.113.2.5, Nov. 1987.
[3] H. Ueda, H. Tsuji, and T.Tsuboi, "New Synchronous Digital Transmission System with Network Node Interface," GLOBECOM '89, pp. 42.4.1-42.4.5,1989.
[4] CCITT Recommendation M.3010.
[5] CCITT Recommendation M.3100.
[6] M. Yamamuro, M. Matsushita, and M. Wakano, "TMN Implementation Strategy Based on OSI Management Standards," This workshop. pp. 1.2.1-1.2.7, 1993.
[7] H. Hara and T. Kunieda, "OAM Function Model for Transport Network with Object-oriented Approach," *NTT R & D*, vol. 41, no.3, pp. 379-90, 1992.
[8] H. Miura, "Special Feature: Construction & Operation of the Transmission Line Network," *NTT Review*, vol. 5, no. 2, pp. 30-58, 1993.
[9] Edited by Tron Association, "Original CTRON Specification Series," Ohmsha, 1989.

SONET Operations in the TINA Context

WILLIAM J. BARR
BELLCORE COMMUNICATIONS RESEARCH
MORRISTOWN, NJ USA

TREVOR BOYD
BRITISH TELECOM
MARTLESHAM HEATH, UK

YUJI INOUE
NTT TELECOMMUNICATION
TOKYO, JAPAN

Abstract—In 1990, the Telecommunications Information Networking Architecture (TINA) Workshops were initiated by major network operators as an open forum for discussing and conducting research on information networking problems with the R&D community from suppliers of both telecommunications and computing technology. The workshops are aimed at promoting a sound worldwide base for the technical advances needed to bring telecommunications and computing together into an overall information networking architecture applicable in the mid-1990s and beyond. The success of these workshops has motivated the establishment of a formal structure — the TINA Consortium (TINA-C). TINA-C was formed in early 1993, and consists of over 40 researchers from the more than 30 member companies and is located in Red Bank, NJ, U.S.

This paper outlines the principles and elements of an information networking architecture as currently conceived by the authors, stimulated by TINA workshop contributions. To illustrate the power of the architecture, we choose an application example — the management and operations of a SONET network — to describe how the architecture concepts could be applied.

TELECOMMUNICATIONS networks are at a juncture comparable with the introduction of stored program control capabilities in the 1960s. Technology breakthroughs are changing the price/performance ratios in fundamental ways. The cost of transmission bandwidth is reducing, providing the impetus for transmission of video and vast amounts of data hitherto thought uneconomical. Telecommunications networks are increasingly using computers, and computers are increasingly being interconnected by telecommunications networks.

Current trends in society, like the distributed office, home banking, video conferencing, home entertainment, and teleworking, are creating new computing applications and services that require closer relationships between the fields of telecommunications and computing. This requirement is forging cross-fertilization of ideas between the computing and telecommunications industries. Furthermore, recent developments in distributed computing and broadband telecommunications systems have increased the feasibility of creating telecommunications services as distributed application programs executing on network-wide computing platforms.

The demand for new sophisticated services, like universal personal communications, virtual network services, mobile and multimedia services, is on the increase. These services require more flexible access, management, and charging regimes than current network infrastructures are capable of providing. Meeting the customer demands for these services requires a network infrastructure into which services can be introduced easily, quickly, and smoothly. Software will play a major role in this infrastructure. Interoperability, portability, and reuse of software components will be of prime importance. The emerging intelligent network (IN) enables the development of a software-based network infrastructure — which we call an information networking architecture.

In 1990, the Telecommunications Information Networking Architecture (TINA) Workshops were initiated by major network operators as an open forum for discussing information networking problems with the R&D community from suppliers of both telecommunications and computing technology. The workshops are aimed at promoting a sound worldwide base for the technical advances needed to bring telecommunications and computing together into an overall information networking architecture applicable in the mid-1990s and beyond. Four international workshops have been held to date. Beginning with TINA'95 in Melbourne, Australia, the workshops were upgraded to full IEEE Conference status. The success of these workshops has motivated the establishment of a formal structure — the TINA Consortium.

1. THE TINA CONSORTIUM

In Spring 1992, Bellcore, United States, BT, United Kingdom, and NTT, Japan initiated a proposal for a TINA Consortium. The proposal was developed over subsequent months in consultation with major telecommunication and computing companies and organizations worldwide, culminating in its formal announcement and invitation at the International Switching Systems Conference (ISS92), Yokohama, Japan, in October 1992. A proposed organization charter and outline technical workplan were presented at this announcement meeting. The objectives briefly stated for the TINA Consortium (TINA-C) are: 1) to define a telecommunications information networking architecture based on advanced distributed processing and service delivery technologies that will enable efficient introduction, delivery, and management of telecommunications and networks; 2) to validate the effectiveness of the architecture through laboratory experiments and field trials; and 3) to promote the use of the telecommunications information networking architecture worldwide.

The Consortium began its research activities in early 1993, with initial architecture specifications developed during 1994. These specifications will be available to the public. They will seek to build on, and maintain compatibility with, relevant emerging standards in distributed processing and telecommunications. Laboratory experiments and field trials will be continued for several years, with completed specifications available during the fifth year.

Most of the technical work is to be carried out by a core team of resident researchers on assignment to TINA-C from Consortium members. For at least the first two years, this team will be located in New Jersey in accommodations arranged by Bellcore. In addition, there will be auxiliary project work, particularly trial and validation of architectural specifications, performed collaboratively by consortium members and coordinated through the core team.

We are on the threshold of a new and exciting era. TINA-C enables vendors and operators to collaborate in research to advance the development of an information networking architecture, and to promote interoperability and reuse of the software for telecommunications services and operations on a worldwide basis. The results produced by the Consortium will positively impact the shaping of our increasingly software-based telecommunications system. This paper outlines the principles and elements of an information networking architecture as currently conceived by the authors, stimulated by TINA Workshop contributions [1]–[3].

2. VISION OF INFORMATION NETWORKING

An information network would provide to the customers of the network the capabilities of on-demand access and management of information any time, any place, in any volume, and in any form. An architecture for information networking would encapsulate reusable network functions and support network-wide interoperability of these functions — enabling flexible construction of services composed of service components distributed over a geographically disperse and technologically heterogeneous telecommunications network.

Telecommunications operations, management, and services functions may be viewed as software-based applications distributed over multiple nodes of a network. The vision of an information networking architecture embodies a distributed application platform for building and executing network-wide applications, where the platform hides the applications from the effects introduced by distribution and from the complexities of the underlying network resources (see Fig. 1).

The Intelligent Network architecture in its present state of development is confined to basic telephony call-control capabilities. An architecture for information networks would embrace the IN and Telecommunications Management Network (TMN) architectures within a framework based on distributed processing principles. To achieve the objectives of information networking, the architecture would include: flexible control of emerging multimedia, multisession, multipoint, broadband

network resources; services interoperability across diverse network domains; rapid creation and flexible deployment of complex services, including the management of service interactions; and coping with increasing sophistication in Customer Premises Equipment (CPE).

3. FROM INTELLIGENT NETWORK TO INFORMATION NETWORK

To date, IN development has been aimed at the construction of a set of interfaces and protocols which would clearly separate the switching from the service aspects of telecommunications networks, so that the application developer has a logical view of call-control, the call-model [4]. IN has yet to seriously address the potentially distributed nature of the service application software itself, and its interoperability with distributed network management, service management, and CPE software. In this respect, it is generally considered that an object-oriented approach should be used for the longer-term IN architecture.

Objects form natural boundaries for distribution, and are fundamental to the development of advanced distributed processing software [5]. In this respect, the IN architecture does not yet generally support the property of distribution transparency for service software components (objects), where service implementation is decoupled from network dependencies allowing, for example, binding of service components across heterogeneous networks, freedom of allocation of components to network nodes, and independence of services from network scale and topology.

An object-oriented connection model for information networking will need to be developed. In the conventional IN call model, control of connection types, other than point-to-point, is complicated. It is difficult to apply the conventional call model directly to broadband service control. With broadband, information traffic by media other than audio will increase, and services accompanied by connection types other than point-to-point, such as unidirectional and multipoint connection, will also increase. For multimedia services, multiple virtual channels would be assigned to the user. Thus, a connection model must provide flexible virtual channel control. A connection model therefore needs to be constructed for these services, with a uniform method of control supporting the range of connection types. It is expected that such a connection model would be object-based.

In order to quickly introduce more complicated services, a uniform distributed processing environment that allows nodes to advertise services (through some sort of registration process) and allows other nodes to access these services (through a binding process) would be very helpful.

For these reasons, and others, the current IN architecture will evolve to become an information networking architecture, with the focus of attention progressively moving from the physical network to the software system, and with the principles of distributed processing increasingly applied to the software architecture.

In the following sections, we outline the elements of an information networking architecture — the major components, the

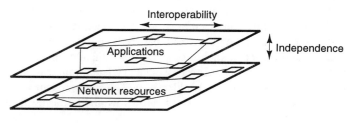

Fig. 1.   Platform model.

key functional separations, the various levels in the architecture — and finally, we consider a control problem (the management and operation of a SONET network) to illustrate the power of the architecture.

An information networking architecture would prescribe an object-oriented model for development of distributed applications, and would support several principles of separation that should be adhered to while packaging these applications into products. These principles are: separation of switching and transmission product technology details from service-related functions, separation of corporate data management from business specific use of those data, and separation of user interface details from other core business functions [6].

4. OBJECTS AND SERVICES

To promote application modularity and reusability, the architecture would prescribe an object-oriented model. Initially, the description is of the logical structure and does not consider issues related to application deployment, such as how the components are packaged into products, how interoperability is ensured, etc. (These issues will be developed next.)

An object is a part of a system which encapsulates data (or information or a resource or some state) and processing. We distinguish between two kinds of objects: computational objects and noncomputational objects. These two kinds differ in the way objects of each kind interact with each other.

Associated with each computational object is a set of operations defined on that object. A computational object interacts with another computational object by invoking operations defined on the other. The data component of a computational object can be accessed and manipulated by other computational objects only through a set of operations defined on the object. Operations that return results are invoked to cause some actions on the called object and obtain the results of these actions. Operations that do not return results are typically used for notification purposes (e.g., notification of exception conditions or alarm conditions). Examples of computational objects are: 1) an object that computes shortest paths between a source and a destination, and 2) an object that provides management operations on some physical entity such as a digital cross-connect.

Noncomputational objects interact with each other in a more general manner by exchanging information which is encoded as a sequence of bits. Examples of noncomputational objects interaction are: 1) bidirectional voice communication, and 2) audio-video communication.

A service is a set of capabilities provided (or offered) by an object (computational or noncomputational) that can be used by other objects. A service may be either a set of capabilities that is offered to an end-user (i.e., customer), such as a conference call, or a set of capabilities that is used in support of these end-user services, such as translation of toll-free numbers to real telephone numbers. The definition of a service provided by an object consists of a service interface and service semantics.

An object may provide several services, each of which may have a different interface. For example, an object may provide

a service that contains its core functions, and may provide another service for management and control of its core functions. Two objects that interact via a service have a client–server relationship between them. The object that provides (or offers) the service is called the server, and the object that uses the service is called the client. Depending on the semantics of a service, concurrent bindings from several clients to the server may or may not be allowed. For example, concurrent bindings to a file server may be allowed, but concurrent bindings to a noncomputational object that interacts via voice communication may not be allowed.

Every service is an instance of a service type. Services that are instances of the same service type have identical service interfaces and semantics (i.e., they provide the same functionality). However, the instances may differ in some nonbehavioral aspects, such as performance, physical resources encapsulated, quality of service parameters, cost of using the service, etc. These nonbehavioral aspects are called service attributes.

5. BUILDING BLOCKS AND CONTRACTS

We now define some concepts and key separations of concern that would be used in the deployment and interoperability of applications. The purpose is to specify guidelines on how application objects can be packaged into products, and how interoperability is to be attained between these products which may possibly be developed and released independently by different vendors.

A building block is a software product that contains one or more computational objects. It is built, installed, and maintained as a unit. A service provided by an object contained in a building block and that is visible outside that building block is called a contract.

Two objects that are in different building blocks interact only via contracts. The infrastructure for this interaction is provided by a Distributed Processing Environment (DPE), outlined later. A building block may provide (or offer) several contracts.

Every building block has the following fundamental properties [7]:

- *Unit of operability* — A building block is installed and administered as a unit. If a building block fails, every contract offered by it becomes unavailable. A building block may be installed and upgraded independently of other building blocks.
- *Unit of distribution* — A building block is a unit that is completely resident in a node of the network. The location of a building block in the network (i.e., the node in which it is resident) may vary over time. To ensure location transparency, every contract is identified by a logical name that is not dependent on the name or location of the building block that provides the contract.
- *Unit of security* — Interactions across building blocks are subject to security checks. Interactions within a building block are not subject to access control checks.
- *Unit of interoperability* — Interactions across building blocks are only via contracts which are specified using well-

defined and standardized models, protocols, and notations (e.g., remote operations, transactions, manager–agent model, etc.). Interactions within a building block may use nonstandard models and protocols.

6. GROUPING OBJECTS INTO BUILDING BLOCKS

A key separation that should be adhered to while grouping objects into building blocks is called the segmentation principle. This principle states that functionalities that are dependent on the technology used in the switching and transmission equipment should be separated from functionalities that are not dependent on these technologies and architectures. Functions that belong to the former category are called delivery segment functions, and functions that belong to the latter category are called service segment functions (see Fig. 2).

Delivery segment functions provide switching and transmission of information. Examples of delivery segment functions are switching equipment control functions, such as port connection/disconnection, bridge connections between ports, report of status information and alarms; administrative functions such as metering and collection of raw traffic data; and hardware-specific operational and maintenance functions.

Service segment functions are concerned with management of network resources and end-user services. Examples of service segment functions are connection setup/release, billing administration, fault management, and so forth.

Service segment building blocks and delivery segment building blocks interact with each other only via contracts. The interaction between service and delivery segment building blocks is based on the manager–agent model that is being promoted by network management standards (e.g., OSI Management Framework) [8]. Thus, each delivery segment building block supports management operations on a collection of network resources which are modeled as managed objects. These operations are grouped into contracts. Service segment building blocks accomplish their management functions by invoking these contracts.

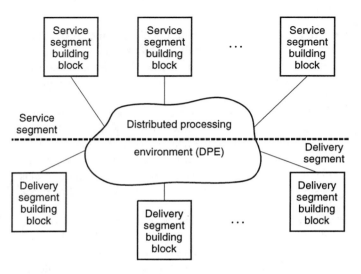

Fig. 2.   Service/delivery segmentation.

Delivery segment building blocks implement their contracts using product-specific technologies.

This separation enables reuse of service segment building blocks across vendor equipment that are based on different product technologies. Only delivery segment building blocks need to be changed in order to introduce a new vendor product. This separation also enables faster introduction and modification of services since they need not be handled on a product-specific technology basis.

7. DISTRIBUTED PROCESSING ENVIRONMENT

The infrastructure required for objects in different building blocks to interact via contracts is provided by a distributed processing environment (DPE). The DPE provides distribution transparencies by hiding from objects the complexities introduced by distribution. In providing distribution transparencies, the DPE builds on the facilities provided by the native computing and communication environment (i.e., operating systems, database management, and protocol stacks) within the computing equipment. However, it hides the heterogeneity of these native environments. Most components of the DPE are built using building block and contract principles. This ensures that DPE components interact with each other via well-known and standard protocols, and that they also have the benefits of release independence and other aspects of interoperability.

The purpose of the DPE is to provide services that enable distributed processing. It does not incorporate within itself policy decisions regarding management and control of network resources, such as bandwidth allocation, traffic management, routing, and billing administration. These business-related functions are accomplished by service segment building blocks.

The various levels in the TINA Architecture are illustrated in Fig. 3. Software functionality within a network node may be divided into these levels. At the bottom, we have the Native Computing and Communication Environment. This consists of the operating system and other related services, such as database management systems that form the computing environment, and transport layer services that provide end-to-end communication services.

At the next higher level, we have the DPE. The DPE runtime environment is composed of two kinds of components, the DPE Kernel and DPE Servers.

The DPE Kernel: The kernel provides basic mechanisms for sending and receiving messages, including request and reply messages needed for remote procedure calls (RPCs), mechanisms for stream-based communication, and mechanisms for establishment and release of application-level associations. It provides support for real-time programs. It also provides support for distributed transactions, including mechanisms for transaction initiation, termination, and atomic commitment service for ensuring the all-or-nothing semantics required for transactions.

A collection of DPE Servers includes servers such as contract trader, authentication server, and authorization server.

The contract trader provides contract registration and contract selection services. Before a contract can be used, it must be

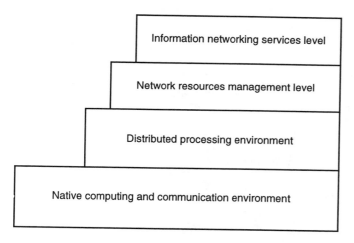

Fig. 3.   Architecture levels.

registered with the DPE via this service. Subsequently, a building block can locate a contract using this service by specifying the contract type and values for the service attributes defined on the contract type. The contract trader may be built on top of an X.500 Directory service.

The authentication server provides services for authentication of customers and building blocks.

The authorization server provides services for setup and modification of access control policies in contract usage.

A notification server allows building blocks to register types of alarms that they may emit. It also allows other building blocks to register the types of alarms they wish to receive. The service provided by the notification server is to allow dynamic binding of emitters and receivers.

DPE servers are built using building block and contract principles. Thus, each server provides one or more contracts that can be used by building blocks and other servers. The DPE kernel is present in every node. DPE servers are placed in the nodes depending on business needs, performance needs, and regulatory considerations.

The next higher level above the DPE (the Network Resources Management level) is concerned with management and control of network resources. Functions provided in these levels are independent of services provided to end-users and are classified into two kinds: systems management functions and network management functions. Systems management functions provide management of individual network resources. Network management functions invoke these functions to enforce policies on network resources management. They are also responsible for managing the relationships between multiple network resources. This functional split is quite similar to the manager–agent paradigm specified in OSI Systems Management Standards.

The highest level in the architecture is the Information Networking Services Level. This is composed only of service segment building blocks. These building blocks provide end-user service-specific functions including the service logic and management functions specific to that service.

8. SONET MANAGEMENT/OPERATION

To illustrate the power of the architecture, we choose an application example — the management and operations of a SONET network — to describe how the architecture concepts could be applied [9].

Network management is especially important in the TINA context, as the architecture is meant to allow for rapid development and introduction of a great variety of new services supported by a flexible and service-independent management platform.

Network management in TINA is intended to be consistent with network management standards, such as ITU-T/TMN and SDH standards, in the sense that it refers to the framework and managed objects (MOs) defined in these standards. SONET is one of the transmission technologies that is likely to be found in a TINA consistent network.

The TINA approach provides benefits to the management and operations of a SONET network in two ways: 1) distribution transparency provided by the DPE can reduce shared management knowledge among applications, thereby increasing flexibility, interoperability, and reusability of applications; and 2) MOs are defined at higher abstraction levels that include managing/controlling aspects of MOs at lower abstraction levels, thereby allowing easy decomposition of management/operation software into manageable components (building blocks). The following paragraphs use a concrete example of the management and operations of a SONET network — reconfiguration of a path layer network triggered by an alarm from SONET equipment — to illustrate how the above benefits are realized.

Notification forwarding supported by the DPE provides an example of benefit 1. Notifications, e.g., X.721 *communication Alarm (cause = "far End Receiver Failure")* from *VC4 Trail Termination Point* MOs [10] in a SONET node, can be filtered and forwarded by the DPE to (contracts offered by) building blocks of the manager role, or simply "managers." The managers inform the DPE of the filtering conditions and the types of notifications they want to receive, without having prior knowledge of MO instances or building blocks of the agent role, or simply "agents," that may emit the notifications. The agents inform the DPE of the types of notifications that they may emit, without knowing which managers may receive the notifications. Thus, managers (agents) need not be concerned about the locations, or in this case even the types and names, of agents (managers). This implies that it is possible for some management tasks, e.g., alarm surveillance, to be "moved around," e.g., from/to a daytime-shift operation center to/from a nighttime-shift operation center daily or even to a new 24-h operation center without modifications to the existing building blocks.

Another example of benefit 1 can be seen in operations on MOs using filtering/scoping [11] in a network-wide manner. Configuration information, e.g., which agent represents what classes of MOs and offers what types of contracts, can be maintained by the DPE. This information may be referred to by the DPE to support filtering/scoping of MOs in agents distributed in a network. With contract selection and message forwarding ser-

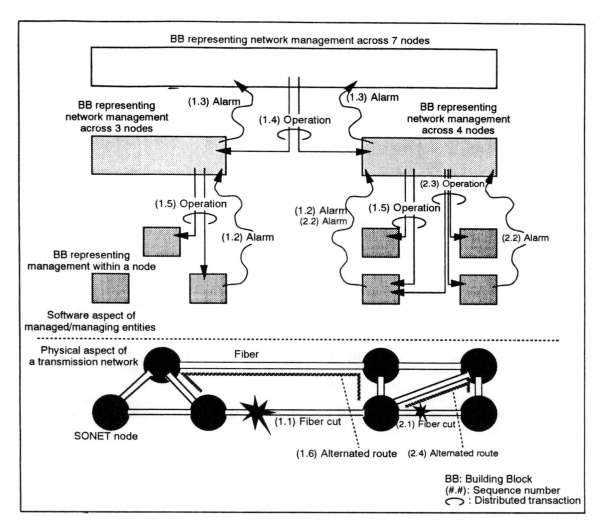

Fig. 4. An example of a SONET network's management and operations.

vices provided by the DPE, it can be sufficient for a manager to issue a single operation, e.g., find *VC4 Trail Termination Point* MOs such that *Operational State = "enabled" AND Administrative State = "unlocked"* (logically equivalent to dynamic generation of a network-wide TP Pool), to poll the status of distributed resources and find out what is available. (The DPE multicasts the operation message to relevant agents in the network.) This indicates that some management tasks, e.g., selecting a back-up (alternative) path and deciding which MOs or agents at which locations should be accessed to switch the path, need not be predefined. They can be performed at runtime, reflecting the current network status, with the help of the DPE.

With respect to benefit 2, an example can be found in an alternative way to implement the network-wide operations on MOs. Figure 4 gives a graphical view of this example. Agents for MOs at the network element layer, e.g., MOs for a SONET node, are implemented by building blocks in the delivery segment. A TINA approach would define MOs at a higher abstraction level, i.e., the element management layer and the network management layer. These MOs are especially useful in managing/controlling distributed network resources as they allow natural decomposi-

tion of resources into abstraction levels as necessary. An agent for MOs at one level of abstraction can be a manager having its (potentially distributed) agents for MOs at a lower level. It hides the details of how to manage/operate distributed MOs at the lower level and provides an abstract view of MOs to managers. Such recursive manager–agent relationships can be terminated, at one end, by a manager whose responsibility covers the whole network and, at the other end, by an agent whose MOs are at the network element layer. As mentioned earlier, distributed transactions required to perform network-wide operation can be supported by the DPE.

9. CONCLUSION

TINA-C is setting out to develop an architecture based on the principles outlined. In defining this architecture, the TINA-C will seek to build on and maintain compatibility with relevant standards in distributed processing and telecommunications. (These standards are not comprehensively referenced or detailed in this outline paper.) Also, an important part of the TINA-C activity will be to validate proposed architectural specifications through collaborative field experiments, prior to any prospective standardization.

To summarize, the TINA architecture is intended to build on current advances in broadband communication and distributed computing technologies, specifying a software-based architecture for future information networks that are required to transport multimedia information and manage multimedia communication. An important aspect of the architecture is that service segment functions are separated from delivery segment functions. This separation is consistent with the manager–agent separation advocated in network management standards.

Another key aspect is the elimination of the rigid division between network applications and operations applications that exists in current-day networks. In the TINA architecture, both kinds of applications execute on a common distributed processing platform. The distributed processing platform hides from applications the effects and complexities introduced by distribution. Most importantly, it supports application interoperability enabling flexible construction of services, composed of service components distributed across network domains, allowing network operators and service providers to cooperatively meet the needs of the user.

References

[1] *TINA 90 Workshop Proceedings*, Lake Mohonk, U.S., June 1990.

[2] *TINA 91 Workshop Proceedings*, Chantilly, France, Mar. 1991.

[3] *TINA 92 Workshop Proceedings*, Narita, Japan, Jan. 1992.

[4] J. J. Garrahan, P. A. Russo, K. Kitami, and R. Kung, "Intelligent network overview," *IEEE Commun. Mag.*, Mar. 1993.

[5] *The ANSA Reference Manual*, Architecture Projects Management Limited, Cambridge, U.K., 1989.

[6] J. J. Fleck, C. C. Liou, N. Natarajan, and W. C. Philips, "The INA architecture: An architecture for information networks," in *TINA 92 Workshop Proc.*, Narita, Japan, Jan. 1992.

[7] W. J. Barr, T. Boyd, and Y. Inoue, "The TINA initiative," *IEEE Commun. Mag.*, Mar. 1993.

[8] Open Systems Interconnection — Basic Reference Model — Part 4, Management Framework, ISO/IEC 7498-4, 1989.

[9] S. Chikara, W. Takita, and H. Kobayashi, "A network management architecture on a distributed processing environment," in *Proc. TINA'93 Workshop*, L'Aquila, Italy, Sept. 1993.

[10] "Synchronous Digital Hierarchy (SDH) management information model for the network element view," ITU-T G.774, 1992.

[11] "Common Management Information Service Definition," ISO/IEC 9595 / ITU-T X.710, 1991.

SDH Management Network: Architecture, Routing and Addressing

Hagay Katz[†]   Greg F. Sawyers[†]   Jeremy L. Ginger[†§]

† Telecom Research Laboratories, 770 Blackburn Road, Clayton, Victoria 3168, Australia
§ Monash University, Wellington Road, Clayton, Victoria 3168, Australia

phone: +61 3 253 6304   fax: +61 3 253 6144   email: h.katz@trl.oz.au

Abstract

CCITT recommendation G.784 provides several options for the implementation of routing and addressing schemes within the Synchronous Digital Hierarchy (SDH) Management Network (SMN). In addition, equipment manufacturers and network operators have identified a hierarchical topology for the SDH network. Based on these routing and addressing options and network topologies, the network operators need to implement an efficient, robust and cost effective management structure. Two alternatives have been identified as possible implementation strategies for the SMN. The first option makes use of an X.25 packet switched network, as well as the SDH Data Communications Channels (DCCs). Here the complexities of implementing a wide area packet switched network are alleviated by making use of the X.25 network with a static routing scheme. The protocol specified in Annex C of G.784 can be used for this implementation, but we identify serious deficiencies in the standards. Hence, its use would require a proprietary solution or considerable effort in the CCITT standardisation arena. The second option uses the DCC exclusively to provide the communications network. The standard protocol stack as specified in CCITT recommendation G.784 is used. Dynamic routing is provided by the Intermediate System to Intermediate System (IS–IS) routing protocol. The first option may be implemented in the initial stages. A gradual migration path can be adopted to take advantage of the second option with the advance of SDH equipment and its management capabilities.

1 Introduction

A Synchronous Digital Hierarchy (SDH) network may be viewed as two logical networks supported by one physical layout. One of these logical networks is a circuit switched network for the transmission of payload information while the other is a packet switched network operating over the Data Communication Channel (DCC) for the communication of management information. In this paper we investigate and recommend the required architecture and routing schemes for the packet switched network, in order to provide an efficient management system for the evolving SDH network.

An outline of the hierarchical SDH network topology is given in Section 2. Section 3 describes the implementation of the SDH Management Network (SMN) [2] based on the network topology. Two options each for implementing a routing scheme for ring and mesh based networks are given in Section 4. Section 5 describes the addressing requirements for the SMN. Sections 6 and 7 describe by means of examples the implementation of these options in the hierarchical SDH network.

2 SDH Network Architecture

In an SDH network, network management information is carried on the DCCs, within the Synchronous Transmission Module (STM) frames, between SDH Network Elements (NEs). In order to implement an effective SMN the physical network topology carrying the STM frames needs to be analysed.

An SDH network is constructed using four types of NEs, namely: Digital Cross–connect Systems (DCSs), Add–Drop Multiplexers (ADMs), Terminal Multiplexers (TMs) and repeaters. These devices support SDH–based network transport services and may also support operation system functions or mediation functions [1].

In large networks such as the one shown in **Figure 1**, different bit–rate transmissions and traffic patterns are present. So to reduce complexity such a network is divided into layers. The national portion of the SDH network may be viewed as a hierarchical network consisting of three levels [10]. The network model in **Figure 1** clearly depicts these three layers. The dashed lines show the interconnection between the layers in the fully integrated SDH network. The *access subnetwork* will mainly consist of ADM rings. The function of the subnetwork at this level is to collect traffic from the access groups. The *regional transit subnetwork* will consist of DCS 4/1s and ADMs in a meshed and/or ring architecture. The *national transit subnetwork* will consist of the links interconnecting the major centres as well as a meshed or ring network within each major centre. DCS 4/4s (mainly) and ADMs (few) will be used at this level. This hierarchical division may be used to reduce the complexity of the network design and dimensioning phases as well as network management and routing of traffic.

Figure 1 illustrates several configurations to allow protection against failure. Protection against node or link failure within a layer is provided by meshed or ring architectures. At higher transmission capacities, where traffic needs to be transferred in bulk, DCSs within a meshed network may provide a reliable and economical solution. At lower bit rates, where traffic needs to be distributed, ring architectures with ADMs may appear to be more attractive. Connectivity between layers under node or link failure is maintained by providing connections in at least two points from a particular layer to a layer above. This is known as *dual parenting*. Some possible interconnections providing protection against a link or node failure down to the switching centres are shown in **Figure 1**. Chain and star networks can be protected against link failures by providing Diverse Route Protection (DRP) mechanisms. Node failure on such a network may isolate a segment of the network. Where protection against failure is not of great concern, star and chain architectures may be used in both the regional transit subnetwork and the access subnetwork to distribute traffic. Interconnections between rings (such as that shown by $ in **Figure 1**), at the regional transit subnetwork and access subnetwork levels are to be avoided. Switching exchanges may be connected to DCS 4/1s, ADMs or TMs at any subnetwork.

Reprinted from *IEEE GLOBECOM '93 Conf. Rec.*, pp. 223–228, Dec. 1993.

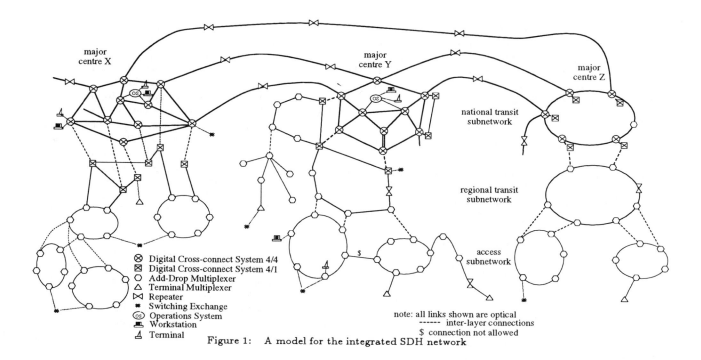

Figure 1: A model for the integrated SDH network

Legend for Figure 1:

⊗ Digital Cross-connect System 4/4
⊠ Digital Cross-connect System 4/1
○ Add-Drop Multiplexer
△ Terminal Multiplexer
⋈ Repeater
✳ Switching Exchange
(os) Operations System
💻 Workstation
△ Terminal

note: all links shown are optical
----- inter-layer connections
$ connection not allowed

3 SMN Architecture and Protocols

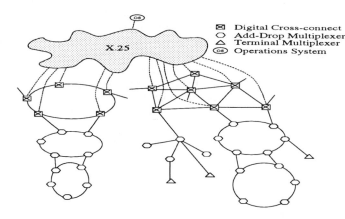

⊠ Digital Cross-connect
○ Add-Drop Multiplexer
△ Terminal Multiplexer
(os) Operations System

Figure 2: Mixture of DCC and X.25

A critical part of the design of the SDH network is its management and control architecture. The SMN needs to be implemented based on the hierarchical physical topology described in the previous section. An SMN is a subset of a Telecommunications Management Network (TMN) [1], responsible for managing SDH NEs. An SMN may be subdivided into a set of SDH Management Subnetworks (SMSs). In the existing Plesiochronous Digital Hierarchy (PDH) networks a separate external network is required for management information transfer. With SDH, provisions have been made to carry the management information within the same STM frame as the information payload by reserving two channels known as the Data Communication Channels, DCC_r (192 kbps) and DCC_m (576 kbps) [2]. The DCCs have been designed so that their bit-rate is independent of the SDH transmission bit-rate. CCITT recommendation G.784 [2] specifies the protocols for a seven

layer OSI stack to be implemented on these channels. The DCC_r has been reserved for SDH NE use. The DCC_m may be used as a wide area, general purpose communications network to support TMN including non–SDH applications [2].

3.1 Options for the SMN Architecture

Figure 2 shows the use of the DCCs within rings and tree structures in the regional transit subnetwork and the access subnetwork for communication of management information. Implementing an efficient packet switched network for communication of management information could be a complicated task. Hence in the initial stages of SDH deployment X.25 connections may be provided to each NE in the national transit subnetwork. One of the disadvantages of using an external network for the management of the SDH network is the large number of external connections and the limited bandwidth compared with the DCCs. The advantages of an external network are its independence from the SDH network and the ability to make use of its existing switching facilities.

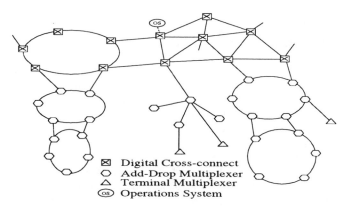

⊠ Digital Cross-connect
○ Add-Drop Multiplexer
△ Terminal Multiplexer
(os) Operations System

Figure 3: Exclusive use of the DCCs

A second option is to make exclusive use of the DCCs for management information transfer. This is depicted in **Figure 3**. Based on past experience, successfully implementing a packet switched network, operating over the DCCs, for the fully integrated SDH network shown in **Figure 1**, could be a complicated and time consuming task.

3.2 Management Protocol Considerations

In order to implement an efficient management network over the DCCs, the specified protocols and their options must be investigated. Here we concentrate on the functionality of the lower three layers of the OSI stack specified in G.784.

The SDH DCCs, carried within the overhead bytes of the STM frame, constitute the *physical layer*. Physical architectures for an SDH network include *mesh, star, chain* and *rings*. The physical media can be an optical system, radio system or electrical system.

G.784 specifies Link Access Procedure – D channel (LAPD) (as specified in CCITT recommendation Q.921) as the protocol to be used at the *data link layer*. This protocol requires a full duplex point–to–point physical layer service. As an alternative, G.784 Annex C specifies a protocol made up from a combination of High level Data Link Control (HDLC), Media Access Control (MAC) and Logical Link Control (LLC) sublayers.

According to G.784, the Connectionless Network Protocol (CLNP) is used at the *network layer* of the OSI stack over the DCCs. This implies that a packet switched network will be implemented over the DCCs for carrying network management information. In a ring configuration the number of paths between any two nodes is limited. In a meshed situation the number of paths between any two nodes could be quite large. In addition, the length, number of hops, cost, etc., of the routes need to be taken into account for an effective implementation of this network.

4 Options for Routing within SDH Networks

This section describes various options for providing a network capable of supporting the OSI connectionless network service. From the perspective of routing architectures, a meshed topology will have different needs compared to a ring topology. Also, the types of equipment used in rings and meshes (ADMs and DCSs respectively) are of different cost and complexity. This may affect the applicability of particular routing architectures to either topology. Routing options for ring and mesh topologies are therefore discussed separately.

4.1 Ring Topology

4.1.1 Option 1: Virtual LAN Subnetwork

(a) The Subnetwork Service.

This option requires the use of the protocol specified in Annex C of G.784. This data link layer protocol uses the unidirectional point–to–point physical links provided by the SDH DCCs. The subnetwork service it provides is that of ISO 8802–2 Logical Link Control class 1 (LLC 1) [5]. This service provides broadcast/multicast transfer and point–to–multipoint connectionless data transfer capabilities. It does this by specifying a MAC protocol for use by the LLC sublayer. When a frame is received on a DCC port, its MAC address is compared with the address of the receiving station. If the addresses match or the received address is a group address, the frame is passed to the LLC sublayer. If the addresses do not match or the received

address is a group address, the frame is forwarded on all DCC ports except the one on which it was received.

Unfortunately, the protocol as specified in Annex C of G.784 has a number of deficiencies. These deficiencies are in the MAC sublayer protocol specification (Section C.1.2 of [2]). The major problem is that the only provision for removing a MAC frame from the ring is when it reaches its destination. This is clearly inadequate in a number of cases. The first of these is for multicast frames. According to the protocol as specified, frames with a group destination address are always forwarded, since the frame must reach many destinations on the ring. In this case, multicast frames will loop around the ring forever. A simple solution to this exists: simply stipulate that when a frame arrives back at its source (determined by examining the source address of the frame), the frame is not relayed any further.

This covers the case of a single isolated ring. However, when rings are interconnected with either other rings or a mesh of DCSs, the protocol requires further enhancement. Obviously, a multicast frame whose source is on another ring cannot be removed from the ring by the source itself. It must therefore be removed by the node where the frame entered the ring, i.e. the interconnecting node. This requires additional functionality in the interconnecting nodes, that of MAC layer bridging. ISO/IEC DIS 10038 (IEEE 802.1D) [8] specifies such a bridging protocol for use in all types of ISO 8802 LAN environments (ethernet, token ring, token bus, etc.). This standard could easily be used to provide the required bridging functionality for ring interconnection (see section below entitled "Ring Interconnection"). It would then be the responsibility of the bridge node to detect frames that have been relayed through it and remove them from the ring. This could be done by determining that the source of a frame is that of a node reachable on one of the bridge's other ports (a standard bridging function). In **Figure 4** the nodes that require bridge functionality are denoted by **B**.

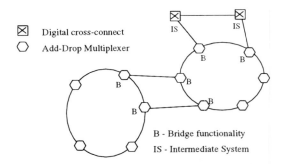

⊠ Digital cross-connect

◯ Add-Drop Multiplexer

B - Bridge functionality

IS - Intermediate System

Figure 4: A single subnetwork showing bridge and IS functionality

(b) Routing within an SDH Ring.

The choice of routing protocol for the architecture described above is a simple one. The OSI End System to Intermediate System (ES–IS) routing protocol, ISO 9542 [6], has been specifically designed to operate over subnetworks which offer broadcast/multicast transmission facilities. Each node in the ring would operate as an OSI End System (ES). An End System is one that acts as a source/sink for Network Protocol Data Units (NPDUs) but does not relay NPDUs. An Intermediate System (IS), on the other hand, relays NPDUs between subnetworks but does not source/sink NPDUs. Each ES operates the ES–IS protocol to dynamically discover the mappings between

Network Service Access Point (NSAP) addresses of other ESs and their corresponding MAC addresses.

The location of the IS will be the node(s) in the meshed DCS network to which the ring is connected. This is illustrated in **Figure 4**. The benefits of operating the ES–IS protocol are as follows. (i) The complexity of routing information and routing algorithms in ESs is minimized. (ii) The amount of a *priori* state information needed by ESs before they can begin communicating is minimized. (iii) Changes in topology, such as node/link failures, new nodes being added to the system, etc., are discovered and compensated for automatically, without the intervention of system management. (iv) ESs can be allocated their NSAP addresses dynamically, without the need for pre-configuration.

(c) Ring Interconnection.

When rings are connected together, or connected to other SDH equipment (e.g. DCSs at another layer), bridging functionality is required at the points of interconnection. ISO/IEC DIS 10038 (IEEE 802.1D) defines the operation of MAC bridges.

Bridge operation is transparent to the LLC sublayer, which means that interconnected rings will appear as one large virtual LAN. Each bridge maintains a table of MAC addresses for each LAN port. The decision as to whether to forward a frame on a particular port is made by consulting the filtering database for that port. If the destination MAC address is found in the filtering database (or the destination address is a multicast address), the frame is forwarded. The filtering database is established and maintained by the learning process. By examining the source MAC address of each received frame, the bridge builds a table of MAC addresses reachable on each port.

When SDH rings are connected together via more than one path, duplicate MAC frames can be generated and loops can occur. To avoid this, MAC bridges operate a protocol which effectively prevents multiple paths between subnetworks (LANs). An algorithm known as the *Spanning Tree Algorithm* is used to reconfigure the bridged LAN topology into a tree structure. The bridge protocol distributes topology information around the bridged subnetwork, causing certain bridge ports to become *blocked*. A blocked port does not forward frames.

The major advantage of this option is that the need for IS functionality in the nodes of a ring is eliminated. The complexity of routing information and the routing protocol within the nodes is kept to a minimum. The major disadvantage is that the data link protocol, as specified in Annex C, G.784, requires enhancement to be made useful.

4.1.2 Option 2: Point–to–Point Subnetwork between Adjacent Nodes

(a) The Subnetwork Service.

This option requires the use of the protocol stack for the DCCs specified in Section 6 of G.784. The only difference between this stack and that required for Option 1 is at the data link layer. The LAPD protocol specified in CCITT recommendation Q.921 is used to provide the OSI data link service (ISO 8886) [9]. This service offers point–to–point connection mode or connectionless data transfer between any two adjacent nodes connected by the DCC. In routing terminology, each data link connection is a point–to–point subnetwork supporting exactly two systems (ES or IS).

(b) Routing.

In this option, each node on the ring operates as both an ES and an IS. Dynamic routing requires that a discrete protocol operates at the network layer to distribute routing information throughout the network. In the case of an SDH ring, the OSI Intermediate System to Intermediate System (IS–IS) routing protocol (ISO/IEC 10589) [7] would be appropriate, as specified in section 6 of G.784. This protocol is a link state based protocol. Each IS reports to the rest of the subdomain all the ESs and ISs that it considers neighbours. A node on a ring would report the two adjacent nodes as being IS neighbours and the ES part of itself as being an ES neighbour. These *Link State PDUs* are *flooded* throughout the routing subdomain, thereby allowing each IS to build up a picture of the topology of the subdomain. This information is then used to build the forwarding database which is used by the Route PDU function of CLNP. The forwarding database for a node on a ring would be very simple, since there are only a small number of paths from which to choose. It is this simplicity of network topology that lessens the benefits gained by operating a sophisticated routing protocol such as IS–IS. One benefit that is gained from using IS–IS is the ability to propagate information automatically about systems that have just come on line.

4.2 Routing in a Meshed DCS Network

4.2.1 Option 1: Extensive Use of the DCC

In a meshed network, the only viable subnetwork service is that provided by the LAPD protocol (see Section 4.1.2). The MAC layer protocol specified in G.784 Annex C is not appropriate for a meshed network due to its highly connected nature.

There are two options for routing in this environment: static and dynamic. Since static routing schemes cannot respond quickly to topology changes, the benefits of having alternate routes between nodes for redundancy are diminished. While it is possible to statically store backup routes for use in case of failure, it is still impossible to guarantee that PDUs will be delivered between nodes even when alternate paths exist. This is not uncommon in meshed networks because there are many distinct paths between any two nodes.

With dynamic routing, changes in network topology are made known throughout the network and alternate routes calculated in real time (in the order of seconds).

4.2.2 Option 2: Mixture of DCC and X.25

As an alternative to the DCC, an X.25 network can be used to transport management information between the OS and the DCSs. Each node would operate CLNP at the network layer using the subnetwork dependent convergence function for ISO 8208 [3] subnetworks. This convergence function is defined in ISO 8473 [4]. Each node would have a X.25 connection to the OS (see **Figure 2**). This connection could be a Permanent Virtual Circuit (PVC), a static Switched Virtual Circuit (SVC) or a dynamically assigned SVC. A static circuit is one that is established and cleared on system management action, whereas a dynamically assigned circuit is established and cleared upon receipt of traffic and expiration of idle timers.

The IS–IS routing protocol distinguishes between dynamically assigned circuits on one hand, and static SVCs and PVCs on the other. No IS–IS routing PDUs are transferred on a dynamically assigned circuit. This means that static routing must be used between ISs on a dynamically assigned circuit. The hierarchical nature of the SDH network can be used to good effect in this case. Each DCS would keep the routing information for its subordinate nodes (rings of ADMs, ADM stars, etc.). In the terminology of IS–IS, each DCS would be a *level 2* IS within one or more *level 1* areas. All subordinate systems connected to the same IS would have the same *Area Address* component of their NSAP address. The OS would then keep a map of which area addresses can be reached over which X.25 ports. The feasibility of such an arrangement depends on the number of X.25 ports required.

Where static SVCs or PVCs are used, routing can be either static or dynamic. The IS–IS protocol treats these circuits the same as any other point–to–point circuit. The division of the addressing domain into areas, as described above, is still desirable for efficient operation of the routing protocol.

5 Addressing Considerations for the SMN

This section recommends an addressing scheme that supports the architectural options canvassed above.

When designing an addressing scheme, consideration must be given to the routing architecture of the network. The IS-IS routing protocol [7] places certain constraints on the addressing scheme chosen. The standard basically states that the NSAP addresses must be split into a high order *Area Address* part, a *System ID* part and a trailing selector part which is not used for routing. The standard also states that the lengths of these parts must be fixed within a particular routing domain (in this case the entire SMN).

The reason for structuring the NSAP addresses in this way is to allow efficient operation of the routing protocols. IS-IS routing is two-level hierarchical. The *Area Address* part of the NSAP address is used as a prefix to identify a single area within a routing domain. The *System ID* part of the address is then used within an area to locate a specific End System. In this way, ISs need only know about routes to destinations within their own area instead of the entire routing domain. Routing tables can be kept as small as possible and routing protocol traffic can be kept to a minimum.

Another addressing consideration is which NSAP address format is used. There are two major categories: subnetwork dependent and subnetwork independent. Subnetwork dependent formats are based on CCITT public network numbering plans and can be used by organizations connected to public networks which use one of these plans. An SMN that used, for example, a public X.25 network for part of its communications network could use such a scheme but any change in this situation (e.g. deciding not to use the X.25 network any more) would require that the NSAP addresses also be changed. To attain maximum flexibility and expandability of the addressing scheme, a subnetwork independent address format should be used. This means that the addressing scheme can be totally independent of the technology used to implement the management network, but introduces complexity in that the point of attachment can not be readily deduced.

6 Example: "Virtual LANs" Connected via X.25

This arrangement is depicted in **Figure 5**. Each ADM in a ring acts as an ES operating the *Annex C* data link protocol. Interconnecting nodes (denoted "B" in **Figure 5**) must also implement IEEE 802.1D bridging functionality. In this manner, each set of interconnected rings operates as a bridged virtual LAN (i.e. a single subnetwork).

All routing requirements for ring nodes are handled by the operation of the ES–IS routing protocol (ISO 9542). The ES–IS protocol will allow each ES to build up a table of (NSAP, SNPA (Subnetwork Point of Attachment address, in this case the MAC address)) address mappings for all other ESs and ISs on its own subnetwork (Virtual LAN). Hence the routing table in each ES will have, at most, one entry for each ES on the virtual LAN. (The actual size will probably be much less than this, since most traffic will be between the ES and the OS.)

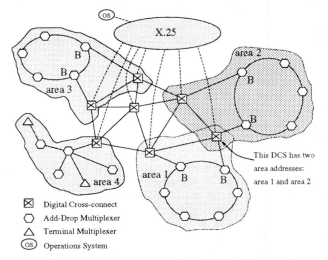

Figure 5: Architecture for virtual LAN subnetworks connected by X.25

Each DCS acts as an IS, forwarding NPDUs between the OS and the ring(s) to which it is connected. A DCS would be required to implement appropriate subnetwork dependent convergence functions to operate CLNP over X.25 and LLC 1. These are defined in ISO 8473. A DCS would also be required to operate the *Annex C* MAC layer protocol to connect to the ring(s). Each DCS would have a permanent or switched X.25 connection to the OS. Through operation of the ES–IS routing protocol, the DCS would maintain a table of (NSAP, SNPA) address mappings for all ESs on the ring(s) to which it is connected. Any NPDUs received from the OS could then be forwarded directly to the destination ES. Also, any traffic received by the DCS from a ring whose destination address does not appear in its routing table would be forwarded to the OS. This covers the case of PDUs destined for the OS and other rings, since inter–ring traffic can be forwarded by the OS.

Note: The management of repeaters may require special attention. Since they will exist in large numbers and remote locations, it is impractical to provide X.25 connections to all of them. Where a number of repeaters exist between two DCSs, the Annex C protocol could be used in exactly the same way as a ring. The same routing methodology as for rings can be used, with each string of repeaters becoming a single area.

The routing tables in the OS can be made fairly simple by choosing an appropriate addressing scheme. The topology of the network is essentially hierarchical, with the OS at the top of the hierarchy and the rings at the bottom. If the addressing scheme is tailored to match this hierarchy, significant simplification of routing information can be achieved. This means that each DCS is considered to be the point of interconnection for one or more *areas*. Each area consists of one or more interconnected rings plus the DCS(s) to which they are connected.

This is illustrated in **Figure 5**. Each area is assigned its own area address. The area address is simply the high order part of the NSAP address. The effect of dividing the NSAP addresses in such a manner is to allow the OS to route only on the area address part of the NSAP address. The routing table in the OS would map between area addresses and X.25 connections. Since there is only likely to be 20–30 DCSs in the network, the table will be of a manageable size.

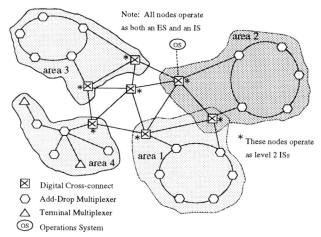

Figure 6: All nodes connected by point–to–point subnetworks

7 Example: Point–to–Point Subnetworks between all NEs

This example is depicted in **Figure 6**. The protocol stack and routing methodology are the same for all types of NEs (DCSs, ADMs, TMs and repeaters). LAPD is used at layer 2 to provide a point–to–point subnetwork service between nodes. Each node will operate the IS–IS routing protocol. The same NSAP addressing and routing architecture as described above can be used. The meshed DCS network would form the level 2 backbone with each DCS acting as the level 2 IS for one or more connected rings. This would lead to maximum routing efficiency by exploiting the hierarchical nature of the SDH network. The OS would simply become another ES, perhaps within its own area.

8 Conclusion

The hierarchical network topology proposed by network operators and equipment manufacturers and the SMN architecture and protocols recommended by the standards have been described. Based on these proposals, two alternative schemes for implementing an SMN have been identified. The first of these schemes, described above in Section 6 (Virtual LANs connected via X.25), makes use of an X.25 network to provide much of the network infra–structure. The advantage of this is that the complexities of routing in a large packet switched network are minimized. Another advantage is that the IS–IS routing protocol is not required at all. However, there are several drawbacks to this option. It requires the use of an as yet unstandardised data link protocol. Since the **protocol specified in Annex C of G.784 is inadequate as specified**, its use would require a proprietary solution or considerable effort in the CCITT standardisation arena.

The second alternative, described above in Section 7 (Point–to–Point Subnetworks between all NEs), uses the DCC to provide the physical network. All routing information exchange

is done by the IS–IS protocol. Routing is totally dynamic, allowing automatic rerouting around node or link failures. One advantage of this scheme is that all equipment supports the same suite of protocols and, from the perspective of the network layer, all network elements appear identical. Another advantage of this approach is that it is a completely standard solution.

The lack of large networks based on IS–IS and the complex nature of the protocol make it difficult to assess the true cost of implementing such a network. It may turn out that the cost saving of using the free transmission capacity provided by the DCC is outweighed by the cost of implementing the packet switching technology.

Whichever alternative is chosen, the addressing scheme will remain the same. The addressing domain should be subdivided into areas to take full advantage of the hierarchical nature of the network. Each area would correspond to a ring or group of interconnected rings and the DCS(s) to which they are connected.

The option incorporating a mixture of the X.25 and the DCC may be implemented in the initial stages. As the management capabilities of SDH equipment advance, the option which makes use of the DCCs exclusively may be adopted.

References

[1] CCITT Draft Recommendation M.3010. *Principles for a Telecommunications Management Network*, 1991. version R5.

[2] CCITT Recommendation G.784. *Synchronous Digital Hierarchy Management*, 1990.

[3] ISO 8208. *Information Processing Systems – X.25 Packet Level Protocol for Data Terminal Equipment*, 1990.

[4] ISO 8473. *Information Processing Systems – Data Communications – Protocol for Providing the Connectionless–mode Network Service*, 1988.

[5] ISO 8802-2. *Information Processing Systems, Local Area Networks – Part 2: Logical Link Control*, 1988.

[6] ISO 9542. *End System – Intermediate System Routing Exchange Protocol, (ES–IS) for ISO 8473*, 1988.

[7] ISO/IEC 10589. *Intermediate System – Intermediate System Routing Exchange Protocol*, 1992.

[8] ISO/IEC DIS 10038. *Local Area Networks – Media Access Control Bridges*, 1990.

[9] ISO/IEC DIS 8886. *Data Link Service Definition for OSI*, 1990.

[10] Mike Sexton and Andy Reid. *Transmission Networking: SONET and the Synchronous Digital Hierarchy*. Artech House, 1991.

Acknowledgements

The permission of the Director of Research, Telecom Australia, to publish this paper is hereby acknowledged.

Section 6
SDH/SONET Future

THE deployment of SONET/SDH is now significantly well advanced in all parts of the transport network. The provision of this facility in the transport network is a key component for evolution toward a Broadband Integrated Services Digital Network, B-ISDN. Such a network can be viewed as an essential component in the provision of an information superhighway, as it will be capable of the near-instantaneous transfer of video, audio, and data among users located anywhere.

The papers in this section are devoted to the role SDH/SONET-based networks will play in the realization of B-ISDN.

In the evolution toward B-ISDN, a key technology is Asynchronous Transfer Mode (ATM). Asynchronous means that there is no strict relationship between time references of the source of information and the network. The first paper, entitled "ATM (Asynchronous Transfer Mode): Overview, Synergy with SDH, and Deployment Perspective" by J. Legras, provides a tutorial introduction to the principles of ATM and interworking with SDH.

The ATM theme is continued in the second article, "An SDH Transmission System for the Transport of ATM Cells" by A. Brosio and A. Moncalvo. This paper discusses the requirements imposed on the SDH-based transmission systems to transport ATM cells; particularly, the mapping of ATM cells in a synchronous transport module is discussed with the aid of equipment functional block diagrams.

"Technologies Towards Broadband ISDN" by K. Murano *et al.* provides a broad overview of three technologies which need to be in place so that a smooth migration from the existing network to B-ISDN may be effected. These technologies are: expansion of fiber networks to the subscriber loop area to provide broadband capabilities everywhere, deployment of SDH transmission systems, and deployment of ATM technologies.

The fourth paper, "Optical Fiber Access — Perspectives Toward the 21st Century" by A. Cook and J. Stern, examines the fiber architectures in the subscriber loop which are being considered and deployed worldwide to meet the emerging demand for broadband services.

Bringing together aspects of the preceding two papers, M. Compton and S. Martin, in "Realizing the Benefits of SDH Technology for the Delivery of Services in the Access Network," emphasize the importance of the "last mile" and the potential role of SDH in delivering low-speed signals to business and residential subscribers. The authors discuss network architectures and network element functionality, and their relationship to both the SDH multiplexing hierarchy and add/drop multiplexing. The authors conclude that until SDH maps out a plan to efficiently offer sub-STM-1 traffic, the role of SDH in access networks could be limited.

The importance of SDH/SONET rings is a major theme of Section 3, where the discussion is limited entirely to synchronous transfer mode (STM). The appearance of ATM, as discussed in the papers above, and the importance of emerging switched data services, require that the role of SONET/ATM rings in the network architecture be considered. These points are discussed in the sixth paper, by T.-H. Wu, "Cost-Effective Network Evolution," including a discussion of network evolution for switched and nonswitched DS-1 and DS-3 services, and operations evolution.

The final paper of this section, "Emerging Residential Broadband Telecommunications" by D. S. Burpee and P. W. Shumate, Jr., is unique in several aspects. Notably, the authors discuss the respective roles of telephone companies and cable operators in delivering broadband services via fiber to residential users, and the possible convergence of these industries. From a services perspective, multimedia information is discussed; from a networking perspective, ATM transport via SDH/SONET is reviewed, along with a survey of the full scope of loop feeder/distribution topologies and technologies.

The above papers, and references contained within, provide an excellent base for readers to appreciate the changes in telecommunication networks which must be made to meet the demands of the future.

ATM (Asynchronous Transfer Mode): Overview, Synergy with SDH, and Deployment Perspective

JACQUES LEGRAS

FRANCE TELECOM CNET

1. INTRODUCTION

SDH is now reaching the stage of large-scale deployment. It enables the construction of a broadband managed transport network, a key component of the broadband ISDN (B-ISDN). In the evolution towards B-ISDN, another key technology is ATM, which stands for asynchronous transfer mode and is considered in CCITT as the target transfer mode of B-ISDN. The concept of "transfer mode" is used in the context of B-ISDN to encompass switching, multiplexing, and transmission, but the use of ATM is possible in combination with other multiplexing and transmission techniques, in particular, SDH. Asynchronous means that there is no strict relationship between the time references of the source of information and of the network. The goal of this section is to give an overview of this technology and to consider the conditions of interworking ATM and SDH. The way in which the deployment of ATM is foreseen is also addressed, although it is difficult to give a complete picture of the situation since new products, experiments, and pilot networks are announced every day.

2. ATM BASIC MECHANISMS

2.1 The Cell Format

The basic element is the ATM cell, which is a short and fixed-length packet of information. It combines flexibility of packet processing and simplicity which enables implementation of basic functions using high-speed hardware and reaching high bit rates.

The cell, which is 53 octets long, is made of two main fields: the header, used for routing, and the information field, which transports user information (Fig. 1). In the information field, the ATM Adaption Layer (AAL) area is used to reduce the effect of impairments of the ATM network (jitter, cell loss), thus restoring user information to its specific requirements. This AAL field is 1 to 4 octets long, depending on AAL type.

The cell header is 5 octets long and contains two identifiers to route two separate logical entities: the virtual channel (VC) and the virtual path (VP). As shown in Fig. 2, the header also contains ancillary indications used by the network and 1 octet (HEC) that protects the header content and recovers cell delineation. At the UNI, the VPI field is 4 bits shorter, these bits being dedicated to generic flow control (GFC).

2.2 The Protocol Reference Model (PRM)

The main functions necessary to implement ATM are usually described in a layered model known as the ATM Protocol Reference Model (PRM), as shown in Fig. 3.

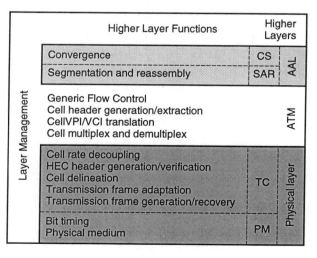

Header structure

VPI Virtual path identifier   R: reserved
VCI Virtual channel identifier
HEC Header error correction

| VPI | VCI | | HEC |

PT   Payload type
CLP Cell loss priority
FCN Forward congestion notification

Fig. 2.   Structure of the cell header.

Higher Layer Functions	Higher Layers
Convergence	CS
Segmentation and reassembly	SAR
Generic Flow Control / Cell header generation/extraction / CellVPI/VCI translation / Cell multiplex and demultiplex	ATM
Cell rate decoupling / HEC header generation/verification / Cell delineation / Transmission frame adaptation / Transmission frame generation/recovery	TC
Bit timing / Physical medium	PM

Layer Management — AAL — ATM — Physical layer

Fig. 3.   Functions of the ATM Protocol Reference Model.

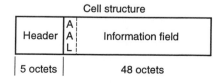

Cell structure

Header	A A L	Information field
5 octets		48 octets

Fig. 1.   Structure of the ATM cell.

The ATM PRM is organized in three layers. These layers should not be confused with the layers of the OSI model. CCITT Recommendations describe the functions of these layers and the data and primitives exchanged between layers and with the outside world. They are briefly described below.

2.3 Adaptation to Services (AAL Functions)

The ATM adaptation layer (AAL) is service-specific and processed in the terminals; it is not seen in the ATM network. The information exchanged with higher layers depends on the service; it may be continuous streams in the case of voice or video or SMDS/CBDS frames for LAN interconnection.

AAL is divided into two sublayers, the convergence sublayer (CS), and the segmentation and reassembly sublayer (SAR). The convergence sublayer implements information-specific functions such as CRC error control or error correction mechanisms for data, or synchronizing processes for real time information, such as voice or video. Up to now, five types of AAL which can accommodate a variety of service classes have been defined in CCITT Recommendation I363. They are shown in Table 1 with their applications. New AAL types could be added if necessary when implementing new services without any impact on the network.

The key feature is the multiservice capability of ATM, combining adaptation tailored to the specific needs of services with a unique network.

2.4 Routing and Multiplexing (ATM Functions)

At the boundary between the AAL layer and the ATM layer, 48 byte blocks containing information and information-related data are exchanged, together with a small number of primitives.

The ATM layer is independent of the service layer and of the transmission medium used to transport ATM flows. It is implemented in all ATM network elements, such as ATM multiplexers, cross-connects, or switches, to route, multiplex, or demultiplex ATM VPs or VCs. The ATM layer adds a cell header to each information field received from the AAL layer, which processes the GFC field to implement traffic control procedures with terminal equipment at the UNI, and generates and processes the VP and VC identifiers (VPIs and VCIs) for multiplexing and routing. OAM flows F4 and F5, which are made of specific cells used to monitor the performance of VPs and VCs, are introduced in the ATM layer. Figure 4 shows ATM multiplexing.

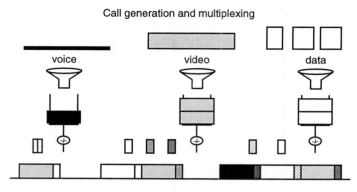

Fig. 4. Cell generation and multiplexing.

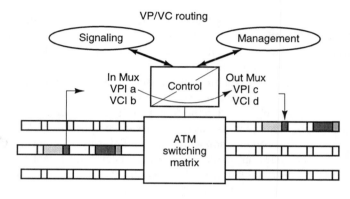

Fig. 5. VP/VC routing.

Buffers are filled at the source clock rate and are read at the network clock rate to produce the information field of each cell. These two clocks are completely independent of each other. The header is added to complete the cell, which is the only service-independent entity in the network. Cellulization generates waiting time jitter, which is generally negligible except for low bit rate sources (e.g., where delay can raise echo problems). Cells of various services are multiplexed to form the ATM stream. Multiplexing jitter (cell delay variation) is a basic performance parameter of ATM networks.

The switching matrix, Fig. 5, routes the cell from the input to the output, depending on the input VPI/VCI values read by the control function, which processes the header of each cell. New

TABLE 1.

AAL Type	1	2	3	4	5
Real Time Constraints	Yes	Yes	No	No	No
Bit Rate	Constant	Constant	Variable	Variable	Variable
Connection Mode	Connection oriented (CO)	CO	CO	Connectionless (CLS)	CO/CLS
Applications	Circuit emulation	Variable bit rate video	Data	LAN interconnection	Data frame relay bearer service

VPI/VCI values are set. The same mechanisms are used in ATM cross-connects, where they are controlled by network management, and in ATM switches, where routing and multiplexing information is obtained from signaling.

2.5 Transmission Aspects (PHY Functions)

At the boundary between the ATM layer and the PHY layer, 53 byte cells containing an information field and routing information in the header are exchanged, together with a small number of primitives.

The physical layer is divided into two sublayers, transmission convergence (TC) and physical medium (PM). PM sublayer and part of the TC sublayer functions are dependent on the transmission medium used, medium adaptation (i.e., electrooptic conversion), bit clock generation and recovery, and generation/recovery of the transmission frame if a transmission frame is used. TC sublayer higher level functions are generic whatever the transmission system below and specific to ATM. They are the following:

- Bit rate adaptation: Bit rate adaptation is similar to the process of justification in plesiochronous multiplexing. It consists of inserting cells with specific identifier and content, which are not used at the ATM level, to the stream of assigned cells to reach the capacity of the transmission system. These cells, which are not seen by the ATM layer since they are added and removed by the physical layer, are called unassigned cells or, more commonly, idle cells.[1]
- Cell delineation and header error detection/correction: These two functions are the result of the same process, based on the HEC octet of the cell header. At the transmit side, the result of the division of the first four octets of the header by the polynomial $1 + X^7 + X^8$ is put in the HEC octet. At the receive side, this redundancy is used to recover cell delineation and to detect loss of cell synchronization, and when synchronization is achieved to detect transmission errors. In the case of single errors, the header is corrected and the cell is given to the ATM layer. In the case of multiple errors, the probability of miscorrection and consequent cell misrouting is too high, and the cell is deleted by the physical layer. To prevent malicious or unintentional imitations of the HEC sequence, the information field of all cells, assigned or unassigned, is scrambled[2] at the physical layer. The result is a self-supporting stream of cells, the bit rate of which corresponds to the capacity of the transmission system, generally octet-aligned with the transmission frame (if any), and which is subsequently coded by lower order functions according to the transmission system-specific format.

Since 1990, the mapping of ATM cells into the SDH/VC4

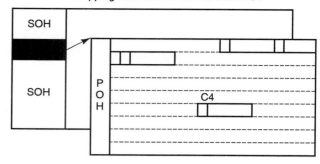

Mapping of ATM cells into theSDH/VC4

Fig. 6.   Mapping of ATM cells into the SDH/VC4.

(Fig. 6) has been specified in CCITT Recommendation G.709. The ATM payload is octet-aligned on the frame and a cell can "cross" the POH boundary. Similar mappings have been introduced more recently for SDH lower-order VCs and PDH bit rates (Recommendation G.804).

Key features are that seamless ATM connectivity can be achieved whatever the structure of the transmission network below, in particular in a mixed PDH/SDH environment, and that bandwidth can be provided exactly to the customer need and easily scalable within the maximum capacity of the physical interface.

3. NETWORK ASPECTS

3.1 Architecture

The architecture of ATM has been defined in the context of B-ISDN studies, and in particular, the relationship of ATM layer with service layers and with the physical layer is represented in the ATM protocol reference model (PRM). Physical layer description in the PRM is not in line with the evolution of the architecture of transport networks due to the capabilities of SDH and is reflected in Recommendation G.803. The PRM and G.803 descriptions should be harmonized to enable ATM equipment and networks to be manageable in a consistent way with equipment and networks implementing SDH and PDH technologies. This harmonization is addressed in ITU-T and ETSI. The first step is to extract from G.803 generic principles which can be applied to all technologies. This is the scope of draft Recommendation Gtna (architecture of transport networks). The second step is to apply these generic principles to ATM transport networks supported by SDH, PDH, and cell-based transport networks. It is the scope of draft Recommendation Gatma (architecture of ATM transport networks).

Some results have already been achieved in ETSI. As shown in Fig. 7, ATM VC and VP networks can be described as layer networks as defined in Recommendation G.803.

The VC layer network is a client of the VP layer network, which is a client of SDH, PDH, or cell-based layer networks. The VP and VC trails are characterized by maintenance flows F4 and F5 (see below) which are equivalent to overhead capacity in SDH POH. They are inserted/extracted in VP or VC trail termination functions. The VC layer network is a server for the

---

[1]This description is complete at the NNI and at the UNI for the SDH-based option. A cell-based transmission system, specific and optimized for ATM, is also defined at the UNI. It makes use of some of the unassigned cells for physical layer OAM purposes.

[2]In SDH and PDH mappings, a $1 + X^{43}$ self-synchronizing scrambler is used. In the cell-based option, a $1 + X^{23} + X^{31}$ DSS (digital self-synchronous scrambler) has been chosen to avoid error multiplication, and two bits of the HEC are used to synchronize the scrambler at the receive side before cell delineation.

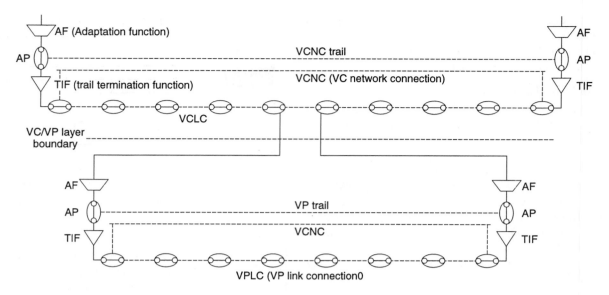

Fig. 7. Application of G.803 concepts to ATM VC and VP layer networks.

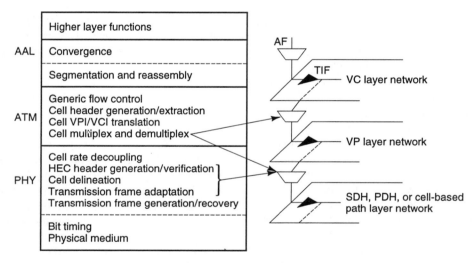

Fig. 8. Correspondence between ATM PRM and G.803 generic architecture.

service layer networks; the adaptation functions are equivalent to AAL functions of the ATM PRM.

An allocation of some of the functions described in the PRM to the generic transport functions (trail termination and adaptation functions) of G.803 is shown in Fig. 8.

This work on functional architecture is a prerequisite to derive information models which will enable ATM equipment interoperability in a multioperator and multivendor environment. The development of such an information model for ATM cross-connects is in progress in ETSI. Equipment standardization should also be done, and should result in an ATM equivalent of Recommendation G.783 for SDH. This is the subject of a question in CCITT SG15.

3.2 OAM

Performance monitoring and maintenance capabilities for the broadband ISDN are defined in Recommendation I 610. In particular, F4 and F5 maintenance flows made of specific cells are defined for ATM VPs and VCs. Maintenance flows may be inserted at network or subnetwork boundaries (F41/F51 flow and F42/F52). The F42/F52 flows are an application of the concept of tandem connection monitoring by sublayering defined in G.803. This structure enables monitoring of a trail end-to-end by the user and by the operators of the subnetworks used to support this trail simultaneously.

Examples of functions of the maintenance flow at the VP layer are:

• Performance monitoring (PM): This is achieved by insertion of recurrent cells in each VP (at least for 16 VPs in a path). The information field of one PM cell contains the result of a BIP-8 calculation on the information field of all cells belonging to the block of cells since the previous PM cell. The number of cells of the block is also given. At the receive side, a comparison between the content of the PM cells and the result of the same calculation made on the cells received indicates errored or lost cells.

- Continuity check: Cells are inserted to check the continuity of a VP in the absence of information.
- Far end receive failure (VP-FERF) and alarm indication signal (VP-AIS): Figure 9 illustrates how these cells are used in case of failure (loss of cell delineation) of one adaptation function in a simple configuration of two VP subnetworks (or VP switch or cross-connect equipment) linked by a 34 Mb/s transmission path.
- Fault location: To achieve fault localization, loopback cells can be sent from a point in a subnetwork with specific identification to be sent back to the source by loopback mechanisms in other subnetworks.

3.3 Traffic Control

Multiplexing ATM streams can be made on the basis of the peak rate. In this case, the maximum capacity required is permanently allocated to a customer. This may be convenient to deal with continuous bit rate services, but leads to a waste of transmission resource in the case of bursty sources such as commonly encountered for data services. For traffic of variable bit rate, the allocation can be made on the basis of the mean bit rate assuming a given distribution. This is generally referred to as statistical multiplexing. However, due to the high bit rates achieved in ATM technology, it is impossible to implement flow control mechanisms, implying a negotiation between source and receiver as used in existing packet networks. Therefore, there is a need to check at the entrance of the ATM network that the traffic shape corresponds to what is really expected. Otherwise, congestion may occur in the ATM multiplexing process.

The traffic is controlled by a function called the usage parameter control (UPC, commonly nicknamed "policing") which checks the conformance of the ATM stream entering the network to the agreed traffic shape and deletes nonconforming cells (in some realizations, the UPC function modifies the CLP bit the header of nonconforming cells to prioritize them for further deletion if congestion occurs in the network; this is called "violation tagging"). First implementations use simple leaky bucket mechanisms based on the control of the average rate and allowing some degree of clustering to accommodate the burstiness of the

traffic. The drawback of this type of algorithm is that cell clustering may also be due to multiplexing, and that it is impossible to distinguish between the two causes. Therefore, a margin has to be allocated which can be used by malicious users to send more traffic than they are allowed. It is now recognized that to be efficient, control of the traffic shape must be combined with a spacing process which removes multiplexing jitter.

4. Deployment of ATM

4.1 Standardization Status

CCITT chose ATM as the B-ISDN target transfer mode in 1988. A first set of 13 Recommendations was issued in 1990, describing basic features and requirements, such as cell size, PRM, header structure, AAL classification, SDH and cell-based options UNIs, networking, and maintenance principles. Another key player in the standardization field is the ATM forum which strives for interoperability between equipment used in private networks, in particular, specifying interfaces and management procedures for permanent virtual circuits compatible with existing private infrastructures and procedures. Standards have now reached a level which enables interoperability between first implementations of VPs of VC cross-connects and using resource allocation based on the peak rate. Standardization work is now devoted to signaling and traffic control. It is a long-term effort which is essential to take full advantage of ATM. A first set of specifications is expected in the near future; the complete set, in particular in the field of signaling necessary to implement various call and connection configurations, will take a few more years. The work to produce standards for ATM equipment has also started.

4.2 Equipment Aspects

Most of the main telecommunications equipment manufacturers have available or are developing ATM equipment for use in the public network. These first products are ATM cross-connect equipment which allow the implementation of VP managed networks and service adaptations or multiplexers which adapt various services to ATM. In the private networks, ATM switches and hubs and ATM interfaces for workstations or PCs are now available to implement ATM-based Local Area Networks using installed metallic or fiber private infrastructures. ATM interfaces for routers are either available or announced by most manufacturers. The fact that ATM is the basic technology for evolution

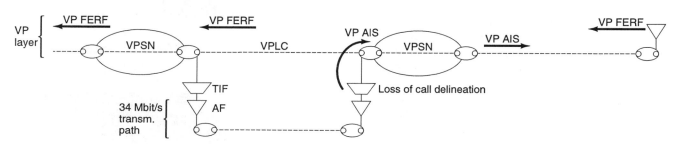

Fig. 9.   Loss of cell delineation in a two–VP-subnetworks configuration, linked by a 34 Mb/s transmission path.

of both public and private networks is of paramount importance since it paves the way to seamless ATM connectivity from terminal to terminal.

4.3 ATM Field Trials and Pilot Networks

RACE 1 programs commissioned by the European Community are coming to an end. They have reached their objective, which was to design and to test components and building blocks of equipment and networks, based on ATM, for integrated broadband communication. RACE 2 programs are now following, with a focus on integration, experimentation, and demonstration. For instance, in the BETEL project, a VP cross-connect located in Lyon (France) was used to test high bit rate data applications among users in Geneva, Lausanne, Lyon, and Sophia-Antipolis.

National R&D projects, such as RECIBA (Telefonica) and BREHAT (France Télécom), were also started a few years ago to design and demonstrate ATM equipment and applications. In the U.S., the gigabit test beds implemented to experiment the evolution of the Internet network are based on ATM. It is also the case of the NTT project VIP (Visual Intelligent and Personal). Recently, most operators have announced ATM field trials and pilot networks, and some operators have already started or announced their plans for the provision of ATM commercial service (Sprint, Time Warner, France Telecom, etc.).

Five operators (BT, DBP Telekom, France Télécom, STET and Iritel acting together, Telefonica) have signed an MOU (memorandum of understanding) to establish an ATM European pilot network based on Eurescom[3] specifications. To date, other operators have joined them to reach a total of 18. The main objectives of this pilot network are to assess that ATM standardization has reached a stage where interoperability can be guaranteed in a multioperator and multivendor environment, and to experiment and demonstrate the capability of ATM to support a variety of services, with the cooperation of pilot users. These "benchmark services" are constant bit rate services, CBDS/SMDS, and frame relay bearer service.

4.4 First Applications

The fact that ATM is considered in CCITT as the target transfer mode of B-ISDN does not justify in itself an early introduction of this technology. The driving factor in the short term is the ability of ATM to support services which are now required by business users. These users are no longer satisfied by existing leased lines services. They require more bandwidth and a more flexible way to exploit this bandwidth. From the network providers' point of view, this is complemented by the multiservice capability of ATM which enables economy of scale in the provision of various services and an easier introduction of new services. The maturity of standardization and the availability of products are such, as shown above, that first implementations are now possible.

Therefore, the first implementation of ATM is likely to be in the field of corporate networks where virtual path networks will interconnect through the ATM public infrastructure and, in

a flexible way, the sites of a customer. The advantages which are expected either for the user or for the provider of this type of network are based on the main following factors:

- Granularity: ATM enables tailoring of the bandwidth to the exact amount required.
- Scalability: Changes in the bandwidth allocation or network configuration are easy to implement, provided a consistent management system is used.
- Multiservice capability: CBR, VBR, connectionless, or connection-oriented data can be integrated on the same network, enabling optimization of transmission capacity and corresponding cost reduction. For connectionless service used for LAN interconnection outside the private environment, the provision of connectionless servers accessible through the VP network is also necessary.
- Enhanced OAM features: The provision of maintenance flows for network and subnetwork enable customer and network provider access to relevant performance and management information.
- Wide area coverage: Overlay ATM networks can be deployed on PDH or SDH infrastructure. This enables decoupling of service offering and evolution of the transport network from PDH to SDH.

It is believed that these features will bring cost effectiveness in the use of ATM, even in the context of relatively low bit rates. Furthermore, the inherent high bit rate capability of ATM will facilitate an increase of the bandwidth where and when necessary. However, the leased line type offering has to be complemented by a more efficient way of sharing the transmission resources among users. For LAN interconnection, which will be the first main application of ATM, the bandwidth optimization is done above the ATM layer by connectionless servers. When available, VC switching will be the basis of a solution applicable to other services.

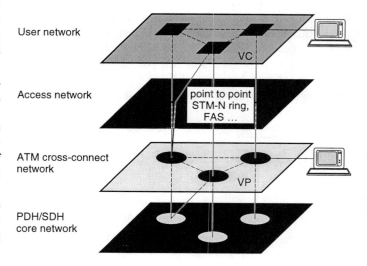

Fig. 10. The BREHAT project.

[3]Eurescom is a research center funded by European telecommunications public network operators.

4.5 An Example: The BREHAT Project

Figure 10 gives an example of the scenario which can be used to introduce ATM technology. It shows the network architecture of the BREHAT project which was launched in 1990 by France Télécom in association with two manufacturers, Alcatel and TRT Philips. It is composed of a VP network made of ATM VP cross-connects and multiplexers supporting a VC network made by service multiplexers which manage the VP bandwidth and adapt non-ATM services. The transport infrastructure used in the access and in the transit portions are PDH or SDH links. VPs are allocated on the basis of the peak rate through management procedures.

This architecture is very close to the architecture of the European ATM pilot. Services offered by the service multiplexer are CBDS/SMDS, providing LAN interconnection by means of an additional router with an HSSI interface, H261 videoconferencing at $n \times 384$ kb/s (n from 1 to 5), ATM VP/VCs through 155 STM-1 or 34 Mb/s interfaces, circuit emulation at $n \times 64$ kb/s (up to 2 Mb/s), and frame relay bearer service through a 2 Mb/s interface.

5. CONCLUSION

This paper contains an overview of ATM main features and a perspective for utilization. In the long term, it is believed that ATM will be used to support the set of services with the wide spectrum of bandwidth and traffic characteristics foreseen in the B-ISDN. In the short term, ATM is needed to enhance services to business users in terms of leased lines optimization, high and flexible bit rates, and multiservice private networks. In both cases, it will rely upon SDH transport infrastructure, and on PDH during a significant period of time. Consistency with the network architecture and management principles defined in the context of SDH is the key for users and operators to take advantage of the complementary features of these technologies.

An SDH Transmission System for the Transport of ATM Cells

A. BROSIO AND A. MONCALVO

CSELT — CENTRO STUDI E LABORATORI TELECOMUNICAZIONI S.P.A.

VIA G. REISS ROMOLI 274 - 10148 TORINO, ITALY

1. INTRODUCTION

THE advent of ATM (Asynchronous Transfer Mode) will require efficient and cost-effective systems to transport cells carrying information with widely differing characteristics in the various portions of the network, from the user premises to the local network node and from node to node. After presenting a general review of the basic ATM concepts, this contribution aims at discussing the requirements imposed on the transmission systems designed for this specific application. Details are given on an implementation based on SDH and intended for application in a broadband optical subscriber link.

2. ATM BASIC CONCEPTS

The development in information technology and in information processing has stimulated, in recent years, an intensive research effort toward the identification of a more efficient methodology for the transfer of information. In the meanwhile, remarkable progress has been achieved as far as switching, transmission, signaling, and network management are concerned. The ATM grew as a synthesis of diversified technical capabilities to handle any kind of information in an integrated manner.

The fundamental idea of the ATM technique [1], [2] consists of the segmentation of a digitally-coded information stream (either discontinuous, like data, or continuous, like voice or video) into a sequence of elementary blocks, called cells, that are transported and routed through the telecommunication network. Cells pertaining to independent connections and with different generation rates can be carried on the same link and are accepted as they are generated, exploiting statistical multiplexing instead of fixed-frame time division multiplexing. When no information is to be transmitted, empty or idle cells are inserted (and discarded at the destination) as the digital channel operates at a fixed bit rate.

The length of the cell has been fixed in 53 octets: the first 5 octets are the cell header and the remaining 48 octets are the information payload. The content and structure of the header are different at the User Network Interface (UNI) and at the Network Node Interface (NNI), but, in both cases, the fifth octet carries the Header Error Control, an error correcting code that protects routing information. In fact, the switching of the ATM cells relies fundamentally on the correct interpretation of complex signaling information that continuously flows between the user and the network.

3. STANDARDIZATION ACTIVITIES RELATED TO ATM

The standardization activities related to the ATM concept were initiated in the second half of the 1980s. In particular, the CCITT Study Group XVIII in the period 1985–1988 set up an "ad hoc" group called the Broad Band Task Group, followed by a regular Working Party (WP 8) since 1989. A first group of Recommendations dealing with the general concepts of ATM was made available at the end of 1990; since then, they increased in number and in scope, covering all aspects of the B-ISDN implementation. As far as transmission aspects are concerned, particular consideration should be given to Recommendation I.432 [3], dealing with UNI specification based on SDH frame and on pure cells, at 155 and 622 Mb/s. At present, ITU-T Study Group 13 is in charge of defining general and network aspects, while Study Group 15 has undertaken the specification of equipment, exploiting a functional block diagram approach, hopefully to be evolved toward the "atomic function" modeling approach. NNI specifications for ATM can be found in Recommendation G.709 for the SDH-based solution and in Recommendations G.832 and G.804 for the interfaces based on new PDH frames [4].

In the framework of the European CEPT (Commission d'Etudes des Postes et Telecommunications), the activity on this subject was initiated by the GSLB (Groupe Speciale Large Bande) since 1984, and continued by ETSI (European Telecommunication Standard Institute) in the Network Aspect 5 (NA 5) group, where the general aspects were dealt with and a Technical Standard on the User Network Interface was prepared. The specifications of UNI can be found in ETSI Technical Standard (ETS) 300 299 [5] and 300 300 [6], while the NNI is described in ETS 300 147 [7] and ETS 300 337 [8].

Much effort, especially concerning the subscriber premises equipment, is carried out within the ATM Forum, that, even if it is not a standardization body, is proposing the adaptation of already available equipment to the transport of ATM cells in a pragmatic way.

4. ATM AND THE TRANSMISSION SYSTEM

A transmission system devoted to the transport of ATM cells

should take into account a number of factors, arising both from the intrinsic characteristics of the ATM (e.g., cell organization, implementation of control and operation mechanisms, etc.) and from the available transmission techniques.

Studies of the adaptability of various transmission systems to ATM cell transport have shown that several approaches may be followed to exploit presently available systems or to design a totally new generation of equipment. In principle, the ATM cell flow can be mapped in any digital stream carried on the available media: metallic coaxial cables or pairs, fibers, radio links, and satellites, provided that a suitable transmission quality is ensured. The effective choice depends chiefly on the type of application and on the time frame in which the application should be available.

The Synchronous Digital Hierarchy (SDH), which was developed in the same time frame during which the ATM concepts were defined, was considered since the beginning a well-suited solution for the cell transport, due to its key features of integrated management, performance monitoring, and transversal compatibility. SDH equipment is going to be introduced in the core network by several Operators on a regular service basis in a medium-term scenario and for high information transfer capacities fitting the B-ISDN perspective well.

The Plesiochronous Digital Hierarchy (PDH) is already extensively deployed, and can cope with the need of business users to access the ATM services in a short time, without waiting for the complete SDH network deployment.

In the distribution network, on the other hand, different solutions could be used in order to take into account the requirements of small business users and eventually of domestic subscribers, and to optimize costs, consumption, and penetration while seeking to limit interworking operations.

The system described in the following paragraphs was developed by CSELT in the framework of the activities related to evaluate the impact of the ATM technique on the distribution network.

All the above-mentioned options for the implementation of the transmission system were carefully analyzed. However, some characteristics of the SDH hierarchy (such as the division of payload into containers of different capacity, the ability to control quality across the ends of the connection, the presence of management and supervision channels, and the division of the connection into different maintenance entities in order to facilitate failure identification) were also considered well suited for exploitation in the distribution network. Another reason lay in the need for carrying on experiments on several "hot topics" in discussion within standardization bodies, while alternative approaches were at that time less interesting, and in the possibility of exploiting the concepts and the technological developments in other network contexts also. On the basis of these considerations, it was decided to use the synchronous hierarchy frame for ATM cell transport (SDH-based solution).

5. ATM CELL ADAPTATION TO TRANSMISSION SYSTEMS

The starting point for the detailed definition of the physical layer is the B-ISDN Protocol Reference Model (PRM) [2]. The model identifies the functions to be performed by the transmission system, and is of crucial importance in permitting cells to be transported independently of the transmission system used.

As will be seen below, the PRM contains a set of general-purpose functions together with other functions which depend on the particular implementation systems used ("SDH-based" or "cell-based" solutions, fiber or copper carriers, etc.). To provide as complete and realistic a view as possible of the problems involved, a number of general aspects will be discussed first.

It must be emphasized that many of the concepts discussed herein are generally applicable, while the parts referring to the specific implementation are included in order to exemplify the complexity of the problems involved.

5.1 Reference Model Physical Layer Definition

The Protocol Reference Model defines the functions required to transfer information between subscribers using the ATM technique. In particular, it describes these functions as regards information structuring into cells and cell transmission, and divides the functions into homogeneous layers where primitives are exchanged in standardized form so that each layer provides the next highest layer with well-defined services.

The Physical Layer (PL) supplies the transmission service for the cells presented by the ATM layer above it. Obviously, transmission may be accomplished using different systems, which must not, however, affect either the ATM layer or the primitives exchanged with the PL.

The independence of ATM cell transport from the transmission system used is considered to be of fundamental importance, both as regards a clearer description of the ATM functions and from the practical standpoint, given that the functions relating to the ATM layer are kept unchanged for different transmission systems.

The PL layer is divided into two sublayers: the Transmission Convergence (TC), which adapts the cells to transmission needs and enters them in the line frame if used, and the Physical Medium (PM), which includes the transceiver functions which depend on the physical medium used (copper cable, optical fiber, etc.).

The first TC function performed on cells coming from the ATM layer is Cell Rate Decoupling. This consists of adding empty cells to the flow of cells from the higher layer (and of removing them in reception). Transfer between the two layers thus applies only to those cells which are effectively used in the connection, leaving the task of completing transmission capacity to the function indicated above. This function is extremely important as it ensures that the transmission system is independent of the requested band within the limits of the maximum available capacity.

The second function, Header Error Control or HEC, consists of entering the error corrector code in the header and checking it in reception for error detection/correction when needed (cells with noncorrectable errors are discarded).

The third function, Cell Delineation (see Section 5.2), is used only in reception and consists of identifying cell boundaries. The currently specified method (see below) is based exclusively on the intrinsic structure of the cells, and is thus independent of the type of Transmission Convergence used (i.e., on the type of frame of the transmission system).

The last two functions concern frame generation and insertion of the cells and vice versa.

The PM sublayer is closely linked to the transmission medium used. Its functions can be compared to those of line terminals, i.e., it must supply the bit flow and the associated timing.

5.2 Cell Delineation

As mentioned above, it is extremely important (particularly from the practical standpoint) to be able to recognize the cell independently of the frame used. Particular attention has thus been devoted to developing a cell delineation method which not only ensures independence from the transmission system, but is also efficient and easy to implement.

One of the first possibilities examined was that of associating cell delineation with the frame structure. Clearly, however, the mechanism which associates the frame structure with cell boundary recognition depends on the transmission system, and in any case does not provide a solution where the transmission of cells is without an external frame. Alternative solutions studied included that of inserting fixed bit sequences which repeat periodically at the beginning of each cell or into dedicated cells. This would make it possible to achieve delineation with procedures similar to those used for the existing system. However, a disadvantage of this solution is that it subtracts part of the information transport capacity.

The possibility was also investigated of using the empty cells, which are identified by predetermined bit configurations. Under high connection usage conditions, however, the limited number of empty cells drastically reduces the chances of detecting delineation losses and of recovering delineation where necessary.

The solution here implemented, which employs the error corrector code (HEC, Header Error Control) provided in the cell header to protect routing information, was incorporated later into Rec. I.432 [3]. The code determines the values of the control bits on the basis of the bits to be protected, using a shortened and expurgated (40, 32) Hamming code with minimum distance 4 which can correct single errors and detect double errors. The generator polynomial is

$$g(x) = x^8 + x^2 + x + 1.$$

This correlation is used to identify the position of the code word, and hence the beginning of the cell. The resulting algorithm (HEC method) thus uses only the information contained in the cell itself, and does not add specific "overhead" for delineation. It maintains the available band without change, and is fully independent of the transmission system (the ATM cell flow is "self-delineating").

As delineation is verified at each cell, attainable performance is better than with methods linked to frame structure. For example, the frames of SDH systems contain an octet every 125 μs which can be used as a pointer to identify the beginning of the

cell. This involves verifying delineation at 155 Mb/s only around every 44 cells, so that out-of-delineation detection times are worse than with the HEC method. Moreover, a fixed sequence of bits was defined which is added to the code word control bits to prevent simulation in the event of out-of-delineation, which would otherwise be possible because of the properties of the error correcting code. For reasons of security, it was also considered advisable to mask the contents of the cell information field. Indeed, without such a procedure which follows rules which are not known to the subscriber, the latter could destroy and simulate the frame transiting on the line. This objective is reached through the use of a scrambling procedure. A number of alternative scrambling procedures were analyzed and compared as regards performance, integrability with delineation, and ease of implementation. For the subscriber–network interface, a scrambler whose generator polynomial is

$$g(x) = x^{43} + 1$$

was standardized.

6. BROADBAND EXCHANGE–SUBSCRIBER CONNECTION

Frame generation and cell insertion/extraction functions as well as the PM sublayer functions depend on the options selected for each individual implementation. To complete the analysis of a complete transmission system in realistic terms, reference will be made to the prototype developed at CSELT. The considerations carried out in Sections 6.1 and 6.2 do not apply only to the specific implementation, but also provide a picture of the general structure of the exchange–subscriber connection, motivate the choice of the SDH-based solution, and indicate its functional characteristics. The remaining sections describe the implementation aspects, and should thus provide some feel for the type of problems which can be encountered as well as for the possible solutions. It should be noted that the optical connection and integrated circuits were implemented as a result of cooperation between CSELT and Italtel in the RACE 1012 Project "Broadband Local Network Technology."

6.1 Functional Blocks

The connection reflects the principles laid down by ITU-T for broadband ISDN systems [9]. As shown in Fig. 1, the connection consists of a number of functional blocks, viz. Exchange Termination (ET), Line Termination (LT), and Network Termination (NT). The latter is in turn divided into two blocks, NT1 and NT2, which are separated by the subscriber–network interface. The functional blocks are separated by reference points V (between ET and LT), U^1 (between LT and NT1), T (between NT1 and NT2), and S (between NT and customer premises equipment).

The main function of the ends of the transmission connection (in this case at the level of blocks ET and NT2) is to enter ATM cells into the SDH hierarchy payload. In practice, this means implementing the PRM model Physical Layer, including cell delineation and scrambling.

The management and control functions in the various reference points are transported in the frame overhead and terminated in the various blocks. In particular, the NT1 block manages the

---

[1] The connection between LT and NT is usually identified as reference point U, although the usage is not contemplated by international standards.

portions dealing with the public network and with the subscriber station. Among other functions, the NT1 block will control activation and deactivation procedures, adapt different clocks which may be present at the two sides of the T interface, etc.

From this standpoint, it should be noted that international standards have been developed for interworking procedures involving the operation and maintenance capabilities offered by the SDH structure (e.g., path overhead) and those envisaged for the ATM technique (e.g., payload type use). Developing these activities is particularly important in order to solve overall serviceability problems (e.g., detection and location of malfunctions on a virtual connection), and thus avoid potential bottlenecks, given that these interworking functions must be provided at the exchange or network termination level.

6.2 Connection Characteristics

The connection characteristics can be summarized as follows:

• ATM cells will be inserted in SDH frame STM-1; the line system will thus operate at 155.520 Mb/s.

• The connection will be implemented using a single monomode optical fiber. Wavelength Division Multiplexing (WDM) will be used to separate the optical carriers in the two directions. The optical carrier in the exchange–subscriber direction will be allocated in the second window (around 1310 nm), while the carrier in the opposite direction will be allocated in the third window (around 1550 nm). These wavelength assignments in the two directions are justified by the possibility of multiplexing broadcast services in the exchange–subscriber direction in the STM-4 frame at 622.080 Mb/s. Comparisons between different solutions show that it is advisable to use a single fiber for both directions, both for economic reasons (comparison between WDM and fiber duplication) [10] and for practical reasons (cable size, underground ducts availability, etc.).

The functional block diagram for the connection developed in accordance with the foregoing considerations is shown in Fig. 1.

6.3 Circuits for Inserting and Extracting ATM Cells in the STM-1 Frame

These circuits insert a continuous flow of ATM cells in the STM-1 frame at the transmitter and extract the ATM flow from said frame at the receiver. In addition, they proceed with cell delineation, scrambling, and cell rate decoupling as described above.

These operations are performed by five functions grouped into four integrated circuits. Figure 2 shows how these functions are used to implement the reference diagram in Fig. 1. The functional characteristics of the three integrated circuits relating to the SDH-based solution will be illustrated briefly below. The fourth IC relates to the other functions (cell delineation, etc.) which have already been illustrated in Section 5.2.

• VC-4 Assembler/Disassembler — This component performs two separate functions.

VC-4 Assembler: The VC-4 Assembler inserts cells in the layer 4 Virtual Container (VC-4) as defined in the SDH hierarchy. To obtain the VC-4, the ATM cells are added to the Path Overhead bytes which follow the VC-4 from the terminal in which it is assembled to the terminal in which the payload is extracted (VC-4 path). The overhead byte functions control the path in terms of error rate monitoring, alarm transfer, payload structure description, etc. A byte (H4) is used to periodically indicate the beginning of each cell. This function (which can be disabled in the component) is envisaged by international standards as an auxiliary to the delineation method based on the corrector code. Incoming cells are presented in parallel byte-aligned form accompanied by the start-of-cell marker and the associated clock.

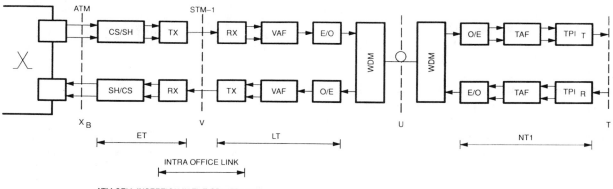

ATM CELL INSERTION IN THE SDH FRAME
ATM CELL EXTRACTION FROM THE SDH FRAME
SIGNAL ADAPTATION BETWEEN REF. POINT V AND REF. POINT U
SIGNAL ADAPTATION BETWEEN REF. POINT U AND REF. POINT T
PHYSICAL INTERFACE AT THE REFERENCE POINT (TRANSMITTING SIDE)
PHYSICAL INTERFACE AT THE REFERENCE POINT (RECEIVING SIDE)
REFERENCE POINTS X$_B$ ARE NOT STANDARD

Fig. 1.   Functional block diagram of the broadband exchange–subscriber connection.

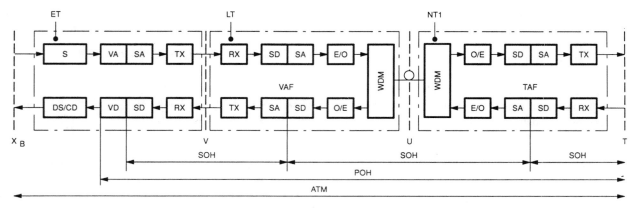

S: SCRAMBLER
DS/CD: DESCRAMBLER/CELL DELINEATION
VA: VC–4 ASSEMBLER
VD: VC–4 DISASSEMBLER
SA: STM–1 ASSEMBLER
SD: STM–1 DISASSEMBLER

Fig. 2. Partitioning of ATM and SDH functions.

VC-4 Disassembler: The VC-4 Disassembler performs the opposite functions to the VC-4 Assembler. In reception, the overhead bytes are extracted from the VC-4 and handled according to the case at hand. If enabled, the H4 byte signals the beginning of the cell. The cells are then emitted with the same format as at the input to the VC-4 Assembler.

• STM-1 Assembler — This function inserts a VC-4 in the STM-1 frame. It performs two basic tasks. The VC-4 phase is adapted to that of the frame through a pointer, whereby it is possible to compensate for any phase difference between VC-4 and STM-1 resulting from different clocks or from jitter and wander. The second task is to insert overhead bytes which accompany the STM-1 frame from the VC-4 assembler to the disassembler; these control the transmission line. In addition to the monitoring and control functions which are also provided for VC-4 Path, it is possible to use groups of bytes to set up message channels for network operation and maintenance (O&M).

• STM-1 Disassembler — This function performs the opposite operations to the STM-1 Assembler. In addition to extracting overhead bytes from the input signal and processing information in accordance with specified procedures, it must identify the beginning of VC-4 through appropriate pointer handling. In addition, it must generate alarms for the rest of the receiver if alarms are detected, e.g., in the event that there is no optical signal, STM-1 frame delineation is lost, nonvalid pointers are received, etc.

For all components described above, the overhead bytes which are not generated internally are taken from outside during transmission, while in reception, they are all accessible. It is thus possible to provide an organ in the terminal equipment which manages control and monitoring equipment and performs the appropriate functions, if necessary cooperating with the remote terminal.

The three components described above were implemented as integrated circuits using 1.2 μm CMOS technology (Fig. 3).

These experimental implementations confirmed the feasibility of transmitting ATM flow in SDH frames, as well as the validity of the ITU-T Recommendations for the various functions envisaged for ATM flow insertion and extraction with SDH frames (both at the Virtual Container level and at the Synchronous Transport Module level).

Fig. 3. Gate arrays VC-4 Ass/Disass, STM-1 Ass, STM-1 Disass. (Developed by CSELT and Italtel in the RACE 1012 Project.)

Fig. 4. Prototype of the 155 Mb/s optical transceiver.

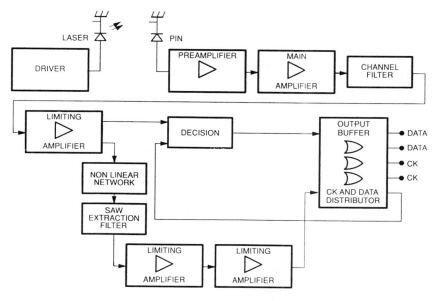

Fig. 5.   Block diagram of the transceiver.

An initial evaluation of the complexity of these functions indicates that implementing them will call, on the one hand, for extensive work in integrating functions (including electrooptic functions), and on the other hand, for careful component selection in order to ensure electrical and optical performance at the minimum possible cost, particularly if the broadband transmission connection is used in the distribution network.

6.4 SDH Optical Link

The prototype 155 Mb/s optical transceiver (see photo, Fig. 4) was implemented according to the block diagram shown in Fig. 5. Semiconductor laser devices operating at 1310 and 1550 nm are used as the optical sources, while the photodetectors consist of ternary P-I-N devices. Prototype transceivers were developed using commercial electronic components; where available, preintegrated functions were preferred in order to make circuit layout easier and limit the subsequent debugging.

The main characteristics of each transceiver are as follows: bit rate of 155.520 Mb/s with NRZ binary format data, bipolar technology monolithic laser driver, automatic optical power output control, optional automatic temperature control, bipolar technology monolithic transimpedance preamplifier, high input dynamic range achieved through a clipping network (without automatic gain control in reception), timing extraction with surface acoustic wave (SAW) filter, alarms for laser bias current, and failure to receive optical pulses. The receiver was fully characterized in the laboratory. The preamplifier transimpedance value is around 1.5 kΩ with a noise of 3.2 pA/Hz$^{1/2}$. Achieving an error rate of 10^{-9} requires a mean input optical power of -31 dBm, and under these conditions, there is a signal-to-noise ratio of 19 dB at the decision point with an eye opening of approximately

75%. The maximum acceptable optical signal is -5 dBm so that the dynamic range is 26 dB.

7. Conclusions

An SDH transmission system developed for the transport of ATM cells has been described in this paper. This system is an integral part of the ATM testbed operating at CSELT Laboratories in Turin.

References

[1] M. De Prycker, *Asynchronous Transfer Mode — Solutions for Broadband ISDN*, 2nd ed., Ellis Harwood, 1994.

[2] J. Legras, "ATM (Asynchronous Transfer Mode): Overview, synergy with SDH, and deployment perspective," in *SONET/SDH: A Sourcebook of Synchronous Networking*, C. A. Siller, Jr. and M. Shafi, Eds. New York: IEEE Press, 1996.

[3] ITU-T Recommendation I.432, "B-ISDN User Network Interface — Physical Layer Specification."

[4] ITU-T Recommendation G.709, "Synchronous Multiplexing Structure." ITU-T Recommendation G.832, "Transport of SDH Elements on PDH Networks: Frame and Multiplexing Structures." ITU-T Recommendation G.804, "ATM Cell Mapping into Plesiochronous Digital Hierarchy."

[5] ETSI — ETS 300 299, "Network Aspects: Cell Based User-Network Access. Physical Layer Interfaces for B-ISDN Applications."

[6] ETSI — ETS 300 300, "Network Aspects: SDH Based User-Network Access. Physical Layer Interfaces for B-ISDN Applications."

[7] ETSI — ETS 300 147, "Transmission and Multiplexing: Synchronous Digital Hierarchy (SDH) Multiplexing Structure."

[8] ETSI — ETS 300 337, "Transmission and Multiplexing. Generic Frame Structures for the Transport of Various Signals (Including Asynchronous Transfer Mode (ATM) Cells) at the CCITT Recommendation G.702 Hierarchical Rates of 2048 kbit/s, 34368 kbit/s, 139264 kbit/s."

[9] ITU-T Recommendations on SDH: G.707, G.708, G.709, G.781, G.782, G.783, G.784, G.957, G.958.

[10] Technical Staff of CSELT, "*Fiber Optic Communications Handbook*," TAB Professional and Reference Books, 1991.

Technologies Towards Broadband ISDN

KAZUO MURANO, KOSO MURAKAMI,
EISUKE IWABUCHI, TOSHIO KATSUKI,
AND HIROSHI OGASAWARA

WAVES of strong driving forces are evident today that point towards the implementation of Broadband Integrated Services Digital Network (B-ISDN) in the early 1990s. In particular, the demand for a high-quality, high-speed intelligent network that enables free exchange of information with various media options is gaining great momentum. High-quality video and high-speed data transmission seem to be of immediate interest to end users, but much more diversified and sophisticated services are expected to emerge as broadband capabilities become widespread. The provisioning of these services also offers new business opportunities for service providers, network operators, manufacturers, and other types of industry involved in the field of telecommunications.

Standards that enable construction of a worldwide integrated network are currently being actively worked out in various international standards bodies, such as the International Consultative Committee for Telephone and Telegraph (CCITT), the International Standards Organization (ISO), and the Institute of Electrical and Electronics Engineers (IEEE). These outcomes provide a positive base on which related organizations can build joint efforts towards early implementation of a new infrastructure that would meet the needs of the 21st century.

In terms of technologies, rapid progress in the area of fiber optics, electronics, and information processing now give a sound basis for economically feasible implementations of B-ISDN. In particular, emerging new transport technologies such as Synchronous Digital Hierarchy (SDH) and Asynchronous Transfer Mode (ATM) [1] provides efficient and flexible means to carry diverse information.

It is vital at this point to form a sound strategy in developing these new technologies and deploying them in the field, so that smooth migration from the existing network to the B-ISDN is accelerated. Three major steps in this context are: to expand fiber networks into the subscriber loop area to provide broadband capabilities everywhere; to construct a universal digital network that facilitates smooth evolution from the existing network to the broadband network of the future through deployment of SDH transmission systems; and to integrate both services and network components through introduction of ATM technologies. These three steps should not be discrete steps, but rather one synergistic, strategic, and coordinated effort, which can perhaps be compared to a "triple jump."

This article is intended to describe development efforts now being carried out at Fujitsu and Fujitsu Laboratories following this line, and to indicate the impact on network construction and service offerings.

ARCHITECTURAL CONSIDERATIONS

In view of the synergistic deployment of user-to-user fiber network and broadband capabilities, laying out a flexible and expandable network architecture is of paramount importance.

Figure 1 describes two primary architectures that share the properties of economic realization and smooth growth. The subscriber loops are divided into feeder loops and distribution loops. In this way, the most cost-sensitive portion (that is, links to the individual users) can be minimized. These two architectures have distinctly different characteristics; the particular choice of one depends on geographical conditions, service needs, and reliability requirements.

Figure 1(a) shows the double-star configuration. In the feeder loop, high-speed SDH transmission systems are used to link Remote Terminals (RTs) to the Central Office (CO). RTs may be equipped with either simple multiplexing functions or line concentration functions for more efficient utilization of the feeder bandwidth. This system can be regarded as the most logical upgrade of existing loop configuration utilizing carrier systems.

Figure 1(b) is the ring-star configuration, where the feeder loop is constructed by a duplicated fiber ring, with RTs inserted along the ring. The fiber ring adopts SDH transmission systems with various bit rates, depending on the traffic requirements. An add-drop function is incorporated in the RTs to provide flexible bandwidth allocation along the ring. Traffic concentration functions or ring access protocols may also be implemented at the RTs for some applications. The duplicated feature of the ring, together with the bypass and loopback functions of the RTs, offer failure-survivable connection.

The distribution loops, in both cases, are constructed in star configuration using Single-Mode Fiber (SMF) to connect each user. Here, a separate fiber may be used for each direction of transmission, or a single fiber with a separate wavelength or other bidirectional multiplexing schemes may be used for each direction of transmission.

For a great many users, simple low-speed bidirectional digital links may suffice to provide Plain Old Telephone Service (POTS) and low-speed data transmission, at least initially. For

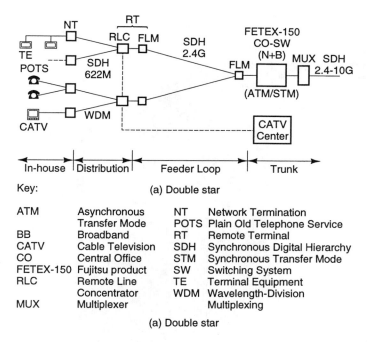

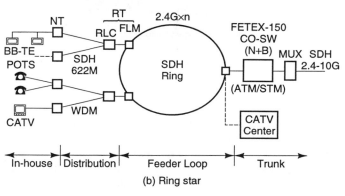

Fig. 1. Network architectures.

Key:

ATM	Asynchronous Transfer Mode	NT	Network Termination
BB	Broadband	POTS	Plain Old Telephone Service
CATV	Cable Television	RT	Remote Terminal
CO	Central Office	SDH	Synchronous Digital Hierarchy
FETEX-150	Fujitsu product	STM	Synchronous Transfer Mode
RLC	Remote Line Concentrator	SW	Switching System
		TE	Terminal Equipment
MUX	Multiplexer	WDM	Wavelength-Division Multiplexing

large business users, much higher capacity may be required, in which case the use of SDH can be justified from the outset. Video distribution capability may also be added on this fiber, either in analog modulated form coupled with use of Wavelength Division Multiplexing (WDM) technique or in digitally multiplexed form integrated with other services.

The architectures described here allow independent development of mass transport technologies in the feeder loops and service-related technologies to be applied in the distribution loops. The following sections describe the development of three key technological steps in implementing the actual network based on these architectures.

EXTENDING FIBERS TO THE USER

The "hop" in the "triple jump" towards B-ISDN is to extend fibers to every user's doorstep and provide broadband connectivity universally. Since a large portion of today's users are content with traditional telephone service and some low-speed data transmission capability for the time being, it is important to pro-

vide these services over fiber in economically feasible terms.

Adopting the architectures of Fig. 1, implementation of the distribution loop, or so-called "last-mile connection," becomes critical. With the rapid decline in the cost of optical fibers and electronics, economically justifiable implementation is now becoming a reality.

Figure 2 describes an example of the transmission arrangement to be applied to the distribution fiber loop. WDM with two wavelengths is used to provide multiple services on a single SMF fiber. A bidirectional digital link is realized with Time Compression Multiplexing (TCM), using a wavelength of 1.3 μm.

The drop bit rate at the Network Termination (NT) is provisionable from 1 B-channel (64 kb/s channel) up to 30 B-channels. Conversion to the standard POTS interface is provided in an NT with an additional adaptor for the RS232C data interface. A 1.55 μm wavelength is reserved for the optional addition of other services (for example, video distribution services).

Battery back-up is provided in the NT to allow eight hours of operation of basic service, in case of AC power failure. Each RT is designed to serve up to 200 lines. The maximum distance between the RT and NT is assumed to be about three miles.

Figure 3 shows an example of some physical implementation images of the NT and RT. Although the above-described configuration assumes direct fiber connection to each home (so-called fiber-to-the-home), slight modification provides fiber-to-the building, fiber-to-the-curb or pedestal arrangements, which have been recognized as economical for initial deployment of narrowband services.

DEVELOPMENT OF SDH SYSTEMS

Standards were established by CCITT in 1988 to provide the first truly worldwide unique SDH [2], [3]. Transmission systems based on SDH have been introduced to form a base for a universal digital fiber network and facilitate smooth migration from the existing network to the B-ISDN.

Exploiting the SDH's synchronous properties and built-in

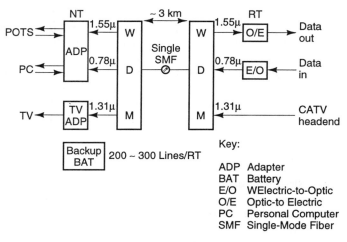

Key:

ADP	Adapter
BAT	Battery
E/O	WElectric-to-Optic
O/E	Optic-to Electric
PC	Personal Computer
SMF	Single-Mode Fiber

Fig. 2. Last-mile distribution.

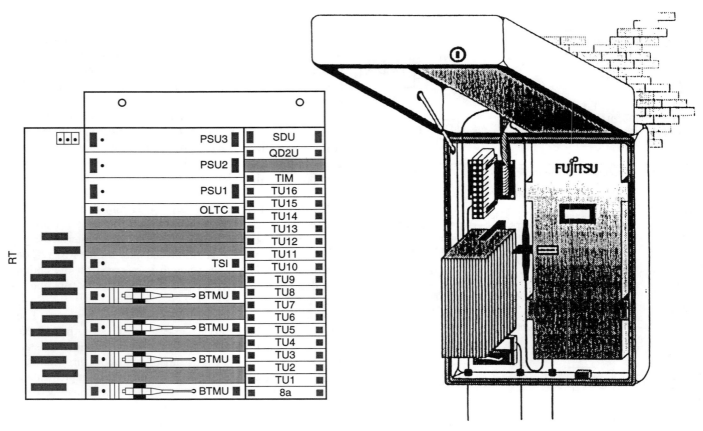

Fig. 3. Physical image.

modularity, flexible network management and self-healing measures can be implemented through the use of cross-connecting and add-drop multiplexing (ADM) functions. Easy upgradability to higher-order systems and economical gains by means of skip multiplexing can also be obtained. Figure 4 shows how SDH systems are expected to gradually penetrate the entire network.

The initial deployment of SDH was introduced into the trunk and feeder loop network in 1990, primarily using STM-1 (155.52 Mb/s) and STM-4(622.08 Mb/s) systems. Figure 5 shows an example of an STM-1 transmission system implementation (FLM 150 ADM). This system maps add-drop 2.048 Mb/s tributaries onto STM-1. A maximum 63 2.048 Mb/s signals can be housed in a single shelf. This system is also equipped with remote maintenance and provisioning capabilities to facilitate unmanned operation.

The SDH system will penetrate the distribution network, thus providing direct user access to broadband channels of about 140 Mb/s, which is the payload portion of the STM-1 signal. An STM-16 (2.48832 Gb/s) system was introduced in 1991 and continues to be deployed widely to support the transport of increasing broadband traffic.

Figure 6 shows a fiber ring configuration based on an STM-16 system with add-drop capability (FLM 2400 ADM). This system serves as the core ring for the network described in Fig. 1(b), but can also be used for a flexible and failure-survivable inter-office link.

As shown in Fig. 4, SDH systems are expected to eventu-

ally penetrate in-house networks. At this stage, terminals will have direct SDH interfaces and will be able to exploit the full broadband and intelligent capabilities of the network. The major optical transport system for the trunk network in this phase is expected to be the 10 Gb/s or a higher bit-rate system.

INTEGRATION OF SERVICES AND NETWORK BY ATM

The third step, or the final "jump," towards the B-ISDN is the network and service integration using the ATM.

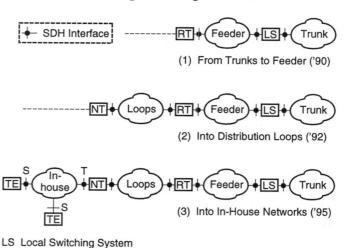

LS Local Switching System

Fig. 4. SDH penetration.

Fig. 5. SDH (SONET) MUX implementation.

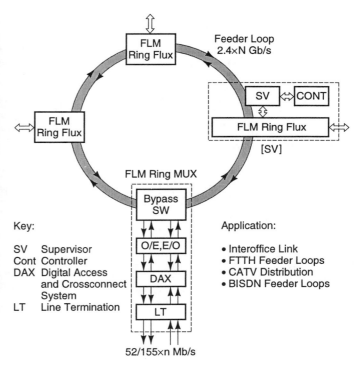

Fig. 6. SDH ring network.

As stated in the CCITT recommendations, B-ISDN is an all-purpose digital network, supporting a wide range of audio, video, and data applications in the same network. The network capabilities include support for variable-bit-rate information transfer from kb/s to Gb/s, both bursty and continuous traffic, and both dialog and broadcast services.

To realize such a network, it should be configured on the basis of a unified idea, which will require a unified architecture for both transmission and switching systems, and also for network management systems.

ATM cells will be packed into the SDH frame as shown in Fig. 7, so that the ATM network can be constructed utilizing the embedded SDH network, deployed earlier in the "step" or second stage described in the previous section. This method will expedite smooth migration from a broadband transport network to a broadband integrated services network.

Fujitsu is now developing technologies for constructing B-ISDN in full conformance to the forthcoming international standards to be recommended by the CCITT.

Figure 8 describes a "standard map" that relates CCITT (now identified as ITU-T) Recommendations, established in 1993 [4], [5], to the major functionalities of the ATM network.

Based on the network architecture of Fig. 1, with SDH transmission systems deployed as described previously, the following features are added:

- Hybrid switching functions at CO and RT to support both ATM and STM switched services

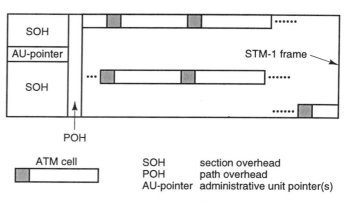

Fig. 7. An example of mapping ATM cells into an SDH frame (reference: CCITT COM XVIII-R 105-E, July 1992).

- Flexible optical bus wiring in the customer premises area based on the Distributed Queue Dual Bus (DQDB) scheme or the proprietary optical passive bus scheme [6].

An RT serves as a concentrator and distributor to make efficient use of the feeder loop. The speed of the user-network interface is either STM-1 level or STM-4 level.

Regarding the switching system to be placed at the CO, the evolution of strategies from the narrowband digital switching system is an important issue. Figure 7 shows a system configuration using the Fujitsu FETEX-150 digital switching system as a platform for the next generation of broadband networks. The system adopts a multiprocessor configuration, where call processors, database processors, and management processors are connected by a processor ring bus. The broadband switch module, which is composed of both ATM and STM switching

333

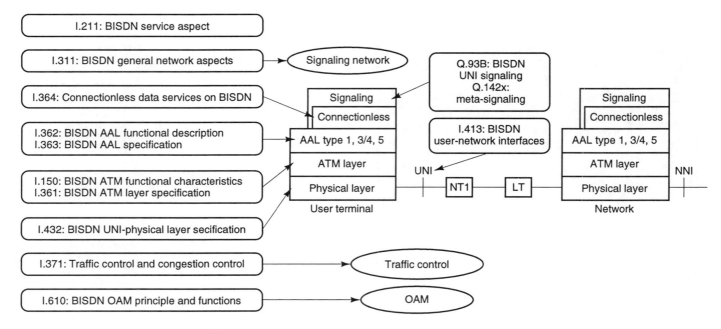

Fig. 8.   ITU recommendations relating to B-ISDN (modified from [5]).

fabrics and a broadband processor, can be added easily to the basic narrowband system by connecting processors via the ring bus.

By adopting such an architecture the broadband traffic can utilize many of the software resources available for the narrowband system. Moreover, the narrowband traffic carried by ATM and destined for the existing narrowband network is easily routed through the internal link via a broadband-narrowband converter.

Based on the above configuration, we developed the Multistage Self-Routing (MSSR) architecture [7] as a promising ATM switching fabric. The MSSR switching is constructed by connecting Self-Routing switch Modules (SRMs) in a three-stage

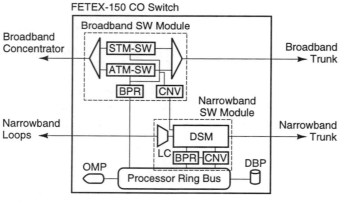

Key:

CPR   Call Processor
BPR   Broadband Processor
CNV   Broadband-to
         Narrowband Converter
OMP Operation and Maintenance
         Processor

DSM   Digital Switch Module
DBP   Database Processor
LC   Line Concentrator

Fig. 9.   B-ISDN switching system.

configuration, as shown in Fig. 8. Multiple routes between the first-stage SRM and the third-stage SRM enable the traffic to be routed efficiently with minimum switching delay.

A Virtual Channel Identifier (VCI) Converter (VCC) is placed at each input highway. This VCC identifies a cell by its VCI, replaces the VCI at the input highway with the VCI of the destination output highway, and generates a tag containing the information about the path within the switching network. Each SRM switches cells to the outlet according to the tags without external software control. MSSR offers several advantages, summarized below:

• Variable-bit-rate switching based on the self-routing principle
• Distributed structure of queuing buffers, simplifying the self-routing algorithm
• Suppression of delays, minimization of delay variations, easy expansion of capacity and functions, and high reliability due to redundant links based on the multistage link configuration
• Modular configuration consisting of small self-routing modules
• Suitability for implementation using Large-Scale Integration (LSI).

By using combined Bipolar (Bi) and Complementary Metal Oxide Semiconductor (CMOS) LSI technology, 1.2-Gb/s highway throughput has already been attained using MSSR, which enables switching capacity of 512 channels, each at 155 Mb/s.

Figure 9 shows the first prototype system for B-ISDN, which was completed in 1988. This system comprises two major subsystems: an ATM-based broadband switching system adopting the MSSR configuration, and an optical fiber subscriber loop network, called "shuttle bus" [8]. The loop has adopted 1.8-Gb/s synchronous transmission, carrying 12 channels of information

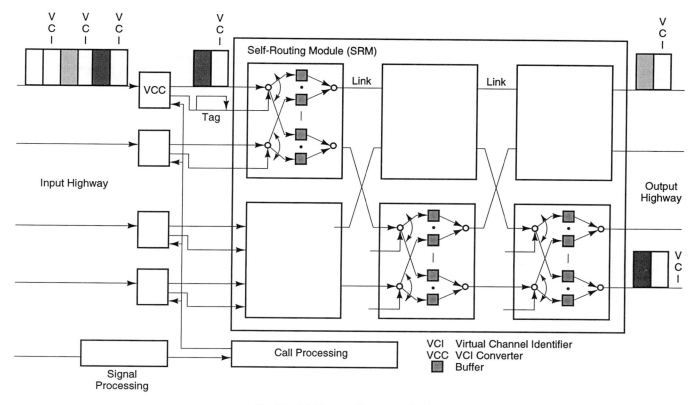

Fig. 10.   Multistage self-routing switching.

at 150 Mb/s. A multimedia terminal, where video, voice, and data are statistically multiplexed using ATM, was implemented to demonstrate the flexible transport characteristics. Digital CATV distribution using STM was also implemented to indicate the full capabilities of an ATM-STM hybrid configuration.

Before these technologies are commercialized, operation and management will be vital. Although not fully implemented in the experimental system of Fig. 10, future commercial broadband systems are expected to be equipped with extensive operation and service provisioning features. A most promising approach is adopting the Information Networking Architecture

TABLE 1. AN OVERVIEW OF B-ISDN SERVICE TIMELINE

- Circa 1994:   Connectionless-type data communication services
  - Frame relay data services ($\sim$ 1.5 Mb/s)
  - Switched multimegabit data services ($\sim$ 45 Mb/s)
  - Workstations LAN ($\sim$ 156 Mb/s)

- Circa 1995:   Permanent Virtual Channel-based broadband services
  - Virtual private network
  - Video on demand
  - CATV distribution
  - Remote education system

- Circa 1996:   Switched Virtual Channel-based broadband services
  - Video telephony
  - Multimedia information retrieval

- Circa 1998:   Multiparty/multiconnection-based broadband services
  - Desktop multimedia conference
  - Remote medical diagnosis system

(INA) [9]. Here, the management systems are implemented based on an object-oriented and distributed-computational framework, allowing for a flexible and reliable modular architecture.

The development of a B-ISDN system should be carried out in a timely manner, keeping pace with the evolution of networks and services. It is important to estimate the expected initial and future service requirements.

Initial service started in 1992 with some field trials. The major goal here was to demonstrate the merits of ATM and its broadband capabilities to business customers as well as residential customers. Commercial services started in 1994, with the first application being high-speed data transport. A wide range of services, including new applications such as Video-on-Demand and Distance Learning, are expected to burgeon from around 1995. Table 1 lists examples of services envisioned and a probable time range for the break-in.

CONCLUSIONS

This paper described Fujitsu's "triple jump" approach in developing technologies towards implementing B-ISDN. Although many efforts are still needed to perfect the technology, define worldwide unique standards, and develop broad user acceptance, the rapid progress over the last few years provides a very positive outlook for initial B-ISDN implementations in the 1990s.

ACKNOWLEDGMENTS

The authors would like to express their sincere appreciation to

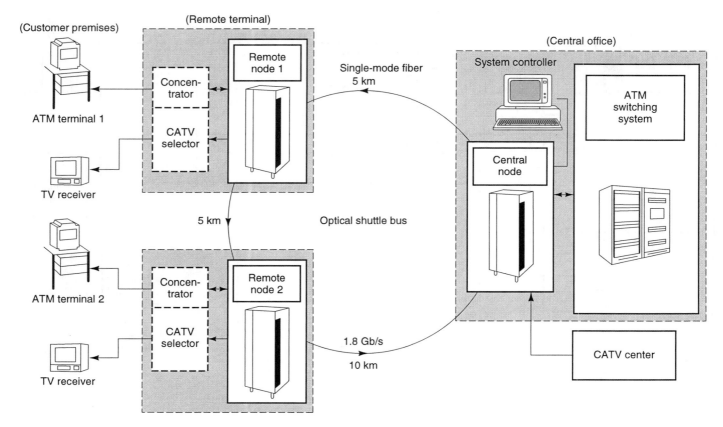

Fig. 11. Experimental B-ISDN system.

H. Takanashi and R. Yatsuboshi of Fujitsu Laboratories Ltd., I. Fudemoto, A. Moridera, and Y. Mochida of Fujitsu Limited, and many other colleagues without whose cooperation this project could not have been accomplished.

References

[1] S. E. Minzer, "Broadband ISDN and Asynchronous Transfer Mode (ATM)," *IEEE Commun. Mag.*, vol. 27, no. 9, pp. 17–24, Nov. 1989.

[2] CCITT Recommendation G.708, "Network Node Interface for the Synchronous Digital Hierarchy," CCITT Blue Book, Vol. III, Fascicle III.4, pp. 109–121, Nov. 1988.

[3] CCITT Recommendation G.709, "Synchronous Multiplexing Structure," CCITT Blue Book, Vol. III, Fascicle III.4, pp. 121–174, Nov. 1988.

[4] ITU-TS Report COM13-R1, "Report of Geneva Meeting 5–16, July 1993)," July 1993.

[5] K. Miyake, "Standardization aspect on B-ISDN," *Trans. IEICE*, vol. J76-B1, Nov. 1993 (in Japanese).

[6] K. Iguchi, S. Amemiya, T. Soejima, and K. Murano, "Optical Passive Bus for Broadband User-Network Interface," *Proc. ISSLS '88*, pp. 235–239, Sept. 1988.

[7] Y. Kato, T. Shimoe, K. Hajikano, and K. Murakami, "Experimental Broadband ATM Switching System," *Proc. GLOBECOM '88*, pp. 1288–1292, Nov. 1988.

[8] N. Fujimoto, T. Ishihara, A. Taniguchi, H. Yamashita, and K. Yamaguchi, "Experimental Broadband Drop/Insert/Cross-Connect System: 1.8 Gb/s Optical Shuttle Bus," *Proc. GLOBECOM '88*, pp. 954–959, Nov. 1988.

[9] G. J. Handler, "Network of the future," in *Proc. Int'l. Switching Symp. '92*, vol. 2, Oct. 1992, B5.2.

Optical Fiber Access - Perspectives Toward the 21st Century

New fiber systems now emerging are expected to provide the basis for large scale deployment of fiber to business and residential customers during the '90s and beyond.

Andy Cook and Jeff Stern

The vision of an all-fiber telecommunications network delivering an ever increasing variety of services to both business and residential customers has been the subject of much speculation for more than a decade. However, although optical fibers are now used routinely for longer haul transmission and for connections to large business customers; fiber to provide the final link to the generality of customers has proved to be a tougher challenge than originally expected.

The extent to which fiber will penetrate the network depends not only on technical capabilities and economic performance, but also on regulation, competition, and the types of service that customers will actually want. In this article, the interplay between these complex factors is discussed, and potential scenarios are described. An endpoint vision of a future broadband infrastructure is described in order to illustrate the major new service delivery and operational benefits, which can flow from the revolutionary new technologies that can now be unleashed.

Current and Near Term System Deployments

The key barrier to widescale deployment of fiber in the local loop is cost. Early ideas for the introduction of fiber (in the late '70s/early '80s) were based on the assumption that a widescale introduction of broadband services — such as cable TV, pay TV, dialup video services, high definition television, etc. — would occur. These would then justify the extra cost of the optical fiber over the copper pair. In reality, however, regulatory restrictions and uncertain demand for new entertainment TV have considerably slowed the pace at which fiber has been deployed.

Most of the fiber deployed for access purposes to date is in the feeder parts of the network. For example, Fig. 1 illustrates how cable companies use feeder fibers to deliver AM-VSB modulated TV channels to nodes serving, typically, 2000 customers. This approach considerably reduces the large numbers of line amplifiers required on earlier all coaxial cable systems. Similarly, Fig. 2 illustrates how

ANDY COOK is with the Access Networks Division at British Telecom Labs.

JEFF STERN is manager of Fiber Access Systems at British Telecom Labs.

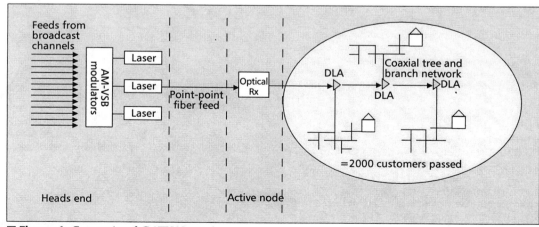

Figure 1. *Conventional CATV Network.*

Reprinted from *IEEE Communi. Maga.*, vol. 32, no. 2, pp. 78–86, Feb. 1994.

the regional Bell operating companies (RBOCs) in the United States use fiber feeders to deliver telephony and other telecommunications traffic to remote electronic (RE) nodes serving groups of customers that are distant from the Central Office.

These feeder links are economically viable because costs are shared over considerable numbers of customers. For links closer to the customer however, costs become much more of a problem and, to this point, fiber has only been routinely deployed in the "local loop," or the last mile ($\approx$ 2 km), for large business customers with high traffic (telephony and/or data) requirements. A network deployment of this kind is illustrated in Fig. 3.

Fiber-in-the-Loop Systems

System Options — During the '80s and the early '90s, continuing uncertainties related to the market for broadband services led telecommunications companies to review priorities for technological development in the local loop. A particular concern of the telcos was to find ways of minimizing the installation of more copper pairs in their networks; a technology that will rapidly become obsolete when the broadband revolution eventually arrives.

Much of the present growth in telecommunications networks, requiring new copper pairs, is for the business sector. Here, the demand is for higher-quality, more responsive networks for the delivery of relatively conventional telephony and data services. In particular, customers are requesting the delivery of tailored packages of services (telephony, private circuits, switched data, etc.) that can be dynamically varied to keep pace with their changing needs — preferably without the need for formal intervention by the telephone company.

These demands have led to a new breed of fiber-in-the-loop (FITL) optical systems, aimed at more flexible delivery of existing narrowband services to small-to-medium business and residential customers. However, these must also be capable of subsequently being upgraded to supply broadband services when the market demand actually arises.

The architectures adopted for these systems are based on double star topologies which are either active, containing intermediate electronics — Active Double Star (ADS); or passive, containing optical splitters — Passive Optical Network (PON). Essentially these systems allow the cost of the network between the exchange and the active or passive splitting node to be shared amongst a number of customers.

The main advantage of the active approach is that the final link to the customer is dedicated, or shared, amongst a small number of customers, and can make use of low-bandwidth, low-performance optical devices. On the other hand, there are problems with locating, powering, and operating remote electronics in the field and with the high upfront costs involved in building such a network.

The key advantage of the PON approach is that initial network costs are particularly low, with the majority of the cost being deferred until revenue-earning customers are actually connected. On the down side, the customer equipment requires higher bandwidth and better performance optical devices, and a multiple access protocol must be implemented to ensure that the returning bit stream from each customer is appro-

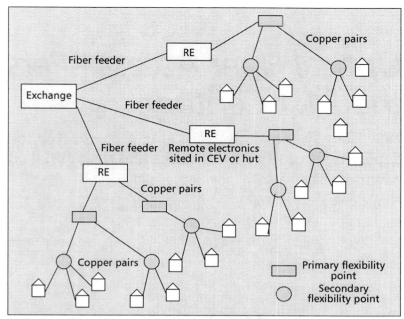

Figure 2. *Use of fiber to feed remote electronics sites in U.S. network.*

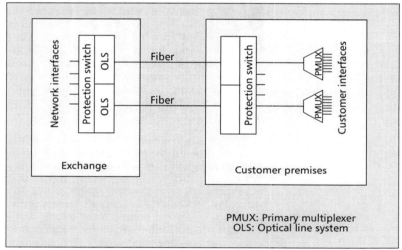

PMUX: Primary multiplexer
OLS: Optical line system

Figure 3. *Conventional delivery technique for large businesses.*

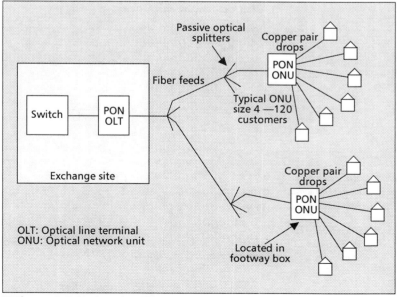

OLT: Optical line terminal
ONU: Optical network unit

Figure 4. *PON-based FTTC system.*

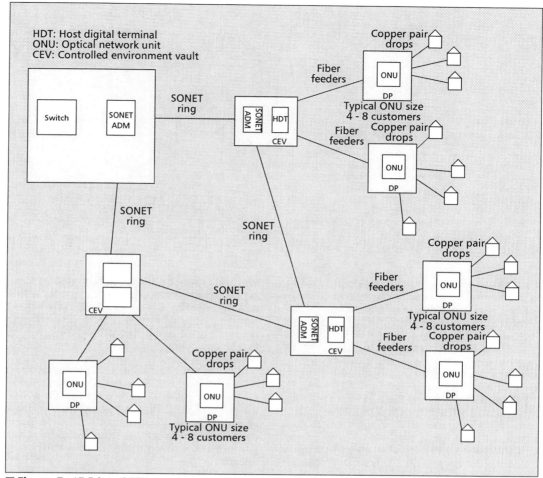

■ Figure 5. *ADS-based FTTC system.*

priately synchronized at the exchange. Privacy and integrity of information must also be safeguarded (since the same signal is broadcast to all locations).

For both ADS and PON systems there are still limitations on the type of customer that can be economically served. Thus it will not be possible in the foreseeable future to take fiber directly to the great mass of one-line telephony customers — such large scale deployment can only be justified via demand for future broadband services (discussed later in this article). For these very small customers, variants to both architectures have been developed in which fiber is extended down to a small active node at the curb, with conventional copper pairs being used for the final customer drops. Figure 4 shows a PON fiber-to-the-curb (FTTC) system. Figure 5 shows the direction in which ADS system are migrating, with the use of SONET/SDH rings to feed the individual CEVs. Figure 6, on the other hand, shows a fiber-to-the-building (FTTB) approach implemented using a PON system. This also shows more details of the constituent parts of the PON system. At the exchange site, the traffic from four PONs is groomed into a set of switch interfaces that consist of a number of tributary units for each service. Some of these tributary units feed the PSTN switch, and some the leased lines network. Each of the PON headends feeds its own splitter networks, and a number of buildings are fed from each PON. In the customer's premises, the traffic is demultiplexed by a core system, which in turn feeds a set of service units for interface to the customer's internal networks.

In an FTTC scenario, essentially, the electronics that were located within the customer's premises for the FTTB system are re-located to an active node in a footway box. Copper pairs are then used for the final drop to the customers.

Deployment Issues and Telco Pans

Both ADS and PON systems have now been extensively tested in technology feasibility trials around the world and operational pilots and initial rollout deployments are currently underway.

United States — The drive has been to develop FTTC systems (ADS and PON) for general deployment in residential areas. Given the relatively low density of U.S. housing developments, this ideally requires very small active nodes to be deployed at the curb each serving only 4 living units (LUs). The problems of engineering, installing, powering and maintaining such large numbers of small nodes, and at the same time achieving acceptable whole life costs, are particularly difficult. These issues have slowed the pace at which operational rollout has occurred up to this point.

Europe — In Europe the drive has centred more on the deployment of fiber systems directly to buildings — thus avoiding the need for active nodes in the network.

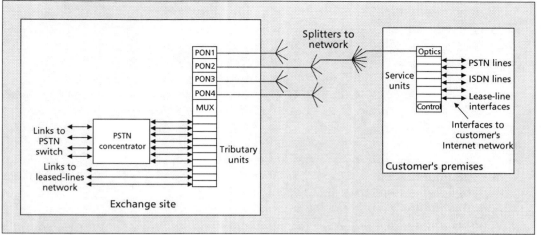

Figure 6. *PON FTTB system.*

Germany —In Germany the need to build a whole new telecommunications infrastructure in former East Germany has created a vast "greenfield" site for the installation of new technology. The lack of existing viable copper pair systems, the number of large apartment buildings and the need to minimize the number of new digital switch nodes [1] are all factors that have helped Deutsche Bundespost Telekom (DBPT) adopt a FTTB approach for narrowband services. DBPT has now placed contracts for the installation of fiber to some 720,000 customers for the period 1993-1994.

PON systems have particular cost advantages for delivering services to relatively small customers (e.g., down to four POTS lines) by direct fiber entry, and, for this reason, have been preferred by both DBPT and BT in their fiber installation programs.

For BT, the drive has been to deploy fiber directly to the small-medium business sector in order to achieve network cost savings and improved responsiveness and quality in service delivery. Deployment plans are now targeted on extending fiber to all such businesses and orders have been placed with a first supplier for the first tranche of rollout PON systems. FTTC PON systems will also be installed, but only in limited situations (typically, new growth housing estates) where they can be economically justified. In these situations the active nodes will be relatively large (e.g., 60 lines).

BT studies have shown that the PON approach has great benefits in minimizing the proportion of the per line cost incurred in deploying equipment at the exchange and in the fiber infrastructure down to close to the customer. This permits capital expenditure to be much more closely linked to actual connection of revenue earning customers than is possible with the copper pair network, thus leading to major benefits in improved financial efficiency and faster service provision. Additionally the PON allows re-allocation of capacity in line with changing demand and the grooming and consolidation of traffic to allow different types of switch port to be efficiently utilized. Examples of the advantages of PON deployment are illustrated in Fig. 7.

Because of their common interest in PON systems, BT and DBPT have converged their specification requirements wherever possible in order to provide a basis for a larger and more focused market for equipment suppliers. The approach adopted has been to lay down a specification framework clarifying all detail necessary from an operational point of view but allowing manufacturers scope to offer competing products with differing detailed designs.

This has led to the development of Joint BT/DBPT Technical Advisory documents describing a generic PON architecture based on a common "core" system. This core system supports a range of sizes/types of optical network units (ONUs) to permit a broad variety of customers to be served [2]. It is recognized that service and network interfaces may differ for each company and these are left to more detailed, company-specific specifications. Premature standardization of items such as the PON line protocol, internal functionality and design, and backplane interfaces has been carefully avoided, although considerable attention has been given to the optical PON interface to ensure manufacturer's equipment can operate satisfactorily over PONs installed by the Network Operator.

The synergy between PON systems and SDH is expected to grow. For some business customers it will be particularly important to have excellent visibility of the performance and status of their circuits. The use of SDH as a delivery mechanism is seen as a rapid route to achieving this goal, and would allow the end-to-end management of 2-Mb/s leased lines. Currently, this would have to be provided by the delivery of an entire STM-1 (155 Mb/s) to the customer. However, if the PON OLT were provided with an SDH interface, VC12 virtual containers could be transported over the PON and terminated at the customer's ONU. This will allow the economic provision of fully-managed 2-Mb/s private circuits to much smaller customers.

Japan — In Japan, aggressive plans have been announced for the creation of a complete nationwide broadband infrastructure between 1995 and 2015. NTT has point-to-point and PON systems for a range of different deployment scenarios and customer types [3]. Avoidance of active nodes in the network is likely to be a major factor in Japan because of severe space limitations in urban situations.

PON systems have particular cost advantages for delivering services to relatively small customers (e.g., down to four POTS lines) by direct fiber entry.

340

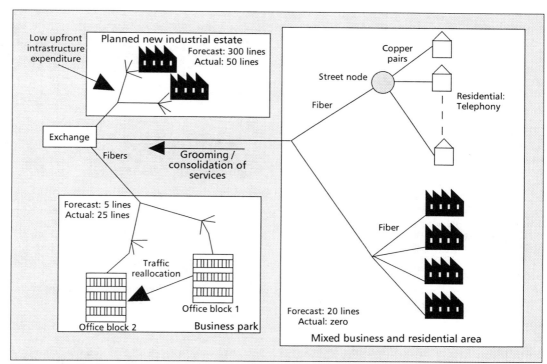

■ **Figure 7**. *Deployment scenarios and features for PON systems.*

Cable TV and Telephony Systems

Figure 8 shows the basic architecture of current fiber-coax/copper pair networks, such as those now being installed by the cable companies and others, such as DBPT, to deliver cable TV and telephony/narrowband services. More details of the constituent parts of the CATV network are shown in Fig. 1. Telephony and narrowband services, where provided, are currently carried over a separate sub-network—sharing occurs only at the cable/duct and active node infrastructure level. ADS or PON systems are used to convey the digital multiplex of narrowband signals to an active node close to the customer (typically 500 customers) with copper pairs used for the final connections.

Future System Options to the Year 2000

Broadband Multimedia Services

Worldwide, there are a number of changes occurring which are serving to blur some of the boundaries between services provided by CATV operators and telcos. Within the United States, there is much discussion of video dial-tone by RBOCs, and cable operators are making plays to be allow to carry narrowband traffic. In the United Kingdom, the rapid expansion of CATV networks has been accompanied by the provision of telephony, and, increasingly, of other narrowband services over these same networks by the new cable operators. In Japan, plans have been in place for a number of years for the installation of a national fiber infrastructure capable of delivery of both narrowband and broadband services.

In the past it has been the case that telephony has formed the vast majority of an operator's service portfolio, particularly to the residential sector. How-ever, this may not remain the situation in the future. The onward march of technology, with the increasing penetration of powerful computers, capable of supporting multimedia applications, paves the way for a demand for new broadband services to emerge. The use of teleworking may be expected to increase, and with it, the demand for high bandwidth services. Advanced teleworking will require remote interconnect to the office LAN, and this will need to be at high speed in order to satisfy the requirements of ever more powerful applications. Access to other remote, perhaps multimedia, databases over the future network will also be a bandwidth-intensive process. This increasing requirement for bandwidth in the residential sector will change the way in which we need to provide service to these areas. Many of the broadband services, such as remote LAN interconnect, and multimedia databases, will also be applicable to the business sector, and the boundaries between business services and residential services will become more blurred.

The major strides made over the last few years in cost reduction and quality improvement in the video coding area have made possible the transport of very large numbers of video channels over a single network. Three years ago, a 1 Gb/s network would have been capable of delivering up to around 20 broadcast TV channels with simple coding techniques. Now, however, approximately 250 channels of similar quality could be fitted into the same bandwidth. ISO standards have been drawn up by the Moving Picture Experts Group (MPEG) for these compression systems. The more recent of these, MPEG2, enables coding of "broadcast quality" video at around 4 Mb/s, although somewhat more bandwidth may be required for material that incorporates substantial amounts of rapid motion, such as sports programs. MPEG2 can also be extended for HDTV at rates in the 20-Mb/s region.

Fiber/Coax Systems

These systems derive from the fiber/coax CATV systems described previously, and are based on a continuation of the usage of AM-VSB carriers.

As demand for switched broadband services develops, the channel capacity of current systems almost immediately becomes inadequate. Assuming that the existing analog AM-VSB channels must be maintained on the network in order to avoid having to upgrade all the customers' CPE, it is necessary to add additional bandwidth to the network. By upgrading the amplifiers, the capacity could perhaps be extended to 850 MHz, allowing around an additional 30 AM-VSB channels to be carried. By using advanced modulation schemes such as 64 QAM (64 level quadrature amplitude modulation) on the AM-VSB carriers, a significant amount of digital data can be transported — typically around 30 Mb/s per carrier. If the video is digitally encoded using a compression scheme such as MPEG2 at around 4 Mb/s, this additional bandwidth is now able to support about an extra 240 channels of digital video.

Video-on-demand is a service which inherently has a much longer call hold-time than telephony, and therefore, at high penetrations, it is likely that even 240 additional channels will not provide adequate quality of service to a 2000-customer area. This implies that, as demand for this service increases, the network needs to be further split and each optical receiver may a much smaller area (as shown in Fig. 9) of perhaps a few hundred customers, thus giving the ability to offer approximately one switched channel per customer. This produces a network with much greater capacity than in Fig. 1, but where the level of equipment sharing has been significantly reduced, with an AM-VSB modulator, and a highly linear optical link being required per few hundred customers. This will have a significant impact upon the cost of the network.

In a new build environment, the starting point is different, and under those circumstances, it is

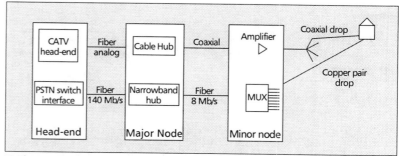

■ **Figure 8.** *CATV Network integrated with telephony.*

not fundamentally necessary to have a network that still carries analog AM-VSB channels. If an 850-MHz system is installed from day one, this gives a capacity of around 800 digital channels which, even after allocating perhaps 100 channels for conventional broadcast services, still leaves sufficient capacity to provide on-demand service to a substantial number of customers, with significantly better level of resource sharing than can be achieved in the upgraded CATV environment.

Current AM-VSB networks are being installed with relatively little in-built capability for providing upstream traffic, with the frequency band 5-30 MHz typically being allocated to traffic in this direction. In order for upstream traffic to be carried, appropriate amplifier modules need to be installed in the DLAs, along with an upstream optical link from the active node up to the head end.

If QAM is also used in this direction, the available bandwidth is sufficient to transport a total of at least 100 Mb/s of upstream traffic, which is enough to give a few Mb/s per customer at reasonable penetration. Where more upstream capacity is required, some could be made available by using the spectrum above the broadcast channels for additional upstream subcarriers. These subcarriers may have a somewhat lower capacity than those in the 5 to 30 MHz band due to reduced carrier to noise (C/N) in the higher parts of the spectrum that may be beyond the original specification of the network.

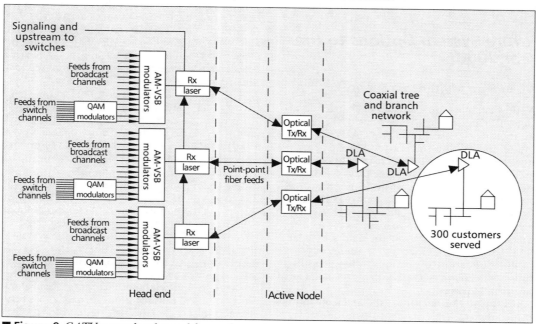

■ **Figure 9.** *CATV network enhanced for on-demand services.*

At the current time, CATV operators' telephony service is being carried over a separate network using fiber-fed, street-based remote multiplexers, with copper pair drops, in a similar way to the U.S. network shown in Fig. 2. However, systems are now emerging that enable telephony to be integrated with the broadband service, by carrying it as a TDM stream modulated onto a subcarrier on the broadband network [4].

FM Systems

An alternative to use of AM-VSB modulation for these networks is to use FM modulation. AM-VSB transmission requires the use of very high performance lasers in order to meet the stringent distortion requirements for this modulation scheme. FM on the other hand is much less sensitive to laser linearity and noise, and this considerably eases the requirements placed upon the device, and hence its cost. Also, FM systems require a much smaller C/N at the receiver, and hence are capable of supporting a very much better power budget than an AM system. This ability to support a greatly increased optical split makes it possible to deploy an FM system as a fiber-to-the-home (FTTH) network if this is desired, although hybrid fiber-coax solutions are also applicable.

The frequency plan of an FM system generally extends from 950 to 1700 MHz, which is compatible with the set-top boxes used within satellite TV systems. This enables the low-cost, mass market CPE produced for the satellite industry to be used in fixed networks. Just as the modulation schemes such as QPSK are being considered for squeezing multiple digital MPEG channels into a standard 27-MHz-wide satellite transponder bandwidth, so the same techniques can be applied to transporting MPEG channels over a fiber-based FM systems.

Digital Baseband Systems

Hybrid Solutions — These system options use a digital baseband rather than an analog subcarrier approach and could prove particularly attractive to telcos with their background in network digitization (switching and transmission links) to gain cost and quality of service benefits.

As an initial step into the broadband arena, telcos are likely to deploy asymmetric digital subscriber loop (ADSL) technology to provide a limited, primarily uni-directional broadband service at 1.5 Mb/s or 2 Mb/s from the local exchange site over the existing copper pair. Solutions using copper pair drops provide a very limited upstream capability, although the discrete multi-tone (DMT) modulation system for ADSL may be able to provide up to 384 kb/s over shorter drops. The use of a second ADSL system over a second copper pair in the reverse direction to obtain higher upstream bit-rates is not viable due to the high levels of near-end crosstalk (NEXT) that will be experienced.

As the bandwidths required by new broadband services increase, the next logical step is to deploy fiber a stage nearer to the customer, by taking it to the cabinet site — perhaps no more than 2 km from the customer. This is often referred to as ADSL3. Current proposals suggest that about 6 Mb/s is possible from the cabinet site over the existing copper infrastructure, although the technique has

the disadvantage that a significant piece of active electronics has now been located within the network infrastructure. This electronics will have costs associated with its installation, maintenance and powering, and these will increase both the capital and current account network costs.

The bit-rate/bandwidth trade-off over copper can be taken a stage further by using very high speed copper drops from the DP site, perhaps achieving 20 Mb/s over 200 m of drop wire, although there are EMC implications to be considered here. This deployment of fiber close to the customer is also broadly compatible with narrowband FTTC solutions. Although the bandwidth available under these circumstances has been significantly increased, the number of locations within the network which require powering has also increased enormously. In spite of the relatively straightforward electronics required at these DP sites, the vastly increased number of active sites will significantly increase the network powering and maintenance costs.

Another option to enhance the bandwidth available beyond that provided by ADSL3, is to replace the copper pair from the cabinet with a coaxial cable to the customer. What we have at this point is essentially the architecture used in CATV networks, although with digital baseband as the transmission method. The use of digital baseband as the modulation scheme is advantageous from the point of view of reducing interface conversion burdens when accessing the core network, and the relaxed requirements for optical device performance will significantly reduce the cost of the optics over the AM-VSB solution. If a tree-and-branch type architecture is chosen, it will be necessary to develop new amplifiers and splitters whose frequency response extends down to low frequencies, however, another option is to terminate the fiber closer to the customer — probably at the DP, and run a dedicated, passive coaxial cable to each customer, or on a very limited sharing basis.

Upstream transmission over the digital baseband coaxial network is possible, and is likely to use similar technology to that suggested for the AM-VSB system in its upper frequency allocation, although a more straightforward alternative may be to use two separate coaxial networks. If a single network is used, the upstream traffic would be likely to be carried to the head-end while still on its subcarrier(s) in order to minimize equipment complexity at the active node site. The overall result of this is that the upstream capacity of the baseband coaxial system is slightly worse than the AM-VSB system for an equivalent level of equipment complexity.

Fiber-to-the-Home — An alternative to the approach of having active nodes for a digital baseband system within the network is to pursue the FTTH concept. In contrast to the fiber/coax hybrid networks, an FTTH solution removes the need for active electronics within the network, and with a large portion of the network costs located within the optical network unit (ONU) equipment, it is possible to defer a significant portion of the network costs until such time as a customer actually requires service. This is much less the case with networks requiring active electronic nodes situated within the network which require a large portion of their costs to be committed up-front.

Achieving network fill in actual deployment of a PON system close to that for which the system was designed has always been critical to the economics of PONs. Where optical split ratios are small, there is a high statistical probability of the achieved fill being significantly different to that for which the network was designed. However, the use of optical amplifiers within PONs enables much larger splits to be achieved, thus reducing the effect of localized poor fill on the overall economics of the PON. The use of amplifiers can also allow the reach of PONs to be significantly extended while maintaining a high optical split. This technique could be used to allow a radical rationalization of the number of switch nodes required in a future broadband network when compared to the present narrowband switching infrastructure. This also enables video servers to be larger in scale and shared over a greater number of PONs without incurring high core transport costs. Additionally, the optical amplifier, which may be located at the current narrowband concentrator site, will be small enough to be located within a piece of street furniture. This gives the operator the key to unlock him from the requirement for a very large number of buildings in the future network.

Telcos are likely to install their new broadband networks as an overlay to their existing network, sharing the same ducts. There could be significant portions of the existing duct network which are too congested to allow coaxial cable to be installed, but into which fiber could be installed.

FTTH is well capable of upstream transmission, although the key to being able to do this economically is to be able to source optical devices at low cost. As already discussed in this article, a number of systems are already being deployed worldwide for narrowband services, and these systems could be used for a low penetration 2Mb/s bi-directional broadband requirement. Beyond this, new systems need to be developed, but a number of systems have been proposed, some particularly well suited to constant bit-rate traffic [5], and some having good capabilities in a bursty data environment [6, 7].

Shape of the Future Network

The shape of the future network in the late '90s and beyond will clearly be the result of a broad range of factors — technical, commercial and regulatory. It may well differ significantly from country to country and is very difficult to predict with any certainty. Nevertheless some broad trends can be discerned.

Telcos will continue to deploy fiber for narrowband services in order to position their networks for future broadband services, and to gain shorter term network cost reductions. Deployments are likely to emphasise FTTB situations (for both business and residential customers) with more restricted use of FTTC systems for new growth housing estates or when co-deployed with broadband delivery systems. Overall, narrowband drivers should lead to extensive deployment of fiber direct to small, medium and large businesses during the '90s but large scale penetration of the residential sector may only be achieved, without the assistance of broadband, in countries with much new growth, and/or large numbers of apartment blocks.

During the '90s the demands of business customers are likely to develop rapidly, particularly in the area of new broadband multimedia services. This will lead to a requirement for higher capacity systems to those initially installed for existing narrowband services. For very large customers, SDH based systems should cope for some time to come, although there is likely to be an increasing trend towards higher bit rates and an ATM (including switching) environment. For small-medium customers, higher capacity PONs are emerging, and also systems based around ATM transport.

In the residential area the demand for broadband services (Cable TV, Pay-per-View, VoD, near VoD, games etc.) is very unpredictable and may vary substantially from place to place. A key requirement will be to develop highly flexible architectures that can cope with a broad range of expected and unexpected service scenarios but without loading costs. Low initial penetrations are likely for many services and this will call for the networks that minimize upfront infrastructure costs and permit 'Just-in-Time'(JIT) provisioning at the customer end.

Analog subcarrier systems, given a new lease of life by the advent of video compression coding, will be the dominant approach in the short to medium term. Cable TV Companies (and their commercial partners) who have an installed base of AM-VSB/coax technology will have an obvious preference to upgrading this approach. In new build situations, however, the FM analogue sub-carrier approach has much to offer, providing synergy with digital satellite TV developments and permitting, if desired, a FTTH approach to be adopted.

Although analogue sub-carrier systems will initially prevail, a key issue is whether this approach will provide the best long term basis for low cost, integrated delivery of services to customers. In particular, next generation digital baseband systems of higher capacity — likely to be deployed first for business — will provide an alternative approach. Resolution of this issue will relate to network economics and in particular the ability of digital baseband solutions to achieve sufficient advantage over sub-carrier approaches to warrant their development and introduction for the residential sector. Key advantages however will relate to:

- Long haul transmission capability ($\approx$ 25 km or more using optical amplifiers), allowing access systems to connect directly back to very remote, highly shared broadband switch/video server nodes.
- Seamless integration of all services into a single bit-stream — avoiding the rigidities imposed by $\approx$ 30 Mb/s granularity associated with AM-VSB or FM channels.
- All digital electronics-permitting complex functions such as traffic concentration, traffic grooming/consolidation, multiplexing and demultiplexing of multiple services to be easily implemented, reducing equipment complexity at all levels in the network.
- Synergy with the introduction of a digital ATM broadband switching fabric capable of handling all types of service.

In the authors' opinion, there is little doubt that the digital baseband approach will prove more efficient in the longer term and eventually dominate

The shape of the future network in the late '90s and beyond will clearly be the result of a broad range of factors — technical, commercial, and regulatory.

ADSL-based solutions will be deployed by telcos to meet the emerging demand for broadband services, but will not provide a long-term solution.

once truly large scale, nationwide broadband networks become prevalent.

Choice of a FTTH, fiber/coax or fiber/copper pair architecture is in principle a separate question to the above. It seems unlikely however that copper pairs will last long as a broadband transmission medium — unless customer demand remains unexpectedly modest in demand for broadband information.

Competition between all-fiber and fiber/coax options could prove slower to resolve. However movement towards digital baseband solutions is likely to encourage the adoption of an all fiber approach in order to avoid the requirement for two, separate, simplex active networks to ensure fully adequate upstream capability for the long term. The latter requirement could increase in importance in the future in the context of creating a truly nationwide infrastructure capable of servicing advanced teleworking/business requirements, e.g., for desktop video publishing, CAD/CAM, very-high-speed data, LANinterconnect/remote access, etc.

Organizations needing to install a broadband pipe could reap other major advantages via the use of fiber including:
• The ability to install in restricted/clogged duct space.
• Maximized JIT provisioning.
• No active nodes to maintain.
• A truly future proof network.
Although today the power of the coax drop seems more than adequate we are laying an infrastructure for the next century and ignoring the potential demand for "futuristic" services that cannot easily be envisaged today — when they might turn out to be the very thing that customers want — would be short sighted in the extreme when the relevant "future proof" technology is at hand.

Conclusions

At the present time, fiber has only penetrated direct to the customer's premises for large business customers. However, systems are now being procured which enable today's protfolio of services to be economically delivered to the small and medium business sector, using FTTB techniques. FTTB systems for the business sector are expected to further evolve to enable wider bandwidths to be delivered, driven by the emergence of broadband services, and at the same time eliminating active nodes from the network.

In the residential sector, hybrid fiber-coax AM-VSB CATV networks will be upgraded, using digital compression technology, to carry a very large number of channels, and will also to have the ability to provide other interactive broadband services. ADSL-based solutions will be deployed by telcos to meet the emerging demand for broadband services, but will not provide a long-term solution. An alternative to these hybrid networks is a FTTH solution, which will enable the network operator to realise a number of benefits.

References

[1] W. Weippert, "The Evolution of the Access Network in Germany," in this issue.
[2] Joint Deutsche Bundespost Telekom/British Telecommunications plc Technical Advisory for a Fiber in the Loop TPON System. Issue 3, May1993.
[3] Y. Wakui, "The Fiber-Optic Subscriber Network in Japan," in this issue.
[4] M. Adams, The Personal Xchange. Cable Television Engineering, vol.15, no.6, Third Quarter 1992, pp. 78-82.
[5] D. W. Faulkner et al., "Novel Sampling Technique for Digital Video Demultiplexing, Descrambling and Channel Selection," Electronics Letters, vol. 26, no. 11, 26th May 1988.
[6] W. Verbiest, G. VanderPlas, and D. J. G. Mestdagh, "FITL and B-ISDN: A Marriage with a Future," IEEE Commun. Mag., vol. 30, no. 6, June1993, pp. 60-66.
[7] J. W. Ballance, P. H. Rogers, and M. F. Halls, "ATM Access Through a Passive Optical Network," Electronics Letters, vol. 26, no. 9, April 26, 1990.

Realising the Benefits of SDH Technology for the Delivery

of Services in the Access Network

Mark Compton and Simon Martin

BT Laboratories
Martlesham Heath, Ipswich, Suffolk IP5 7RE, UK.
Tel: +44 473 643476 Fax: +44 473 646445

ABSTRACT The deployment of Synchronous Digital Hierarchy (SDH) networks has gained increasing momentum world-wide as network operators seek to reap the benefits of low operating costs, high flexibility, and high quality of service. Commercially available SDH equipment is primarily based on deployment in core networks and is therefore unsuitable for locating at most customers premises. This paper describes how SDH technology and the advantages it offers may be extended into the access network and customer premises. Described are proposals using equipment available now and modified future options such that service delivery to customers is optimised.

1. Introduction

The Plesiochronous Digital Hierarchy (PDH) has for many years been the mainstay for network operators, being used in core and access networks as well as at customers' premises. However, customers are increasingly demanding greater flexibility of service routing and bandwidth allocation with higher resilience and protection requirements. For the operator this necessitates a high degree of network control and visibility of small granularity channels. The PDH is inherently very inflexible, restricts the level of monitoring, control and management and is therefore becoming less able to meet both present day and future network requirements.

The advent of synchronous transmission standards, culminating in CCITT Synchronous Digital Hierarchy (SDH) standards in 1989 (1), heralded an end to these restrictions, and in due course, PDH networks. These standards have assured the development of SDH equipment by manufacturers, thus facilitating widespread deployment by network operators.

SDH equipment to date has been deployed almost exclusively to construct core networks with a view to replacing the existing PDH equipment. The deployment of SDH technology in this way affords many advantages to both network operators and customers, but restricted to the core networks. By using SDH technology in access networks and at customer premises, these benefits may be extended to the point of service delivery and thus provide full end to end management and flexibility, benefiting both customer and network operator. This paper describes how these advantages may be realised by using a number of options based on current and future SDH equipment in conjunction with various service delivery technologies.

This paper does not present BT's future policy for the use of SDH, but proposes some ideas that are either being considered or are worthy of consideration.

2. The Motivation For Using SDH

Many business customers require to interconnect a number of geographically separated offices which may be spread across national and international boundaries. In order to make the most cost effective use of their networks, there is an increasing demand for flexible and dynamic routing while supporting high protection levels for telecommunication services many of which are 2Mbit/s and below. Within the PDH domain, 2Mbit/s channels are multiplexed into increasing hierarchical data rates, from 2Mbit/s to 8Mbit/s, 34Mbit/s, 140Mbit/s and so on. At any hierarchical level greater than 2Mbit/s, it is not possible to locate and thus retrieve individual 2Mbit/s channels due to the bit interleaving and justification process employed during the PDH multiplexing procedure. If it is required to extract a 2Mbit/s channel from a PDH frame, a number of multiplexing stages must be encountered, which usually requires numerous separate multiplexer/demultiplexer equipment. This operation is thus very expensive, inflexible and inherently less reliable than if the same result was achievable using a single piece of equipment. The introduction of SDH standards and subsequent

Reprinted from *IEEE ICC '94 Conf. Rec.*, pp. 1071–1076, May 1994.

production of SDH equipment has enabled these issues to be resolved.

The SDH data structure is based on multiples of the synchronous transport module STM-1 (which has a data rate of 155.520Mbit/s). Within an STM-N frame, the lowest granularity channel is 2Mbit/s (1.5Mbit/s US)which, with its SDH path overhead, enables individual 2Mbit/s channels to be identified and managed. This therefore means that it is possible to add, remove or switch a particular 2Mbit/s channel at any STM-N level directly, using a single piece of network equipment. By deploying bi-directional rings to interconnect switching nodes, a number of protection options yielding high service availability are also possible.

The flexibility, management and protection provided by SDH, allied with the ability to purchase equipment with open interfaces in a world market and thus enable mid-span meet, has led to the widespread deployment of SDH equipment (and SONET equipment in the US which is based on a sub-set of the CCITT standards). However, to date this has been almost exclusively for the interconnection of core network nodes. In the UK, BT is continuing to conduct SDH trials and will begin deployment in 1994, with volume roll-out scheduled for 1995. In order to realise the full benefits available, SDH functionality and interfaces should be extended as close to the customer as possible. If an SDH interface is used to connect customers directly to the network, single 2Mbit/s channels may be flexibly routed and managed end to end between remote sites. This has the additional advantage that the same network management platform may be used for both the core and access networks with seamless operation.

3. Implementation Options Using Currently Available Equipment

The key equipment building block for SDH networks is the add-drop multiplexer (ADM), Figure 1.

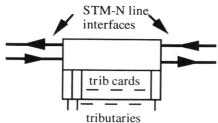

Figure 1 *The Add - Drop Multiplexer (ADM)*

The ADM may be configured as a terminal multiplexer for point to point connections, or in standard mode when deployed in rings. For core networks, nodes will be interconnected by rings supporting an STM-N channel where N will depend on the total bandwidth requirements of the customers connected to the ring. In rural locations STM-1 (155Mbit/s) may be sufficient, but in densely populated locations or business districts, STM-4 (622Mbit/s) or even STM-16 (2.4Gbit/s) rings may be used.

For business customers with large bandwidth requirements at a single site (e.g. ≥34Mbit/s), it will be cost effective to locate an ADM at the customer's premises and provide connection to the serving site using an STM-1 path. This connection may be point to point or via a ring shared by a number of users within the same locality. At sites where the bandwidth requirement is sufficiently less than this, a reduced bandwidth delivery multiplexer or sharing of an ADM would be a more cost effective solution. In the UK, the average distance between customers premises and the serving site in the access network is approximately 2.5km. For this reason, the STM-1 connection to large bandwidth sites will be either optical, or in some instances radio.

For the point to point option, Figure 2, the fibre connection between the customer premises and serving site may be duplicated and diversely routed to the serving site in order to provide very high network resilience.

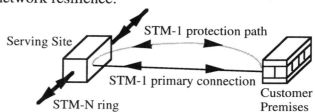

Figure 2 *Point to Point Connection*

This configuration can utilise the multiplex section path protection (MSP) function of the terminal multiplexer, or path protection using an ADM. In each case, both paths are constantly monitored in order that the path with the best performance may be selected. The medium for the protection path may be fibre or radio, since radio standards also support STM-1 interfaces. A further enhancement on network performance may be achieved by connecting the customer to two separate nodes, Figure 3.

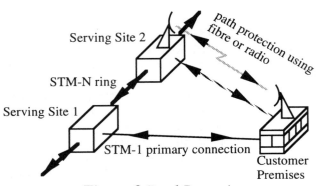

Figure 3 *Dual Parenting*

The protection path for this architecture (known as dual parenting or dual homing) again may be via an optical or radio connection.

The smallest bandwidth granularity tributary currently supported on most ADMs is 2Mbit/s presented as plesiochronous interfaces. In order for these ADMs to support telecommunication services at the sub 2Mbit/s level, additional service multiplexers will be required. Alternatively, sub 2Mbit/s interfaces may be integrated with the ADM thereby relinquishing the need for separate service multiplexers. A further option for supporting various bandwidth services is to use a SDH delivery for 2Mbit/s and above, and a passive optical network (PON) for the less than 2Mbit/s services (2).

For small bandwidth business users, a street located terminal multiplexer could be deployed providing connection to a number of customers sites, Figure 4.

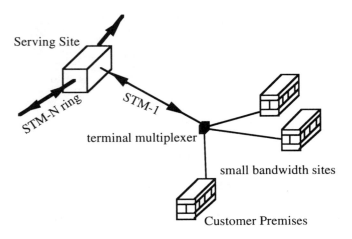

Figure 4 *Street Located Terminal Multiplexer*

The physical connection between the customers and the terminal multiplexer could be via coaxial cable. This method extends the SDH functionality to the street and helps to rationalise the number of physical connections to be routed to the serving site.

Although it is possible to interconnect a number of business customers on a ring, for large bandwidth users this is likely to be undesirable due to the limitations on bandwidth upgrade since the STM channel is shared. There is also an issue regarding data security if the ADMs are located at customer sites. However, rings may be usefully deployed by connecting business sites with small bandwidth requirements to street located ADMs, Figure 5, using the same techniques as described for the street located terminal multiplexer.

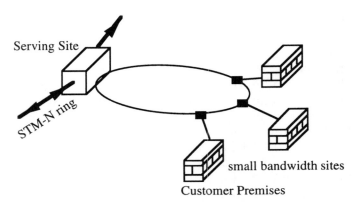

Figure 5 *Street Located Add - Drop Multiplexers*

4. Future Implementation Options

The network connection options discussed enable SDH functionality to be extended to customer premises for large bandwidth users, and possibly to the street for others. For both applications, the advantages that SDH offers end at the location of the ADM or terminal multiplexer. This therefore limits the extent to which the SDH potential may be exploited. In order to fully realise all the benefits, SDH needs to be extended to the sites of lower bandwidth users, provide flexibility at 64kbit/s and be integrated with existing and new service delivery platforms. There are a number of ways in which these requirements may be met, some of which are now addressed.

As previously described, SDH enables the switching and management of circuits using a minimum granularity of 2Mbit/s (1.5Mbit/s US), which means that sub 2Mbit/s services have to be switched as an entire 2Mbit/s block in the SDH network. With the increasing use of data compression techniques and virtual private networks, many current and future telecommunication services may be supported using 64kbit/s channels. The inability for SDH to manage and configure at this level is therefore somewhat restrictive and limits the flexibility afforded to these services. If in the future the standards bodies produce recommendations for the management of SDH 64kbit/s channels, the flexibility available will be immense.

Commercially available SDH equipment is not at present well tailored for locating at the premises of a customer requiring substantially less than 32Mbit/s since it tends to be expensive, very large and supports more bandwidth and functionality than is required. The only possible concession for cost reduction being a limited number of tributaries supported by an ADM. Typically, SDH equipment manufacturers provide 2Mbit/s interfaces on tributary cards which have at least 16x2Mbit/s ports. For customers requiring substantially less than 32Mbit/s, this is a very expensive option. In order to increase the cost effectiveness, smaller granularity cards supporting only a few 2Mbit/s interfaces in a reduced size multiplexer are needed. Path protection could be supported in the same way as described earlier. The advantages with this option are that STM-1 interfaces are standardised, many customers could be connected to a single serving site ADM and there is ample spare capacity to enable large future bandwidth upgrades. The disadvantages however are that the use of an STM-1 radio path for protection of low bandwidth will be very expensive, and the serving site ADM may need to support many very low utilised STM-1 channels, which is a very expensive way to provide service. The caveat to this is that the cost of an STM-1 interface will not be such a limiting factor in the future as it is now.

An alternative and more efficient approach to the delivery of an SDH path for small bandwidth users is to use a sub STM-1 connection between the customer's premises and the serving site, Figure 6.

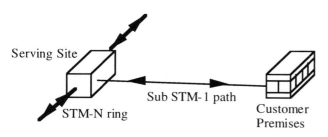

Figure 6 *Sub STM-1 Connection*

In order to maintain the SDH path management, the data structure should be an integer number of low order SDH multiplex units. Within the SDH

multiplexing hierarchy, Figure 7, the smallest unit is the level 1 virtual container (VC-12 for 2Mbit/s and VC-11 for 1.5Mbit/s).

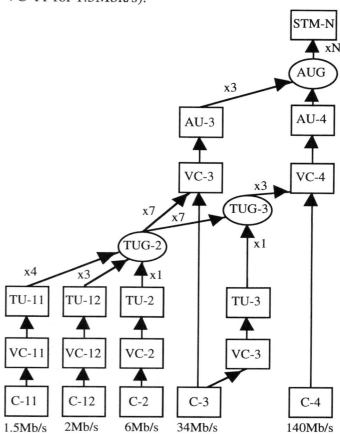

Figure 7 *SDH Multiplexing Hierarchy*

The VC-12 is a 2Mbit/s channel with a path overhead which is used to configure and monitor the performance of the channel. A pointer is added to the VC-12 to create a tributary unit (TU-12), 3 TU-12s are combined to form a tributary unit group (TUG-2), and 7 TUG-2s are combined to form a TUG-3. The payloads supported by the TUG-2 and TUG-3 are 6Mbit/s and 42Mbit/s respectively. The provision of a sub STM-1 connection using a TUG-2 or TUG-3 between the serving site and the customers' premises would be less expensive than the STM-1 option and would more optimally deliver the required bandwidth. The TUG2 or TUG3 could be assigned a proportion of the STM section overhead in order to maintain MSP and section management between the customer's and serving site equipment.

Sub STM-1 delivery is a very attractive option enabling SDH channels to be cost effectively delivered to small bandwidth users while maintaining full SDH management and functionality. At present there are no standards to support this, although in the interim a proprietary solution may be used, but at the expense of relinquishing mid-span meet.

For the delivery of services using optical fibre to small business customers, a PON will be the most likely solution (3), whereas customers with larger bandwidth requirements will have a dedicated point to point delivery, which may be SDH. If a PON could be integrated within an ADM or SDH terminal multiplexer, by street locating the equipment it will be possible to serve a mixture of small and large bandwidth users from the same equipment, Figure 8.

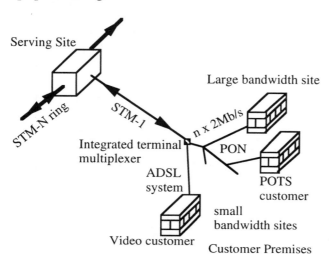

Figure 8 *Integrated Street Located Multiplexer*

It is also possible for PONs to support VC-12s so that the SDH path management channel may be extended across the PON to a remote optical network unit at the customer's site. This arrangement not only rationalises the number of connections and equipment required at the serving site, but it also simplifies network management, since one element will effectively support many bandwidth delivery rates and a range of customers. As indicated in Figure 8, this option may also support the delivery of POTs and entertainment services to residential customers using fibre or copper delivery. Copper delivery of entertainment services may be achieved by interfacing to an asymmetrical digital subscriber loop (ADSL) system (4).

One of the disadvantages of using SDH is that in order to support the services of most customers, connection to non SDH equipment is required. To manage this remote equipment at present requires a

separate management platform to that for SDH. This problem may be overcome by extending the data communications channel (DCC), Figure 9, which is embedded in the STM overhead.

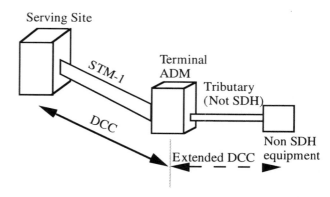

Figure 9 *Extending the Data Communications Channel (DCC)*

There is the facility to use parts of the DCC for management of remote equipment and thus extend monitoring to non SDH equipment. This will be particularly useful for performance monitoring and fault location. There are number of possible implementation techniques to support this, none of which have yet been recommended by the standards forums.

5. Conclusions

This paper has described how the emerging SDH technology, initially designed for core network deployment, may be utilised in the access network and at customers' premises. By doing so, end to end routing flexibility of 2Mbit/s channels is possible with high resilience supported by a single network management platform.

Commercially available equipment at present does not generally support <32Mbit/s, so will only be cost effective for high bandwidth applications, or if it is street located and shared by a number of users. To optimally support bandwidth requirements of substantially less than 32Mbit/s, the most likely cost effective solution will be to use a sub STM-1 delivery, although in Europe there is currently no sub STM-1 standard. Until this issue is resolved, it is unlikely that equipment supporting this will be available in large volumes. Further issues requiring the attention of the standards bodies in order to fully exploit SDH are 64kbit/s switching and the management of remote equipment connected to the SDH multiplexers.

Some of the benefits available with SDH technology are already being realised in core networks. This is only the beginning. By implementing the options described in this paper and with the help of evolving standards, the vast potential that SDH offers may be fully realised to the very point of service delivery.

6. References

1 CCITT Recommendations G.707, G.708 and G.709.

2 Marshall J and Guyon R, "Towards an Integrated Access Network", IEE Colloquium on "Customer Access - the last 1.6km", 1 June 1993.

3 Stern J R et al, "Passive Optical Networks for Telephony Applications and Beyond", Electronics Letters, Vol 23 No 24, 19 Nov 1988.

4 Waring D W, "The Asymmetrical Digital Subscriber Line (ADSL): A New Transport Technology for Delivering Wideband Capabilities to the Residence", Proc. Globecom'91, Phoenix, AZ, pp 1979-1986, 25 Dec., 1991.

Cost-Effective Network Evolution

A three-phase path from todays SONET/STM ring transport to a SONET/ATM VP ring transport could facilitate the network evolution for broadband services.

Tsong-Ho Wu

TSONG-HO WU is with the Network Control Research Department at Bellcore, where he is responsible for broadband fiber network design, survivable network architectures, and emerging technology applications for SONET and ATM virtual-path-based networks.

[1] *STM is a multiplexing and switching process in which a connection (or channel) is represented by time slots, which appear in the same location within a frame occurring periodically (e.g., every 125 μs for SONET systems). Figure 1a depicts this STM concept. Examples of STM for present SONET systems include time slot assignment (TSA), used in SONET add-drop multiplexers (ADMs) for chain and ring applications; and time slot interchange (TSI), used in SONET ADMs and digital cross-connect systems (DCSs) for ring and mesh networks.*

ynchronous optical network (SONET) rings are being deployed in many local exchange carriers' (LECs') intra-local-access-transport-area (intra-LATA) networks due to their standard signal interfaces, economic high-speed signal add-drop capability, fast self-healing capability, and simple network operations. Present SONET rings using synchronous transfer mode (STM)[1] for signal multiplexing and switching (see Fig. 1a) only implement the virtual tributary (VT) and STS-1 features to support existing nonswitched DS1 and DS3 services. The transfer mode is referred to as the integrated multiplexing and switching process. Several network plans have been reported by LECs (e.g., [1]) that use existing or SONET self-healing rings to provide reliable transport for high-speed switched data services (e.g., frame relay service, switched multimegabit data service — SMDS — and fiber distributed data interface — FDDI — interconnection). This strategy reduces the initial capital investment cost for providing high-speed data services in the initial service introduction stage. However, it is conceivable that some of the network infrastructure should be updated to accommodate these new data services due to inefficient use of bandwidth for burst-type data services (e.g., SMDS) when they are carried on the present non-SONET/STM or SONET/STM ring.

Recently, a SONET/asynchronous transfer mode (ATM)[2] ring architecture, called SONET/ATM ring using point-to-point virtual paths (SARPVP) [2], has been proposed to overcome the problem of inefficient and expensive use of SONET bandwidth for SMDS service and other high-speed switched data services. The SARPVP is a SONET ring that uses ATM virtual path (VP) technology to lower transport cost. The ATM VP-based network architecture [3, 4] is essentially a compromise between the SONET/ STM architecture and ATM virtual channel (VC)-based network architecture: it takes a system simplicity concept from the SONET/STM network and keeps the flexibility of ATM technology (see Fig. 1b for the ATM con-

cept). The proposed SARPVP ring architecture can evolve from the present SONET self-healing ring to carry existing nonswitched DS1 and DS3 services, and SMDS and other types of high-speed data services as well, with a minimum cost investment.

The purpose of this article is to show how to use the proposed SARPVP ring architecture to develop a cost-effective and efficient evolution path from today's SONET/STM environment, supporting DS1 and DS3 services, to a hybrid STM/ATM ring architecture for additionally supporting SMDS and other high-speed data services over the STM network of today, and finally to a SONET/ATM VP-based ring architecture for cost-effectively supporting both switched and nonswitched DS1 and DS3 services.

In this article, the author first briefly reviews network architectures used to support SMDS service, and then the SARPVP ring architecture and some necessary information. A cost-effective three-phase network evolution path for supporting both the switched and nonswitched DS1 and DS3 services, and then a possible operations system (OS) evolution path under the proposed three-phase network evolution scenario is described. Conclusions and summary information are then presented.

Network Architectures for Supporting SMDS Service

Figure 2 depicts a logical network configuration that supports SMDS service. A variety of customer premises equipment (CPE) are supported by an SMDS switch, and each type of CPE accesses the SMDS switch via a dedicated subscriber-network interface (SNI) with the SMDS interface protocol (SIP).[3] CPE and a serving SMDS switch form an SMDS access network, as depicted in Fig. 2. If the SMDS demand requires two or more SMDS switches, these SMDS switches are interconnected via the inter-switching system protocol (ISSI) [6]. The inter-SMDS

Access class	SIR* (Mb/s)	Network type
1	4	IEEE 802.5 LAN
2	10	IEEE 802.3 LAN
3	16	IEEE 802.5 LAN
4	25	**
5	34	++

* SIR (sustained information rate) is the rate of trans- fer of user information that CPE could sustain over long periods using that particular access class.
** Class 4 allows for an SIR intermediate between 16 Mb/s and 34 Mb/s. This class allows for configurations of equipment that are capable of transferring data at moderately high sustained rates but that still do not require the full DS3 bandwidth.
++ Class 5 represents the case where there is no enforcement applied — the SIR is the maximum effective access bandwidth achievable using SIP protocol across a DS3-based SNI (approximately 34 Mb/s).

■ **Table 1.** *SMDS access class and bandwidth requirement.*

switch network is sometimes referred to as the SMDS backbone network. The SMDS access network will be the primary target network of interest for the proposed SARPVP ring architecture.

Figure 3 depicts two possible transport structures of the SMDS access network for a single SMDS

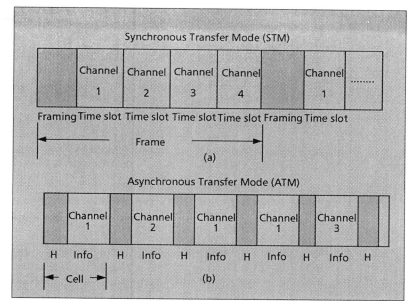

■ **Figure 1.** *(a) STM concept versus (b) ATM concept.*

switch configuration. The first scenario, depicted in Fig. 3a, is to use dedicated T1 or multiplexed DS3 links with the SIP to access the serving SMDS switch. The second scenario, as proposed by [1] (also see Fig. 3b), is to use a ring as a reliable transport architecture to access the serving SMDS switch. Note that initially, the high-speed data switches are not expected to be deployed at every central office (CO) due to capital cost considerations and the relatively small initial service demand level. Thus, only one SMDS switch

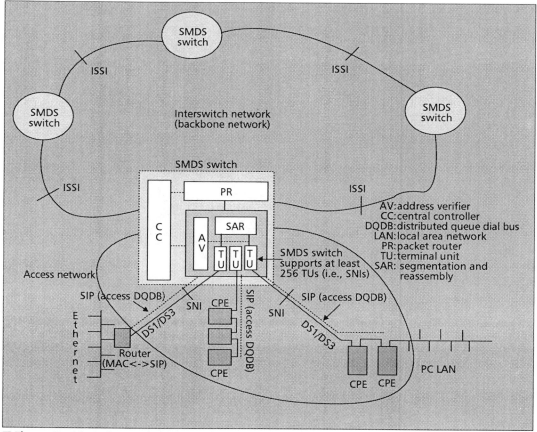

■ **Figure 2.** *A logical network configuration for supporting SMDS.*

AV: address verifier
CC: central controller
DQDB: distributed queue dial bus
LAN: local area network
PR: packet router
TU: terminal unit
SAR: segmentation and reassembly

[2]ATM is a target technology for the broadband integrated services network (B-ISDN). In ATM, information to be transferred is segmented to a set of fixed-size cells. Cells are identified and switched by means of a label in the header. Cells belonging to the same connection may exhibit an irregular recurrence pattern, as cells are filled according to actual demand) (Fig. 1b).

[3]The IEEE 802.6 standard distributed queue dual bus (DQDB) protocol [5] is used as the basis of the SIP.

353

may be needed in each LATA for supporting initial SMDS service in that region.

Due to different bandwidth requirements associated with each SMDS access class (1.5 Mb/s, 4 Mb/s, 10 Mb/s, 16 Mb/s, 25 Mb/s, and 34 Mb/s), as shown in Table 1, and the burst characteristics of each access class, a significant portion of SONET bandwidth will be wasted[4] when these high-speed bursty data services are carried on the present STM network. For example, suppose an SMDS subscriber has an Ethernet that runs at 10 Mb/s. This subscriber needs to request Access Class 2 (Table 1), and the SONET network has to dedicate an STS-1 path (51.84 Mb/s) to that subscriber. Also, the average Ethernet utilization is typically about 20 percent; thus, the STS-1 path utilization is only about 4 percent.[5] Therefore, in the near future the network provider may look for another transfer technology to replace the STM technology for efficiently and cost-effectively supporting SMDS service.

An ATM/SONET Ring Architecture Using the VP Concept

ATM VP Transport Concept

Two switching concepts — VC and VP switching — have been proposed for use in ATM switched network architectures. The VC-based switched network architecture, as depicted in Fig. 4a, utilizes ATM VC switches as key network elements and processes ATM VC connections on a VC-by-VC basis. The VC-based switched network architecture is sometimes referred to as the "full" ATM switched network architecture. This architecture is very flexible and efficient in terms of bandwidth management and allocation, since these capabilities are distributed throughout the network. However, this bandwidth management flexibility comes at the expense of complex ATM VC switches.

To simplify the signal transport complexity at intermediate nodes while preserving some degree of the bandwidth management flexibility of ATM technology, the VP-based network architecture, as depicted in Figure 4b, was introduced [3, 4]. The ATM VP concept is a hybrid concept that combines SONET system simplicity with ATM system flexibility. In the ATM VP-based network architecture, VPs are processed at two end nodes of a VP connection using ATM switches or add-drop multiplexers (ADMs) and are transported transparently via an ATM VP-based transport network. An ATM VP transport network is made up of ATM DCSs for the mesh-type network or ATM ADMs for the ring topology. This transport nodal equipment is simpler than equipment at two ends of the VP connection, since no call control and bandwidth management functions are needed at intermediate nodes. (Please refer to [4] for more discussions on the ATM VP-based transport network architecture.)

SARPVP Ring Architecture

Figure 5 depicts an SARPVP. A point-to-point VP is a VP that contains VCs terminating at the same two end points. The SARPVP ring architecture was proposed to cost-effectively and efficiently support both switched and nonswitched DS1 and DS3 services [2]. The signal add-drop on the SARPVP ring is performed at the VP level. In this SARPVP architecture, each ring node pair is preassigned a duplex VP. For example, in Fig. 5, VP #2 and VP #2'

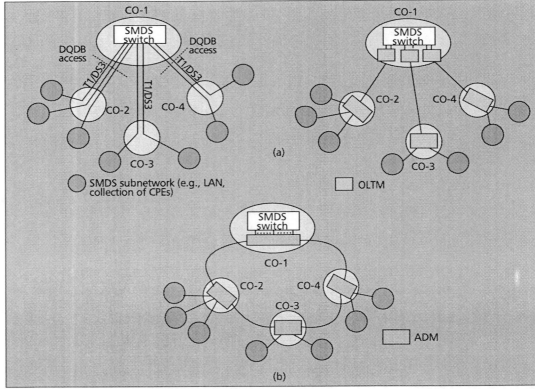

■ **Figure 3.** *Transport alternatives for SMDS access: (a) dedicated point-to-point transport for SMDS access; (b) reliable ring transport for SMDS access.*

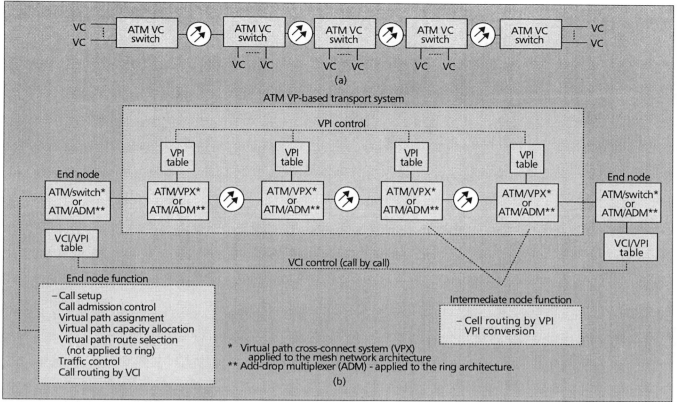

(a)

(b)

Figure 4. *(a) ATM VC-based and (b) VP-based network architectures.*

(not shown in the figure) carry all VC connections from nodes 1 to 3 and from node 3 to node 1, respectively. The physical route assignment for the VP depends on the type (unidirectional or bidirectional) of the considered SONET ring. If it is a unidirectional ring, two diverse routes that form a circle are assigned to each VP. For example, in Fig. 5, two physical routes, 1-2-3 and 3-4-1, are assigned to VP #2 and VP #2' if the considered ring is a unidirectional ring. If it is bidirectional, only one route is assigned to each duplex VP (e.g., route 1-2-3 is assigned to both VP #2 and VP #2'), and demands between nodes 1 and 3 are routed through route 1-2-3 bidirectionally. (Please refer to [7] for more details on SONET unidirectional and bidirectional ring architectures.)

In order to avoid VP translation at intermediate ring nodes of a VP connection, the VP identifier (VPI) value is assigned on a global basis. The ATM cell add-drop or pass-through function at each ring node is performed by checking the cell's VPI value. Since the VPI value has global significance and only one route is available for all outgoing cells, VPs need not be cross-connected at each intermediate ring node. Thus, no VP cross-connect capability is needed for the ATM ADM of this SARPVP ring architecture. The implementation of the ADM for this SARPVP ring architecture is discussed below.

The global VPI value assignment presents no problem here, since only one route exists for all outgoing ATM cells, and the number of nodes supported by a ring is usually limited (primarily by traffic and reliability requirements). For example, the 12-bit VPI field in the network-to-network interface (NNI) ATM cell represents 4096 VPI values available for use. Thus, the maximum number of ring nodes is 91,[6] which is sufficient to

support LECs' interoffice and loop rings.

If the point-to-point VP ring is used to support present nonswitched DS1 services (via circuit emulation), each DS1 comprises a VC connection and is assigned a VPI/VC identifier (VCI) based on its addressing information and the relative position of the DS1 within all the DS1s terminating at the same source and destination on the ring. For example, VPI = 2 and VCI = 3 represent a DS1 that is the third DS1 of the DS1 group terminating at Node 1 and Node 3.

The VPI/VCI assignment for SMDS transport is different from that of point-to-point non-switched DS1s/DS3s described above. A two-stage switching function is needed here for SMDS packets, and VPI/VCI translation is also needed, which is performed at the SMDS switch. The details of the VPI/VCI assignment for SMDS service are discussed later.

[6]*Let* n *be the number of ring nodes. The maximum number of ring nodes is the number satisfying the equation*

$$\frac{n \times (n-1)}{2} = 4096$$

which is 91.

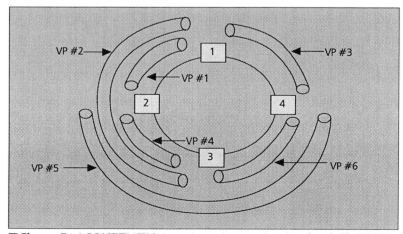

Figure 5. *A SONET/ATM ring using point-to-point virtual path (SARPVP).*

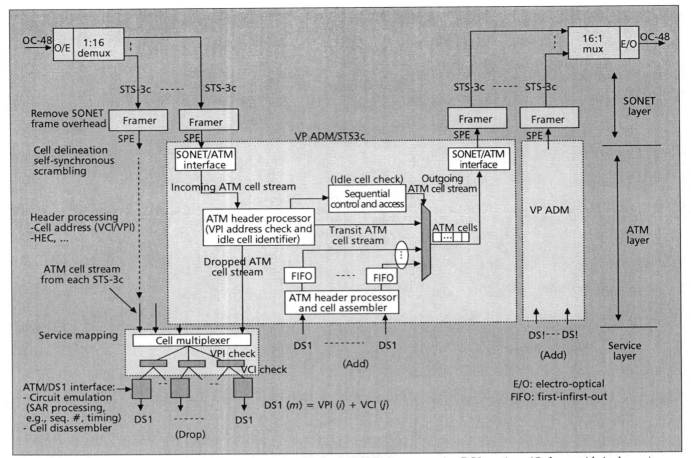

Figure 6. *An STS-3c add-drop hardware configuration for the SARPVP ring supporting DS1 services. (Only one side is shown.)*

Add-Drop Multiplexer (ADM) Design for the SARPVP Ring

The ATM ADM for the SARPVP architecture (see above) can be implemented in different ways depending on physical SONET STS-*N*c terminations. The most common proposed ATM STS-*N*c terminations are STS-3c, STS-12c, and STS-48c, although only the STS-3c ATM termination has been specified in current International Consultative Committee for Telephone and Telegraph (CCITT) Recommendations.[7] When ATM cells are extracted from the STS-*N*c frame, dropping or passing through of the cells occurs by examining the VPIs of the cells; these VPIs indicate the destination nodes of the cells on the ring.

Figure 6 depicts a possible ADM configuration with STS-3c terminations for an SARPVP ring implementation that supports DS1 services via circuit emulation. In this architecture, the hardware design for signal terminations above STS-3c is the same as that for SONET/STM ADMs[8] for ring applications [8]. The ATM VP add-drop function, which is performed at the STS-3c level, requires three major modules. The first module is the ATM/ SONET interface, which converts the STS-3c payload to an ATM cell stream and vice versa. The functions performed in this module include cell delineation, self-synchronization, and scrambling. The scrambling process here is to increase the security and robustness of the cell delineation process against malicious users or unintended simulations of a correct header error control (HEC) in the information field.

The second module is to perform header processing, which includes cell addressing (VPI in this case) and HEC. In order to perform cell add-drop/pass-through, this module checks the VPI value of each cell to determine if it should be dropped or passed through and identifies an idle cell, which can be used for cells from the considered office (i.e., signal adding) via a simple sequential access protocol. This sequential access protocol can be implemented by the third module, which passes through each nonidle cell and inserts the added cells into outgoing idle cells in a sequential order.

The third functional module is a service mapping module, which maps ATM cells to their corresponding DS1 cards based on VPI/VCI values of ATM cells. This service mapping module first multiplexes all ATM cells from different STS-3c payloads into a single ATM cell stream, and then distributes ATM cells to corresponding DS1 groups according to their VPI values. For each DS1 group, the ATM cells are further divided and distributed to the corresponding DS1 cards by checking their VCI values. This service mapping module essentially performs a simple VPI/VCI comparison function.

Note that DS1 services that may be supported by this VP ring architecture include existing nonswitched DS1/DS3 services and new switched DS1/DS3 services (e.g., frame relay service, SMDS, and FDDI interconnection). Note that, in Fig. 6, we show a pure STS-3c termination configuration for an OC-48 ATM VP ADM. If STS-3c ATM cards supporting the new high-speed data services do not replace all original STS-3 STM cards (e.g., these STS-3 STM

[7]*The ATM cell mapping at the STS-12c termination has been specified in American National Standards Institute (ANSI) T1S1/91-634, "Draft for Broadband ISDN User-Network Interfaces: Rates and Formats Specifications."*

[8]*The present SONET/STM ADM only implements VT and STS-1 features to support nonswitched DS1 and DS3 services.*

Evolution phase	Non-switched DS1/DS3 access	SMDS		Ring Transport (SARPVP) supporting SMDS			
		Access	Year	Access port	Transfer mode	Motivation for each phase	ADM availability
I	DS1/DS3	DS1/DS3	1992	VT/STS-1	STM	Offers SMDS with minimum capital investment	1992
II	DS1/DS3 or VT/STS-1	DS1/DS3 or VT/STS-1	1993-94	VT/STS-1 for non-switched DS1/DS3; STS-3c ATM cards for SMDS	STM/ATM	Cost-effectively support SMDS, and gain ATM operations experiences	1995
III	VT/STS-1	VT/STS-1 or ATM (Class D)	1994-95	STS-3c (ATM)	ATM	cost-effectively support all services and provide an integrated operations system	1995–6

■ **Table 2.** *Three-phase network evolution using SARPVP ring transport.*

cards are still used to support nonswitched DS1 and DS3 services), the ring transport architecture forms a hybrid ATM/STM transfer mode at each ADM. The desirability of a hybrid architecture has been recognized as a practical solution by some LECs [9].

For ATM termination at the STS-3c level, the ATM/ADM can be upgraded from the present SONET SONET ADM by replacing the SONET STS-3 line cards with ATM line cards, thus minimizing the initial capital costs for evolving from the SONET STM ring to the hybrid STM/ATM ring and, eventually, to the ATM VP ring. The ATM technology needed to implement the SARPVP ring architecture is available today. Of course, operations for processing ATM cells will be different from those for processing SONET channels. The requirements and design for SONET rings require no change. Thus, the ATM ADM with STS-3c terminations could be the first candidate for the transition from the SONET STM ring to the ATM ring for early deployment of ATM technology on the SONET infrastructure.

Self-Healing Capability for SARPVP Ring

The self-healing function of the ATM VP ring with STS-3c or STS-12c terminations can be performed at the path (STS-3c or STS-12c) layer[9] [10] or the line layer[10] [11, 12] in an identical way to those functions for SONET ring architectures.

To ensure a reliable transport for switched data services, dual SMDS switches may be used to mitigate a serving SMDS switch failure. Like the dual central controller protection in the centralized SONET digital cross-connect system (DCS) networks [7], these two identical SMDS switches are placed in different COs, as depicted in Fig. 7, but only one SMDS switch is in control at any given time. Both SMDS switches are on-line and synchronized so that the standby controller can instantly assume the restoration control function. Alternatively, if the SMDS subscriber requires two or more DS3s across its SNI interface, this subscriber may split demands to both SMDS switches, just like the present dual-homing SONET ring architecture [9]. Both SMDS switches are connected to OSs via redundant-control data networks or SONET DCC channels.

A Cost-Effective Access Network Evolution for Broadband Services

Table 2 shows a three-phase network evolution path that allows a cost-effective evolution from the present SONET/STM ring transport architecture supporting nonswitched DS1 and DS3 services to a near-future SONET/ATM ring transport supporting both switched and nonswitched DS1 and DS3 services. Note that, in Table 2, the initial transmission interface is asynchronous DS1 and/or DS3. Later on, when the ATM access technology matures (expected to be 1994-95), the SMDS access can be directly converted to the ATM cell form at the CPE. Figure 8 depicts SNI mappings for SMDS access in this three-phase evolution path, and Fig. 9 shows a transfer mode evolution within the ADM. Please also note that, although this section discusses an evolution path for SMDS, the same evolution path is applied to other high-speed data services.

[9]A SONET self-healing ring is called a path protection switching ring if the ring's self-healing function is triggered by the path alarm indication signal (AIS), which resides in the path overhead.

[10]A SONET self-healing ring is called a line protection switching ring if the ring's self-healing function is triggered by the line overhead (e.g., K1 and K2 bytes).

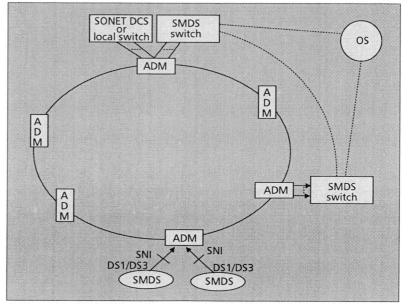

■ **Figure 7.** *A SONET ring transport with dual SMDS switch configuration.*

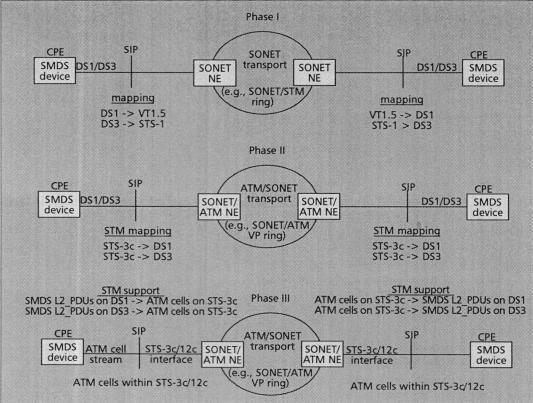

■ Figure 8. *SMDS SNI mapping in a three-phase evolution path.*

Phase I Evolution

Phase I is to add SMDS service over the existing SONET/STM ring,[11] which only supports present nonswitched DS1 and DS3 services. As shown in Table 2 and Figs. 8 and 9, the SMDS DS1 and DS3 are mapped to SONET VT1.5 and STS-1 paths,

respectively, and are then transported to the ADM on the ring, where the SMDS switch is located. Customers not served by COs with SMDS switches can be hauled on the SONET ring network to a CO with an SMDS switch. In this STM architecture, VTs and STS-1s are dedicated to each SNI. In other words, if there is no traffic sent from that SMDS customer, no other SMDS customers can access that time slot. This architecture can use existing SONET provisioning, OSs, and the self-healing system to support SMDS without any change. Thus, it may be the best strategy to initially provide SMDS service [1].

Phase II Evolution

Phase II, as shown in Figs. 8 and 9, is to upgrade the ring transport from the STM format to a hybrid STM/ATM format by replacing the STS-3 line cards carrying high-speed data services with STS-3c line cards implementing the ATM VP add-drop scheme, as described earlier or in [2]. In this phase, the STS-3 cards carrying existing nonswitched DS1 and DS3 services remain unchanged (i.e., still use the STM format), and new STS-3c cards carry high-speed data services. The SMDS DS1/DS3s access the network via the cell format of the ATM adaptation layer (AAL) Class D[12] (Fig. 10), to support the format conversion between the service and the signal transport. Thus, the ring architecture forms a hybrid STM/ATM architecture. The self-healing ring system remains unchanged, as in Phase I. It is expected that the Phase II ring transport will provide a much more efficient and cost-effective transport for high-speed data services than the SONET/STM ring in Phase I due to the inherent characteristic of ATM technology to

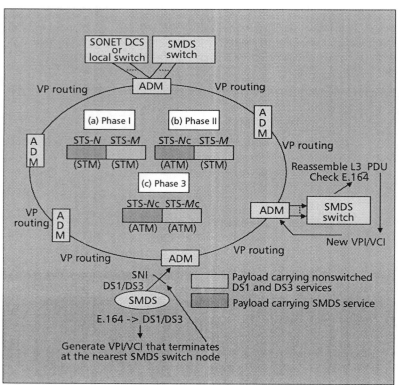

■ Figure 9. *Three-phase evolution for transfer mode in ADMs.*

share bandwidth and the simple ADM design for the SARPVP ring architecture [2]. In this phase, Oss may need to upgrade from SMDS-type OSs to ATM VP-based OSs. The savings on transport for supporting high-speed data services in this phase may justify the investment of OS upgrade. This phase also offers a good opportunity for network operations personnel to gain experience on ATM VP-based network operations, which will make the last phase transition to the fully ATM VP ring easier.

The routing for SMDS packets in Phase II is a two-stage VP routing, just like fiber-hubbed routing. SMDS packets (i.e., layer 3 protocol data unit — L3_PDU) are first segmented to a series of layer 2 protocol data unit (L2_PDU) packets of fixed size (53 octets), and these L2_PDUs are routed to the central (or main) SMDS switch via the ring transport using the point-to-point VPI assigned to those L2_PDUs; at the SMDS switch, incoming L2_PDUs are reassembled to a L3_PDU, and a new VPI/VCI value is regenerated for this L3_PDU based on the E.164 address information carried in the L3_PDU. This L3_PDU is then segmented into a set of L2_PDUs, which are then routed to their destination nodes and SNIs using this new VPI/VCI value.

Phase III Evolution

When sufficient operations experience has been gained from the high-speed data services using the ATM VP technology in Phase II, Phase III (Fig. 9) can be introduced, replacing the remaining STM portion (which supports nonswitched DS1 and DS3 services) with the ATM VP add-drop scheme, as described earlier. In this phase, nonswitched DS1s and DS3s are converted to ATM cells via circuit emulation (i.e., AAL Class A).

A case study based on a five-node LEC's ring model network indicated that the SARPVP ring had either significant cost savings or more spare capacity[13] compared with an STM ring [2]. An integrated OS for the SARPVP ring for both

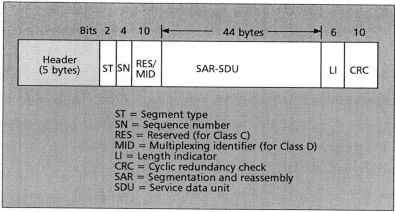

■ Figure 10. *Cell format for AAL Class D.*

switched and nonswitched broadband services in this phase should further reduce the operations, administration, management, and provisioning (OAM&P) costs. Thus, the network evolution from Phase II to Phase III may be justified by its potential economical and operations merits. However, these merits still require more detailed studies. Again, like the evolution from Phase I to Phase II, the self-healing system remains unchanged when the ring evolves from Phase II to Phase III.

In Phase III, when the ATM technology becomes mature, the SMDS access in CPE can be converted to the ATM cell format, which can then be directly placed into the STS-Nc payload in the ADM of the ring. This can further reduce the network cost, since the DS1/DS3-to-ATM (or VT/STS-to-ATM) conversion is no longer needed. After completing this three-phase network evolution, the underlying ring transport system is a true ATM B-ISDN transport system.

Table 3 summarizes a comparison of three ring transport architectures in this three-phase network evolution path.

Operations System Evolution

Figure 11 depicts an operation evolution scenario for the proposed three-phase network evolution path. The network OSs are used to manage ser-

Factors	SONET/STM ring	Hybrid (STM-DS1,ATM-SMDS)	SONET/ATM VP ring (SARPVP)
Bandwidth use efficiency for non-switched DS1s	fair	fair	better*
Bandwidth use efficiency for high-speed switched data service (e.g., SMDS)	poor	better	better
Provisioning for non-switched DS1s/DS3s	yes	yes+	no
Provisioning for high-speed switched data service	yes	no++	no
Delay for SMDS	best	fair	fair

* See analysis in [2].
+ Nonswitched DS1 and DS3 services are still supported by the STM network, which requires time slot provisioning.
++ Switched DS1 and DS3s here are supported by the ATM technology, which does not require service provisioning as long as these STS-3c/STS-12c frames used to carry ATM traffic have been provisioned.

■ Table 3. *Comparison among ring transports in three-phase evolution.*

[13]*A case study based on an LEC ring network suggested that the proposed SARPVP ring architecture can support existing non-switched DS1 services more economically (savings of up to 26 percent to 39 percent) than present SONET/STM rings [2]. A sensitivity analysis indicated that when all compared rings need the same high-speed line rate, the SARPVP ring cost approximates the cost of the best SONET ring bandwidth management scheme, but with much more spare capacity available for supporting other services.*

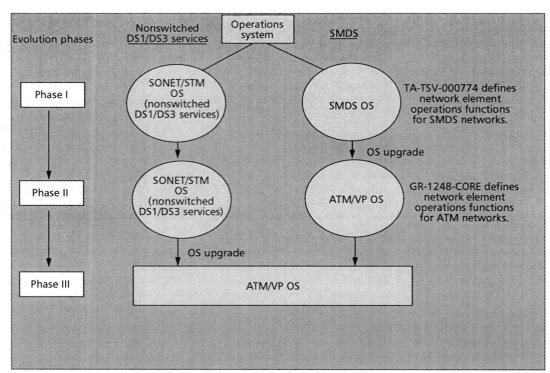

■ **Figure 11.** *An OS evolution scenario.*

vices (e.g., SMDS), and to manage and maintain the network[14] via network elements' operation modules (interfaces) and a data communication network. In Phase I, we assume that two separate OSs exist to support both switched and nonswitched DS1 and DS3 services. The operations modules of network elements for supporting SMDS should meet the operations criteria of TA-TSV-000774 [14]. In Phase II, these SMDS network-element operations modules are expected to upgrade to support ATM VP technology, which has been specified in GR-1248-CORE [13]. The cost savings in transport facilities in Phase II may help justify the operations upgrade from the SMDS OS to the ATM VP-based OS. The ATM VP-based OS has the long-term potential to provide an integrated OS for both switched DS3 and DS1 services in Phase III. The experiences gained from operating the ATM VP ring for supporting SMDS service may simplify the process of upgrading the network from Phase II to Phase III.

Note that efforts for changing OSs from one type of network to another are always significant. It may be difficult to evaluate the cost for the OS transition from one phase to another. However, an operations impact analysis for each network evolution phase can be used to estimate the complexity of the OS transition. For example, the provisioning methods and administration for various mappings as they relate to the functionality of individual network elements and the visibility of the proper performance of VPs (e.g., surveillance in support of maintenance) should be addressed. The operations impact analysis for this three-phase evolution path is currently under study.

Summary and Remarks

The author has discussed a cost-effective three-phase network evolution path by using the proposed SONET/ATM VP-based ring (SARPVP)

architecture. This evolution path evolves from the present SONET/STM ring transport to a hybrid SONET STM/ATM ring transport and, eventually, to a SONET/ATM VP ring transport for both switched and nonswitched voice and data services. The potential economical and operations benefits associated with each phase transition may facilitate the network evolution. The ADM for the proposed SARPVP ring architecture can evolve from today's SONET ADM by replacing the STS-3 termination cards with ATM STS-3c line cards in each phase. Thus, the capital investment needed for each phase of network evolution can be minimized. The self-healing system for present SONET/STM rings can be applied to the proposed SARPVP ring without any modification. Considering the factors of economic merit and smooth ADM evolution, the investment savings for the proposed SARPVP ring may result in a potential early deployment opportunity from the present SONET/ STM infrastructure to the future ATM/SONET infrastructure; this evolution can also help to support future broadband network services.

The network operations impact on the present SONET OS is certainly one of the key factors for the smooth transition in this three-phase evolution path. A possible OS evolution scenario has also been discussed to make the OS upgrade easier. Generic criteria for SARPVP and its operations impact have been studied by Bellcore and documented in GR-2837-CORE [16].

The SARPVP ring discussed in this article uses STS-3c terminations for ATM cells due to the consideration of easy evolution from the present SONET/STM rings. It may be desirable to design ADMs with higher STS-Nc terminations, such as STS-12c or STS-48c, for ATM traffic due to cost-effective support of a mixture of DS1 and DS3 services. The higher-speed ATM terminations may also reduce the network operations and maintenance complexity with fewer components to be controlled and maintained. Also, the present OC-48 ring capacity may not be enough to carry both the switched and non-switched DS1 and DS3 services on the same ring in the near future. It may be desirable that ultra-high-speed OC-192 technology for the ring application or some alternate ring architectures (e.g., a wavelength-division multiplexing — WDM — based switch-consolidated ring architecture [15]) be made available in Phase III to alleviate this capacity constraint problem.

Acknowledgments

The author would like to thank Joe Sosnosky for providing the information on an LEC network plan; and Joe E. Berthold, Richard H. Cardwell, Bob W. Klessig, Fred Klapproth, Wayne Tsou, and Bill Rubin for their valuable comments on the draft.

References

[1] M. Walter, "US West Targets '92 for SMDS/Frame Relay," *TE&M News*, pp. 26, Nov. 1, 1991.

[2] T.-H. Wu, D. Kong, and R. C. Lau, "An Economic Feasibility Study for Broadband Virtual Path SONET/ATM Self-Healing Ring Architecture," *IEEE J. Sel. Areas in Commun.*, vol. 10, no. 9, pp. 1459-1473, Dec. 1992.

[3] CCITT Study Group XVIII, Report R 34, COM XVIII-R 34-E, June 1990.

[4] K.-I. Sato, S. Ohta, and I. Tokizawa, "Broadband ATM Network Architecture Based on Virtual Paths," *IEEE Trans. Commun.*, vol. 38, no. 8, pp. 1212-1222, Aug. 1990.

[5] IEEE P802.6/D15, "Distributed Queue Dual Bus (DQDB) Subnetwork of a Metropolitan Area Network (MAN)," Oct. 1990.

[6] "Inter-Switching System Interface Generic Requirements in Support of SMDS Service," Bellcore Tech. Adv. TA-TSV-001059, issue 1, Dec. 1990.

[7] T.-H. Wu, Fiber Network Service Survivability: (Artech House, May 1992.)

[8] T.-H. Wu and M. Burrowes, "Feasibility Study of a High-Speed SONET Self-Healing Ring Architecture in Future Interoffice Fiber Networks," *IEEE Commun. Mag.*, vol. 28, no. 11, pp. 33-42, Nov. 1990.

[9] J. K. Conlisk, "The Graceful Transition of the Public Networks," *Telephony*, Sept. 30, 1991.

[10] Bellcore Generic Requirement, "SONET Dual-Fed Unidirectional, Path Switched Ring (UPSR) Equipment Generic Criteria," GR-1400-CORE, March 1994.

[11] "SONET Bidirectional Line Switched Rings Standard Working Document," T1X1.5/92-004, Feb. 1992.

[12] T.-H. Wu and W. Way, "An Enhanced Novel Passive Protected SONET Self-Healing Ring Architecture," *IEEE J. Lightwave Tech.*, vol. 10, no. 9, pp. 1314-1322, Sept. 1992.

[13] Bellcore Generic Requirement, "Generic Requirement for Operations of Broadband Switching Systems," GR-1248-CORE, Aug. 1994.

[14] "SMDS Operations Technology Network Element Generic Requirements," Bellcore Tech. Adv. TA-TSV-000774, issue 3, plus supplements, Feb. 1991.

[15] S. Wagner and T. E. Chapuran, "Multiwavelength-Ring Networks for Switch Consolidation and Interconnection," *Proc. IEEE ICC '92*, Chicago, Ill., June 1992.

[16] Bellcore Generic Requirement, "ATM Virtual Path Functionality in SONET Rings—Generic Criteria," GR-2837-CORE, Dec. 1994.

Efforts for changing OSs from one type of the network to another are always significant.

Emerging Residential Broadband Telecommunications

DAVID S. BURPEE AND PAUL W. SHUMATE, JR., FELLOW, IEEE

Invited Paper

Although near-term benefits are driving both telephone companies and cable operators to deliver residential services over optical fiber, in the longer term fiber enables new interactive, multimedia, switched "Information Age" services. We address residential fiber deployment, and opportunities for these two industries to work together in providing new services.

I. INTRODUCTION

In the mid-1970's, while optical-fiber systems were being developed for telephone trunking, experiments were already being planned to extend fiber directly to homes. The ever-growing need for bandwidth was widely recognized even in the very early 1970's, and the bandwidth potential, small size, flexibility, ruggedness, corrosion resistance, immunity to signal leakage and EMI, and other advantages of fiber over metallic media were seen as the enabling factor for delivering broadband services.

Analogous to the way new software concepts for personal computers emerged only after PC's had become established, many new broadband services will take form only after versatile broadband platforms have become widely available. One reason is that an identifiable market is needed to justify the investment in software development, as well as in advanced hardware to support the market needs. Nevertheless, service concepts are crystallizing. Video entertainment, of course, leads the list, offering interactive program guides and on-demand selections from vast libraries of material. This is followed by a rich spectrum including multimedia games and bulletin boards, interactive shopping and financial services, telecommuting, education, workforce retraining, healthcare, demand-side energy management, and so on [1]–[6].

Interestingly, many of these services have been envisioned since the earliest proposals for trials using either fiber or coaxial cable. For example, in 1977, Warner Amex Cable initiated their QUBE experiment, an advanced two-way coaxial-cable cable television system in Columbus, OH

Manuscript received August 13, 1993; revised October 4, 1993.
The authors are with Bellcore, Morristown, NJ 07962.
IEEE Log Number 9215565.

[7], where entertainment- and community-oriented service concepts were tested. On the fiber front, the first fiber-to-the-home (FTTH) trial began in 1978, serving residents in the Japanese community of Higashi-Ikoma[1] [8], [9]. Here, all the services mentioned above were tested except telecommuting. Canada [10], [11], Germany [12], the United Kingdom [13], France [14], [15] and the United States [16] soon followed with various fiber trials, generally emphasizing either switched two-way telephony services augmented with video, or cable television services using fiber to replace coaxial cable.

Notable early video-on-fiber installations in the United States included replacing coax with fiber for video transport in a building in New York City by Manhattan Cable Television, use of fiber drops in Alameda, CA, to deliver cable service to the home (albeit from a coaxial-cable feeder) by Times Fiber using their Mini-Hub system [17], and delivery of switched video services in fiber-to-the-home trials by Southern Bell in Orlando, FL [18]–[20], and Bell Atlantic in Perryopolis, PA [21]. All of these trials demonstrated that fiber could indeed replace coaxial cable for delivering high-quality video, but at high cost. Also, the technical requirements for services and optimal transmission formats were undefined at the time, so efforts on system and device development were unfocused. Nevertheless, both the telephone and cable-television sectors recognized the potential of optical fiber, and research and some development continued unabated.

II. BROADBAND SERVICE REQUIREMENTS

Visual information—both still images and full-motion video—has always been central to concepts of broadband services, as has been the concept of interacting with the information itself. This is increasingly termed "multimedia" for voice, video, and data. Multimedia services are assumed

[1] This trial was known as the Hi-OVIS experiment, for "Higashi-Ikoma Optical Visual Information System" for its location in the Higashi-Ikoma area of Japan near Osaka, but later for "Highly Interactive Optical Visual Information System."

Reprinted from *Proc. IEEE*, vol. 82, no. 4, pp. 604–614, April 1994.

to require broadband bit rates in excess of 1.5 to 2 Mb/s, and this has influenced standards directions during the last decade[2]. Also, for services other than teleconferencing or sending stored images from the home, residential service demands are often considered to be asymmetric: Downstream bit rates for movies, games, databases, libraries, etc., will exceed those upstream which control the incoming information.

Broadband ISDN standardization within CCITT focused on meeting existing and future video bit-rate requirements as well as being maximally compatible with national digital hierarchies. The basic international broadband access rate is 155.52 Mb/s (called STS-3 in North America and STM-1 elsewhere), and higher rates are byte-interleaved synchronous multiplexes of this basic rate [22]. In North America, there is a synchronous subrate at 51.84 Mb/s (STS-1) which is compatible with the existing 44.736-Mb/s asynchronous transmission rate. Also in North America, the emerging synchronous hierarchy is known as SONET (Synchronous Optical Network) [23], while elsewhere it is the SDH (Synchronous Digital Hierarchy).

A. Asynchronous Transfer Mode

A wide range of bit rates for both isochronous and anisochronous services[3] must be accommodated for "bandwidth on demand." Wide recognition of this need is leading to the adoption of a packet-like multiplexing and switching technique known as asynchronous transfer mode or ATM [24], [25]. ATM uses short, fixed-length packets or cells (53 bytes) transmitted over low-error-rate channels so that simple protocols can be used to minimize transmission delays. This is quite unlike conventional packet protocols such as X.25 which cannot support isochronous services.

In ATM, the term "asynchronous" refers to the lack of alignment between the packets or cells associated with any particular channel or service and the transmission frame. This is illustrated in Fig. 1(a), where cells associated with particular services are designated with letters. Service "A" (e.g., low-speed data) requires fewer bytes per frame than services "B" or "C" (e.g., video), so "B" and "C" cells appear more frequently, in no particular alignment with

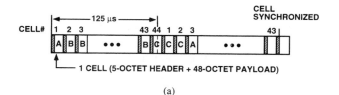

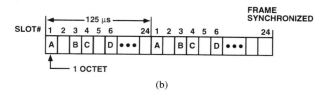

Fig. 1. Comparison of asynchronous (cell-synchronized) and synchronous (frame-synchronized) transfer modes. The structure shown in (a) portrays current SONET-based ATM, and (b) portrays standard T1 framing. The shaded area on the left side of each cell in (a) represents the header. The conventional framing bit in T1 is not explicitly indicated in (b).

the 125-μs frame. Thus both isochronous services like video and voice can be multiplexed with anisochronous services like data. This is different from the synchronous transfer mode (STM) shown in Fig. 1(b). Figure 1(b) indicates a standard "T1" circuit-switched transmission frame consisting of 24 64-kb/s channels. Bytes for channel "A" are allocated synchronously, in the same positions in the frame and for the duration of the call, whether or not there is content to fill them. Unused channels are blank in the figure. In addition to restricting services to the basic synchronous rate or multiples thereof, STM also results in switches being optimized for this basic rate, greatly limiting flexibility.

Either ATM or STM can be transferred via a network that itself is either synchronous with a master network clock, like SONET, or asynchronous like most of today's networks. Timing differences for the latter are made up for using "stuffing" bits inserted and removed during transmission.

B. Advances in Video Technologies

In parallel with the international broadband standards activities during the last decade, dramatic progress in video transmission and compression has been made: analog transmission of 80 or more AM-VSB (amplitude-modulated vestigial sideband) channels is now possible over many kilometers of fiber while maintaining signal-to-noise ratios above 50 dB, and compression of digitized video signals by factors of between 50× and 100× are recently possible [26]. This latter accomplishment means that video-recorder quality is possible using only 1.5 Mb/s. At 3 to 4 Mb/s, off-air broadcast quality can be achieved, while studio quality is possible at about 6 to 8 Mb/s. "Advanced television" (e.g., high-definition TV) can be similarly compressed, from a raw bit rate as high as 1.2 Gb/s to near 20 Mb/s. This factor of about 60× is six times lower than the best HDTV compression only five years ago. Compressed digital video is having a major impact on multimedia services, including storage, retrieval, and transmission. More broadly, the relentless progress in fiber technologies and transmission, and

---

[2] During the last decade, the diferences between narrowband, wideband, and broadband have been widely discussed. Narrowband was generally understood to mean bit rates less than 1.5 (1.544) to 2 (2.048) Mb/s, which could be transported via ISDN (Integrated Services Digital Network). Wideband was understood to include rates between narrowband and 32 (32.064) to 45(44.736) Mb/s, although some place the cutoff at 10 Mb/s. (The specific values in parentheses relate to standard bit rates in international telecommunications hierarchies.) Broadband simply spanned bit rates above wideband. With the progress in digital video compression, however, traditional multimedia "broadband" services can now be transmitted at much lower rates, and some multimedia applications are compatible with ISDN. Therefore less attention is paid to assigning precise distinctions among these terms.

[3] Isochronous services are characterized by a continuous bit rate (e.g., full motion video) and traditionally have required a circuit-switched connection; i.e., a dedicated circuit maintained for the duration of a session. Anisochronous services, on the other hand, are bursty (e.g., keyboard entries, file transfers) and can be transmitted as packets or cells via a packet-switched or "connectionless" network. For further discussion, see M. Schwartz, *Telecommunications Networks: Protocols, Modelling and Analysis.* Reading, MA: Addison Wesley, 1987.

in the digital-video arena, will have fundamental effects on residential services beginning this decade.

III. Telco Fiber to the Home

Numerous FTTH trials conducted throughout the 1980's were largely technology exercises to assess how well traditional narrowband services could be delivered and to test the delivery of video services. Most trials utilized a two-fiber single- or double-star approach with active optoelectronics dedicated to each subscriber. Single- and double-star topologies are illustrated in the upper and lower parts of Fig. 2, respectively. In the double-star topology, the number of transmission fibers is reduced by multiplexing customers' channels to higher bit rates for transmission over the "feeder" portion of cable between the central office and a remote terminal site (RT). An example of such a system today is digital loop carrier (DLC). Since by the late 1980's, it was widely assumed that 155 to 622 Mb/s would be provided to each home, feeder rates were assumed to be in the vicinity of 2.5 Gb/s. Fiber-to-the-home installations demonstrated that the cost of directly substituting digital fiber for analog copper at that time was prohibitive. Widespread deployment of FTTH would require either greater sharing of the optics or significantly lower optical technology costs.

During the late 1980's, studies of passive optical networks (PON's) [27], [28] and bus networks suggested alternative architectures that shared optical fiber and components. These studies showed the potential for achieving cost parity with traditional copper delivery. However, emphasis remained primarily (but not exclusively) on delivering narrowband services, such as "plain old telephone service" or POTS.

Original PON concepts eliminated the need for active multiplexing equipment in the loop, but still terminated fiber at the home. The bus approach was the first to introduce the concept of fiber-to-the-curb (FTTC), where the optoelectronics interface is backed out from a dedicated point at the home to shared equipment near the street [29]. This curbside interface is designated an Optical Network Unit (ONU)[4], and final connections (the drop or service cables) are made using metallic media. Fiber-to-the-curb is illustrated in Fig. 3 along with the PON concept, shown in the figure as a "passive distribution network" or PDN. Also shown in the figure is the optional use of a digital-loop-carrier platform for longer-loop applications. If the PDN is utilized in the distribution portion of the network, splitting is generally achieved using passive multiport splitters, although wavelength multiplexing is also an option.

Even though copper drop media remain, FTTC achieves a 99% replacement of metallic conductors with fiber, greatly simplifying FTTH upgrades when they may be warranted.

---

[4] The ONU is often portrayed as a curbside above-ground unit mounted in a pedestal enclosure, but an ONU can also be housed in a pole-mounted enclosure, in an aerial strand-mounted enclosure, in a below-ground enclosure in a handhole, or even inside a multifamily unit such as an apartment complex or small business. The ONU also need not strictly be located at a curb, but may also be placed near utility accesses in rear-lot-fed installations.

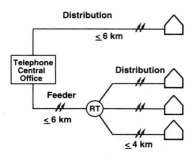

Fig. 2. Single- and double-star telephone fiber-to-the-home loop networks asenvisioned circa 1985.

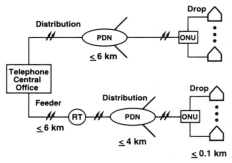

Fig. 3. Single- and double-star telephone fiber-to-the-curb loop networks with additional provisions for passive distribution networks using optical splitters.

Curbside electronics provide greater opportunities for in-service testing, for scheduled maintenance, and for automated provisioning of new lines and services, and possibly even for enabling self-healing features to survive accidental cable cuts. New features such as these are expected to reduce ongoing (life-cycle) costs of operating the network, continuing the historical improvements in network efficiency. Life-cycle savings will be discussed shortly.

By early 1990, numerous fiber architectures and first-generation products were available and being tested throughout the telephone network. It is undesirable to deploy multiple architectures lacking a common physical infrastructure, particularly in the distribution plant where the cost of future changeouts is highest. The distribution plant as identified in Figs. 2 and 3 is the part of the loop that provides the final connection to the end user, which may entail placing buried cable along residential streets and buried drops across customers' lawns. Consequently, telephone companies want this part of their network to be free from the need constantly to refurbish or add new cable to expand capacity.

To support the Bell Operating Companies in establishing a common direction for fiber access systems, Bellcore developed generic system requirements known as "TR-909"[5]

---

[5] Bellcore Technical Reference, "Generic requirements and objectives for fiber in the loop systems," Tech. Ref. TR-NWT-000909, Issue 1, Dec. 1991. This document describes distribution architectures, minimum service specifications for 12-, 24-, and 48-line optical network units, operations and maintenance specifications, optical spectrum allocations, power limits, etc., but leaves unspecified the technical details of the system itself to promote development innovations among the manufacturers. The document was reissued in December 1993, to include additional service and system options.

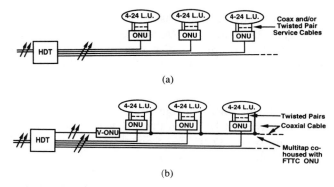

(a)

(b)

HDT = Host Digital Terminal, located at remote site or central office
ONU = (Curbside) Optical Network Unit
V-ONU = Video Optical Network Unit
L.U. = Living Unit

Fig. 4. Fully integrated delivery of narrowband through broadband services over fiber-to-the-curb (a), compared with a dual-system hybrid in (b). Also shown in (b) is additional sharing of the video ONU across multiple narrowband ONU's.

which apply to a physical fiber-to-the-curb architecture capable of supporting either point-to-point or point-to-multipoint (i.e., PON) transmission systems. The basic concept underlying this architecture would be to place distribution fiber on a point-to-point basis from the last location in the network beyond which future reinforcement would be undesirable or cost-prohibitive. This location could be a conventional remote electronics site or a PDN coupler/splitter site[6]. Additionally, terminals near customers were defined to allow sharing of the optoelectronics among end users. This architecture has become the industry norm for FTTC deployment.

Although the basic FTTC architecture has the inherent capability to support broadband services, initial requirements defined by Bellcore concentrated on providing existing narrowband services. The primary thrust of this activity has been to accelerate deployment and quickly reach installed-first-cost parity with copper-based digital-loop-carrier alternatives. In many applications, this objective has been achieved or exceeded. Equally important are life-cycle costs, mentioned earlier, which quantify ongoing expenses associated with operating, administering, maintaining and provisioning (OAM&P) service over the life of a system. Possible annual maintenance savings on the order of 23% to 26%, and provisioning savings as high as 90% have been predicted [30]–[32]. These are significant enough so that, even with the higher cost of electrical powering for FTTC, a net life-cycle savings approaching $200 per line over 20 years appears possible [33]. As first-costs continue to decline, and as life-cycle savings are verified, fiber will be used more and more to replace existing copper service as part of continuing maintenance. Increasingly, now, attention is turning to adding broadband capabilities to these new fiber systems.

With the recent FCC decision on video dialtone [34], namely, to describe two levels of video service that local-exchange carriers may now offer[7], the evolution of FTTC systems to support broadband services (or the equivalent) has sparked widespread interest. Initial directions for providing a long-term plan for delivery of switched digital video services are being re-evaluated as cable companies increasingly review the opportunities to offer telephone services over their networks.

For positioning FTTC for video dialtone service delivery, two popular approaches include integrated digital delivery of both narrowband and video services on a single system, and hybrid delivery of digital narrowband services on one system combined with either analog and/or digital video services on another.

IV. INTEGRATED NARROWBAND/BROADBAND FTTC

A fully integrated approach to narrowband plus video/broadband is straightforward and is eminently upgradable to a wide range of switched services [35]. Furthermore, by incorporating ATM to provide bandwidth flexibility, services with different bit rates (from telemetry and voice to high-definition video) and traffic characteristics (continuous or bursty) are easily accommodated. In an integrated approach such as illustrated in Fig. 4(a), all services for the customers served from an ONU are time-division-multiplexed into a bitstream which directly modulates a light source at the HDT, to be transmitted over a single fiber to the ONU. (This system is denoted as a *baseband* system, to distinguish it from a *passband* system where different services, although also carried over a single fiber, are subcarrier-multiplexed rather than time-division multiplexed.) Telephony signals are generally demultiplexed from the bitstream at the ONU, analog-converted, and augmented with the usual battery and ringing voltages, on-hook/off-hook supervision functions, etc. Telephony is delivered via conventional copper twisted-pair drops to each home, and is indistinguishable from service provided today over copper.

Baseband video signals are partially demultiplexed from the bitstream arriving at the ONU so that each home receives only its intended channels, but in a digital format delivered over coax drops. Each television set or other video equipment requires a set-top interface for digital-to-analog conversion and for upstream communications to control the video. Other approaches, of course, are possible such as decoding at the ONU and remodulating as AM-VSB, but this may compromise the video and/or sound qualities otherwise attainable, as well as being expensive and requiring additional electrical power. (The increased power needed for video or broadband is an important issue. If provided over the network, it can be costly. The most

---

[6] A passive optical coupler, splitter, or wavelength-division multiplexer affords the same one-to-many topological and multiplexing flexibilities that an electronic multiplexer site offers, but achieves it without the need for active electronics or powering. Such devices are also transparent to upgrades that involve higher bit rates or a change in the modulation format. Only the equipment at the ends of the network need be changed.

[7] As a "level-one platform," a local-exchange carrier can provide video services on a nondiscriminatory, common-carrier basis, including end-to-end transmission plus basic directory and routing functions. As a "level-two gateway," the carrier can provide advanced video gateway and related services (except programming), unregulated and subject to competition.

economical approach is to minimize the power consumed at the ONU and carry out more functions in the home where the cost of providing power is as much as ten times less [36]. This issue applies for either advanced telephony or cable networks.)

Should it prove desirable to eliminate the coax drop, extra copper pairs in the telephone service cable may have the capability to deliver digital video at aggregate bit rates up to 52 Mb/s or higher over the short drop distances. [37] This area is currently being examined [38]–[40].

Chief attractions of integrated baseband delivery are its robust signal and its similarity to ordinary telephone transmission (affording commonality in testing procedures and equipment, and craft training), and its inherent bidirectional transmission capability over fiber which allows for unlimited interactivity and upstream bandwidth for future services.

For the *passband* integrated approach, Fig. 4(a) still applies, but multiplexing is carried out in the frequency domain. Different services first modulate electrical subcarriers (e.g., using QPSK or QAM for digital services and AM-VSB for analog services). Both microwave and conventional television subcarriers have been proposed [41]. After electrically combining the RF subcarriers, the composite signal modulates the light source for optical transport to the ONU. At the ONU, telephony services can be demultiplexed and delivered to the appropriate customer over copper pairs, providing today's level of privacy and security, or they can remain on one of the RF carriers delivered over coax to be processed in the home. The video in the FTTC passband approach is essentially an AM-VSB broadcast system where the ONU provides the O/E conversion. The ONU could provide additional functionality, such as channel interdiction to prevent theft of service. In this approach, interactivity is limited to the narrow bandwidth available for telephony services.

The chief advantage of a passband system is that it can preserve the AM-VSB format for compatibility with existing television/VCR equipment, eliminating the need for individual set-top boxes for signal processing. It also maintains the capability for adding digital signals on other subcarriers which, however, would need signal processing in the home. In a passband approach, the subcarrier modulation requires additional analog test equipment, procedures, and practices which are new for the telephone craft, however.

V. HYBRID NARROWBAND/BROADBAND FTTC

As we define the hybrid approach, *separate physical systems* deliver narrowband and broadband services. Telephony is carried via an FTTC system designed to meet current TR-909 generic requirements. Video services are often carried using AM-VSB over a separate fiber either directly to each narrowband ONU or to a more central node (a "video ONU" or V-ONU) for subsequent distribution to several narrowband ONU's via a coaxial cable bus.

A hybrid delivery system is illustrated in Fig. 4(b). In this figure, the video fiber terminates on such a shared video ONU. Economic studies suggest that at least 64 homes total should share the V-ONU if used in low-density single-family living-unit areas. (This is similar to the cable television fiber/coax approach described below, but serving fewer total subscribers.) If four homes derive service from each narrowband ONU, then the V-ONU might serve up to 16 narrowband ONUs fed from a tapped coax bus. If the upper limit of 24 homes suggested in the figure applies, then the V-ONU would serve only three narrowband ONUs. With only approximately 64 homes served from a V-ONU, access to dedicated video dialtone channels with relatively low blocking probability will be possible, depending on the number of channels delivered, usage characteristics, etc.

Initial studies suggest that this approach provides cost-effective deployment of broadcast capability plus limited interactivity. Again, a key advantage is the possibility of carrying the AM-VSB format directly, meaning converters are not needed at each set except as might ordinarily be required for descrambling. Another advantage is the flexibility to deploy the video and telephony systems together for new construction, or only the video system if the in-place telephony plant is in satisfactory condition.

An all-digital hybrid system has been analyzed which, although requiring converters at each set, reduces optoelectronics costs and the sensitivities to noise and reflection characteristic of analog transmission. One-hundred forty compressed NTSC broadcast channels (4 Mb/s each) plus 28 switched channels are carried in 42 6-MHz channels using 16-QAM modulation [42].

Contrary to occasional misconceptions, the cost of either a fully integrated FTTC system, or a hybrid voice-plus-video system, is less than the sum of the costs for separate networks; i.e., one placed for telephony and another separately placed for cable service. This suggests possibilities for cooperation between cable and telephone companies, discussed later in the paper. More importantly, the numerous options fiber offers for transporting broadband services, at costs becoming more similar with time, make it attractive to accelerate installing such networks to capitalize on new business opportunities.

VI. CABLE FIBER TO THE NODE

Fiber's potential advantages over coaxial cable were also immediately recognized by the largest user of coax—the cable television industry. Work on multichannel analog video transmission over fiber began in the mid-1970's [43]–[45]. Laser properties, specifically noise, modal instabilities, and distortion at that time limited the number of channels, but progress was made in reducing light-source nonlinearities using linearization techniques [46]–[49].

Robust modulation techniques such as subcarrier FM or digital PCM could overcome the noise sensitivities of optical analog transmission. But since both required significant spectrum (36 to 40 MHz for FM and 100 Mb/s per NTSC video channel at that time), as well as

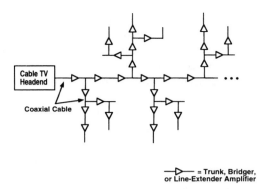

Fig. 5.   Today's cable-television tree-and-branch network based on coaxial cable.

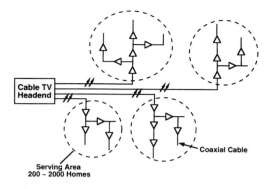

Fig. 6.   A tree-and-branch cable network repartitioned into small serving-area nodes using fiber trunks followed by short sections of amplified, tapped coaxial cable.

added cost for modulation and coding, neither format was extensively used except in the cable supertrunk network between head-end facilities. The ultimate goal remained to transmit conventional analog video in 6-MHz channels.

During the early 1980's, research and development led to very-high-speed, narrow-linewidth lasers for long-distance, high-bit-rate transmission. The distributed-feedback (DFB) and dynamic-single-mode (DSM) lasers that emerged by the mid-1980's were the first devices that promised to work for analog transmission. Early experiments with 20 to 30 channels were extended quickly to 40 and more. Currently, both directly modulated and externally modulated lasers can transmit up to 80 channels [50]–[52] and over 100 appears on the horizon.

At the same time, a new fiber topology was proposed for repartitioning the cable television tree-and-branch network into small groups (nodes) of subscribers, each node served by dedicated fiber. In a tree-and-branch network (Fig. 5), many thousands of subscribers can be served from a cascade of as many as 50 amplifiers, limited in extent by noise and distortion accumulation [53]. The signal-to-noise ratio in decibels decreases linearly with the number of amplifiers. In addition, a single amplifier failure can simultaneously affect thousands of subscribers.

In the repartitioned network (Fig. 6), first called a "fiber backbone, [54]" fiber deployed analogous to telco feeder leaves from a minimum of one (at the fiber-to-coax transition) to only a few amplifiers between any customer and the originating signal. This reduction in cascade depth of $10\times$ to $20\times$ reduces unavailability attributable to amplifier failures by a similar amount, from over 1000 min/year (0.997 availability, a typical figure) to 100 min/year (0.9998 availability) or less.[8] Signal-to-noise ratio and bandwidth potential are also improved, the former by less accumulation of noise and distortion, and the latter by shortening the length of coaxial cable which has an inherent

[8] Annual unavailability (h/year) can be defined as 8760 times the system mean-time-to-repair (MTTR_s) divided by the sum of the system mean-time-to-failure (MTTF_s) and MTTR_s. For n equivalent amplifiers in series, MTTF_s is the amplifier MTTF_A divided by n. Since $\text{MTTF}_s \gg \text{MTTR}_s$, the unavailability is approximately $\text{MTTR}_s \cdot n/\text{MTTF}_A$. If MTTF_A is 4 years and MTTR_s is 2 h, then a 40-amplifier cascade has an annual unavailability of 20 h, whereas a 2-amplifier cascade has an annual unavailability of 1 h.

Unavailability figures are taken from various articles in the trade press.

bandwidth·distance[2] product (an alternate statement of the $\sqrt{f}$ dependence of attenuation) [55].

Numerous variations on this original concept have followed, such as fiber-to-the-feeder[9], fiber-trunk-and-feeder, etc., or collectively "fiber-to-the-serving-area" (FSA). All minimize or eliminate cascades of amplifiers.

By sizing the node appropriately, the cost of building, rebuilding, or upgrading cable plant with FSA can be held to approximately the cost of traditional construction alternatives. The operational benefits are a significant improvement in downtime (unavailability), a typical increase in received quality of 8 to 10 dB in signal-to-noise ratio, and an improvement in the potential bandwidth of several hundred megahertz (if utilized by respacing amplifiers and adding channels, all at additional cost). For example, today's 35- to 80-channel systems (300 to 550 MHz) can be raised to as many as 150 channels, or about 1 GHz, using emerging amplifier technology. In combination with digital compression and spectrally efficient modulation formats such as 16-QAM (3.3 b/Hz), 64-QAM (5 b/Hz) or 16-VSB (7 b/Hz), these 150 analog channels could, in principle, carry as many as 1500 downstream digital video channels. As with the previously described FTTC alternatives, the use of set-top boxes for processing digital signals received at the home is required.

Figure 7 shows two proposals for utilizing this new capability. Both utilize 54 to 450 MHz for conventional AM-VSB broadcast video delivery, and both allow for limited upstream transmission using the 5- to 30-MHz "subsplit" band. The AT&T concept in Fig 7(a), known as Cable Integrated Services Network [56], adds 30 SONET STS-1 channels (51.84 Mb/s) downstream between 450 and 750 MHz, to be shared by the customers served from the node. The use of 64-QAM fits each STS-1 into a 10-MHz channel. The spectrum above 750 MHz is reserved for 75 channels of upstream voice and wideband data. (To use this spectrum, additional costs would be incurred for equipment at the home and in the network, beyond the cost to upgrade the plant to FSA, and is necessary to expand the initial capability to support higher levels of interactivity.)

[9] The cable television industry denotes as trunk plant what the telephone industry calls feeder plant, and as feeder what telephone calls distribution. Both designate the last connection as the drop cable.

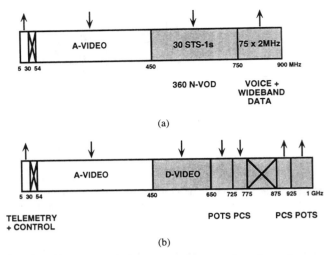

Fig. 7. Two proposed frequency assignments for advanced 1-GHz cable television coaxial plant. (a) AT&T/Optical Networks International's "Cable Integrated Services Network [66]." (b) Time Warner's "Full Service Network" as proposed for testing in Orlando, FL [57].

The Time Warner plan in Fig. 7(b), part of their Full Service Network [57], adds 200 MHz of downstream compressed digital video channels between 450 and 650 MHz, to augment the 54- to 450-MHz analog channels. These, of course, can be allocated among the customers served by the node for dedicated "dial-up" services. The spectrum above 650 MHz is reserved for two-way telephony (POTS) services and wireless personal communications services (PCS). The PCS channels would be terminated at the fiber/coax nodes and connected to PCS microcell transmitting and receiving equipment.

Neither proposal relies extensively on use of the subsplit band for two-way services, although many uses for this portion of the spectrum have been proposed, including telephony, Ethernet, and Personal Communication Service. To increase the capacity of the subsplit band and to avoid interference sources such as Citizens Band radio, there is current thought to expanding it include 5 to 42 MHz [58]. The limitations of the subsplit band have been discussed, as well as possible disadvantages of locking in the upper end ("supersplit" region) of the spectrum prematurely [59].

Should it prove undesirable to use the coax plant at frequencies approaching 1 GHz, the top half of the spectrum (i.e., above 450 MHz in Fig. 7) could be placed on a second cable. So-called "double-cable" systems have been used for a decade to expand capacity beyond 80 channels.

Key to both proposals is the ability to use the two-way capability of the coax plant. Although in principle available in larger systems since the early 1970's, two-way communications using the 5- to 30-MHz (42-MHz) subsplit band for upstream messages was rarely provisioned because of cost, limited need for the capability compared with alternative means (such as the telephone network), and problems with noise ingress and accumulation in the upstream direction [60], [61]. In an FSA topology, two-way capability is significantly enhanced because of the smaller number of amplifiers. In fact, for nodes smaller

than about 200 homes, line-extender amplifiers between the optoelectronic interface and all subscribers can be eliminated. Such "passive" nodes remove the need to provide 60 V_{ac} power over the coax (unless supplied for interdiction equipment), the need to provide downstream equalization, slope or gain compensation, and the need to install diplexers at each amplifier for upstream signaling. The disadvantages of small, passive nodes are the need for more optical receivers, associated power supplies, and more upstream lasers. Currently, although node sizes usually range from 500 to 2000 homes passed[10], passive nodes as small as 200 homes are being engineered [62]–[64]. It is our belief that, as optoelectronics and fiber costs continue to fall, nodes will become generally smaller to increase the available bandwidth per subscriber. Extension of fiber to the tap (curb) or home has previously been considered for cable television delivery systems [65]–[70].

With compression, appropriate modulation schemes, small node sizes, and two-way operation, it is clear that fiber/coax networks offer possibilities never envisioned a decade ago. These capabilities are central to providing broadband services over networks relying on coaxial cable in the distribution part of the network.

While cable's FSA nodes move to smaller sizes, telco FTTC systems are being tested for larger nodes for reducing near-term costs, particularly for rehabilitation in dense areas. At some point, the system architectures of the two industries will become very similar: fiber-rich nodes serving tens of homes, both capable of interactive video and voice services. On the other hand, we feel fiber-to-the-home will once again become economically attractive with falling costs. Technology costs continue downward, closing the gap between FTTC and FTTH. Life-cycle savings from eliminating metallic drop testing (both copper pair and coax) and curbside powering may close or eliminate the first-cost gap. The result is a growing belief that FTTH will become a factor in the next few years.

VII. THE FUTURE

Many telephone companies and cable operators are on fast tracks to deploy fiber close to the subscriber: Telcos with FTTC, hybrid fiber/coax, or even cable-like FSA, the first two possibly evolving to include FTTH within the decade, and cable operators with FSA, with smaller and smaller node sizes, but probably not FTTH in the foreseeable future.

How fast might these transformations occur? Telephone companies rehabilitate or rebuild about 3% of their lines annually as part of routine maintenance. Therefore, it would take on the order of 30 years to convert to fiber if, for every situation, fiber were the most cost-effective alternative. Recognizing the growing needs of business customers for advanced services, the business opportunities that video dial tone may offer [71], [72], and positive effects on the national economy [73]–[76], several Regional Bell

---

[10] Current practices are to engineer nodes for upgrades, which have the lowest cost target, at about 2000 homes, and for rebuilds and new-builds at abut 500. These vary depending on the specific details of each situation.

Operating Companies (RBOC's) have announced accelerated infrastructure plans. The changeover to fiber (and accompanying switches, feeder, and interoffice trunks), reaches "completion" generally between 2010 and 2020. A standard S-curve (e.g., Fisher–Pry) representing such schedules suggests that 10% to 20% of homes will have switched broadband access by the end of the decade, and 50% three or four years later. Examples of these accelerated plans include New Jersey Bell (2010), Bell of Pennsylvania (2015), Pacific Bell (2015), BellSouth (2011), and US WEST (2020).

Cable operators typically rebuild or upgrade from 8 to 10% of their plant annually, so a conversion to fiber could be completed in about 10 to 12 years. Motivating such a changeout is the fact that fiber's benefits are significant and immediate, in terms of quality and reliability. There are also the business opportunities in advanced services and targeted advertising. Conveniently, FSA technology is available now, coinciding with a peak in rebuild activity, after earlier construction around 1980 to address demand for then-emerging premium services. This construction gives cable an opportunity to accelerate FSA deployment. The National Cable Television Association has estimated that $14B will be spent over the next ten years in upgrading plant and equipment, and that 75% of existing systems will be rebuilt [77]. More specifically, TCI, the largest multiple-system operator (MSO) in the U.S., expects 90% of its subscribers to be connected to FSA networks by the end of 1996 [78]. The recent alliance between US WEST and Time Warner, the second-largest MSO, will allow Time Warner to accelerate construction of its Full Service Network [79].

As telcos and cable rapidly deploy similar networks, both with an eye on similar if not the same markets, there is clearly an opportunity for competition. The current situation is often referred to as the "telco-cable battle." Indeed, there are examples where telcos have announced plans for delivery of video services which, although cooperating with one service provider, compete for the same subscribers with another, an "overbuild" situation [80], [81]. Similarly, cable operators have announced plans for offering voice services in competition with telcos, usually planning to provide interexchange access, bypassing the local-exchange carrier's access charges [82]–[84].

A. Telco/Cable Cooperation

Equally clear, though, are opportunities for cooperation or alliances. Recent examples include Sammons arranging with New Jersey Bell to provide video dialtone service using FTTC facilities in lieu of rebuilding an aging cable plant [85], [86], Time Warner partnering with US WEST for funding to accelerate deployment of the Full Service Network outside of US WEST territory [87]-[89], and Time Warner working with NYNEX to test video dial-tone service in New York [90], [91]. The authors do not profess to know how the issue of cooperation versus competition will play out over the next few years. The regulatory environment will more likely steer developments than advances in technology. Perhaps the most significant of these

is the recent ruling [92], [93] that telephone companies may provide programming within their service areas, thus setting aside restrictions from the 1984 Cable Act [94]. Many future changes are anticipated as Federal policy emerges to promote building of a national information infrastructure. Nevertheless, we do see several areas where the two industries have complementary strengths, and we feel that cooperation will stimulate growth of advanced networks capable of broadband service delivery.

B. Construction Costs

As noted earlier, the cost to build two independent networks—one for telephony and one for cable—is higher than integrating them. This obviously suggests scenarios for cooperation to reduce costs where both service providers are faced with either new construction or extensive rebuilding of old plant.

C. Operations Support Systems

Telcos have highly developed OSS's in place, and processes and procedures for developing and implementing new ones. Telcos have depended on these "management information systems" increasingly since the 1960's, to deal with increasingly complex networks. Although new OSS's must be developed to deal with switched, transactional multimedia/multiparty services, telcos are in an excellent position to work with cable operators in designing, implementing, and supporting such software.

D. Programming and Packaging

Cable operators have broad experience in programming and program packaging, making it an ideal video information provider for video dialtone services. (Broadcasters have this strength as well, but are limited in the number of programs they can produce by their transmitting facilities or other outlets. As video information providers, broadcasters can become "multiple-channel programmers" with an almost unlimited number of outlets for new material [95].)

E. Quality and Pricing

Telcos have a reputation for high-quality service and stable pricing. In a recent survey, telephone service was rated eighth out of fifty products and services.[11] Although prices have risen with time and inflation, they have trailed the Consumer Price Index since the 1935 [97]. Working with cable operators as partners, this reputation may stimulate new subscriptions to cable service delivered as video dialtone (or simply to new cable service for cable systems that may be owned by RBOC's outside their serving areas)[12] at

[11] In a 1993 survey of 50 consumer products and services by the Conference Board, telephone service and electric service were ranked 8th and 10th, respectively. The only service ranked higher was videotape rental (2nd).[96].

[12] As of 1990, 231 independent telcos (15% of the total number) provided cable television service through over 500 systems within their serving areas. These companies were exempt from the ownership restrictions of the Cable Communications Policy Act of 1984 ("1984 Cable Act") through the rural exemption or other waiver provisions in Section 613.

a time when new subscriptions are inevitably approaching saturation.[13]

F. Standards and Open Interfaces

Telcos, cable, and consumers will all benefit from having standardized equipment and open interfaces, particularly on the customer's premises, so that, like software for the IBM PC, the development of new services becomes widespread and competitive. Telcos have a long history of working with standards bodies, but common belief now is that in a growing number of cases this process results in untenable delays. In cooperation, telcos and cable may be able to speed such processes, with industry buy-in.

G. Ubiquitous Networks

Cable operators currently pass more than 90% (and serve 64% [98]) of homes with a coaxial plant having the capability of providing early broadband services. New service offerings could be accelerated through a cooperation in which telcos and cable work to upgrade this plant, install switching, develop new transactional billing systems, perhaps use the coax plant to facilitate powering of telco FTTC ONU's, and perhaps leverage the underlying narrowband telephone network to integrate or coordinate the two networks, simplifying the upstream control needed for multimedia services. This could be further enhanced using ISDN (Integrated Services Digital Network) which is becoming widely available during in the next few years.

H. Switched Backbone

Telcos offer an advanced, ubiquitous switched backbone network which is rapidly evolving to broadband ISDN capabilities including SONET, ATM, and network intelligence. This backbone is crucial for achieving widespread connectivity to information providers as well as additional customers who will use multiparty services such as teleconferencing and telecommuting.

I. Customer Experience

Cable has broad experience working with residential customers in providing entertainment services, whereas telcos have complementary experience in working with business customers and services. These areas need coupling to move into many of the advanced broadband services alluded to at the beginning of the paper.

J. Common Goals

Finally, both industries are anxious to see the development of a National Information Infrastructure, where both also have major roles. We feel that cooperation will accelerate this goal, whereas competition will probably delay it.

---

[13] Fifty percent cable penetration was reached during 1987, at which point new subscribers were theoretically being added at the highest rate—the slope at the inflection point. Subscription rate will decline in subsequent years unless stimulated by some new service offering. For example, the emergence of premium channels distorted the usual S-curve during the 1980's.

VIII. CONCLUSION

We have reviewed the major aspects of telco and cable activities over the last two decades in learning to use fiber and exploit its advantages over metallic transmission media. Although fiber (or fiber combined with coax) is only one of several important media for delivering interactive broadband services (existing copper pairs operated digitally at up to 6 Mb/s, and new terrestrial- and satellite-based microwave options are being proven in even now, and should all be very important options during this decade), fiber will eventually be the most important. Cable may have a majority of subscribers served from FSA networks in as few as ten years, whereas telcos may only reach 50% of subscribers in the same time with FTTC and/or FTTH. But working together, the networks that *are* built can be open, interworkable, switched, and of highest quality, all requirements for "Information Age" services and for being leading members of the 21st Century international telecommunity.

REFERENCES

[1] S. B. Weinstein and P. W. Shumate, "Beyond the telephone—New ways to communicate." *The Futurist*, pp. 8–12, Nov./Dec. 1989,
[2] D. J. Wright, "Strategic impact of broadband telecommunications in insurance, publishing and health care," *IEEE J. Selected Areas Commun.*, vol. 10, pp. 1369–1381, Dec. 1992.
[3] J. Rosenberg *et al.*, "Multimedia communications of users," *IEEE Commun. Mag.* (Special Issue on Multimedia Communications), pp. 20–36, May 1992.
[4] G. Lawton, "Utility tests fiber network," *Lightwave*, pp. 1, 24–26, Mar. 1993.
[5] S. McCarthy *et al.*, "Project ThinkLink—Educational applications for FITL," in *Proc. Nat. Fiber Optics Engr. Conf.*, vol. 2, pp. 255–261, 1993.
[6] C. Holliday and V. Junkmann, "Broadband services begin to shape up," *Telephony*, pp. 26–29, Apr. 26, 1993.
[7] S. B. Weinstein, *Getting the Picture.* New York: IEEE Press, 1986, pp. 52–55 and pp 87–89.
[8] M. Kawahata, "Hi-OVIS (Higashi Ikoma Optical Visual Information System) development project," in *Proc. Integrated Optics and Optical Commununications (IOOC'77)*, pp. 467–471, 1977.
[9] M. Kawahata, "Hi-OVIS in the new media era," in *Proc. 1983 NCTA Conv. (Cable'83)*, pp. 195–201, 1982.
[10] K. Y. Chang and E. H. Hara, "Fiber-optic board-band integrated distribution–Elie and beyond," *IEEE J. Selected Areas Commun.*, vol. SAC-1, pp. 439–444, 1983.
[11] G. A. Tough and J. J. Coyne, "Elie—An integrated broadband communication system using fiber optics," in *Proc. 1983 NCTA Conv. (Cable'83)*, pp. 202–206, 1982.
[12] J. Kanzow, "BIGFON: Preparation for the use of optical fiber technology in the local network of the Deutsche Bundespost," *IEEE J. Selected Areas Commun.*, vol. SAC-1, pp. 436–439, 1983.
[13] E. J. Powter, "Milton Keynes and beyond, The next-generation systems," in *Proc. 1983 NCTA Conv. (Cable'83)*, pp. 185–190, 1982.
[14] P. Touyarot, *et al.*, "First lessons from the Biarritz trial network," in *Proc. SPIE*, vol. 585, pp. 138–145, 1985.
[15] C. Veyres and J. J. Mauro, "Fiber to the home: Biarritz (1984)...Twelve Cities (1988)," in *Proc IEEE Int. Commun. Conf. (ICC'88)*, vol. 2, pp. 874–878, 1988.
[16] P. W. Shumate, "Optical fibers reach into homes," *IEEE Spectrum*, pp. 43–47, Feb. 1989,
[17] M. F. Mesiya *et al.*, "Mini-hub addressable distribution system for high-rise application," in *Proc. 1982 NCTA Conv. (Cable'82)*, pp. 37–42, 1982.
[18] A. M. Flaherty, "Will video do it? Justifying fiber to the home," *Fiber Optics Sourcebook*, Phillips Pub., 1988, pp. 119–123.

[19] A. Mathur *et al.*, "Fiber-to-the-home: Heathrow system," in *Proc. FOC/LAN'88*, pp. 174–179, 1988.

[20] M. Balmes *et al.*, "Fiber to the home: The technology behind Heathrow," *IEEE LCS Mag.*, pp. 25–29, Aug. 1990.

[21] J. Parker, "Penn Bell to test multimode fiber in the home," *Lightwave*, pp. 10–11, June 1988,

[22] K. Asatani, "CCITT standardization of network node interface of synchronous digital hierarchy," *IEEE Commun. Mag.*, pp. 15–20, Aug. 1990.

[23] R. Ballart and Y-C. Ching, "SONET: Now it's the standard optical network," *IEEE Commun. Mag.*, pp. 8–15, Mar. 1989.

[24] S. E. Minzer, "Broadband ISDN and Asynchronous Transfer Mode (ATM)," *IEEE Commun. Mag.*, pp. 17–24, Sept. 1989.

[25] For a complete treatment of ATM in transport and switching, see M. de Prycker, *Asynchronous Transfer Mode–Solution for Broadband ISDN.* New York: Ellis Horwood, 1991.

[26] P. H. Ang *et al.*, "Video compression makes big gains," *IEEE Spectrum*, pp. 16–19, Oct. 1991,

[27] J. R. Stern *et al.*, "Passive optical local networks for telephony applications and beyond," *Electron. Lett.*, vol. 23, pp. 1255–1257, 1987.

[28] D. W. Faulkner *et al.*, "Optical networks for local loop applications," *J. Lightwave Technol.*, vol. 7, pp. 1741–1751, 1989.

[29] C. Wilson, "Raynet rolls out first commercial FTTC," *Telephony*, pp. 18, 22, Apr. 2, 1990.

[30] W. E. Ensdorf *et al.*, "Economic considerations of fiber in the loop plant," in *Proc. Int. Symp. on Subscriber Loops and Services (ISSLS'88)*, pp. 291–296, 1988.

[31] M. O. Vogel and L. F. Garbanati, "Estimating equipment and facility repair costs for fitl architectures," in *Proc. Nat. Fiber Optics Engr. Conf. (NFOEC'90)*, paper 1.4, 1990.

[32] H. Sinnott and W. Macleod, "Fiber access maintenance leverages," in *Proc. SPIE—Fiber Optics in the Subscriber Loop*, vol. 1363, pp. 196–198, 1990.

[33] P. W. Shumate, "Economic considerations for fiber to the subscriber," in *Proc. European Conf. on Optical Communications (ECOC'93)* vol. 1, pp. 120–123, 1993.

[34] FCC: CC Docket No. 87-266, First Report and Order, Nov. 22, 1991.

[35] P. W. Shumate and R. K. Snelling, "Evolution of fiber in the residential loop plant," *IEEE Commun. Mag.*, pp. 68–74, Mar. 1991.

[36] K. Mistry, "Video dial tone powering issues," to be presented at *Intelec'93*, Sept. 1993.

[37] P. W. Shumate and R. K. Snelling, "Evolution of fiber in the residential loop plant," *IEEE Commun. Mag.*, p. 71 Mar. 1991.

[38] J. Terry, "Alternative technologies and delivery systems for broadband ISDN access," *IEEE Commun. Mag.*, pp. 58–64, Aug. 1992.

[39] W. E. Stephens *et al.*, "Transmission of STS-3c (155 Mb/s) SONET/ATM signals over unshielded and shielded twisted pair copper wire," in *Proc. GLOBECOM'92*, pp. 170–174, 1992.

[40] P. Kothary and C. Look, "Bandwidth characterization of residential copper drops," in *Proc. Nat. Fiber Optics Engr. Conf. (NFOEC'93)* , vol. 2, pp. 127–136, 1993.

[41] See, for example, *IEEE J. Selected Areas Commun.* (Special Issue on Applications of RF and Microwave Subcarriers to Optical Fiber Transmission in Present and Future Broadband Networks), vol. 8, Sept. 1990.

[42] S. S. Wagner *et al.*, "Technical and economic analysis of a digital fiber/coax video-distribution system," in *Proc. Nat. Fiber Optics Engr. Conf. (NFOEC'93)*, vol. 2, pp. 113–125, 1993.

[43] E. H. Hara and T. Ozeki, "Optical video transmission by fdm analogue modulation," *IEEE Trans. Cable Television*, vol. CATV-2, pp. 18–34, 1977.

[44] S. Akiyama *et al.*, "3-channel TV transmission on optical fiber," in *Proc. Int. Conf. on Integrated Optics and Optical Fiber Communication (IOOC'77)*, pp. 481–484, 1977.

[45] F. W. Dabby *et al.*, "Twelve video channel transmission over a single optical fiber," in *Int. Conf. on Integrated Optics and Optical Fiber Commununication (IOOC'77)*, postdeadline paper, Tokyo, 1977.

[46] J. Strauss and O. I. Szentesi, "Linearisation of optical tranmitters by a quasifeedforward compensation technique," *Electron. Lett.*, vol. 13, pp. 158–159, 1977.

[47] K. Asatani and T. Kimura, "Linearization of LED nonlinearity by predistortions," *IEEE J. Solid-State Circuits*, vol. SC-13, pp. 133–138, 1978.

[48] K. Asatani, "Nonlinearity and its compensation of semiconductor laser diodes for analog intensity modulation systems," *IEEE Trans. Commun.*, vol. COM-28, pp. 297–300, 1980.

[49] M. Bertelsmeier and W. Zschunke, "Linearization of light-emitting and laser diodes for analog broadband applications by adaptive predistortion," in *4th Conf. on Integrated Optics and Optical Fiber Communication (IOOC'83)*, paper 30C2-3, 1983.

[50] H. Blauvelt *et al.*, "Trends in output power and bandwidth of 1310 nm, DFB lasers for FM video," in *Proc. 42nd Ann. NCTA Conv.*, pp. 385–390, 1993.

[51] R. J. Plastow, "80-channel AM-VSB CATV transmitters utilizing external modulation and feedforward error correction," in *Proc. IEEE Broadband Analog & Digital Optoelectronics Topical Meet.*, pp. 10–11, 1992.

[52] M. Nazarathy *et al.*, "Externally modulated 80-channel AM CATV fiber-to-feeder distribution system over 2 × 30 km total fiber span," in *Proc. IEEE Broadband Analog & Digital Optoelectronics Topical Meet.*, pp. 12–13, 1992.

[53] K. A. Simons, "The optimal gain for a CATV line amplifier," *Proc. IEEE*, vol. 58, pp. 1050–1056, 1970.

[54] J. A. Chiddix, "Fiber backbone-multichannel AM video trunking," in *Proc. 1989 NCTA Conv. (Cable'89)*, pp. 246–253, 1989.

[55] *Reference Data for Engineers: Radio, Electronics. Computer, and Communications.* Indianapolis, IN: Howard W. Sams & Co., 1985, pp. 29–13. *Technical Note 1019–Attenuation Versus Center Conductor Diameter*, Times Fiber Communications, Inc., June 1992.

[56] "AT&T courting cable industry with plan to develop 1.5 Gbit interactive network," *The Cable-Telco Report*, pp. 1 and 5–6, Oct. 1991.

[57] "Time Warner accelerates network expansion, eyes cooperation with telcos," *The Cable's Telco Report*, pp. 1, 11–13, Jan. 1993.

[58] "Chiddix on cable's full-service network," *Multichannel News*, pp. 3 and 54–55, Aug. 23, 1993.

[59] L. Ellis, "The 5-30 MHz return band: A bottleneck waiting to happen?" *Commun. Eng. and Des.*, pp. 40–44, Dec. 1992.

[60] A. S. Taylor, "Coaxial cable—The hostile medium," in *Proc. Cable'83 (NCTA)*, pp. 40–43.

[61] J. W. Ward, Jr. "Ingress—Sources and solutions," in *Proc. 34th. Ann. Conv. and Expos.*, pp. 11–17, 1985.

[62] G. Kim, "Syracuse rebuild brings passive networks closer," *Lightwave*, p. 1, May 1993.

[63] P. Lambert and F. Dawson, "Adelphia plans 200-home fiber nodes," *Multichannel News*, pp. 3, 31, Mar. 8, 1993.

[64] J. Selvage and V. Kurjakovic, "Passive cable network topology," *CED: Commun. Engrg. & Des.*, pp. 36–39 and 53, Sept. 1993.

[65] D. E. Robinson and D. Grubb, "A high-quality switched FM video system," *IEEE LCS Mag.*, pp. 53–59, Feb. 1990.

[66] D. E. Robinson, "Switched star fiber optic architectures for cable TV," in *Proc. 38th. Ann. Conv. and Expos.*, pp. 134–140, 1989.

[67] J. Speidel *et al.*, "New fiber network in the local loop for telecommunication and TV distribution services," *Philips Telecommun. Rev.*, vol. 49, pp. 21–28, Dec. 1992.

[68] R. E. Chalfont *et al.*, "Taking fiber to the curb via optical repeating," *Commun. Engrg. and Des.*, pp. 30–35, Aug. 1992.

[69] G. Kim, "Cable TV fiber-to-subscriber development: Recent developments," in *Proc. Nat. Fiber Optics Engr. Conf. (NFOEC'92)*, vol. 2, pp. 575–583, 1992.

[70] D. Raskin *et al.*, "1 GHz fiber-to-the-pedestal system for CATV," in *Proc. Cable'92 (NCTA)*, pp. 337–343, 1992.

[71] D. L. Wenner, "Are you ready for residential broadband?" *Telephony*, pp. 84–103, May 22, 1989.

[72] "Residential services planning program," *Noel Dunivant & Assoc.*, Feb. 1991.

[73] "Impact of broadband communications on the U.S. economy and on competitiveness," Economic Strategy Institute, 1992.

[74] "The economic impact of Bell Operating Company participation in the information services industry," Wharton Econometric Forecasting Assoc. (WEFA), 1992.

[75] "Can telecommunications help solve America's health care problems?"Arthur D. Little Co, July 1992.

[76] "An Infostructure for all Americans: Creating economic growth in the 21st century," prepared by Ameritech, Bell Allantic, BellSouth, NYNEX, Pacific Telesis, Southwestern Bell and US WEST, Apr. 1993.

[77] "Cable television and America's telecommunications infrastructure," Nat. Cable Television Assoc., Apr. 1993.

[78] P. Lambert, "TCI's $1.9B pledge for superhighway," *Multichannel News*, pp. 1, 49, Apr. 19, 1993.

[79] F. Dawson, "Time Warner goes west for partner—Executives see more cable-telco alliances," *Multichannel News*, pp. 1, 65, May 24, 1993.

[80] M. L. Carnevale, "Bell Atlantic, cable-TV operator reach pact to offer service over phone lines," *Wall St. J.*, p. B10, Dec. 16, 1992,

[81] M. L. Carnevale, "Bell Atlantic takes on cable in wireless pact," *Wall St. J.*, p. B1, Aug. 5, 1993,

[82] M. Fahey, "Cable TV finds new revenues in fiber," *Lightwave*, pp. 1, 21, Jan. 1992.

[83] C. F. Mason, "CATV operator gets into access," *Telephony*, p. 3, Apr. 27, 1992,

[84] A. Kupfer, "The race to rewire America," *Fortune*, pp. 42–61, Apr. 10, 1993.

[85] G. Kim, "Sammons dials up NJ Bell agreement," *Multichannel News*, pp. 1, 48, Nov. 23, 1992.

[86] M. L. Carnevale, "Bell Atlantic's phone network to carry cable," *Wall St. J.*, p. C15, Nov. 17, 1992.

[87] J. M. Higgins, "US WEST to buy 25% of TW cable," *Multichannel News*, pp. 1, 65, May 24, 1993.

[88] J. L. Roberts *et al.*, "US WEST and Time Warner to form strategic alliance," *Wall St. J.*, pp. A3, A10, May 17, 1993.

[89] M. L. Carnevale *et al.*, "Cable-phone link is promising gamble," *Wall St. J.*, pp. B1, B10, May 18, 1993.

[90] P. Lambert, "TW joins NYC foe in NYNEX dial tone trial," *Multichannel News*, p. 44, Aug. 2, 1993.

[91] "Time Warner, in turnaround, to join test of cable TV services with NYNEX," *Wall St. J.*, p. B5, July 26, 1993.

[92] E. L. Andrews, "Ruling frees phone concerns to offer cable programming," *New York Times*, pp. A1 and D5, Aug. 25, 1993.

[93] A. Lindstrom, "Ruling could boost broadband," *Commun. Week*, p. 27, Sept. 6, 1993.

[94] "Cable Communications Policy Act of 1984," Public Law 98-549, sec. 613(b), Oct. 30, 1984.

[95] "Broadcasters and telephone companies: Risks and opportunities in a fiber future," Shooshan & Jackson, Inc., report prepared for the National Assoc. of Broadcasters, Jan. 5, 1989.

[96] See F. Linden, "What MDs and cable TV have in common," *Across the Board*, pp. 14–15, Jan./Feb. 1993.

[97] "Equal access now at 90%," *Telephony* Aug. 1989, pp. 23–24. "Telecommunications: Numbers behind the numbers," *Wall St. J.*, p. R10, Oct. 4, 1991.

[98] "Cable reaches 59M households" (Arbitron survey results), *Multichannel News*, p. 95, June 17, 1993.

Author Index

Ballart, Ralph, 19
Barr, William J., 301
Bars, G., 91
Blau, G. L., 117
Blume, Johan, 281
Boehm, Rodney, 28
Bourdeau, D., 91
Boyd, Trevor, 301
Brosio, A., 324
Burpee, David S., 362
Burrows, Maurice E., 140

Callaghan, J., 117
Campbell, L. H., 275
Cardwell, Richard H., 200
Carlisle, Chris J., 47
Ching, Yau-Chau, 19, 28, 110
Chum, Stanley, 62
Compton, Mark, 346
Cook, Andy, 337

Doverspike, Robert D., 172

Eaves, John, 39
Egawa, Takashi, 192
Engel-Smith, B., 103
Everitt, H. J., 275

Fleury, B., 211

Gersht, Alexander, 235
Ghosal, Dipak, 182
Ginger, Jeremy L., 308
Green, Mike, 226
Gruber, J., 211
Gruber, John, 226

Hägg, Peder, 281
Hansson, Leif, 281

Hasegawa, Satoshi, 191
Holter, Rony, 259
Hwu, Cannon, 62

Inoue, Yuji, 301
Iwabuchi, Eisuke, 330

Joncour, Gilles, 39

Katsuki, Toshio, 330
Katz, Hagay, 308

Kheradpir, Shaygan, 235
Klein, Michael J., 242
Kobrinski, Haim, 182
Kunieda, Toshinari, 294

Lakshman, T. V., 182
Lau, Richard C., 151
Lazzaro, P., 84
Leeson, J., 211
Leeson, Jim, 226
Legras, Jacques, 317
Leland, Will, 172

McEachern, James, 123
Mackenzie, J. H., 117
Maki, Kazumitsu, 97
Manley, A. M., 117
Martin, Simon, 346
Miura, Hidetoshi, 97
Moncalvo, A., 324
Morgan, Jonathan A., 172
Murakami, Koso, 330
Murano, Kazuo, 330

Nagaraj, K., 211
Nishihata, Kazuhiro, 97
Nunn, Robert O., 251

Oei, W. S., 53
Ogasawara, Hiroshi, 330
Okanoue, Yasuyo, 191

Parente, F., 84
Portejoie, J. F., 266

Reid, Andy, 75
Richman, Geoff, 47
Ritchie, G. R., 103

Sakauchi, Hideki, 191
Sandesara, N. B., 103
Sasaki, Noriyuki, 294
Sawyers, Greg F., 308
Say, H. Sabit, 110
Shafi, Mansoor, 1, 47, 218
Shulman, Alexander, 235
Shumate, Paul W., Jr., 362
Siller, Curtis A., Jr., 1
Smith, Peter J., 218
Sosnosky, Joseph, 162
Steinberger, Michael, 39
Stern, Jeff, 337
Stinson, Willis, 235

Sugimoto, Satoru, 294
Sundin, Leif, 281

Tamboli, S., 53
To, Michael, 123
Tremel, J. Y., 266
Urbansky, Ralph, 242

Wasem, Ondria J., 200
Whitt, S., 117
Wright, Tim, 39
Wu, Tsong-Ho, 140, 151, 182, 200, 352

Subject Index

NOTE: Bold page numbers indicate figures and illustrations.

A

abstract syntax notation 1 (ASN.1), 9
access
    multiple, 56
    to optical fiber, 337–345
    transport alternatives for, **354**
    unauthorized, 281–282
access network
    SDH for services delivery in, 346–351
    in SONET, 106
    See also networks
access points
    connected by trail, **266**
    in transport network architecture, 76–77
adaptation layer
    in ATM, 318
    See also layer network
add/drop multiplexer (ADM)
    relative cost ratios, **206**
    for SARPVP ring, 356–357
    in SDH network, 85, 87, 308, **347**
    in SONET, 110, 140, **143**, 163, 172, 191
    street located, **348**
    in survivable network, 173–**174**
    See also multiplexing
addressing, and SDH management network, 312–316
ADM. *See* add/drop multiplexer
administrative unit (AU), in CCITT standards, 31–**32**
ADS-based FTTC system, **339**
ADSL. *See* asymmetric digital subscriber loop
ADSL3, effect on fiber optic bandwidth, 343
agent, interaction with manager and objects, **270**, **283**
AIS. *See* alarm indication signal
alarm, autonomous, in NTT, 101–102
alarm indication signal (AIS)
    parameters for, 261
    recommendations for, 42
    required defect indications, **272**
    in response to LOP primitive, **228**
    in SDH/SONET, 262, 267
    See also fault management
alarm surveillance, and fault management, 271–272
algorithms
    distributed control, 193–194
    distributed self-healing, 192–197
    multiperiod capacity expansion, **202**
    multiperiod demand bundling, **202–203**
    for network restoration, 182–184
    restoration, 182–184
    ring fiber routing, 203–**204**
    ring selection, **203**
    routing, 79–**80**
    SONET control, 191–199
American National Standards Institute (ANSI), 1

ANSI. *See* American National Standards Institute
apportionment
    for G.821, **218**
    for G.826, **219**, **222**
APS. *See* automatic protection switching
architectural issues
    for broadband ISDN, 330–336
    for SDH network, 56, 75–**76**
architecture
    add/drop comparisons, **135**
    conventional compared to new, **98**
    defined by TMN, 228
    diverse protection (DP), 140
    double star, 330–**331**
    dual homing (DH), 140, **141**
    fiber access, **86–87**
    hubbed, 140
    hubbed and distributed, **126**
    layered, 260
    network management, 101
    ring, **6–8**
    ring-star, 330–**331**
    and SDH management network, 308–316
    for SDH networks, 75–83
    self-healing, 131–139, 140–150, 151–161, **152**
    for SONET-based metro network, 123–128
    TMN architecture, **10**
    See also ring architectures; transport network architecture
ASN-1. *See* abstract syntax notation
asymmetric digital subscriber loop (ADSL), in fiber optic system, 343–345
async. *See* asynchronous transport systems
asynchronous digital hierarchy bit rates, **20**
asynchronous transfer mode (ATM)
    adaptation layer functions, 318
    architecture, 319–**320**
    ATM virtual connection monitoring, 231–232
    basic mechanisms, 317–319
    and B-ISDN, 332–335
    BREHAT project, **322–323**
    broadband exchange-subscriber connection, 326–329
    cell header structure, **317**
    cell structure, **317**
    cell transport system, 324–329
    compared to synchronous transfer mode, **363**
    field trials and pilot networks, 322
    mapping in, 2, 34, 52
    overview and deployment, 317–323, 324
    permanent virtual connection (PVC) for, 108
    protocol reference model functions, **317**
    and SDH, 13
    and SONET, 108–109, 110–111, 113, 115, 166, 215–216, 230–231, 235
    for SONET pipe congestion management, **240**
    SONET/ATM ring, 352
    standards, 324

three-phase evolution of, **358**
and transmission system, 324–325
vs. STM, **353**
asynchronous transport systems (async)
availability and error performance, compared to SONET,
211–217
outage data for DS3 circuits, **214**
AT&T, role in SONET standards, 1, 28
ATM. *See* asynchronous transfer mode
ATPC. *See* automatic transmitter power control
attributes, and modeling, 270
AU. *See* administrative unit
automatic protection switching (APS), 26, 34
in SONET self-healing ring, **157**–158
in survivable network, 173, 200
See also protection switching; switching
automatic transmitter power control (ATPC), for SDH network, 51
availability
G.821 and G.826 recommendations, 218, 219
objectives for, 214
simulating for networks, 80
of SONET, compared to asynchronous transport systems,
211–217

B

background block error (BBE), parameters for, 229
background block error ratio (BBE), parameters for, 273
backlogs, avoiding, 294–300
bandwidth
hubbed bandwidth management via DCCs, **123**
SMDS requirement for, **353**
bandwidth efficiency, in SDH network, 53
bandwidth management, via ADM chains or rings, **124**
base signals
layering of, 30
nested signals, **33**
for SONET, 29
BBE. *See* background block error; background block error ratio
BCCs. *See* Bellcore Client Companies
B-DCS, *See* broadband digital cross-connect system
behaviors, and modeling, 270
Bellcore, 2, 19, 110
generic SONET requirements, **111**
Bellcore Client Companies (BCCs), SONET deployment by,
103, 106
BER
relationship with G.826 EPOs, 220–**222**, **223**
requirements for, **223**
bidirectional line-switched ring (BLSR), 7–**8**
ring interworking for, **136**–137
in SONET, 113, 132, **133**–134
two-fiber compared to four-fiber, 134
bidirectional self-healing ring (B-SHR), 151–153, **152**, 158–159
capacity arrangement for, **155**
versus U-SHR, **156**
See also self-healing rings
BIP. *See* bit interleaved parity
B-ISDN. *See* broadband ISDN
bit interleaved parity (BIP), in detection of primitives, 228
BLSR. *See* bidirectional line-switched ring

British Telecom
SDH/SONET networks in, 117–122
trunk network, **118**
broadband
and SDH, 281
and SONET, 108–109, 113–114
broadband digital cross-connect system (B-DCS)
use in SONET, 172
See also digital cross-connect system
broadband exchange-subscriber connection, in ATM, 326–329
broadband ISDN (B-ISDN), 2, 12, 76, 163, 165–166, 191, 218
and ATM, 333–335
cost-effectiveness of, 357–360
development of, 330–336, 352–361
functional blocks for, 326–**327**
integrated with narrowband, **365**–366
physical image, **332**
residential telecommunications using, **362**–370
service timeline overview, **335**
and SONET standards, 35, 62
and transmission performance, 230–231
broadband multimedia services, 341
broadband networks, transmission performance in, 230–232
broadband payload transport, with payload pointers, **24**–25
broadband services, 164, 165–166
requirements for, 362–364
B-SHR. *See* bidirectional self-healing ring
buffer approach, to frame alignment, 243
building blocks
and contracts, 303–304
grouping objects into, 304
properties of, 303
bundle restoration, 187–188. *See also* network restoration
business
conventional delivery technique for, **338**
SDH connections for, **347**–348
services for, 368

C

C. *See* container
cable cut. *See* fiber cut
cable systems, 69
BER requirements, 222–224
See also fiber optic cable
cable television
optical fiber access for, 341
proposed frequency assignments, **368**
and telephone companies, 369
tree-and-branch network, **367**
cablehead, for SDH, 65–**66**, 69
call level control, 239. *See also* control
carrier group alarm (CGA)
and call dropping threshold, 164
See also alarm
CATV network
conventional, **337**
integrated with telephony, **342**
with on-demand services, **342**
CCITT. *See* Consultative Committee on International Telegraphy
and Telephony

cell adaptation, to transmission systems, 325–326
cell delineation
    in ATM, 326
    loss of, **321**
cell format, for AAL, Class D, **359**
cell generation, and multiplexing, **318**
cell transport, with SDH transmission system, 324–329
cell-level control, 239–**240**. *See also* control
central office (CO)
    in SDMS network architecture, **235**–236
    See also office
central processor, in SDH management, **285**
CGA. *See* carrier group alarm
chain network, link failure protection for, 308–**309**
characteristic information
    and access points, 76
    See also information
client-server relationship, in SDH network, 78, **79**, **267**
clock characteristics, for network synchronization, 242–243
CM. *See* configuration management
CMIP. *See* common management information protocol
CMISE. *See* common management information service element
coax cable, and fiber optic cable, 342–343
coding violations, parameters for, 261
common management information protocol (CMIP), 9–10.
    See also network management
common management information service element (CMISE)
    and message communication function, 267, 271
    in SONET, 112
    See also network management
compatibility, in SONET, 105
configuration management (CM), 274
    in SDH management, 282, 284
    See also network management; performance management
connectivity
    mutual, 296
    and network cost, **175**, 281
Consultative Committee for Telephone and Telegraph (CCITT)
    G.707 to G.709, 39–40
    G.781, 259
    G.783, 244
    G.784, 259, 308, 310
    G.803, G.831, 44–46
    G.821 to G.826, 218–220
    G.822, 242
    G.823, 824, 243
    G.826, 218–225
    G.957, G.958, 42–44
    and SONET/SDH standards, 1, 2, 19–21, 28, 31–33, 110
CONT minicomputers, performance parameters of, **298**
container (C)
    and administrative units, **32**
    in multiplexing structure, 5, 39–**40**
    See also virtual container
contracts, and building blocks, 303–304
control
    call level control, 239
    cell-level control, 239–**240**
    DCS control architecture, **162**–163
    distributed control algorithms, 193–194

integrated self-healing, **198–199**
levels of network control hierarchy, **237**
on network elements in SDH network, 281–293
network management control layers, **275**
network-level control, 236–238, **237**
SONET network traffic management and control, 235–241
in SUCCESS, **298**
See also protection switching; traffic management and control
control system
    architecture and mapping of, **287**
    SDH-based architecture for, 284–290
cost-effectiveness, for broadband services evolution, 357–360
CPE. *See* customer premise equipment
customer
    restoration time impact on, **167**
    SDH/SONET benefits to, 8
    See also user
customer layer, in layered network, **275**–276
customer premise equipment (CPE), in SONET network, 115
cyclic demand pattern, cost/capacity comparison for, **155–156**

D

data communication networks, protocol stacks of, **299**
data communications channel (DCC), 2, 9, 26-27, 262–**263**
    extending, **350**
    and network management, 68
    protocol stack, **34**
    in SDH network, 308
    in SONET, 108, 110, 264–**265**
    standards, 33–**34**, 37
database management systems, 288
DCC. *See* data communications channel
DCN management, 283
DCS. *See* digital cross-connect system
DCX. *See* digital cross-connect system
DDF. *See* digital distribution frame
defect indications, required, **272**
defects, in networks, 11–12
delivery of services
    in access network, 346–352
    See also services
delivery technique, conventional, **338**
demand patterns, cost/capacity comparison for, **155–156**, **177**
deployment
    of ATM, 317–323
    for optical fiber, current and near term, 337–341
    of PON-based systems, **341**
    of SDH/SONET, 260
    of SONET, 103–109, **104**, **105**, 110–116, 132
    of SONET metro network, 123–128
    See also implementation
design methodology
    multiperiod network growth assumptions, 202
    for network survivability, 200–207
DFB. *See* distributed-feedback lasers
DH. *See* dual homing
digital baseband systems, in fiber optic system, 343–344
digital cross-connect system (DCS)
    applications, 162–171
    broadband DCS, 172

control architecture, **162**
  and distributed protection control, 169–170, 192
  and distributed restoration, 182–190
  and distributed self-healing algorithm, 192–197
  hubbed bandwidth management via DCCs, **123**
  in network topology, 6, **66**
  parallel architecture for, 185–186, 188, 189
  and restoration time, **169**
  routing in, 311–312
  in SDH network, 75, 85, 87, 282, 308
  self-healing, 131, 137–138
  in SONET, 110, 111, 113–**114**, 191, 235
  in survivable network architectures, 172–181
digital distribution frame (DDF), compared to DCSs, 75
digital network
  digital switched network, **85**
  evolution, 84–**85**
  synchronization of, **232**–234
  wander in, 243
  See also networks
digital private lines traffic, on SDH/SONET gateway, 67–68, **70**
digital radio-relay system (DRRS)
  applications of, **49**
  regenerator, **48**
  for SDH networks, 47–52
digital switched circuits, for SDH/SONET gateway, 69
dispersion-limited systems, recommendations for, 43, 44
distributed control algorithms, 193–194
  See also control
distributed processing environment (DPE), in TINA, 304–305
distributed protection control, 169–170
  See also network protection; network restoration
distributed restoration
  DCS network restoration time, 184–187
  with DCSs, 162–171
  self-healing rings as, 191
  SONET DCS systems for, 182–190
  See also network restoration; self-healing rings
distributed-feedback (DFB) lasers, 367
diverse protection (DP) architecture, 140–**141**
  in network survivability, 200
  See also architecture
diverse routed protection (DRP), 131
  for link failure, 308
  in self-healing architecture, 137
  See also network protection; protection switching
diverse routing
  in point-to-point network, 172–173
  See also point-to-point network; routing
double star network architecture, 330–**331**, **364**
DP architecture. *See* diverse protection architecture
DPE. *See* distributed processing environment
DRP. *See* diverse routed protection
DRRS. *See* digital radio-relay system
DS1 paths, availability and performance of, **216**–217
DSM. *See* dynamic-single-mode lasers
dual homing (DH), in SDH network, **348**
dual homing (DH) architecture, 140–141, 145, 201
  in interoffice network, 149
  See also architecture

dual parenting. *See* dual homing
duplex cross-connection
  model of, **283**
  See also digital cross-connect
dynamic-single-mode (DSM) lasers, 367

E

ECC. *See* embedded control channel
ECSA. *See* Exchange Carrier Standards Association
EDFA. *See* erbium doped fiber amplifier
embedded control channel (ECC)
  in SDH management, 291–292
  See also control
equipment
  costs for, 180
  functional reference block derivation for, 80–**81**
  See also synchronous equipment
equipment configuration, 282
equipment management, 274
  See also network management; performance management
erbium doped fiber amplifier (EDFA), recommendations for, 45
error block, and performance monitoring, 12
error detection code, **273**
error performance
  G.821 and G.826 recommendations, 218–220
  of SONET, compared to asynchronous transport systems, 211–217
error seconds (ES), parameters for, 261
errored second ratio (ESR)
  parameters for, 273
  vs. BER for random errors, **221**
ES. *See* error seconds
ESR. *See* errored second ratio
ETSI. *See* European Telecommunications Standards Institute
Europe, FTTC systems development in, 339
European Telecommunications Standards Institute (ETSI)
  international SDH/SONET gateways, 62–74, **66**
  SDH definition, 75
evolution. *See* network evolution
Exchange Carrier Standards Association (ECSA), 28

F

failure
  generation of in performance monitoring, 228–229
  protection against, 308
  See also network survivability; performance monitoring
far-end error block (FEBE), in performance monitoring, 229–**230**
fault management (FM), 271–272, 281, 282. *See also* alarm
    indication signal; fiber cut
FDDI. *See* fiber distributed data interface
FDF. *See* fiber distribution frame
FEBE. *See* far-end error block
fiber access network, architecture of, 86–**87**
fiber cut, 7, 140, 162, 281
  in unidirectional path-switched ring, **132**–134
  See also alarm; fault management
fiber distributed data interface (FDDI)
  as MAN interface, 165
  mappings for, 2

fiber distribution frame (FDF), for HRDP, **212**–213
fiber optic cable
    and coax cable systems, 342–343
    compared to other media, 92–93
    costs for, 180
    in local and regional networks, **85**
    submarine, 69
    to home and office, 86, 140, 343–344, 362
    to nodes, 366–368
    to remote sites, **338**
    See also optical fiber
fiber optic transmission system (FOTS), and radio-relay
    systems, 50
fiber-in-the-loop systems, 338–339
fiber-to-the-curb (FTTC) system, **364**–365
    ADS-based, **339**
    hybrid narrowband/broadband, 366
    PON-based, **338**
fiber-to-the-home (FTTH), 86, 140, 343–344, 362, **364**–**365**
fixed satellite service (FSS)
    design and application of, 53–54
    and SDH networks, 53–59, **55**
flow-through operation, in Japan NTT, **100**–101
FM modulation, in fiber optic system, 343
FM. *See* fault management
folded ring architecture, with APS, **157**–158. *See also* ring
    architecture
formats
    G.707 to G.709, 39–40
    ITU-T recommendations, 39–**46**
    for SONET/SDH, 2–5, **3**–**4**, 21
FOTS. *See* fiber optic transmission system
FR. *See* frame relay
frame
    alignment of with pointers, 25
    fiber distribution frame (FDF) for HRDP, **212**–213
    mapping into, **333**
    severely errored frame (SEF) primitive, 228
frame alignment for time multiplex switching, 243
frame formats, **3**
    SONET frame, **29**
    and standardization, 28
    STS-1 frame, **21**
frame relay (FR), and SDH networks, 76
frame structure
    and pointer processor, 245
    and SDH multiplexing, 40
    of SDH and SONET, compared, 62
France Telecom
    deployment of SDH, 91–96
    existing transmission network, **91**–92
    junction networks, **94**
    local exchange networks, 92, **93**, 94
    long distance network, 92, **93**, **94**
FSS. *See* fixed satellite service
FTTC. *See* fiber-to-the-curb
FTTH. *See* fiber-to-the-home
function assignments, in SUCCESS CONT control equipment, **298**
functional blocks
    for broadband systems, 326–327
    derivation of for equipment, 80–**81**

of DRRS regenerator, **48**
recommendations for, 41–42, 46
and SDH radio equipment, 47

G

gate arrays, in STM-1, **328**
gateway interconnection, of SDH and SONET, 62–74, **66**
gateway network element (GNE), 10
    connected to OS, **291**
    See also network element
Germany, FTTC systems development in, 340
glossary
    for SDH, 61
    for SONET, 15–16
GNE. *See* gateway network element
graphical user interface (GUI), in SDH management, 284
grooming. *See* traffic grooming
GUI. *See* graphical user interface
G.XXX recommendations. *See* Consultative Committee for
    Telephone and Telegraph

H

hitless switching, **99**–100. *See also* switching
home
    fiber optic delivery to, 86, 140, 343–344, 362, **364**–**365**
    See also fiber optic cable; optical fiber
hubbed architecture
    compared to distributed architecture, **126**
    cost model, 142–**145**
    high-speed self-healing, 140–150
    See also architecture
hypothetical reference digital paths, 211–**213**, **212**
    and CCITT recommendations, **220**

I

ICCF. *See* Interexchange Carrier Compatibility Forum
IDT. *See* international dedicated transit
implementation
    by British Telecom, 119
    cost-effective network evolution, 352–361
    with currently available equipment, 347–348
    and international networks, 62
    of pointer processors, 248–250
    recommendations for, 41, 44, 110–116
IN. *See* intelligent network
information
    characteristic, and access points, 76–77
    derivation of model for, 81
    network management information traffic, 68–69
    relation to network overhead, 260
    in TMN architecture, 228
information model
    for network modeling, 268–269
    in SDH management, 283–284, 288
information network
    and intelligent network, 302
    vision of in TINA, 302
information payload
    floating, 3
    See also payload
intelligence, for remote units, 235

intelligent network (IN)
    and information network, 302
    and TINA, 301–302
intelligent services, 164, 301. *See also* services
INTELSAT, 56
Interexchange Carrier Compatibility Forum (ICCF), 1
interfaces
    for cable TV, 370
    CONT-NE interface, 299–300
    craft interface, 52
    fiber distributed data interface (FDDI), 2, 165
    and functional blocks, 42
    NE to OS, 36, **296**
    for network management, **51**
    network node and satellite equipment, 54
    for operations support, 38
    optical, 88, 226
    performance monitoring at, 229–**230**
    physical (SPI), 48
    Q3-interface, 291
    recommendations for, 44
    for SDH, 48, 110, 281
    for SDH/SONET, 19, 27, 28, 35, 104, 110–**111**, 114, 259
    and SONET standards, 30–31
    on SUCCESS workstation, **299**
    user-to-network interface, (UNI), 108
internal communication network, in SDH management, 285–286
international dedicated transit (IDT), for SDH/SONET
        interconnection, **67**, **70**
International Telecommunication Union-Telecommunication
        Standardization Sector (ITU-T), 1
    B-ISDN recommendations, **334**
    SDH/SONET recommendations, 39–46, 75, 259
Internet, 76
interoffice networks, 140–150
    dual homing, 149
    fiber-hubbed, **141**
    single homing, 147–149
    SONET ring placement in, 200
    See also networks
interoffice transport network, 106–**107**
ISDN services, 164, 165
Italian transmission network, SDH systems in, 84–**90**
ITU-T. *See* International Telecommunication
        Union-Telecommunication Standardization Sector

J

Japan
    FTTC systems development in, 340
    SDH mapping, 63
    SDH network evolution, 97–102
jitter
    allocation of, 252–253
    defined, **251**–252
    generated by pointer processor, **248**
    and network elements, 233
    PDH compatible, 244
    and pointer adjustments, 253–255
jitter interworking, SONET requirements for, 251–255

L

LAN. *See* local area network
lasers, distributed-feedback (DFB) and dynamic-single-mode
        (DSM), 367
last-mile distribution, **331**
LATA. *See* local access transport area
layer network, 76–78, 131
    network management control layers, **275**, 308–**309**
    partitioning of, **77**–**78**, **267**
    and telecommunications management network, 275–280
    topology of, **266**
    in year 2000, 81–83, **82**
    See also networks
layered overhead structure, **260**–**261**
layering
    in ATM, 318
    of base signals, 30
    independence of topology in different layers, **79**
    path layer protection, **278**
    in SDH network, **266**, 286–287
    in TINA-based SDH network, 304–**305**
    of transport network, 45, **54**–55, **120**, 227
    in transport network architecture, 77–**78**, 260
LEC. *See* local exchange carrier
line layer, 260. *See also* layer network
line network element (LNE), 260. *See also* network element, (NE)
line overhead (LOH), and SONET standards, 30–31
line protection switching, 151
    SONET self-healing ring as, 357n10
    See also protection switching; switching
line restoration, **192**
    and range of restoration time, **196**
    See also network restoration
line systems
    ITU-T recommendations, 39–46
    in SDH network, 87, **227**
linear add/drop configuration, **6**
linear add/drop network, **131**
link connections, and partitioning, 77, **79**
link protection, 308–**309**
    compared to network protection, **120**
LNE. *See* line network element
local access transport area (LATA) network
    backbone architecture example, **163**
    component outage times, predicted, **215**
    and single-homing architecture, **144**
local area network (LAN), virtual, connected in SDH management
        network, **312**–313
local communications network (LCN)
    deployment of, **36**–37
    in TMN architecture, **10**
local exchange carrier (LEC) networks
    in France Telecom, **92**
    in SONET network, 103
    See also networks
logical configuration layer, in layered network, **275**–276
logical network, for supporting SMDS, **353**
LOH. *See* line overhead
LOP primitive. *See* loss-of-pointer primitive
loss-of-pointer (LOP) primitive, 228. *See also* pointers

M

maintenance signals, in SDH/SONET, 262
maintenance thresholding, in performance management, 273–274
MAN services, 164, 165
managed object (MO)
    allocation of in NE and OS, **297**
    properties of, 270
    See also objects
managed transmission network (MTN), in British Telecom,
        117, **118**
management. *See* network management
management network. *See* network management
manager, interaction of with agent and objects, **270, 283**
mapping
    of ATM cells, **319, 333**
    bit-stuffed compared to fixed location, 23
    of CSA software, **287**
    for multiplexing structures, 3–**4**
    of SDH, 41, 45, 52
    of SDH and SONET, compared, 62, **63, 64, 66**
    SMDS SNI mapping in 3-phase evolution, **358**
    of SONET, 34, 104–105
    sub-STS-1, 26
MCF. *See* message communication function
MCM. *See* message communications module
mediation device (MD)
    for element management, 264
    in TMN architecture, **10**
medium routes
    FSS-SDH network element scenario for, 57–59, **58**
    See also routing
mesh networks, 8, 155
    protection for, 172, 173
    restorable, 162
    See also networks
message communication function (MCF), in SDH network,
        267, 271
message communications module (MCM), in SUCCESS CONT
        control equipment, 298
metro SONET network, 123–128, **125**, 211
    HRDP component counts, **213**
    HRDP error performance, **216**
    hubbed and distributed, compared, **126**
    hypothetical reference digital path, 211–**213**
    transition to target, **126**–127
    See also synchronous optical network
MO. *See* managed object
modeling
    and network management, 268–269
    recommendations for, 41
monitoring. *See* performance monitoring
MSOH. *See* multiplex section overhead
MSP. *See* multiplex section protection
MST. *See* multiplex section termination
MTN. *See* managed transmission network
multidestination, in SDH network, 53–54
multimedia services
    broadband, 341
    See also video
multiperiod demand bundling algorithm, 202–203

multiple access
    in SDH network, 56
    See also access
multiplex section overhead (MSOH)
    and frame format, 3
    in STM-1 frame, 47
multiplex section protection (MSP)
    in radio-relay system, 48, 49
    in SDH network, 347
multiplex section termination (MST), in radio-relay system, 48
multiplexer
    optical line termination multiplexer (OLTM), 142
    street located, **350**
multiplexing
    and cell generation, **318**
    compared to ATM, 111
    and frame alignment, 243
    optical wavelength division multiplexing (WDM), 76
    recommendations for, 42
    SDH multiplexing hierarchy, **349**
    SDH multiplexing tree adapted to FSS, 55
    of SDH and SONET, compared, 62, **63, 66**
    for SONET/SDH, 3-5, **4**, 22–23, 29, 39–**40**, 242–250, 333
    sub-STM-1 multiplexing routes, **50**
    and synchronization, 23–**24**
    T1 MUX network, 197
    See also add/drop multiplexer; terminal multiplexer
multivendor environment problems, avoiding, 294–300, **295**

N

naming convention, in SONET, **260**
naming tree, in SDH, **270**
narrowband, integrated with broadband FTTC, 365–366
national research and education network (NREN), SONET trials
        by, 115
NE. *See* network element
nested signals, **33**. *See also* base signals
network connections, and partitioning, 77
network element (NE)
    configuration of, 299
    control of, 284
    DCSs as, 163
    gateway network element, 10
    ITU-T recommendations, 39–46
    in Japan NTT system, 98–**99**, 101
    line network element, 260
    and managed objects, **297**
    management of, 264
    monitoring process, **228**
    in point-to-point configuration, **37**
    recovery of, 282
    in SDH network, 281–293, **292, 295**, 296, 308
    section network element, 260
    in SONET, 111, 163, 191, 211, 226, 227
    standards for, 2
    termination functions, 260
    in TMN architecture, **10**
    transport facilities provided by, 267
    See also gateway network element (GNE)
network element (NE) layer, **275**. *See also* layer network

network evolution
    cost-effective, 352–361
    future of, 368–370
    phase I, 358
    phase II, 358–359
    phase III, 359
    with SARPVP ring transport, 357
    See also deployment; implementation
network layer
    in layered network, **275**
    See also layer network
network-level control (NLC), 236–238, **237**
    and operations systems, 240–241
    optimization, 238–239
    in SONET network, 235
    See also control
network management (NM), 8–10, 31
    in British Telecom, **121**–122
    control layers for, **275**
    hubbed bandwidth management via DCCs, **123**
    information model derivation, 81
    information traffic, 68–69
    interaction among manager, agent, objects, **270**
    interfaces for, **51**
    in Japan NTT, **100**–101
    layered approach to, 275–280
    layers for, 82–83
    protocols for, 267
    remote, 76
    for satellite SDH systems, 60
    for SDH, **51**–52, 54, 55–56, 76, 88, 96, 281–293, 294–300,
        308–316
    for SDH/SONET, 107–**108**, 259–265, 266–274, **268**, **306**
    SONET distributed bandwidth management, **124**
    SONET network traffic management and control, 235–241
    standards for, 259
network node interface (NNI), 2
    and satellite equipment interface, 54
network product layer
    in layered network, **275**–276
    See also layer network
network protection
    compared to link protection, **120**
    diverse routed protection (DRP), 131
    ring protection applications, 45, **134**
    in SDH network, 308–**309**
    See also protection switching
network restoration
    algorithms for, 182–184
    in British Telecom SDH/SONET, 119–120
    bundle restoration, 187–188
    DCS distributed restoration, 162–171
    enhanced with parallel processing, 188–**189**
    impact on customer, **167**
    near-time, **99**
    queuing model for, 186–187
    range of restoration time, **196**
    with self-healing control, **198**–199
    time objectives for, **167**–168
    See also distributed restoration; network survivability

network service groupings, for SONET, 164
network survivability, 151, 191
    DCS use for, 172–181
    design methodology for, 200–207
    and network redundancy, 175
    network survivability measure, 145
    for operating telephone companies, 140
    via 1+1 diverse routed system, **124**
    via DCCs and centralized intelligence, **124**
    worst-case survivability (WCS), 179–180
    See also network restoration; protection switching;
        self-healing rings
network topology, 5–**8**
    in British Telecom, 118–**119**
    designing and sizing, 78–79
    independence of topology in different layers, **79**
    linear add/drop, **6**
    mesh networks, 8
    and network survivability, 204–205
    performance issues, 10–12
    ring architectures, **6–8**
network traffic management optimization, **279**
    See also network management; traffic
networking, standards development for, 1
networks
    access networks, 106, 346–351
    availability simulation, 80
    CEPT, 31–33
    chain network, 308–**309**
    impact on of standards evolution, 37–38
    interoffice network, 140–150
    Italian transmission network, 84–90
    ITU-T recommendations, 39–46
    layer networks, 76–77
    metro SONET network, 123–128
    network-level control, 236–238, **237**
    North American, 33
    performance parameters derivation, **80**
    recommendations for, 44–46
    sectorial, 92
    self-healing network, 281
    SONET network, **38**
    synchronization, 242–243
    transport network architecture, 76–78
    ubiquitous, 370
Nippon Telephone and Telegraph (NTT), SDH use, 97, **294**
NLC. *See* network-level control
NM. *See* network management
NNI. *See* network node interface
nodes
    fiber optic cable to, 366–368, **367**
    15-node network, **176**–177
    53-node network, **177**
    9-node network, **173**
    point-to-point subnetwork between, 311, **313**
    response to failure of, **137**
    for rings interconnect, **126**
    in SONET, 131
North America
    network standards, 33, 36
    SDH standards in, 62
NTT. *See* Nippon Telephone and Telegraph

O

OAM&P. *See* operations, administration, maintenance, and
  provisioning
object-oriented approach
  in intelligent network, 302
  to network modeling, 268
objects
  grouping into building blocks, 304
  interaction with manager and agent, **270**, **283**
  properties of, 270
  and services, 303
  See also managed object (MO)
OFA. *See* optical fiber amplifier
office
  fiber routing to, 86
  in generic metro network, **125**, 138
  in self-healing ring connection, 151
  See also business; central office
OLS. *See* optical line system
OLT. *See* optical line terminal
OLTM. *See* optical line termination multiplexer
ONR model
  and software backlog, 297
  for transmission line, **296**
open system specification, in SDH network, 54
open systems interconnection (OSI), 1, 231
  in layered networks, 78, 111
  for SDH management, 282
operation support systems (OSSs), 9
  evolution of, 359–**360**, 369
  interfaces for, 38, 281
  and network level control, 240–241
  and SDH/SONET management, 259, 267, 294
  for SONET, 111–112
operations
  in Japan NTT, **100**–101
  modeling of, 270
  of SDH network elements, 281–293
  in SONET, 107–**108**
operations, administration, maintenance, and provisioning
  (OAM&P), 1, 2, 12
  in ATM, 320–321
  satellite-specific services for, 55
  and SDH networks, **51**–52, 53, 63
  in SDH/SONET networks, 117, 226, 228, 232
  See also network management
optical fiber
  access to in 21st century, 337–345
  and coax cable, 342–343
  fiber-in-the-loop systems, 338–339
  to home and office, 86, 140
  See also fiber optic cable
optical fiber amplifier (OFA), 45
optical line system (OLS)
  SDH recommendations, 42–44
  See also synchronous optical line system
optical line terminal (OLT), in radio-relay system, 50
optical line termination multiplexer (OLTM), 142. *See also*
  multiplexing
optical parameters, 1–2, 27
  and standardization, 28, 45

optical transceiver, **328**–**329**
optical wavelength division multiplexing (WDM), 43, 76.
  See also multiplexing
orderwire channels, 262
OSI. *See* open systems interconnection
overhead
  layered, 260
  for SONET/SDH, 19, 22, 34, 63, **64**
  specification for, 35
  termination of, **261**
overhead bytes, in multiplexing, 40
overhead layers, **260**–**261**. *See also* layer network

P

packet-switched services, 164–165
partitioning
  of ATM and SDH functions, **328**
  of layer network, **77**–**78**, **267**
  in SDH/SONET interconnection, 71
  of trail, **79**
passive optical network (PON)-based system
  advantages of, 340
  deployment scenarios and features, **341**
  for FTTC, **338**, **340**
  with FTTH, 364
path layer, 260
  See also layer network
path layer protection, **278**
  See also protection switching
path network element (PTE), 260
  See also network element
path overhead (POH)
  functions supported by, 3, 9
  in SDH/SONET interconnection, 65, 68
  and SONET standards, **31**
path protected ring (U-SHR/PP), **159**
path protection switching, 151
  SONET self-healing ring as, 357n9
  See also protection switching; switching
path restoration, **192**
  and range of restoration time, **197**
  See also network restoration
path switched ring, **6**, 132
  ring interworking for, 135–**136**
  See also ring architectures; unidirectional path switched ring
path-level restoration, 183
  See also network restoration
paths
  DS1 paths, **216**–217
  hypothetical reference digital paths, 211–**213**, **212**
payload
  broadband payload transport, 24–25
  on IDT, 67
  management of, 260–261
  payload jitter analysis, 253
  and pointers, 23–**24**
  of SONET/SDH, 19, 35, 62, 63, **64**
  in STM-1 frame, **47**
  sub-STS-1, 25–27
  synchronous payload envelope, 30
  See also pointers
PDH. *See* plesiochronous digital hierarchy

performance assessment, in SONET/SDH networks, 226–227, 233–234

performance issues, 10–12, **11**, 216

performance management (PM)
  maintenance thresholding for, 273–274
  parameters for, 273
  required defect indications, **272**
  for SBMS, **241**
  and SDH management, 282–283
  for SONET-based multiservice networks, **235**–241
  See also network management

performance monitoring
  in ATM, 320–321
  and error block, 12
  history of, 273–274
  in-service, 52
  in Japan NTT, 101–102
  near end, and far end, 229–**230**
  at network interfaces, 229–**230**
  parameters and failures generation, 228–229
  provided by VT/VC path, 261
  for satellite SDH systems, 59
  in SONET/SDH networks, 227, 261, 267, 281
  transmission convergence monitoring, 231

performance parameters, in SONET/SDH networks, 261–262

performance primitives. *See* primitives

phase-locked loop, **251**

photonic layer, 260. *See also* layer network

physical configuration sublayer, in layered network, **275**

physical layer (PL)
  divisions in, 319
  reference model description, 325–326
  See also layering

PJE. *See* pointer justification event

PL. *See* physical layer

plain old telephone service (POTS), adequacy of, 330–331

planning
  for SDH network, 89
  of SONET metro network, 123–128

plesiochronous digital hierarchy (PDH)
  in multiplexing, 5, 40
  and SDH/SONET, 42, 47, 53, 75-76, 84, 89, 91, 97, 117, 242, 243, 279–280, 346

PM. *See* performance management

POH. *See* path overhead

point-to-point network
  between adjacent nodes, 311, **313**
  network element in, **37**
  protection for, 172
  SDH technology for, **347**

point-to-point virtual path SONET/ATM ring, **355**

pointer adjustments, and jitter, 253–255

pointer alignment, in SDH/SONET, 243

pointer justification event (PJE), effects of, 243–244, **249**

pointer processor (PP)
  advanced, 246–247, 248
  classical, 244–**245**, 247–248, **249**
  designs for, 245–**248**
  implementation, 248–250

pointers
  pointer justification events (PJE), 243–244
  in SDH multiplexing structure, **40**, **66**
  synchronization of, 23–**24**
  See also payload

PON. *See* passive optical network

POTS. *See* plain old telephone service

PP. *See* pointer processor

primitives, detection of, 228

private line services, 164

private lines
  on SDH/SONET gateway, 67–68, **70**
  SONET access scenario, 115

PRM. *See* protocol reference model

programming and packaging, for cable TV, 369

protection architecture
  defined, 172
  See also network protection; protection switching

protection path, for SDH network, 347–**348**

protection. *See* network protection; protection switching

protection switch count (PSC), parameters for, 229

protection switching
  and distributed restoration, 162
  multiplex section protection (MSP), 48
  and network topology, 7
  radio protection switching, **49**
  for satellite SDH systems, 59–60
  in SDH network, 281, 282
  in SONET self-healing ring, 151, 152, **157**
  See also automatic protection switching; network protection

protocol reference model (PRM), functions in ATM, **317**–318

PSC. *See* protection switch count

PSTN. *See* public-switched telephone network

PTE. *See* path network element

public-switched telephone network (PSTN), 67, 164
  architecture, 75–**76**
  See also voice services

pulse stuffing, and jitter, 253

Q

Q3-interface, 291

QAx protocol, 267, **269**. *See also* network management

Qecc protocol, 267, **269**, **270**. *See also* network management

quality and pricing, for cable TV, 369–370

queuing model, for network restoration, 186–187

R

radio, terrestrial, BER requirements, 222–224

radio protection switching (RPS), **49**
  early warning activation for, 51
  See also protection switching

radio-relay systems
  for SDH networks, 47–52, 54, 87
  to transport STM-1, 48–**50**, **49**

radio-relay terminal (RRT), functional block diagram, **48**

rates
  G.707 to G.709, 39–40
  ITU-T recommendations, 39–46
  for SONET/SDH, 2, 3, 19, 20, 91, 104–105
  See also transmission rates

RBOCs. *See* Regional Bell Operating Companies

RDI. *See* remote defect indication
reach parameters, and SONET standards, 35
regenerator section overhead (RSOH)
    in frame formats, 3
    in SONET/SDH interconnection, 68
    in STM-1 frame, 47, 63
regenerator section termination (RST), in radio-relay systems, 48, 50
regenerators
    and optical amplifiers, 45
    optical regenerator (OR), 50
Regional Bell Operating Companies (RBOCs), role in standards development, 28
remote defect indication (RDI), 262
remote failure indication (RFI), 262
remote sites, fiber optic service to, **338**, 340
remote unit (RU), intelligent, in SONET network, 235
residential telecommunications, broadband services for, 362–370
restoration. *See* network restoration
RFI. *See* remote failure indication
ring architecture
    bidirectional line-switched ring (BLSR), 113
    comparison of approaches to, 134–**135**
    folded ring architecture, **157–158**
    in metro network, **125**
    in network topology, **6–8**
    protection for, 172
    ring add/drop network, **131**
    ring interworking, **135–137, 136**
    SARPVP ring architecture, 354–356
    self-healing rings, 131–139, 140–150, **193**
    in SONET, 107, 113, 114, **125, 160**
    and submarine cable, **69**
    unidirectional path switched ring (UPSR), 113
    See also self-healing rings
ring fiber routing algorithm, 203–204
ring network, in SDH, 333
ring protection applications, 45
    See also network protection
ring selection algorithm, 203
ring topology, in SDH management network, 310–**311**
ring transports, comparison among, **359**
ring-star network architecture, 330–**331**
routing
    in meshed DCS network, 311–312
    in SDH management network, 310–316
    VP/VC routing, **318**
routing algorithms, developing and defining, 79–**80**
RPS. *See* radio protection switching
RRT. *See* radio-relay terminal
RSOH. *See* regenerator section overhead
RST. *See* regenerator section termination
RU. *See* remote unit

S

SA. *See* section adaptation
SAR. *See* SONET/ATM ring
SARPVP. *See* SONET/ATM ring with point-to-point virtual paths
satellite equipment interface (SEI)
    and network node interface, 54
    See also fixed satellite services; interfaces

satellite systems, BER requirements, **224**
satellite systems integration
    in SDH transport networks, 53–61
    See also fixed satellite services
SBMS. *See* SONET-based multiservice networks
SDH. *See* synchronous digital hierarchy
SDXC control system architecture, 285
    software for, 286
section adaptation (SA), in radio-relay system, 49
section layer, 260
    See also layer network
section network element (STE), 260
    See also network element
section overhead (SOH)
    in SDH, capabilities supported by, 3
    in SDH multiplexing structure, **40**
    and SONET standards, 31
    in STM-1 frame, 47
sectionalization, and trouble locating function, 227
sectorial hubs, using DCC, **123**
sectorial networks
    in France Telecom, **92**
    See also networks
security management (SM), in SDH management, 283, 284
SEF. *See* severely errored frame
self-healing capability, for SARPVP ring, 357
self-healing networks, 281
    See also networks
self-healing rings (SHR), 131–139, 151–161, **152**
    bidirectional and unidirectional, 151
    comparisons among, **160**
    control algorithms of, 191–199
    conventional, 191–192
    cost/capacity tradeoff, 153–156, **155**
    distributed, 183
    in high-speed SONET, 140, 141, **142, 143**
    integrated, 197–199
    See also bidirectional self-healing ring; unidirectional self-healing ring
SEMF. *See* synchronous equipment management function
sensitivity
    cost sensitivity, 179
    demand sensitivity, 177–178
    to network connectivity, 180–181
serial processing, serial cross-connect, **183**
    See also digital cross-connect
service providers, SDH/SONET benefits to, 8
service/delivery segmentation, **304**
services
    and objects, 303
    packet-switched, 164–165
services delivery, in access network, 346–351
SES. *See* severely errored seconds; severely errored second
severely errored frame (SEF) primitive, 228
    parameters for, 229
    See also frame
severely errored second (SES), parameters for, 219, 228–229, 261, 273
SH. *See* single-homing architecture
SHR. *See* self-healing rings
signal. *See* base signal

signaling networks, 164, 166
single-homing (SH) architecture, 140
    in interoffice network, 147–149
    and LATA network, **144**
    See also architecture; dual homing
SLS. *See* synchronous optical line system
SM. *See* security management
SMAEs. *See* system management applications entities
SMDS. *See* switched multi-megabit data service
SMUX control system architecture, **288**–289
software
    backlog for, **296**–297
    for SDXC control system, **286**
SOH. *See* section overhead
SONET. *See* synchronous optical network
SONET-based multiservice (SBMS) networks, 235–241
    performance management procedure, **241**
    three-level control strategy, 236–239, 237
    user service attributes, **236**
SONET/ATM ring (SAR), in SONET network, 352
SONET/ATM ring with point-to-point virtual paths (SARPVP)
    self-healing capability for, 357
    in SMDS service, 352–356, **355**
    three-phase network evolution using, **357**
source arrival process, in network restoration, **187**
SPE. *See* synchronous payload envelope
speech encoding, 63
SPI. *See* interfaces, physical
Sprint Communications, 62, 69–71, **70**
standards
    for cable TV, 370
    for CCITT, 31–33
    related to ATM, 324
    for SONET/SDH, 1–2, 13–15
star network, link failure protection for, 308–**309**
STE. *See* section network element
STM. *See* synchronous transfer mode
STM-1. *See* synchronous transport module level-1
STS-1 frame, **21**
STS-3. *See* synchronous transport signal level-3
STS/VT. *See* synchronous transport signal/virtual tributary
submarine cable systems, 69
    See also fiber optic cable
subnetwork connection
    in France Telecom, 92
    and partitioning, 77, **79**
    See also networks
subscriber lines networks
    in France Telecom, 96
    See also customer
SUCCESS
    configuration of, **297**
    CONT functions, **298**
    relationship to TEAMS, **300**
Surcap, 174
survivability. *See* network survivability
switched multi-megabit data service (SMDS)
    access class and bandwidth requirement, **353**
    logical network for supporting, **353**
    network architecture for supporting, 352–354

and SDH networks, 76
and SONET, 165
and SONET ring transport, **357**
transport alternatives for access to, **354**
switching
    in B-ISDN system, **334**
    digitalization of, 84–**85**
    frame alignment for, 243
    hitless switching, **99**–100
    line protection switching, 151
    multistage self-routing, 335
    path protection switching, 151
    in SDH network, 93
    in SONET, 132–133, 151
    virtual channel vs. virtual path, 354
    See also protection switching
synchronization
    of digital network, **232**–234
    of pointers and multiplexing, 23–**24**
    priority master-slave synchronization, **232**–233
    for satellite SDH systems, 60
    for SDH/SONET, 242–250, 282
    standards for, 1, 5, 34, 47, 226, 250
synchronous digital hierarchy (SDH)
    cablehead, 65–**66**
    capabilities of, 1, 9, 346–347
    compared to ATM, 75, 317–323
    compared to PDH, 47, 53
    compared to SONET, 28, 62–65
    development of, **331**–335
    evolutionary strategy, 92–96
    international gateway interconnection with SONET, 62–74
    ITU-T recommendations, 39–46
    multiplexing structure, **4**
    network recommendations G.803, G.831, 44–46
    optical line systems recommendations G.957, G.958, 42–44
    optical link for, 329
    and PDH, 279–280
    physical interface (SPI), 48
    and PON systems, 340
    rates and formats G.707 to G.709, 39–40
    standards, 1–2, 14–15, 84, 250
    STM-1 frame structure, **3**
    and TMN, 259, 264
    traffic types, 65
    transmission rates, **3**
    transmission system for ATM cells, 324–329
synchronous digital hierarchy (SDH) networks
    architecture, **308**–**309**, 309–310
    British Telecom system, 117–122, **118**
    deployment strategy, 88–89
    first-phase equipment, **294**
    France Telecom system, 91–96
    integrated, **309**
    Italian transmission system, 84–**90**
    Japan NTT system, 97–102, **100**
    management, 88, 282–284, 285–293, 294–300, 308–316
    network architecture, 75–83, 85–**86**
    network element control and operation, 281–293
    network scenarios, 56–59
    radio-relay systems, 47–52

requirements, 87–88
ring network, **333**
satellite systems integration, 53–61
second-phase network management system, 297–300
synchronization for, 242–250
transmission performance, 226–234
virtual synchronous digital hierarchy networks, 70–71
See also networks
synchronous equipment
recommendations G.781, G.782, G.783, 40–42
See also equipment
synchronous equipment management function (SEMF)
in Japan NTT, **102**
and network elements, 267
for SDH network, 52, **271**
synchronous optical line system (SLS)
recommendations, 43
See also optical line system
synchronous optical network (SONET)
applications, **108**, 138
architectures, 106–**108**
British Telecom system, 117–122
capabilities, 9
compared to asynchronous transport systems, 211–217
compared to SDH, 28, 62–65
cost characteristics, 175–176
deployment, 103–109, **104**, **105**, 110–116
design methodology, 200–207
discussed, 1, 12–13, 235, 251
frame, **29**
glossary of terms, 15–16
history, 19–21
implementation, 110–116
initial document release, 29–30
international gateway interconnection with SDH, 62–74
interoffice networks, 140–150
jitter interworking requirements, 251–255
management/operation, 305–**306**
multiplexing structure, **4**
network (typical), **38**
overhead channels, 22
performance issues, 10–12
phase II, 33–**34**
self-healing rings, 131–139, 140–150, 151–161, **152**
signal standard, 21–25
SONET island, **252**
SONET logical network, **198**
SONET physical network, **198**
SONET/ATM ring, 352
standards, 1–2, 13–15, 19–21, 28–38, 110, 156–160
STS-1 frame, **3**, **21**
STS-3c frame, 63–**64**
STS-N frame, **22**
synchronization for, 242–250
in TINA context, 301–307
and TMN, 259, 264
transmission performance, 226–234
transmission rates, **3**
transport and control architectures, **108**
transport and path overhead byte designations, **22**

transport products list, **112**
See also metro SONET network; self-healing rings
synchronous payload envelope (SPE), 30
See also payload
synchronous transfer mode (STM)
and SONET, 166
vs. ATM, **353**
synchronous transport module level-1 (STM-1)
and administrative units, 32
ATM cell insertion and extraction, 327–328
frame structure, **47**
radio-relay systems for, 48–50, **49**, 87
for SDH systems, 2, 63, **64**, **69**, 87
and SONET standards, 35
STM-N frame, **41**, 47
sub-STM-1 connections, **349**
sub-STM-1 multiplexing routes, **50**
synchronous transport mode (STM), in SDH management network, 308
synchronous transport signal level-1 (STS-1)
bandwidth assignment in, 195
and CCITT standards, 31–32
development of, 31–32
frame, **21**–22
payload pointer, 23–**24**
synchronous transport signal level-3 (STS-3), 2
add/drop hardware configuration for with SARPVP ring, **356**
bandwidth assignment in, 195
synchronous transport signal/virtual tributary (STS/VT), and restoration, 163
system management applications entities (SMAEs), and message communication function, 267

T

T1 multiplexer
and network restoration, 196–**197**
See also multiplexing; network restoration
T1M1 standard, 33–34, 36
T1X1 standard, 1–2, 19–21, 28, 30, 34, 36, 110, 115, 191, 260
TBE. *See* terrestrial baseband equipment
TEAMS. *See* telecommunications engineering and management systems
telecommunication restoration algorithms for network survivability (TRANS), 192–197, 198, 199
telecommunications, residential, 362–370
telecommunications engineering and management systems (TEAMS)
relationship to SUCCESS, **300**
role in network design, **295**
telecommunications information networking architecture (TINA)
information networking vision, 302
and SONET, 301–307
TINA consortium, 301–302
telecommunications management network (TMN), 9–**10**, 226
architecture, **228**, **263**, **296**
in British Telecom, **121**–122
described, 276–277
entities for, 263–264
layered approach to, 275–280, 287
message communication function within, 267

network traffic management optimization, **279**
    in SDH, 277–279, 282, 294
    and SDH/SONET, 259, 264
    service restoration in, 277
    standards for, 259
    and TINA, 302
    See also network management; performance management
telephone companies, and cable companies, 369
telephony systems, optical fiber access for, 341
terminal multiplexer (TM)
    in network topology, 6
    in SDH network, 308
    in SONET, 106
    street located, **348**
    See also multiplexing
termination function, network element function for, 260
termination point provisioning, 282
terrestrial baseband equipment (TBE)
    functional blocks in, **57**
    segregated FDMA-SDH, **58**
    See also fixed satellite services
thick route, FSS-SDH network element scenario for, **58**–59
thin/medium route
    for FSS-SDH network element scenario, 56–57
    See also routing
threshold mechanism, in performance management, 273–274
timing
    for satellite SDH systems, 60
    in SDH/SONET path, **244**
    standards for, 250
TINA. *See* telecommunications information networking architecture
TL1. *See* transaction language 1
TM. *See* terminal multiplexer
TMN. *See* telecommunications management network
toll HRDP, 211–**212**
    component counts, **213**
    error performance, **216**
topology. *See* network topology
traffic
    and DCS use, 113–**114**
    digital private lines traffic, 67–68
    international, types of, 65, **66**
    network management information traffic, **68**–69
    network traffic management optimization, **279**
    point-to-point arrangement, **152**
traffic configuration sublayer, in layered network, **275**–276
traffic grooming, in SDH/SONET network, **121**, 163
traffic management and control
    in ATM, 321
    characteristics of, **237**
    in SONET networks, 235–241
trail
    defined, 266
    in layer network, **77, 266**
    partitioning of, **79**
    in transport network architecture, 76–77
trail termination, and partitioning, 77, **79**
TRANS. *See* telecommunication restoration algorithms for
    network survivability

transaction language 1 (TL1), in SONET communications, 112
transfer modes, synchronous vs. asynchronous, **363**
transmission, digitalization of, 84–85
transmission convergence monitoring, 231. *See also* performance
    monitoring
transmission network architecture, **98**
transmission performance
    in broadband networks, 230–232, 281
    in evolving SONET/SDH networks, 226–234
transmission rates, **3**, 22
    of SDH and SONET, compared, 62, **63**
    See also rates
transmission resources management, 274. *See also* network
    management; performance management
transmission system, for ATM cells transport, 324–329
transmission-line network ONR model, **296**
transport function
    functional layers of SDH transport networks, **54**
    as network description, 44–45, 52
    provided by SDH network element, 267
    in SDH/SONET interconnection, 68
transport network
    interoffice, 106–**107**
    layered model of, **120**
    SDH-based, 281, 284
    SONET-based, 119
transport network architecture
    applications of, 78–81, 110
    asynchronous transport systems, compared to SONET,
        211–217
    generic, 76–**78**
    layer network, access points, trails, 76–77
    in year 2000, 81–83, **82**
    See also architecture
transport network layered model, SDH-based, **278**
transport overhead (TOH)
    and frame construct, 3
    and SONET standards, 30
transport systems, and SDH/SONET, 259–263
tree-and-branch network, **367**
tributary
    on IDT, 67
    See also virtual tributary
tributary group unit (TUG), in multiplexing structure, 5
tributary unit (TU), and networks, 33
TU. *See* tributary unit

U

U-SHR. *See* unidirectional self-healing ring
ubiquitous networks, 370. *See also* networks
unavailability periods, in performance monitoring, 274
unavailable seconds for line, parameters for, 261
UNI. *See* user-to-network interface
unidirectional path switched ring (UPSR), **6**–7, 113
    in SONET, 131–133, 132
    See also ring architecture
unidirectional self-healing ring (U-SHR), 151, **152, 153**
    protection for, 173
    See also self-healing rings
unit processors, in SDH management, 285
United States, FTTC systems development in, 339

UPSR. *See* unidirectional path switched ring
user
    extending fibers to, 331
    See also customer
user channels, 262
user-to-network interface (UNI), 108

V

VC 4-layer network topology, **79**
VC 12-layer network topology, **79**
VC. *See* virtual container
video
    effected by impairments, 227
    provided by SONET, 115, 211
video technology, and broadband services, 363–364
video-on-demand, 342, 365
virtual container (VC)
    in multiplexing structure, 5, **40**
    payload independence for, 44
    See also container
virtual container (VC) overhead, **9**
"virtual LAN", connected in SDH management network, **312**–313
virtual path (VP)
    ATM/SONET ring architecture with, 354–357
    SONET/ATM ring with point-to-point virtual paths, 352–356
    See also path
virtual synchronous digital hierarchy networks, 70–71
    See also synchronous digital hierarch network
virtual tributary synchronous payload envelope (VTx-SPE), in multiplexing structure, 5

virtual tributary (VT)
    and networks, 33
    in SONET, **105**, 235
virtual tributary/container (VT/VC), in layered overhead, 260
voice services
    sensitivity to impairment, 226–227
    See also public-switched services
VP. *See* virtual path
VT. *See* virtual tributary
VT/VC. *See* virtual tributary/container
VTx-SPE. *See* virtual tributary synchronous payload envelope

W

wander
    in digital networks, 243
    generated by pointer processor, **248**, **249**
    and network elements, 233–234
wavelength division multiplexing (WDM) systems, 43, 76
WCS. *See* worst-case survivability
WDM. *See* optical wavelength division multiplexing
WDM systems. *See* wavelength division multiplexing (WDM) systems
wireless services, 164, 166
workstations, in SUCCESS CONT, 298–299
worst-case survivability (WCS), 179–180
    See also network survivability

X

X3T9 Committee, 35

Editors' Biographies

Curtis A. Siller Jr. received the B.S.E.E. (with Highest Honors), M.S., and Ph.D. degrees from the University of Tennessee, Knoxville. He is a Distinguished Member of Technical Staff at AT&T Bell Laboratories. His career spans exploratory studies in analytic antenna design, microwave radio propagation, signal processing, communication theory, system engineering for private wide-area networks, definition of SONET network element functionality, asynchronous transfer mode applications and networking for multimedia communications, and medium-access control layer protocol design. He has published extensively, given international presentations, and holds four patents in the above areas.

Dr. Siller is a member of Phi Eta Sigma, Phi Sigma Phi, Eta Kappa Nu, Tau Beta Pi, and Sigma Xi honorary societies. He was awarded a Distinguished Technical Staff Award and named a Fellow by AT&T Bell Laboratories in 1984 and 1989, respectively. In 1991 he was elected a Fellow of the IEEE. He has organized and chaired numerous IEEE Communications Society-sponsored conference sessions, served on related Technical Program Committees, and helped organize several international workshops. In 1992 Dr. Siller was elevated to the grade of IEEE Fellow. From 1988 to 1992, he chaired the Society's Signal Processing and Communication Electronics Technical Committee. In 1987, he was appointed a Technical Editor of IEEE COMMUNICATIONS MAGAZINE, and was appointed Editor-in-Chief in 1993. His term expired in mid 1995, and he now serves as a Senior Technical Editor. He is a voting member of the IEEE P802.14 Working Group, "Standard Protocol for Cable-TV Based Broadband Communication Network," and is their designated liaison to the ATM Forum Residential Broadband Working Group.

Mansoor Shafi (S'69–A'70–M'82–SM'87–F'93) was born in Multan, Pakistan, on May 5, 1950. He received the BSc (Eng) degree from the University of Engineering and Technology, Lahore, Pakistan, and the Ph.D. degree from the University of Auckland, New Zealand, in 1979, both in electrical engineering.

He was awarded the Gold Medal for standing first in the preengineering examinations of 1966 from Government Emerson College Multan, and won a four-year Government scholarship to study engineering at the Lahore University. From 1971 to 1974, he worked as a Lecturer at the University of Engineering and Technology, Lahore, and from 1975 to 1979, he worked as a Junior Lecturer at the University of Auckland and was engaged in research in the fields of radio astronomy and signal processing. Since 1979, he has been with the New Zealand Post Office (mother organization for Telecom New Zealand). His research interests are in the fields of fixed and mobile transmission systems.

Dr. Shafi has published widely in the various subjects relating to transport systems, and in 1988 he coedited an IEEE Press book, *Microwave Digital Radio*. Recent activities have also included the planning of telecommunications in developing countries. Dr. Shafi was awarded the 1992 IEEE Communications Society Public Service Award in Communications. He serves as a New Zealand delegate to various meetings of the ITU-R sector.